A COURSE IN
THEORETICAL
STATISTICS

OTHER GRIFFIN BOOKS ON STATISTICS AND MATHEMATICS

Descriptive brochure available from Charles Griffin & Co. Ltd.

A COURSE IN THEORETICAL STATISTICS

FOR SIXTH FORMS · TECHNICAL COLLEGES
COLLEGES OF EDUCATION · UNIVERSITIES

N. A. RAHMAN

M.A.(Alld), M.Sc.(Stat.) (Calc.)
Ph.D.(Stat.) (Camb.)

Senior Lecturer in Mathematical Statistics
University of Leicester

GRIFFIN   LONDON

0852640084

732,550

First published in 1968

Set by E. W. C. Wilkins & Associates Ltd., London
Printed in Great Britain by J. W. Arrowsmith Ltd., Bristol 3.

Dedicated to

MY MOTHER

and

THE MEMORY OF MY FATHER

PREFACE

The presentation of yet another introduction to statistics requires some explanation. In general terms, currently available introductory texts fall into one of two categories: either they assume a certain level of mathematical knowledge and maturity on the part of the reader and are restricted to the mathematics of statistics, with particular emphasis on probability and sampling distribution theory, or they present the methods of statistics with the minimum amount of mathematics and a correspondingly greater stress on the applications of statistical theory. This text avoids both these approaches, principally because it is planned not simply as a book on statistical mathematics or the methods of statistics but as an introduction to the statistical profession. In other words, the essence of this approach to statistics lies in the view and the use of mathematics as an indispensable tool in the analysis of observational data, for which the present spectacular development and use of electronic computers is bringing innumerable further opportunities.

In the first place, this work is planned for sixth-formers specialising in mathematics, to show them some of the possibilities of a new dimension in applied mathematics and thereby stimulate their interest in the statistical profession. However, the text is written to no set syllabus but addressed flexibly to the needs of instruction at the sixth-form and first-year undergraduate stages. To achieve this dual purpose, the work has been planned to allow for some selection. Furthermore, in thinking of the content and its presentation at A-level, the needs of the extension of statistics into S-level have also been kept in view. In particular, it is believed that the first ten chapters (with the possible omission of some mathematically difficult sections) *could* provide a complete and ordered A-level course, with suitable practical work requiring the use of hand-operated desk calculators, and the last seven chapters a similarly coordinated S-level course. This, in the view of the author, is a plan for a desirable sixth-form syllabus, and it is hoped that the publication of the work will encourage a development in this direction.

Next, so long as statistics is initiated largely at the university, this text provides a complete and ordered coverage for a one-year course within the framework of a mathematics honours degree. This course should be particularly suitable for first-year undergraduate mathematicians who have not studied statistics at school or have had only a brief introduction. In fact, the text has grown out of such an undergraduate course given at the University of Leicester over a number of years. A considerable measure of success with undergraduates might well be an indicator of a more general appeal of this approach to the statistical profession in other universities. Besides, in conformity with existing syllabuses, a selective use of the text will prove suitable for relatively restricted but professionally orientated teaching in sixth forms, technical colleges, and colleges of education.

One of the crucial decisions in the detailed planning of this text pertained to the level of pure mathematics to be assumed. With a few carefully considered exceptions, the general line adopted was to restrict the pure mathematics to the present A-level syllabuses. However, it was felt that the scope of statistical ideas included in the text should not be completely dependent upon the limitations of the pure mathematics taught in sixth forms, since the appropriate relation should be one of interdependence between pure and applied mathematics generally. Moreover, the pure mathematics required for statistics needs an early introduction to some ideas which are either largely excluded from the present A-level syllabuses (for example, finite differences, simple polynomial interpolation, and difference equations), or relatively neglected (for example, summation of series and elementary properties of the gamma and beta functions). Accordingly, bearing in mind statistical requirements, introductions to such topics are included to ensure a systematic development.

Chapters 1 – 3 are devoted to elementary numerical methods and the planning of computations. This unconventional beginning is regarded by the author as essential for the proper orientation of students towards the computational discipline required for the analysis of any substantial data. Chapter 4 is concerned with the summation of finite and infinite series of particular interest in the study of discrete probability distributions. Part I (Chapters 1 – 4) thus includes such introductory mathematical ideas as classroom experience has shown to be particularly helpful for later work. Part II (Chapters 5 – 10) deals with the usual preliminary ideas of statistics and probability which are related to practical applications. However, statistical inference is deliberately excluded, although the evaluation of probabilities in a diversity of examples should help beginners to think in terms of uncertain categories. The material covered should also provide ample scope for computational and mathematical work well within the range of sixth-formers. Part III (Chapters 11 – 17) is devoted wholly to statistical inference. However, no attempt is made to prove all the basic mathematical results in normal sampling distribution theory which underlie the methods of statistical analysis.

A considerable number of solved examples are included in the text. These illustrate mathematical methods and computational techniques, and should, it is hoped, help beginners to understand clearly the concepts of a novel subject. Nevertheless, it must be stressed that the ultimate method of learning the subject is by practice with theoretical exercises and analysis of data. It was decided not to follow the conventional procedure of including a limited number of unsolved problems in the text. The systematic presentation of theoretical and practical exercises requires a more extended treatment. Accordingly, it is suggested that users of this text supplement their study by the theoretical exercises in the author's *Exercises in Probability and Statistics* issued separately by the present publishers. It is hoped that a companion volume of practical exercises will follow.

For permission to reprint tables, I am indebted to the following: Professor A. Hald and John Wiley and Sons, Inc., New York (Table 5.9 from *Statistical Theory with Engineering Applications*); Professor M.S. Bartlett, F.R.S., and the Department of Statistics, University College London (Table 13.3 from Tracts for Computers No. 24);   Professor G.W. Snedecor and the Iowa State University Press, Ames (Table A.4 from *Statistical Methods*); and the Literary Executor of the late Sir Ronald A. Fisher, F.R.S., Cambridge, and Oliver and Boyd, Ltd, Edinburgh (Tables A.2 and A.3 from *Statistical Methods for Research Workers*).

It is a pleasure to record my thanks to colleagues and friends who have helped me in various ways. I should like to thank Dr M.G. Kendall, who initially proposed my writing this work and generously gave advice in its planning, and Professor H.E. Daniels for his unfailing encouragement and for suggesting improvements in the presentation. To this I would add my thanks to Mr H.F. Downton, since the approach adopted in this work owes not a little to his past pioneering efforts in undergraduate teaching at this university and to the intangible effects of our earlier professional association. My gratitude to Dr R.O. Davies is deep — for the conscientiousness with which he read parts of the manuscript and sought to improve them by many constructive suggestions. I am obliged to Professor R.L. Goodstein for giving me the opportunity for experimentation with different groups of students. I have learnt much through this contact, and I hope that something of my experience has passed into this text and now will be more generally acceptable. My thanks are also due to the publishers for their consideration and efficiency in publishing this work.

I should be grateful to readers of this work for pointing out any mistakes or obscurities found. For these the author alone is responsible.

Leicester
31st March, 1967

N.A.R.

CONTENTS

PART I

MATHEMATICAL INTRODUCTION

PART II

STATISTICAL MATHEMATICS AND PROBABILITY

PART III

STATISTICAL INFERENCE

PART I

MATHEMATICAL INTRODUCTION

CHAPTER 1

ELEMENTARY COMPUTATIONAL TECHNIQUES

1.1 Planning of computations

All practical applications of mathematics require numerical calculations. Occasionally these computations may be simple, but in general they are involved and time-consuming, with a genuine risk of making mistakes which cannot be completely eliminated even with the greatest care and concentration. The time required and the ever-present possibility of mistakes are, in fact, the two most important considerations which need the attention of a beginner learning the computational discipline. Accordingly, we may state that the basic aim of the art of computation is two-fold: to economise in time and to minimise the risk of mistakes; and in order to realise this objective, it is essential to plan the computations before starting on the calculations. This concept of computational planning is a comprehensive notion, and we shall first explain its important constituents in some detail.

To begin with, a good computational plan must take into account all the devices and aids which are available to the computor. These may be broadly classified as *published tables*, *calculating machines*, and *computing paper*. And the quality of the plan depends upon the degree of efficiency achieved in using these aids.

1.2 Published tables

The reader of this book is almost certainly familiar with the commoner mathematical tables of logarithms and trigonometric functions; and, in principle, all numerical tables are of this kind, that is, they give in tabular form the values of a mathematical function, $f(x)$, to a certain specified degree of accuracy, corresponding to usually equispaced values of the variable x. At present a large number of mathematical tables are available, and these give extensive tabulations of simple and highly complex mathematical functions. It is therefore advisable to find out the tables which may be helpful in any piece of computation. Besides, it is particularly important to understand clearly the instructions for the use of tables by noting any special signs and abbreviations used in tabulation. A judicious use of tabulated values is one of the best ways of reducing the time spent on computation and also thereby of decreasing the risk of mistakes. Some of the simpler tables which will be found useful in the kind of numerical work encompassed by this book are:

(i)   *Chambers's Four-figure Mathematical Tables*
        by L.J. Comrie
(ii) *Barlow's Tables of Squares, Cubes,* etc.
(iii) *Cambridge Elementary Statistical Tables*
        by D.V. Lindley and J.C.P. Miller
(iv) *Tables of Random Sampling Numbers* — Cambridge Tract for
        Computers No.24
        by M.G. Kendall and B. Babington Smith

As a simple illustration of the use of tabulated functions in reducing computational work, we may consider the following elementary numerical exercise.

Example 1.1

Tabulate, correct to six places of decimals, the function

$$f(x) = \frac{11 \cdot 8\, x^7 + 6 \cdot 7\, x^5 - 12 \cdot 4\, x^3 + 1 \cdot 92}{6 \cdot 8\, x^6 - 13 \cdot 7\, x^4 + 2 \cdot 5\, x^2 - 0 \cdot 58}$$

for values of x in the range $1 \cdot 5 \leqslant x \leqslant 3 \cdot 0$ at intervals of $0 \cdot 1$.

It is immediately obvious that the simplest way of evaluating $f(x)$ is first to determine the numerator and denominator of $f(x)$ for each value of x in the specified range. The values of x^2, x^3, and x^4 can be obtained from *Barlow's Tables*, and the corresponding values for the higher powers x^5, x^6, and x^7 are then derived by appropriate multiplications of the values written down from the tables. For example, x^7 is obtained as the product of x^3 and x^4. In the same way, x^5 is deduced as the product of x and x^4, or of x^2 and x^3. The appropriate powers of x are then multiplied by the stated coefficients to determine the numerator and denominator of $f(x)$ for each x. Finally, simple division gives the required values of $f(x)$.

However, even with the help of *Barlow's Tables*, the computations needed to determine $f(x)$ are so laborious that they can hardly be carried out by the usual methods of school arithmetic. And this simple sequence of numerical processes reveals the importance of calculating machines. Indeed, modern applied mathematics would be impossible without them.

1.3   Calculating machines

The idea of using some aid in numerical calculation is not new. We know from the researches of archaeologists that even in the earliest stages of numeration man used devices to help him in his simple calculations. The English tally-stick and the counting board or abacus found in many different countries may rightly be regarded as perhaps two of the earliest forms of computing machines. The development from these primitive devices to the present-day spectacular electronic computers is a long and human story. The details of this story do not concern us here, but the central idea is clear: each of these mechanical developments simplified or even made possible the increasingly complicated and extensive calculations which were needed in the ever-widening applications of mathematics in the quantitative study of natural and social phenomena. It is also important to remember that the development of computing machines, in itself a remarkable account

of scientific and technological progress, was, in the main, made conceptually possible by the demands on human ingenuity to create the tools for the
application of mathematics to solve practical problems.

At present we have access to a wide variety of computing machines,
and even the simplest of them makes possible calculations in reasonable
time which would have defeated human calculators a century ago. It is this
power of the computing machine which has added a new dimension to all applied mathematics and, in particular, helped greatly in the rapid development
of the science of statistics. To anticipate matters, we may state that one
of the important aims of this book is to show how the computing machine has
brought about many new and interesting applications of mathematics; and
these are the applications which are leading us now to a better understanding of our complex society and a more purposeful use of our natural re-
sources. This positive creativity is one of the greatest attractions of modern applied mathematics, and its secret lies in the computing machine.

The simplest of the mechanical aids to numerical calculation is the
slide-rule, which is much used by scientific workers. This is a small instrument of considerable flexibility and can provide quick answers, but its accuracy is limited and so also is its use in extensive numerical work . At
the other extreme, the most elaborate calculating machines are the *electronic computers*. These have an almost fabulous speed for calculations of
enormous complexity. However, the planning of the calculations with the
appropriate instructions for the computer (known as *programming*) may take
a long time to work out. Thus, for example, in the British general election
of October 1964, it took many months to make the computer programme used
for forecasting results by the B.B.C. on election night. Between the simple
slide-rule and the big electronic computer, there are a variety of computing
machines which concern us more closely. These machines are the compact
desk calculators, and they represent a special class of *digital calculating
machines*.

In general, desk calculators are of three main types, distinguishable by
the way they are operated to perform the four main arithmetical operations —
addition, subtraction, multiplication, and division. Some of these machines
are also designed to extract square roots by a simple continuous machine
operation. All these machines, however, are built on the rotary principle,
and may be divided into three main categories of *hand-operated, semi-
automatic*, and *fully automatic* desk calculators. It is not our purpose here
to explain in detail the instructions for operating a desk calculator because
they depend considerably upon the actual model, but some general observations are worth making.

The hand-operated calculator, as its name implies, is operated by the
turning of a crank handle, and hand manipulation is required for the four
arithmetical operations. The semi-automatic and fully automatic machines
are operated electrically, though numbers are set on them manually. The
main difference between them is that a semi-automatic machine does the
basic operations, except multiplication, automatically once the numbers to
be operated upon have been set, whereas a fully automatic machine also

does multiplication automatically. Some of the more elaborate fully auto-matic machines are also equipped with one or more features like automatic clearance, automatic return to a fixed decimal point, or automatic subtrac-tion of products from a previously recorded total.

Two of the commonest models of desk calculators found in this country are of the type of "Facit" and "Brunsviga" machines. The ability to use such a desk calculator efficiently and precisely requires practice and con-centration. This is particularly so with the cheaper hand-operated machines. It is therefore essential for a beginner to understand the manufacturer's special instructions for the use of a model, and to develop facility in using the machine with the maximum economy of time and accuracy. Thus, for ex-ample, machines like the hand-operated "Brunsviga" have an *automatic transfer device*, and a simple routine operation for square roots. An effec-tive use of the potentialities of a machine is essential for good computation-al programmes. It is also important to remember the limitations of a machine, and one of the commonest of these is the restricted digital capacity of the machine. Accordingly, care has to be exercised in doing near-capacity mul-tiplications and divisions. Despite all the qualities of a machine, it should be remembered that it is only an aid and it does not eliminate the human de-sign and thought which are the basis of efficient numerical calculations. Finally, a desk calculator is a precision instrument with absolutely smooth functioning, and it should invariably be handled with care strictly according to the manufacturer's elaborate instructions provided with each machine.

1.4   Computing paper

In planning a piece of computation with the aid of a desk calculator, the first important step is to visualise all the distinct stages of the required calculations as an ordered sequence of processes each of which could be carried out on the machine. This presupposes a facile use and understand-ing of all the computational potentialities of the particular machine used. This ordering of the operations is necessary to avoid duplication of work and to ensure an organic development of the entire computational programme. A particular machine operation should, in general, only be performed when all the quantities needed for it have been obtained either by machine opera-tions or from already tabulated values.

As observed earlier, numerical computation is usually extensive and there is always a possibility of mistakes despite the greatest care. Besides, a machine can perform a number of operations cumulatively to give a result that may either be the required one, or an intermediate value needed for further operations. Accordingly, there will be many stages of the calcula-tions when numbers will have to be recorded from the machine on to paper, and subsequently from paper to machine. Some of the commonest mistakes occur in such transcription of figures. There is, for example, the risk of transposition of two consecutive digits and, of course, there is the abundant possibility of misreading a badly written digit. It is, therefore, important to reduce transcription to the absolutely essential, and to ensure that necessary calculations are written down in an orderly form. To help in this objective, computations are invariably done (and also finally presented) on what is known as *computing paper*. This is special foolscap paper ruled into squares by rulings which are approximately one-fourth of an inch apart. The rulings are very faint and do not catch the eye more than is necessary to guide the alignment of the calculations. It is usual to record all the required intermediate calculations and transfers of figures from the machine on to computing paper in an ordered, preferably tabular, form. Each square should take two digits, well placed centrally with a digit allowance for a decimal point.

The ability to plan a computational layout is an art which is learnt by both care and practice. The aim is to obtain a pleasing arrangement without cramming the figures. This is less taxing in reading, and it is always a positive advantage in checking for mistakes. Every effort should be made to print the digits evenly with a well-pointed medium-soft pencil. A well-written page of computations should have the even quality of a page typed out by a competent secretary. It is never advisable to do computations of any length on odd bits of paper. Unfortunately, this is advice which beginners too often fail to observe till harsh experience teaches them the importance and relevance of this simple maxim. In the opinion of this author it is also desirable not to do computational work in ink. Opinions differ on this point, and it is perhaps true that working in ink would, in general, encourage more deliberate concentration; but, on balance, the advice of this author is firmly in favour of pencil. Such work can be erased easily and this is a helpful point for beginners. It is also useful to remember that work is done on only one side of computing paper. This makes possible the simple continuity of the calculations and also saves the recorded computations from being smudged. Furthermore, it is essential for future reference to have enough written explanation on a computational page for it to be easily understood at a later date. Finally, it is good practice to follow the convention of preserving computation sheets in a flat file.

1.5 Checks

Despite all reasonable care, mistakes will occur in computations, and so the checking of calculations is invariably necessary. The most relevant check is first to make sure that the formulae and processes to be used are free from any theoretical inaccuracy. Nothing is more disheartening than to find that a laborious piece of computation is wrong because of some theoretical slip. Next, the actual computations should also be checked, preferably by some independent means. Sometimes, the only possible check is recalculation. However, in any case, it is essential to use the available checks, even if this apparently means some more computation. An incorrect computation is of no use except as a salutary lesson, and the consistent use of checks is, indeed, a long-run economy. It is also well worth emphasising that the neater the computational layout, the easier it is to detect mistakes. And in this context it is pertinent to quote the wise and comforting caution of Professor Hartree:

"No one, and no machine, is infallible, and it may fairly be said that the ideal to aim at is not to avoid mistakes entirely, but to find all mistakes that *are* made, and so free the work from any unidentified mistakes."

Example 1.1 (concluded)

We now illustrate the application of some of our computational principles by presenting the tabulated values of $f(x)$ in the example cited above. The values of x^2, x^3, and x^4 are obtained from *Barlow's Tables*. Given these values, the simplest computational sequence for x^5, x^6, and x^7 is to multiply x^3 successively by x^2, x^3, and x^4. The values of $P_1(x)$ and $P_2(x)$ in Table 1.1 are obtained by cumulative machine operations, whence, finally, division gives the required values of $f(x)$. The presentation of these values correct to six decimal places, involves the principle of *rounding off* which we shall consider in the next chapter.

1.6   Extraction of square root

We assume that the reader is by now familiar with the use of a desk calculator like the "Facit" or the "Brunsviga", and knows how to carry out the four basic arithmetical operations. The extraction of a square root is generally not regarded as a basic operation, though it occurs quite frequently in statistical work. *Barlow's Tables* give an extensive table of square roots, and in many cases the required square root may be obtained simply from these tables. However, especially when greater accuracy is required, it is necessary to use some machine operation to evaluate the square root. It is therefore appropriate to consider the theoretical reasoning which justifies the usual methods used on desk calculators for the extraction of square roots, namely, the method of *successive approximation* and the method of *subtraction of odd numbers*. On machines of the "Brunsviga" type the two methods can be combined to give a highly efficient and simple process for finding a square root.

(i) *Method of successive approximation*

This method depends upon obtaining a first approximation $\sqrt{N} + \epsilon$ of the number N, where ϵ is small, and may be positive or negative. Such an approximate square root can in most cases be read off from *Barlow's Tables*. Divide N by $\sqrt{N} + \epsilon$, and the average of the quotient and $\sqrt{N} + \epsilon$ gives the second approximation. This process of division and averaging is repeated till stability is established to the desired number of digits. Theoretically, the first division gives

$$\frac{N}{\sqrt{N} + \epsilon} = \sqrt{N}\left[1 + \frac{\epsilon}{\sqrt{N}}\right]^{-1} = \sqrt{N}\left[1 - \frac{\epsilon}{\sqrt{N}} + \frac{\epsilon^2}{N} - \cdots\right].$$

Therefore the average of $\sqrt{N} + \epsilon$ and $N/(\sqrt{N} + \epsilon)$ is

$$\sqrt{N}\left[1 + \frac{\epsilon^2}{2N} - \cdots\right].$$

Now ϵ is small compared to $\sqrt{N}$, and the successive terms of the alternating series converge to zero very rapidly. The second approximation has a smaller error than the first approximation $\sqrt{N} + \epsilon$ as the term in ϵ is not present in the series. In general, this method is convenient as in many

Table 1.1   Showing the values of $f(x)$, correct to six decimal places, for a range of values of x

x	x^2	x^3	x^4	x^5	x^6	x^7	$P_1(x)$	$P_2(x)$	$f(x)$
1·5	2·25	3·375	5·0625	7·59375	11·390625	17·0859375	24·6187500	13·1450000	1·872718
1·6	2·56	4·096	6·5536	10·48576	16·777216	26·8435456	42·8902848	30·1207488	1·422907
1·7	2·89	4·913	8·3521	14·19857	24·137569	41·0338673	68·9563128 4	56·3566992	1·223569
1·8	3·24	5·832	10·4976	18·89568	34·012224	61·2220032	105·1818 5856	94·9860032	1·107341
1·9	3·61	6·859	13·0321	24·76099	47·045881	89·3871739	154·2767 7212	149·8172208	1·029767
2·0	4·00	8·000	16·0000	32·00000	64·000000	128·0000000	219·520 00000	225·4200000	0·973827
2·1	4·41	9·261	19·4481	40·84101	85·766121	180·1088541	304·805 45028	328·3756528	0·928222
2·2	4·84	10·648	23·4256	51·53632	113·379904	249·4357888	414·7 2677504	461·5726272	0·898508
2·3	5·29	12·167	27·9841	64·36343	148·035889	340·4825447	554·6 7021676	635·9068752	0·872251
2·4	5·76	13·824	33·1776	79·62624	191·102976	458·6471424	730·9 1592192	858·7871168	0·851103
2·5	6·25	15·625	39·0625	97·65625	244·140625	610·3515625	950·7 4812500	1140·0450000	0·833957
2·6	6·76	17·576	45·6976	118·81376	308·915776	803·1810176	1222·57460608	1490·8901568	0·820030
2·7	7·29	19·683	53·1441	143·48907	387·420489	1046·0353203	1556·05582524	1924·0901552	0·808723
2·8	7·84	21·952	61·4656	172·10368	481·890304	1349·2928512	1962·24413696	2453·8553472	0·799658
2·9	8·41	24·389	70·7281	205·11149	594·823321	1724·9876309	2453·73348772	3096·3286128	0·792465
3·0	9·00	27·000	81·0000	243·00000	729·000000	2187·0000000	3044·82000000	3869·4800000	0·786881

$$P_1(x) = 0·8\,x^7 + 6·7\,x^5 - 12·4\,x^3 + 1·92$$
$$P_2(x) = 6·8\,x^6 - 13·7\,x^4 + 2·5\,x^2 - 0·52 \quad \text{and} \quad f(x) = P_1(x)/P_2(x)$$

cases no more than one division is required to attain sufficient accuracy in the square root. Indeed, if the first approximation of the square root is correct to m decimal places, the second approximation will be found to be correct to $2m$ decimal places.

This method is known as an *iterative process*, and it can be used on any desk calculator. We next consider a simple illustration of the process.

Example 1.2

Calculate the square root of $163 \cdot 7893$ correct to four decimal places.

From *Barlow's Tables* an approximate value of the square root is $12 \cdot 80$, which is correct to two decimal places.

Division gives $\dfrac{163 \cdot 7893}{12 \cdot 80} = 12 \cdot 7960$, so that the second approximation of the square root $= \frac{1}{2}(12 \cdot 80 + 12 \cdot 7960)$
$$= 12 \cdot 7980.$$
This square root is correct to four decimal places, since $(12 \cdot 7980)^2 = 163 \cdot 7888$, with an error of -5 in the fourth decimal place.

(ii) *Method of odd numbers*

This method is based on the fact that the sum of the first n odd numbers is n^2. Since
$$x^2 - (x-1)^2 = 2x - 1,$$
we have, on addition for $x = 1, 2, 3, \ldots, n$,
$$n^2 = \sum_{x=1}^{n} (2x - 1).$$
Accordingly, if N is a perfect square, then to extract its square root it is only necessary to subtract the successive odd numbers, beginning with 1, since the number of odd numbers subtracted equals N. For example, if $N = 49$, then $\sqrt{49} = 7$, since $49 - 1 - 3 - 5 - 7 - 9 - 11 - 13 = 0$, and seven odd numbers are subtracted.

When N is a number with more than two digits, then the square root is obtained by combining the subtraction of odd numbers with digiting. Since the first ten odd numbers total 100, they may all be subtracted by subtracting 100. In using this principle on a desk calculator like the "Brunsviga", the subtraction of 100 ensures that a "1" (representing 10) is recorded in the revolutions counter. The eleventh odd number, 21, is subtracted and the process continues in this way. As an example, suppose $N = 196$. Then
$$\sqrt{196} = 14, \text{ since } 196 - 100 - 21 - 23 - 25 - 27 = 0.$$
It is to be noted that in subtracting
$$100; \quad 400 = 100 + 300; \quad 900 = 100 + 300 + 500;$$
$$1600 = 100 + 300 + 500 + 700; \quad \text{etc.}$$
we are subtracting the first 10, 20, 30, 40, etc. odd numbers. The next odd numbers are 21, 41, 61, 81, etc., respectively. Hence the following method, usable on the "Brunsviga" or a similar machine with setting levers.

(i) Place the number N at the left of the products register.

(ii) Arrange the digits in groups of two as measured from the decimal point.

(iii) The successive odd numbers are subtracted from the extreme left group until negative numbers are introduced in the products register (bell rings).

(iv) Go back one revolution in the positive direction.

(v) Change the last odd number subtracted to the preceding even number.

(vi) Shift carriage one step to the right and carry on.

The revolutions counter at any stage of this process gives the approximate square root in which the decimal point is placed appropriately. In this method the square root of N is $\sqrt{N} - \epsilon$, where $\epsilon > 0$. At any stage, the revolutions counter has $\sqrt{N} - \epsilon$, the setting levers have $2(\sqrt{N} - \epsilon)$, and the products register has the remainder $N - (\sqrt{N} - \epsilon)^2$. If ϵ is sufficiently small, then we divide out and the additional fraction obtained in the revolutions counter is

$$\frac{N - (\sqrt{N} = \epsilon)^2}{2(\sqrt{N} - \epsilon)} = \frac{2\epsilon\sqrt{N} - \epsilon^2}{2\sqrt{N}(1 - \epsilon/\sqrt{N})}$$

$$= \epsilon\left[1 - \frac{\epsilon}{2\sqrt{N}}\right]\left[1 + \frac{\epsilon}{\sqrt{N}} + \frac{\epsilon^2}{N} + \ldots\right] \sim \epsilon + \frac{\epsilon^2}{2\sqrt{N}}.$$

This correction is automatically added to the approximate square root $\sqrt{N} - \epsilon$ already in the revolutions counter to give

$$(\sqrt{N} - \epsilon) + \epsilon + \frac{\epsilon^2}{2\sqrt{N}} = \sqrt{N}\left[1 + \frac{\epsilon^2}{2N}\right],$$

which is a much better approximation than $\sqrt{N} - \epsilon$. For increasing the accuracy of the square root by this method, it is necessary to make ϵ small, which can be done by evaluating the approximate square root to a greater degree of accuracy by subtracting the successive odd numbers. This is a *non-iterative* method. If the square root is desired correct to $2m$ decimal places, then the approximate square root should be evaluated correct to m decimal places.

As a check, it is essential to verify the accuracy of a square root obtained by either of the two methods by actually squaring the square root.

CHAPTER 2

ERRORS AND ACCURACY OF NUMERICAL COMPUTATIONS

2.1 Approximate numbers

We have already made a reference to the principle of rounding off a digital number to obtain an approximate value which is correct to a certain number of digits. This is a notion of great practical importance in numerical computations, and it arises simply because of the difference between algebraic symbols and their digital representations. All algebraic symbols are, by definition, exactly what they represent, but numbers can only contain a finite number of digits. This limitation is a practical necessity, though we can make such numbers as accurate as we please. For example, 1·4142 is a digital approximation of the number denoted by $\sqrt{2}$. Similarly, 3·14159 is a digital approximation of the numerical value of the mathematical constant π, and 2·7183 that of the Napierian constant e. These three digital numbers are *approximate numbers* or, more correctly, *approximate values* of the numbers denoted symbolically by $\sqrt{2}$, π, and e respectively. It is clear that there is no single digital approximation of such a number represented by a symbol, and different approximations for the same quantity may be written down by retaining a varying number of digits. We may thus consider 3·14, 3·142, and 3·1416 as different digital approximations of π obtained by retaining two, three, and four digits after the decimal point with appropriate rounding off. Such approximations have different degrees of accuracy, and our first concern is to consider methods of determining the accuracy. However, the accuracy of a digital approximation is properly defined only when the rounding off is done according to definite rules.

2.2 Significant figures and decimal places

There are two ways of defining the accuracy of a digital number, namely, by the number of *significant figures*, or by the number of *decimal places*.

A significant figure is any one of the digits from 1 to 9; and 0 is a significant figure except when it is used to fix the decimal point or to take the place of discarded or unknown digits. According to this definition, the number of significant figures in a digital number is the number of digits obtained by counting from the first non-zero digit on the left.

Example 2.1

    (i) 768·59 contains five significant figures.

    (ii) 0·00053096 contains five significant figures. The zeros used to fix the decimal point are not significant.

(iii) 4906 contains four significant figures.

(iv) As written, there is nothing to indicate whether the zeros in the number 56400 are significant figures. This ambiguity can be removed by writing the number in one of the following ways:

(a) $5 \cdot 64 \times 10^4$, which contains three significant figures;

(b) $5 \cdot 640 \times 10^4$, which contains four significant figures;

(c) $5 \cdot 6400 \times 10^4$, which contains five significant figures.

In each case the significant figures are indicated by the factor on the left.

The number of decimal places used to indicate the accuracy of a digital number is the number of digits after the decimal point.

Example 2.2

(i) $768 \cdot 59$ contains two decimal places and five significant figures;

(ii) $0 \cdot 00053096$ contains eight decimal places and five significant figures; and

(iii) $0 \cdot 6492$ contains four decimal places and four significant figures.

It is to be observed that the number of decimal places can be more, equal to, or less than the number of significant figures. The two methods for indicating the accuracy of a digital number are different, but they are both based on the principle of rounding off.

2.3 Rounding off

The number $872 \cdot 50$ contains five significant figures and two decimal places only if it implies that the number has been measured to this accuracy. For example, suppose the number represents road mileage between two places. This means that if the mileage were recorded by using a more precise measuring technique to determine the third figure after the decimal point, then the number would lie between $872 \cdot 495$ and $872 \cdot 505$. In other words, correct to five significant figures or two decimal places, the number $872 \cdot 50$ represents any number lying between $872 \cdot 495$ and $872 \cdot 505$. This implies the concept of rounding off a number.

To round off a number is to retain a certain number of digits counted from the left, and discard the others. For example, rounding off π to three, four, five, and six significant figures respectively gives

(i) $3 \cdot 14$, correct to three significant figures or two decimal places;

(ii) $3 \cdot 142$, correct to four significant figures or three decimal places;

(iii) $3 \cdot 1416$, correct to five significant figures or four decimal places; and

(iv) $3 \cdot 14159$, correct to six significant figures or five decimal places.

Numbers are rounded off so as to cause the least possible error. This is done by the following rule.

Rule

To round off a number to n significant figures (or decimal places) discard all digits to the right of the nth place. Then

(i) if the discarded number is less than half a unit in the nth place, leave the nth digit unchanged;

(ii) if the discarded number is greater than half a unit in the nth place, add one to the nth place;

(iii) if the discarded number is exactly half a unit in the nth place, leave the nth digit unaltered if it is an even number, but increase it by one if it is an odd number. In other words, round off to leave the nth digit an even number.

Example 2.3

(i) $34 \cdot 526548 = 34 \cdot 5265$ rounded off to six significant figures or four decimal places;

(ii) $0 \cdot 67982 = 0 \cdot 680$ rounded off to three significant figures or three decimal places;

(iii) $482 \cdot 745 = 482 \cdot 74$ rounded off to five significant figures or two decimal places;

(iv) $1089 \cdot 7635 = 1089 \cdot 764$ rounded off to seven significant figures or three decimal places;

When a number has been rounded off according to the above rule, it is said to be *correct to n significant figures or decimal places*. It is important to note that this means that there is an error of approximation in the nth place retained. The correctness implies that the rounding off has been done correctly by taking due account of the magnitude of the discarded digits.

2.4   Errors and mistakes

In everyday usage, the words *error* and *mistake* are synonymous, but in the technical use of the terms in this book an important distinction is made between them. A mistake is a chance inaccuracy which may be due to human or machine fallibility. Thus if a number 739 is by transposition recorded as 793, then the inaccuracy is a mistake. In the same way, a technical flaw in a computer may lead to an inaccurate sum of certain numbers, and in this case also the inaccuracy is a mistake. On the other hand, an error is an inherent element in approximate numbers which represent digital approximations of mathematical constants or quantitative measurements obtained to a certain degree of precision. In this sense we make an approximation error in using $3 \cdot 14$ as a digital approximation for the constant π. Similarly, the inaccuracy of recording a mileage between two places as $872 \cdot 50$ is an error of measurement since we have adopted the convention that any distance between $872 \cdot 495$ and $872 \cdot 505$ will be denoted indifferently by $872 \cdot 50$ correct to two decimal places. An error of measurement or of rounding off a mathematical constant can be decreased but it can never be completely eliminated. And we may rightly state that numerical computation of applied mathematics is the arithmetic of approximate numbers. This is a totally different concept from that of school arithmetic as an "exact" science. It is because a beginner does not understand the notion of error implicit in any measurement that he is led into the unfortunate position of deducing results with six-figure accuracy from data which are correct to only two figures! As a general rule, no result can be more accurate than the numbers from which it is obtained. However, this does not imply that all calculations should be rounded off to the significant figures (or decimal places) of the data at each step of the computations. Indeed, the correct procedure is just the opposite since it is always advisable to carry more figures in the intermediate stages of the computations. A good working procedure is to adopt the following rule.

Rule

During intermediate stages of the calculations retain one more figure than that given in the data, and round off after the final operation has been carried out.

This rule may be regarded as generally safe, but it must be emphasised that it is not infallible. Discarding figures in intermediate stages of the calculations is a risky operation, and it can only be carried out with reasonable assurance when the computor has a clear understanding of how errors will accumulate in subsequent stages of the calculations. As an example, suppose a sequence of operations leads to a number $1 \cdot 0074362$ rounded off to eight significant figures. If the final answer is required correct to four significant figures, it may seem reasonable to round off this number to $1 \cdot 0074$ correct to five significant figures. But if at the next stage one has to be subtracted from this number then the difference $0 \cdot 0074$ has only two-figure accuracy! In fact, to obtain a four-figure accuracy we need to retain the number as $1 \cdot 0074362$, which then, after subtraction, gives $0 \cdot 007436$, a result correct to four significant figures.

The moral of this simple example is clear: although it is an undesirable waste to carry extra digits in intermediate stages of computation, the temptation to throw away digits should be resisted by an understanding of all the implications of the *subsequent* calculations. Numerical computation requires judgement based on experience. In this connection, it is some help to know how errors accumulate in the basic numerical processes which we consider next.

2.5 Errors in basic numerical processes

(i) *Addition*

The algebraic sum of x_1 and x_2 is $x_1 + x_2$. But if x_1 and x_2 are approximate numbers, then they will usually be given correct to, say, d_1 and d_2 decimal places (or significant figures) respectively. Besides, in view of the rule for rounding off, these numbers may be in error by as much as a half in units of the last digit. Therefore the number of reliable digits in the sum $x_1 + x_2$ is $\min(d_1, d_2)$, where this means the lesser of d_1 and d_2.

Example 2.4

Suppose $x_1 = 769 \cdot 5$ with a maximum error of $\pm 0 \cdot 05$

and $x_2 = 0 \cdot 0000543$ with a maximum error of $\pm 0 \cdot 00000005$.

Then the sum $x_1 + x_2$ is $769 \cdot 50$ with a maximum error of $\pm 0 \cdot 05$ and *not* $796 \cdot 5000543$.

In general, if ϵ_r is the maximum possible positive or negative error in x_r $(r = 1, 2, \ldots, n)$, then the maximum possible positive or negative error in the sum

$$\sum_{r=1}^{n} x_r \quad \text{is} \quad \sum_{r=1}^{n} \epsilon_r.$$

It is to be understood that in this context error is defined as the difference between the true value and the observed value represented by a digital approximation. It also follows from the rounding-off rule that numerically the maximum possible positive error in an approximate number is equal to the maximum possible negative error. The following example illustrates the way approximate numbers of different accuracies should be added.

Example 2.5

$$
\begin{aligned}
\text{Let} \quad x_1 &= 482 \cdot 56, & \epsilon_1 &= \pm 0 \cdot 005, \\
x_2 &= 791 \cdot 492, & \epsilon_2 &= \pm 0 \cdot 0005, \\
x_3 &= 882 \cdot 3475, & \epsilon_3 &= \pm 0 \cdot 00005, \\
x_4 &= 92 \cdot 6, & \epsilon_4 &= \pm 0 \cdot 05, \\
x_5 &= 532 \cdot 234, & \epsilon_5 &= \pm 0 \cdot 0005.
\end{aligned}
$$

The observed sum is $\sum_{r=1}^{5} x_r = 2781 \cdot 2335$, and the maximum possible value of the sum is

$$\sum_{r=1}^{5} (x_r + \epsilon_r) = 2781 \cdot 28955.$$

Therefore the maximum possible error in the sum is $\pm 0 \cdot 06$.

Hence $\sum_{r=1}^{5} x_r = 2781 \cdot 23$, with an error of $\pm 0 \cdot 06$.

(ii)   *Averages*

An important case in the addition of approximate numbers is that of averaging. As we have observed, the entries in all numerical tables and the results of all measurements are expressed in terms of approximate numbers in which the error is not greater than half a unit in the last significant figure. In general, these errors are equally likely to be positive or negative so that their algebraic sum is never large. Usually we may expect it to be less than half a unit in the last figure. Thus the error in the average

$\frac{1}{n} \sum_{r=1}^{n} x_r$ is $\frac{1}{n} \sum_{r=1}^{n} \epsilon_r$. Hence if $n \geqslant 10$ and x_r are correct to d significant figures, then their average is usually correct to $d + 1$ significant figures. This is an important instance when the accuracy of a derived figure is more than the accuracy of the numbers on which it is based.

(iii)   *Subtraction*

Suppose x_1 and x_2 are approximate numbers with maximum possible errors ϵ_1 and ϵ_2 respectively. The difference is $x_1 - x_2$; but since the errors may be either positive or negative, therefore the maximum positive error of x_1 is to be combined with the maximum negative error of x_2 to obtain the maximum positive error of the difference $x_1 - x_2$. Hence the maximum positive error in $x_1 - x_2$ is $\epsilon_1 + \epsilon_2$, and similarly the maximum negative error is $-(\epsilon_1 + \epsilon_2)$.

Example 2.6

$$\text{Let} \quad x_1 = 254 \cdot 56, \quad \epsilon_1 = \pm 0 \cdot 005,$$
$$\text{and} \quad x_2 = 189 \cdot 0972, \quad \epsilon_2 = \pm 0 \cdot 00005.$$

The observed value of the difference $x_1 - x_2 = 65 \cdot 4628$, and the maximum value of the difference is $(x_1 + \epsilon_1) - (x_2 - \epsilon_2) = 65 \cdot 46785$. Therefore the maximum possible error in the difference is $\pm 0 \cdot 005$. Hence $x_1 - x_2 = 65 \cdot 463$, with an error of $\pm 0 \cdot 005$.

The most important point in the subtraction of two approximate numbers has already been referred to. It arises when two nearly equal approximate numbers are subtracted. For example, consider the difference of $74 \cdot 3958$ and $73 \cdot 9824$ which are both correct to six significant figures. The difference is $0 \cdot 4134$ with an error of $\pm 0 \cdot 0001$.

Errors arising from the disappearance of the most important figures on the left of two approximate numbers are quite frequent. It is important to plan these calculations carefully to eliminate such loss of accuracy, for otherwise a whole piece of computation can be rendered quite useless. The general rule is simple: if the difference of two approximate numbers is desired to d significant figures, and if it is known that k of the figures on the left will disappear on subtraction, then $d + k$ significant figures are required in each of the numbers initially.

Example 2.7

Calculate the difference of $\sqrt{52 \cdot 02}$ and $\sqrt{52}$ correct to five significant figures.

Now $\sqrt{52 \cdot 02} = 7 \cdot 2124892$ correct to eight significant figures,
and $\sqrt{52} = 7 \cdot 2111026$ correct to eight significant figures.
Hence the difference $\sqrt{52 \cdot 02} - \sqrt{52} = 0 \cdot 0013866$ correct to five significant figures.

(iv) *Multiplication*

Let the maximum possible error in x_r be ϵ_r $(r = 1, 2, ..., n)$. Then the observed product is $\prod\limits_{r=1}^{n} x_r$, and its maximum value could be as much as

$$\prod_{r=1}^{n} (x_r + \epsilon_r) = \prod_{r=1}^{n} x_r \left[1 + \frac{\epsilon_r}{x_r} \right]$$
$$= \prod_{r=1}^{n} x_r \times \prod_{r=1}^{n} \left[1 + \frac{\epsilon_r}{x_r} \right]$$
$$\sim \prod_{r=1}^{n} x_r \left[1 + \sum_{r=1}^{n} \frac{\epsilon_r}{x_r} \right], \quad \text{since } \frac{\epsilon_r}{x_r} \text{ is small.}$$

Therefore the maximum possible error in the product is

$$\prod_{r=1}^{n} (x_r + \epsilon_r) - \prod_{r=1}^{n} x_r \sim \prod_{r=1}^{n} x_r \times \left[\sum_{r=1}^{n} \frac{\epsilon_r}{x_r} \right].$$

In particular, for $n = 2$, the maximum error in the product is

$$(x_1 + \epsilon_1)(x_2 + \epsilon_2) - x_1 x_2 \sim x_1 \epsilon_2 + x_2 \epsilon_1.$$

Example 2.8

(i) Let $x_1 = 7 \cdot 24, \quad \epsilon_1 = \pm 0 \cdot 005,$
and $x_2 = 4 \cdot 673, \quad \epsilon_2 = \pm 0 \cdot 0005.$

Then $x_1 x_2 = 33 \cdot 83252$, and the maximum possible value of the product is

$$(x_1 + \epsilon_1)(x_2 + \epsilon_2) = 33 \cdot 8595075.$$

Hence the maximum possible error in the product is $0 \cdot 03$, so that

$$x_1 x_2 = 33 \cdot 83 \text{ with an error of } \pm 0 \cdot 03.$$

We are not justified in writing the product with more than two decimal places.

(ii) If $x_1 = 7 \cdot 240$, $\epsilon_1 = \pm 0 \cdot 0005$,

and $x_2 = 4 \cdot 673$, $\epsilon_2 = \pm 0 \cdot 0005$,

then the maximum possible error in the product is $0 \cdot 006$. We are therefore entitled to write the product as $33 \cdot 833$ with an error of $\pm 0 \cdot 006$.

(v) *Division*

If ϵ_1 and ϵ_2 are the maximum possible errors in x_1 and x_2, then the maximum possible error in the quotient x_1/x_2 is

$$(x_1 + \epsilon_1)(x_2 - \epsilon_2)^{-1} - \frac{x_1}{x_2} = \frac{x_1}{x_2}\left[1 + \frac{\epsilon_1}{x_1}\right]\left[1 - \frac{\epsilon_2}{x_2}\right]^{-1} - \frac{x_1}{x_2}$$

$$= \frac{x_1}{x_2}\left[1 + \frac{\epsilon_1}{x_1}\right]\left[1 + \frac{\epsilon_2}{x_2} + ...\right] - \frac{x_1}{x_2}$$

$$\sim \frac{\epsilon_1 x_2 + \epsilon_2 x_1}{x_2^2}, \text{ ignoring second-order terms.}$$

In the same way, the maximum negative error in the quotient x_1/x_2 is

$$(x_1 - \epsilon_1)(x_2 + \epsilon_2)^{-1} - \frac{x_1}{x_2} \sim - \frac{\epsilon_1 x_2 + \epsilon_2 x_1}{x_2^2}, \text{ ignoring second-order terms.}$$

Example 2.9

$$\text{Let} \quad x_1 = 16 \cdot 837, \quad \epsilon_1 = \pm 0 \cdot 0005,$$
$$\text{and} \quad x_2 = 9 \cdot 5492, \quad \epsilon_2 = \pm 0 \cdot 00005.$$

$$\text{Therefore } \frac{x_1}{x_2} = 1 \cdot 763184 ... \quad \text{and} \quad \frac{x_1 + \epsilon_1}{x_2 - \epsilon_2} = 1 \cdot 763245 ... .$$

Hence the maximum possible error in $\frac{x_1}{x_2}$ is $0 \cdot 00006$, so that we can write the quotient as $1 \cdot 76318$ with an error of $\pm 0 \cdot 00006$.

2.6   Absolute, relative, and percentage errors

We have so far considered error as the difference between the observed and true value of a number or a measurement. Thus if x is the observed value and $x + \epsilon$ the true value, then

$$\text{Error} = (x + \epsilon) - x = \epsilon.$$

This error is known as the *absolute error*, and it is dependent upon the unit of measurement of x. The *relative error* is absolute error divided by the true value. Thus the relative error of x is $\epsilon/(x + \epsilon)$. The *percentage error* is $100 \times$ relative error.

The relative and percentage errors are both independent of the unit of measurement, and therefore provide the true index of the accuracy of a measurement or a calculation.

Example 2.10

A three-inch steel shaft is measured to the nearest thousandth of an inch, and a mile of railway line to the nearest foot. The absolute errors of the two measurements are $0 \cdot 0005$ of an inch and six inches respectively. But the corresponding relative errors are

$$\frac{0 \cdot 0005}{3} = \frac{1}{6000} \quad \text{and} \quad \frac{0 \cdot 5}{1760 \times 3} = \frac{1}{10560}.$$

Therefore the measurement of the railway line is more accurate than that of the steel shaft, although the absolute error of the former measurement is 12,000 times as great as that of the latter measurement.

We recall that if x_1 and x_2 are approximate numbers with maximum possible errors ϵ_1 and ϵ_2 respectively, then the maximum possible errors in the sum $x_1 + x_2$ and the difference $x_1 - x_2$ are both equal to $\epsilon_1 + \epsilon_2$, that is, the sum of the absolute errors in x_1 and x_2. On the other hand, the maximum possible relative error in the product $x_1 x_2$ is

$$\frac{\epsilon_1 x_2 + \epsilon_2 x_1 + \epsilon_1 \epsilon_2}{(x_1 + \epsilon_1)(x_2 + \epsilon_2)} = \frac{\epsilon_1}{x_1 + \epsilon_1} + \frac{\epsilon_2}{x_2 + \epsilon_2}\left[1 + \frac{\epsilon_1}{x_1}\right]^{-1}$$

$$\sim \frac{\epsilon_1}{x_1 + \epsilon_1} + \frac{\epsilon_2}{x_2 + \epsilon_2}, \text{ ignoring second-order terms.}$$

Thus the maximum possible relative error in the product $x_1 x_2$ is the sum of the relative errors in x_1 and x_2. In the same way, the maximum possible relative error in the quotient x_1 / x_2 is

$$\left[\frac{x_1 + \epsilon_1}{x_2 - \epsilon_2} - \frac{x_1}{x_2}\right] \Big/ \left[\frac{x_1 + \epsilon_1}{x_2 + \epsilon_2}\right] = \frac{(\epsilon_1 x_2 + x_1 \epsilon_2)(x_2 + \epsilon_2)}{(x_1 + \epsilon_1)(x_2 - \epsilon_2) x_2}$$

$$= \frac{\epsilon_1}{x_1 + \epsilon_1}\left[1 - \frac{\epsilon_2}{x_2}\right]^{-1} + \frac{\epsilon_2}{x_2 + \epsilon_2}\left[1 + \frac{\epsilon_1}{x_1}\right]^{-1}\left[1 + \frac{\epsilon_2}{x_2}\right]^{2}\left[1 - \frac{\epsilon_2}{x_2}\right]^{-1}$$

$$\sim \frac{\epsilon_1}{x_1 + \epsilon_1} + \frac{\epsilon_2}{x_2 + \epsilon_2}, \text{ ignoring second-order terms.}$$

Hence the maximum possible relative error in the quotient x_1 / x_2 is also the sum of the relative errors in x_1 and x_2.

We therefore conclude that absolute error is connected with decimal places and is important in addition and subtraction, whereas relative error is associated with significant figures and is important in multiplication and division.

2.7 Maximum errors and random variation

The error analysis based on the maximum possible errors considered in Section 2.5 is useful in avoiding particular sources of error. However, in any extensive calculation an error analysis made on the basis of the maximum possible error will exaggerate the errors which are likely to occur in practice. For example, suppose $x_1, x_2, \ldots, x_n$ are n numbers, each having a maximum possible error ϵ. Then the maximum possible error in the sum $\sum_{r=1}^{n} x_r$ is $n\epsilon$, but this is hardly ever likely to be realised. The essential point is that if the x_r are unrelated numbers, then their errors will have a "random variation". This means that, in general, the errors will tend to have the possible different values in the interval $\pm\epsilon$ with equal frequency, and so the actual error in $\sum_{r=1}^{n} x_r$ will be less than $n\epsilon$. Of course, since the actual errors in the x_r are unknown, we cannot know for certain the actual error in $\sum_{r=1}^{n} x_r$. Nevertheless, an argument based on statistical theory indicates that there is approximately a five per cent chance that the actual error in $\sum_{r=1}^{n} x_r$ will exceed $2(n/3)^{\frac{1}{2}}\epsilon$. This is considerably less than $n\epsilon$ for large n. To anticipate matters, we may also state that conclusions of this kind, which have a measure of uncertainty associated with them, are typical of statistical inference.

CHAPTER 3

INTERPOLATION AND FINITE DIFFERENCES

3.1 The problem of interpolation

One of the basic ideas in mathematics is that of *correspondence*. To illustrate this concept by a very simple example, suppose x is a variable which can take the positive integral values $1, 2, 3, \ldots$. Then to every value of x there corresponds a value of x^3 so that

$$x = 1, 2, 3, 4, 5, \ldots,$$
$$x^3 = 1, 8, 27, 64, 125, \ldots .$$

It is usual to call one of the variables between which correspondence holds as the *argument*, and the other variable is said to be a function of the argument. Accordingly, in our example x is the argument and x^3 is the function of x.

In general, if y is a function of an argument x, then the defining relation may be formally denoted by the equation $y = f(x)$. The forms of $f(x)$ can of course be very diverse, and for purposes of numerical computation we may subdivide them into two broad classes. Firstly, suppose $f(x)$ is an algebraic expression involving the basic arithmetical operations. Then, given x, we can determine accurately the corresponding value of $f(x)$. Secondly, $f(x)$ may be of the type $\log x$, $\tan x$ or

$$\frac{1}{\sqrt{2\pi}} \int_0^x e^{-\frac{1}{2}t^2} \, dt,$$

so that it is not possible to calculate the value of $f(x)$ for any x by using a *finite* number of arithmetical operations. Such functions are of frequent occurrence in mathematics, and one of the important tasks in numerical analysis is to determine techniques for calculating readily the values of such functions. For this purpose, we are obliged to have recourse to a *table* which gives the values of $f(x)$ corresponding to certain selected values of the argument x. Indeed, this is exactly what mathematical tables provide. For example, Table 3.1 overleaf gives the values of $\log_{10} x$ for all integral values of x in the range $1500 \leqslant x \leqslant 1510$. The given values of $\log_{10} x$ are approximate numbers which have been rounded off to nine decimal places. Given this table, we are concerned with finding out some method for determining the values of the function $\log x$ for values of the argument x which are intermediate between the tabulated values. For example, we may wish to determine $\log_{10} x$ for $x = 1500 \cdot 78$ to the same

21

Table 3.1   Showing the values of $\log_{10} x$ for $x = 1500(1)\,1510$

x	$\log_{10} x$	x	$\log_{10} x$
1500	3·176091259	1506	3·177824972
1501	3·176380692	1507	3·178113252
1502	3·176669933	1508	3·178401341
1503	3·176958981	1509	3·178689239
1504	3·177247836	1510	3·178976947
1505	3·177536500		

degree of accuracy as the tabulated values. The solution of this problem is provided by the theory of *interpolation*, which in its most elementary aspect may be described as the science of "reading between the lines of a mathematical table".

It is to be noted that the argument in Table 3.1 changes by equal intervals. This is not essential to the theory of interpolation, but it is a convenient property of most published tables. As such, much of the theory of interpolation has been developed for *equispaced* arguments, and this is also the convention that we shall follow here. Furthermore, we shall be concerned with only such elementary ideas of the theory as are relevant to the scope and nature of the numerical work necessary to the study of this book.

Suppose the numerical values of a function $y = f(x)$ are given only at an enumerable set of values of the argument x, namely,

$$x = a, \quad a + h, \quad a + 2h, \quad a + 3h, \quad a + 4h, \quad \ldots$$

and nowhere else. We may denote the corresponding values of y as y_0, y_1, y_2, y_3, y_4, ... With this specification, our problem is to determine the value of $f(x)$ when x lies between two consecutive values of the argument for which $f(x)$ is tabulated. However, it is important to note that for the theory of interpolation to be applicable, it is *not* essential that the functional form of $f(x)$ be known. The only information needed, apart from the *smoothness* of $f(x)$, are the values of the function given for $x + rh$ $(r \geqslant 0)$. Indeed, some of the interesting applications of the theory of interpolation in applied mathematics are associated with empirical functions whose functional forms cannot be specified. In principle, the method of interpolation consists in fitting a polynomial to the given set of points (x,y) and finding the intercept y when $x = a + \theta h$, say, where $0 < \theta < 1$. This method was first introduced by James Gregory in 1670. The formal justification of this method is partly based on the following theorem proved by Weierstrass in 1885.

Weierstrass's theorem

Every function $f(x)$ which is continuous in an interval (a, b) can be represented in that interval, to any specified degree of accuracy, by a polynomial; that is, it is possible to find a polynomial $P(x)$ such that

$$|f(x) - P(x)| < \epsilon$$

for every value of x in the interval (a, b), where $\epsilon > 0$ and is arbitrary.

We do not prove this theorem here, but it is pertinent to indicate briefly its importance and its one major limitation. The theorem implies that in practice, when the tabulated function values are given to a limited degree of accuracy, we can replace the continuous function $f(x)$ over the range by a suitable approximating polynomial $P(x)$ without appreciable loss of accuracy. The drawback lies in that the theorem gives no indication of the degree of the polynomial $P(x)$. This is a fundamental practical point and, as we shall see, the degree of $P(x)$ affects the accuracy of our interpolation procedure. In essence, the underlying assumption about $f(x)$ which justifies the method of interpolation is more than the continuity of $f(x)$ over the specified range. The assumption refers to the smoothness of $f(x)$. Roughly, this signifies that the shape of the curve $y = f(x)$ changes gradually over the specified range so that the tabulated y values give a good indication of its shape. It is then possible to obtain for $f(x)$ a polynomial representation of specified degree to the required accuracy.

The interpolation based on such an approximate representation of $f(x)$ is known as *polynomial interpolation*. However, it is not necessary to calculate the polynomial coefficients as they can all be expressed in terms of *finite differences*.

3.2 Finite differences

Suppose a function $y = f(x)$ has the tabulated values $y_0, y_1, y_2, \ldots, y_n$ corresponding to the values of the argument $a, a + h, a + 2h, \ldots, a + nh$. Then the *first differences* of the y values are

$$y_1 - y_0, \quad y_2 - y_1, \quad \ldots, \quad y_n - y_{n-1},$$

and it is conventional to denote them as $\Delta y_r = y_{r+1} - y_r$ for all $0 \leqslant r \leqslant n-1$. In this notation Δ is known as the *difference operator* which operates on the function values y_r. In the same way, the *second differences* of the y_r are defined as the differences of the first differences. The second differences are denoted by the power superscript 2 on Δ so that

$$\Delta^2 y_r = \Delta(\Delta y_r) = \Delta y_{r+1} - \Delta y_r.$$

The process is continued to define the *third differences* as

$$\Delta^3 y_r = \Delta^2 y_{r+1} - \Delta^2 y_r,$$

and, in general, for kth differences

$$\Delta^k y_r = \Delta^{k-1} y_{r+1} - \Delta^{k-1} y_r.$$

The difference operator Δ is analogous to the operator $\dfrac{d}{dx}$ in the calculus, and it is always placed before the function being differenced. Again, although each higher difference is expressed in terms of the preceding differences, it is possible by continuous substitution to obtain higher differences in terms of the function values y_r. For example,

$$\Delta^2 y_r = \Delta y_{r+1} - \Delta y_r = (y_{r+2} - y_{r+1}) - (y_{r+1} - y_r)$$
$$= y_{r+2} - 2y_{r+1} + y_r.$$

Similarly,

$$\Delta^3 y_r = \Delta^2 y_{r+1} - \Delta^2 y_r = (y_{r+3} - 2y_{r+2} + y_{r+1}) - (y_{r+2} - 2y_{r+1} + y_r)$$

or $$\Delta^3 y_r = y_{r+3} - 3y_{r+2} + 3y_{r+1} - y_r.$$

The formation of these higher differences suggests the following theorem.

Theorem

For any non-negative integer k,

$$\Delta^k y_r = \sum_{i=0}^{k} (-1)^{k-i} \binom{k}{i} y_{r+i}.$$

Proof

We prove this result by induction. The theorem holds for $k = 1, 2,$ and 3. Suppose, then, it also holds for some integer m. Then

$$\Delta^m y_r = \sum_{i=0}^{m} (-1)^{m-i} \binom{m}{i} y_{r+i},$$

so that

$$\Delta^{m+1} y_r = \sum_{i=0}^{m} (-1)^{m-i} \binom{m}{i} \Delta y_{r+i}$$

$$= \sum_{i=0}^{m} (-1)^{m-i} \binom{m}{i} (y_{r+i+1} - y_{r+i})$$

$$= \sum_{i=0}^{m} (-1)^{m-i} \binom{m}{i} y_{r+i+1} - \sum_{i=0}^{m} (-1)^{m-i} \binom{m}{i} y_{r+i}$$

$$= \sum_{j=1}^{m+1} (-1)^{m-j+1} \binom{m}{j-1} y_{r+j} + \sum_{i=0}^{m} (-1)^{m-i+1} \binom{m}{i} y_{r+i}$$

$$= y_{r+m+1} + \sum_{i=1}^{m} (-1)^{m-i+1} \left[\binom{m}{i-1} + \binom{m}{i} \right] y_{r+i} + (-1)^{m+1} y_r$$

$$= y_{r+m+1} + \sum_{i=1}^{m} (-1)^{m-i+1} \binom{m+1}{i} y_{r+1} + (-1)^{m+1} y_r,$$

or $$\Delta^{m+1} y_r = \sum_{i=0}^{m+1} (-1)^{m-i+1} \binom{m+1}{i} y_{r+i}.$$

Therefore the result also holds for $m + 1$, whence the theorem by induction.

The totality of the differences of $y = f(x)$ are known as *finite differences*.

3.3   Difference tables

It is usual to arrange the differences of a function in the form of a table, and the most commonly used scheme is that of the *diagonal difference table* presented in Table 3.2.

A table of this kind is made out by writing the initial pairs $(a + rh, y_r)$ on alternate lines of computing paper. The differences are then written on the intermediate lines between the two numbers for which they represent the differences. Differences with the same suffix in Table 3.2 are called *forward differences*. The first term y_0 in the table is called the *leading term*, and the differences of y_0, namely, $\Delta y_0, \Delta^2 y_0, \Delta^3 y_0, \ldots,$ are called *leading differences*. These lie along the top diagonal of the table. In the same way,

Table 3.2 Showing a schematic representation of the forward differences of $f(x)$

x	$f(x)$	Δ	Δ^2	Δ^3	Δ^4	Δ^5
a	y_0					
		Δy_0				
$a+h$	y_1		$\Delta^2 y_0$			
		Δy_1		$\Delta^3 y_0$		
$a+2h$	y_2		$\Delta^2 y_1$		$\Delta^4 y_0$	
		Δy_2		$\Delta^3 y_1$		$\Delta^5 y_0$
$a+3h$	y_3		$\Delta^2 y_2$		$\Delta^4 y_1$	$\Delta^5 y_1$
		Δy_3		$\Delta^3 y_2$		
$a+4h$	y_4		$\Delta^2 y_3$		$\Delta^4 y_2$	$\Delta^5 y_2$
		Δy_4		$\Delta^3 y_3$		
$a+5h$	y_5		$\Delta^2 y_4$		$\Delta^4 y_3$	$\Delta^5 y_3$
		Δy_5		$\Delta^3 y_4$		
$a+6h$	y_6		$\Delta^2 y_5$		$\Delta^4 y_4$	
		Δy_6		$\Delta^3 y_5$		
$a+7h$	y_7		$\Delta^2 y_6$			
		Δy_7				
$a+8h$	y_8					

the differences of any y_r lie along a diagonal.

A simple but useful check for such a difference table is that the sum of the entries in any column of differences is equal to the difference between the first and last entries in the preceding column. For example,

$$\Delta y_7 - \Delta y_0 = \sum_{r=0}^{6} \Delta^2 y_r.$$

Example 3.1

There are some further considerations in the making of such a forward difference table which are best illustrated by a numerical example. For this purpose we difference the values of $\log_{10} x$ given in Table 3.1 above.

Notes

The following points in the formation of Table 3.3 deserve attention.

(i) The function $\log_{10} x$ is not a polynomial function of the argument x and its values are approximate numbers which have been rounded off correct to nine decimal places.

(ii) In writing down the differences the decimal points and all zeros appearing on the left have been omitted. This is standard practice, and it is clear from the accuracy of the function values that all differences contain nine decimal places. Thus,

$$\Delta \log_{10} 1500 = 0 \cdot 000\,289\,433,$$
$$\Delta^2 \log_{10} 1500 = -0 \cdot 000\,000\,192, \text{ etc.,}$$

although all the non-significant zeros and the decimal points are not written in Table 3.3.

(iii) Since $\log_{10} x$ is an increasing function of the argument x, it is more convenient on a desk calculator to start the differencing from the bottom. This ensures that each particular function value can be kept

Table 3.3 Showing the differences of $\log_{10} x$ tabulated for $x = 1500(1)1510$

x	$\log_{10} x$	Δ	Δ^2	Δ^3
1500	3·176091259			
		289433		
1501	3·176380692		−192	
		289241		−1
1502	3·176669933		−193	
		289048		0
1503	3·176958981		−193	
		288855		2
1504	3·177247836		−191	
		288664		−1
1505	3·177536500		−192	
		288472		0
1506	3·177824972		−192	
		288280		1
1507	3·178113252		−191	
		288089		0
1508	3·178401341		−191	
		287898		1
1509	3·178689239		−190	
		287708		
1510	3·178976947			

on the setting levers for two successive operations. The first differences of $\log_{10} x$ are decreasing, and so these should be differenced
by starting from the top of the table. Such simple planning reduces the
computational labour considerably.

(iv) The second differences are almost equal, and the third differences
indicate only rounding-off errors. This implies that although $\log_{10} x$
is not a polynomial function of x, still, to the accuracy to which the
function values are given, it can be adequately represented by a second-
degree polynomial.

(v) In the tabulated range not all functions will exhibit the steady increasing or decreasing characteristics revealed by $\log_{10} x$. In such
cases the function values and differences will usually fluctuate in magnitude and sign. Hence differencing will require particular care, and
the simple check mentioned above should necessarily be used.

3.4   Differences of a polynomial

As another instructive example, consider the differences of the polynomial $x^4/10^7$ given in Table 3.4. We observe that the fourth differences of
the function are equal. This is a characteristic result for polynomials of all
degrees. Hence the following theorem.

Theorem

For equispaced argument, the nth differences of a polynomial of degree
n are constant. Conversely, if the nth differences of a function tabulated
for equispaced intervals of the argument are constant, then the function may

Table 3.4 Showing the differences of the polynomial function $x^4/10^7$ tabulated for $x = 150(1)155$

x	$x^4/10^7$	Δ	Δ^2	Δ^3	Δ^4
150	50·6250000				
		1363560 1			
151	51·9885601		27 3614		
		13909215		3636	
152	53·3794816		277250		24
		14186465		3660	
153	54·7981281		280910		24
		14467375		3684	
154	56·2448656		284594		
		14751969			
155	57·7200625				

be represented by a polynomial of degree n.

Proof

Let $f(x) = \sum\limits_{r=0}^{n} a_r x^r$, where the a_r are constants and $a_n \neq 0$. If the interval of the argument is h, then

$$\Delta f(x) = f(x + h) - f(x) = \sum_{r=0}^{n} a_r \left[(x + h)^r - x^r \right]$$

$$= \sum_{r=1}^{n} a_r \left[\binom{r}{1} x^{r-1} h + \binom{r}{2} x^{r-2} h^2 + \ldots + \binom{r}{r} h^r \right].$$

which is a polynomial of degree $n - 1$.

Again, $\Delta^2 f(x) = \Delta [\Delta f(x)]$

$\qquad\qquad = \Delta [\text{a polynomial of degree } n - 1]$

$\qquad\qquad = \text{a polynomial of degree } n - 2.$

Hence successive differencing gives

$$\Delta^n f(x) = \text{a constant,}$$
$$\Delta^{n+k} f(x) = 0, \text{ for } k \geqslant 1.$$

Conversely, suppose $f(x)$ is a polynomial of degree $m \neq n$ and $\Delta^n f(x)$ is constant. But we know that if $m < n$, then $\Delta^n f(x) = 0$. Also, if $m > n$, then $\Delta^n f(x)$ is a polynomial of degree $m - n$. Hence $\Delta^n f(x)$ is a constant and non-zero only if $m = n$.

3.5 Tabulation of polynomials

The property that the nth differences of a polynomial of degree n are constant can be used to tabulate a polynomial for equispaced intervals of the argument. This is done by building up successively the lower differences from the higher ones by repeated addition, using the relation

$$\Delta^{r-1} y_{k+1} = \Delta^r y_k + \Delta^{r-1} y_k. \tag{1}$$

Thus it is necessary to evaluate $n + 1$ values and the initial n differences to start the construction of the difference table in reverse. However, it is advisable as a check to compute directly one or two more values of the function than the minimum $n + 1$ required to start the table.

Example 3.2

We illustrate the computational plan of the method by using it to tabulate the polynomial

$$y = 6x^3 - 25x^2 + 17x + 89, \quad \text{for } x = 0(1)12.$$

The notation $x = 0(1)12$ means that all values of x between 0 and 12 at unit interval have to be considered in the tabulation. The computations are presented in Table 3.5.

Table 3.5   Showing the tabulation of the values of the polynomial
$y = 6x^3 - 25x^2 + 17x + 89$ for $x = 0(1)12$

x	$6x^3$	$-25x^2$	$17x$	89	y	Δ	Δ^2	Δ^3
0	0	0	0	89	89	-2		
1	6	-25	17	89	87	-16	-14	
2	48	-100	34	89	71	6	22	36
3	162	-225	51	89	77	64	58	36
4	384	-400	68	89	141	158	94	36
5					299	288	130	36
6					587	454	166	36
7					1041	656	202	36
8					1697	894	238	36
9					2591	1168	274	36
10					3759	1478	310	36
11					5237	1824	346	36
12	10368	-3600	204	89	7061			

The values above the unbroken line in Table 3.5 have been calculated directly, and are necessary to start the building up of the y values. The values between this line and the broken line are also calculated directly but only to provide a check on the constancy of the third differences. The result of the table is built up by summation of the differences implied by (1). The value of the polynomial for $x = 12$ is obtained directly to provide another check on the accuracy of the tabulation.

Caution

It is necessary in using this process to keep all figures without rounding off, even though the final values of the polynomial are not required to this accuracy.

3.6   Checking by differences

Another simple but useful application of differences is that of checking for random (but not systematic) mistakes in tabulated values of a function. Such mistakes are usually due to some human fallibility though, less frequently,   they may be the outcome of an undetected technical flaw in a desk calculator. Suppose the tabulated values of a function are y_{-4}, y_{-3}, y_{-2}, y_{-1}, $y_0 + \epsilon$, y_1, y_2, y_3, y_4 corresponding to equispaced values of the argument $x = a - 4h$, $a - 3h$, ..., $a + 4h$, where ϵ is the undetected inaccuracy in y_0, the correct value of the function for $x = a$. Table 3.6 below gives the difference table formed from these values.

Table 3.6   Showing the effect of a mistake in a function value in the successive differences of $f(x)$

x	y	Δ	Δ^2	Δ^3	Δ^4
$a - 4h$	y_{-4}				
		Δy_{-4}			
$a - 3h$	y_{-3}		$\Delta^2 y_{-4}$		
		Δy_{-3}		$\Delta^3 y_{-4}$	
$a - 2h$	y_{-2}		$\Delta^2 y_{-3}$		$\Delta^4 y_{-4} + \epsilon$
		Δy_{-2}		$\Delta^3 y_{-3} + \epsilon$	
$a - h$	y_{-1}		$\Delta^2 y_{-2} + \epsilon$		$\Delta^4 y_{-3} - 4\epsilon$
		$\Delta y_{-1} + \epsilon$		$\Delta^3 y_{-2} - 3\epsilon$	
a	$y_0 + \epsilon$		$\Delta^2 y_{-1} - 2\epsilon$		$\Delta^4 y_{-2} + 6\epsilon$
		$\Delta y_0 - \epsilon$		$\Delta^3 y_{-1} + 3\epsilon$	
$a + h$	y_1		$\Delta^2 y_0 + \epsilon$		$\Delta^4 y_{-1} - 4\epsilon$
		Δy_1		$\Delta^3 y_0 - \epsilon$	
$a + 2h$	y_2		$\Delta^2 y_1$		$\Delta^4 y_0 + \epsilon$
		Δy_2		$\Delta^3 y_1$	
$a + 3h$	y_3		$\Delta^2 y_2$		
		Δy_3			
$a + 4h$	y_4				

Notes

The following points indicate how the growth of the mistake spreads out in the successive differences of the y values, and the considerations involved in correction.

(i)   The mistake spreads out fanwise through the higher differences.

(ii)   The coefficients of the ϵ's in the rth differences are the rth-order binomial coefficients with alternating signs.

(iii) The binomial coefficients multiplied by ϵ are centred in each column of differences about the horizontal line through the inaccurate function value $y_0 + \epsilon$.

(iv) If the mistake is $y_0 - \epsilon$, then the contribution of the ϵ's in the difference columns will change in sign.

(v) If the function differenced is a polynomial of degree three, and the inaccurate value is $y_0 + \epsilon$, then the fourth differences will be $\epsilon(1, -4, 6, -4, 1)$.

(vi) A random mistake in a polynomial function can thus be clearly shown to exist and located by differencing the tabular values.

(vii) If the tabulated function is not a polynomial then, in general, its values will be approximate numbers affected by rounding-off errors. This will also be the case if the function is an exact polynomial but its table values have been rounded off to a certain degree of accuracy. In such cases the mistakes will be confounded with the random error contributions to the differences, and the binomial coefficients will not show up with full accuracy as implied in the formal algebraic scheme presented in Table 3.6. However, if ϵ is numerically large compared with the rounding-off errors, the progressively increasing contributions of the mistake will show up quite distinctly.

(viii) If two or more near function values are inaccurate with inaccuracies ϵ_1, ϵ_2, ... respectively, then the fanwise contributions of the ϵ's will overlap, and the binomial coefficients will not apparently appear as distinctly as in Table 3.6. In such cases the existence and location of the mistakes will be a matter of some judgement; and in particular cases, especially when the function values are approximate numbers, it may not be possible to correct the mistakes through adjustment of the differences. The appropriate procedure in these instances is to recalculate the inaccurate function values.

We next illustrate these points in the correction of the mistakes in two numerical examples. In the first of these the tabulated function is an exact polynomial, and the second example pertains to a function whose tabulated values are approximate numbers affected by rounding-off errors.

Table 3.7   Showing the correction of a mistake in an exact polynomial function $y = f(x)$

x	y	Δ	Δ^2	Δ^3	Δ^4	Δ^5	Corrections to Δ^5
0·10	11·5000						
		5191					
0·11	12·0191		354				
		5545		−24			
0·12	12·5736		330		24		
		5875		0		360	−360
0·13	13·1611		330		384		
		6205		384		−1800	1800
0·14	13·7816		714		−1416		
		6919		−1032		3600	−3600
0·15	14·4735		−318		2184		
		6601		1152		−3600	3600
0·16	15·1336		834		−1416		
		7435		−264		1800	−1800
0·17	15·8771		570		384		
		8005		120		−360	360
0·18	16·6776		690		24		
		8695		144			
0·19	17·5471		834				
		9529					
0·20	18·5000						

Example 3.3 (see Table 3.7)

The fifth differences of the function clearly indicate that the y value corresponding to $x = 0 \cdot 15$ is incorrect. These differences are $360 \times (1, -5, 10, -10, 5, -1)$, and so the mistake is an increase in the function value by 360. Therefore the correct value of y is $14 \cdot 4375$.

This correction illustrates another important point. We note that the numerical correction in the function value is 9×40. Such multiples of 9 almost invariably show up when two consecutive digits are transposed. In our particular example, we have in fact interchanged the digits 3 and 7 and so increased the function value to $14 \cdot 4735$. The figure 40 indicates that the error is in the second digit from the right and it is due to an interchange of two consecutive digits which differ by 4.

It is also easily seen that the corrections to the fifth differences are $360(-1, 5, -10, 10, -5, 1)$.

Example 3.4

Table 3.8   Showing the correction of two mistakes in a function whose tabular values $y = f(x)$ are also affected by rounding-off errors

x	y	Δ	Δ^2	Δ^3	Corrections to Δ^3	Corrected Δ^3
1	8·7799					
		6151				
2	9·3950		−101			
		6050		1		1
3	10·0000		−100			
		5950		−719	720	1
4	10·5950		−819			
		5131		2160	−2160	0
5	11·1081		1341			
		6472		−2160	2160	0
6	11·7553		−819			
		5653		686	−720 + 36	2
7	12·3206		−133			
		5520		109	− 108	1
8	12·8726		− 24			
		5496		−110	108	−2
9	13·4222		−134			
		5362		39	− 36	3
10	13·9584		−95			
		5267		0		0
11	14·4851		−95			
		5172		0		0
12	15·0023		−95			
		5077				
13	15·5100					

Notes

(i) It is not too difficult to see from the third differences in Table 3.8 that the y values corresponding to $x = 5$ and $x = 8$ are incorrect, there being some overlap in the two fanwise contributions of the mistakes. The magnitudes of the mistakes require some more thought. It is clear that the contributions to the third differences due to the incorrect function value for $x = 5$ are $720(1, -3, 3, -1)$, and $720 = 9 \times 80$. Hence the correct function value for $x = 5$ is $y = 11 \cdot 1801$. There is, on the other hand, some doubt about the magnitude of the mistake in the y value corresponding to $x = 8$. But it is seen that the best overall reduction in the third differences is obtained when the contributions of the second mistake are taken to be $36(1, -3, 3, -1)$. Since $36 = 9 \times 4$, the correct value of the function for $x = 8$ is $12 \cdot 8762$. Hence the stated correct values of the third differences, which only reveal rounding-off errors.

(ii) In this example the two mistakes were numerically large compared with the rounding-off errors in the function values, and so we could convincingly determine the magnitudes of the two mistakes made. However, there is always a risk that the cumulative effects of rounding-off errors may be regarded in the higher differences as indicative of a small random mistake. As such, the correction of function values through the adjustment of differences should not be carried out unless there is almost certain evidence of the magnitude of the mistake. If the function is known, it is advisable to recalculate from first principles the suspected function value. Alternatively, if the function is not known or the y values are measurements of an empirical function, then the suspected function value could be calculated by interpolation using double the initial spacing of the argument. Thus, in the above example, the function value corresponding to $x = 8$ could be interpolated by using the difference table formed by differencing the function values for $x = 7, 9, 11$, and 13. After the correct value for $x = 8$ is obtained, the value for $x = 5$ is again found by interpolation using the function values for $x = 6, 8, 10$, and 12.

(iii) There is another practical consideration in the location of mistakes which needs mention. It is clear that if there are no mistakes in the function values being differenced, then the successive differences should decrease numerically. On the other hand, when mistakes are present then differences tend to decrease first and later begin to increase and also, generally, to alternate in sign. Differencing should not be carried out automatically, but each column should be carefully studied for possible indication of the binomial coefficients before the next column is obtained. Experience is the best teacher to indicate when to stop differencing.

(iv) The preceding analysis of the behaviour of difference tables leads us to a final caution. Apart from polynomials calculated to full accuracy, in most tables every function value will be in error due to rounding off. Although the rounding-off error in such approximate numbers is never more than half a unit in the last decimal place, the effect of such error is necessarily increased in the higher differences. Consequently, these differences do tend to become somewhat irregular, and the more so the higher the order of the differences. It is important to realise this, otherwise irregularities in higher differences, which are solely due to the accumulative effects of rounding-off errors, may be wrongly regarded as indicative of mistakes. In fact, such irregularities can only be eliminated (or lessened) by taking more significant figures in the initial tabulation of the function values. And in this context, it is also useful to remember that rounding off and then differencing is not the same as differencing and then rounding off. As a safe procedure, it is advisable to carry a couple of extra digits in all intermediate calculations to ensure the required accuracy of the final results.

3.7   Interpolation

We have considered some of the simple but useful applications of finite differences and difference tables, and we now revert to the basic problem of interpolation. We recall that we are given that the function $y = f(x)$ takes the values $y_0, y_1, y_2, \ldots, y_n$ corresponding to the values of $x = a$, $a + h$, $a + 2h$, $\ldots$, $a + nh$. The function $f(x)$ may be a polynomial, a non-polynomial function whose form is known, or an empirical function which has no known analytical form. Our basic assumption is that over the given range the function is smooth and can therefore be represented by a polynomial to any desired degree of accuracy. We shall use this representation to determine $f(a + \theta h)$, where $0 < \theta < 1$. The simplest method for interpolation of this kind is based on the forward difference table of the given values, and we shall consider this procedure only. Besides, it is instructive to consider the interpolation in three stages, that is, when

(i) the approximating polynomial is linear in x, which leads to *linear interpolation*;

(ii) the approximating polynomial is quadratic in x, which gives the *quadratic interpolation* formula; and

(iii) the approximating polynomial is of degree n in x. This gives the general *Gregory–Newton forward difference interpolation* formula (p.35).

(i) *Linear interpolation*

Suppose $y = c_0 + c_1 x$ is assumed to be an adequate representation of the function $y = f(x)$, where c_0 and $c_1 \neq 0$ are unknown constants. Then

$$\Delta f(a) = [c_0 + c_1(a + h)] - (c_0 + c_1 a) = c_1 h.$$

Therefore
$$f(a + \theta h) = c_0 + c_1(a + \theta h)$$
$$= (c_0 + c_1 a) + \theta\, c_1 h,$$
or
$$f(a + \theta h) = f(a) + \theta\, \Delta f(a),$$

which is the required linear interpolation formula. An alternative way of expressing this formula is

$$f(a + \theta h) = (1 - \theta) f(a) + \theta f(a + h). \tag{1}$$

This is a somewhat more useful form for machine calculations. Besides, this also shows that the interpolated value only depends on the two adjoining tabulated values of $f(x)$ for $x = a$ and $a + h$.

Example 3.5

Suppose $y = 1302$ for $x = 6$ and $y = 2408$ for $x = 7$. We wish to interpolate for the value of y for $x = 6 \cdot 2$.

Here $a = 6$, $h = 1$, and $\theta = 0 \cdot 2$.

Therefore, using (1),

$$f(6 \cdot 2) = 0 \cdot 8 \times 1302 + 0 \cdot 2 \times 2408 = 1523 \cdot 2.$$

(ii) *Quadratic interpolation*

Here we assume that $y = c_0 + c_1 x + c_2 x^2$ is a suitable representation of
the function $y = f(x)$, where the c's are unknown constants and $c_2 \neq 0$.

Then

$$\begin{aligned}
\Delta f(a) &= [c_0 + c_1(a + h) + c_2(a + h)^2] - (c_0 + c_1 a + c_2 a^2) \\
&= c_1 h + 2c_2 ah + c_2 h^2.
\end{aligned}$$

Similarly,

$$\begin{aligned}
\Delta^2 f(a) &= [c_1 h + 2c_2(a + h)h + c_2 h^2] - (c_1 h + 2c_2 ah + c_2 h^2) \\
&= 2c_2 h^2.
\end{aligned}$$

Hence

$$\begin{aligned}
f(a + \theta h) &= c_0 + c_1(a + \theta h) + c_2(a + \theta h)^2 \\
&= (c_0 + c_1 a + c_2 a^2) + \theta(c_1 h + 2c_2 ah) + \theta^2(c_2 h^2) \\
&= (c_0 + c_1 a + c_2 a^2) + \theta(c_1 h + 2c_2 ah + c_2 h^2) + (\theta^2 - \theta)(c_2 h^2),
\end{aligned}$$

or $f(a + \theta h) = f(a) + \theta \, \Delta f(a) + \dfrac{\theta(\theta - 1)}{2!} \Delta^2 f(a),$

which is the required quadratic interpolation formula.

Example 3.5 (continued)

Suppose the data of the example are extended to give the following
forward difference table:

Table 3.9   Showing the forward differences of the function $y = f(x)$
of Example 3.5, for $x = 6(1)8$

x	y	Δ	Δ^2
6	1302		
		1106	
7	2408		590
		1696	
8	4104		

We can now use the additional information given by this table to apply the
quadratic interpolation formula to obtain

$$f(6 \cdot 2) = 1302 + 0 \cdot 2 \times 1106 + \frac{0 \cdot 2(-0 \cdot 8)}{2} \times 590$$

$$= 1302 + 221 \cdot 2 - 47 \cdot 2 = 1476 \cdot 0.$$

(iii) *General polynomial interpolation*

As an obvious generalisation of the results for linear and quadratic
interpolation, we now assume that $y = f(x)$ can be represented by a polynomial of degree n. We then have the following theorem.

Theorem

If $f(x)$ may be represented by a polynomial of degree n (≥ 1) in x, then

$$f(a + \theta h) = \sum_{r=0}^{n} \frac{\theta^{(r)}}{r!} \, \Delta^r f(a), \quad \text{where } 0 < \theta < 1, \quad \text{and}$$

$$\theta^{(r)} \equiv \theta(\theta - 1)(\theta - 2) \dots (\theta - r + 1), \quad \text{for all } r.$$

This is the celebrated Gregory–Newton [James Gregory (1638–1675); Isaac Newton (1642–1727)] forward difference interpolation formula. We observe that this formula is certainly correct for $n = 1$ and 2. The general result is proved simply by using the following notational lemma.

Lemma

The *n*th *factorial power* of x is defined as

$$x^{(n)} = x(x - 1)(x - 2) \dots (x - n + 1), \quad \text{for all integral } n \geq 0.$$

Further, writing the difference of one unit in $x^{(n)}$ as

$$\Delta_1 x^{(n)} = (x + 1)^{(n)} - x^{(n)},$$

we have

$$\Delta_1 x^{(n)} = [(x + 1) x(x - 1) \dots (x - n + 2)] - [x(x - 1)(x - 2) \dots (x - n + 1)]$$
$$= x(x - 1)(x - 2) \dots (x - n + 2)[(x + 1) - (x - n + 1)]$$
$$= nx^{(n-1)}.$$

Similarly, $\Delta_1^2 x^{(n)} = n^{(2)} x^{(n-2)}$ and, more generally, for $r \leq n$,

$$\Delta_1^r x^{(n)} = n^{(r)} x^{(n-r)}.$$

Hence, in particular,

$$\Delta_1^n x^{(n)} = n^{(n)} = n!,$$
$$\Delta_1^{n+k} x^{(n)} = 0, \quad \text{for all } k > 0.$$

and

It is to be noted that the results of differencing $x^{(n)}$ are entirely analogous to those obtained by differentiating x^n. The proof of the theorem now follows by using this lemma.

Proof

By our assumptions,

$$f(a + \theta h) = \sum_{r=0}^{n} c_r (a + \theta h)^r,$$

where c_r are constants and $c_n \neq 0$. In general, the right-hand side is a polynomial of degree n in θ and may also be written formally as

$$f(a + \theta h) = \sum_{r=0}^{n} a_r \theta^{(r)} \equiv F(\theta), \quad \text{say}, \tag{1}$$

where a_r are constants and $a_n \neq 0$.

Now the difference of h units in $f(a + \theta h)$ is

$$\Delta_h f(a + \theta h) = f(a + \theta h + h) - f(a + \theta h)$$
$$= F(\theta + 1) - F(\theta)$$
$$= \Delta_1 F(\theta).$$

But

$$\Delta_1 F(\theta) = \Delta_1 \sum_{r=0}^{n} a_r \theta^{(r)}$$

$$= \sum_{r=1}^{n} r a_r \theta^{(r-1)} . \qquad (2)$$

Again,

$$\Delta_h^2 f(a + \theta h) = \Delta_h f[a + (\theta + 1)h] - \Delta_h f(a + \theta h)$$
$$= \Delta_1 F(\theta + 1) - \Delta_1 F(\theta) = \Delta_1^2 F(\theta).$$

But

$$\Delta_1^2 F(\theta) = \sum_{r=2}^{n} r(r - 1) a_r \theta^{(r-2)}. \qquad (3)$$

The differencing operations can be continued to give a sequence of equations (1), (2), (3), ... Hence, putting $\theta = 0$ in these equations, we have

$$a_0 = f(a); \quad a_1 = \Delta_h f(a); \quad a_2 = \frac{1}{2!} \Delta_h^2 f(a); \ ... \ ; \quad a_n = \frac{1}{n!} \Delta_h^n f(a).$$

Hence substitution in (1) and writing Δ for Δ_h gives the stated theorem.

Example 3.5 (concluded)

We are now given the y values for all integral values of x in the range $6 \leqslant x \leqslant 12$. The differences of these function values are presented in Table 3.10.

Table 3.10   Showing the forward differences of the function $y = f(x)$ of Example 3.5, for $x = 6(1)12$

x	y	Δ	Δ^2	Δ^3	Δ^4
6	1302				
		1106			
7	2408		590		
		1696		180	
8	4104		770		24
		2466		204	
9	6570		974		24
		3440		228	
10	10010		1202		24
		4642		252	
11	14652		1454		
		6096			
12	20748				

Since the fourth differences are constant we immediately infer that y is exactly a polynomial of degree four. Hence the fourth degree interpolation formula should give the exact value of y for $x = 6 \cdot 2$. To obtain the interpolated value of y, we compute systematically as follows:

$$a = 6; \quad h = 1; \quad \theta h = 0 \cdot 2; \quad \theta = 0 \cdot 2.$$

$$\tfrac{1}{2}(\theta - 1) = -0 \cdot 4 \qquad\qquad \frac{\theta^{(2)}}{2!} = -0 \cdot 08$$

$$\tfrac{1}{3}(\theta - 2) = -0 \cdot 6 \qquad\qquad \frac{\theta^{(3)}}{3!} = 0 \cdot 048$$

$$\tfrac{1}{4}(\theta - 3) = -0 \cdot 7 \qquad\qquad \frac{\theta^{(4)}}{4!} = -0 \cdot 0336$$

Therefore

$$f(6) = 1302$$

$$\theta\,\Delta f(6) = 221 \cdot 2$$

$$\frac{\theta^{(2)}}{2!}\,\Delta^2 f(6) = -47 \cdot 2$$

$$\frac{\theta^{(3)}}{3!}\,\Delta^3 f(6) = 8 \cdot 64$$

$$\frac{\theta^{(4)}}{4!}\,\Delta^4 f(6) = -0 \cdot 8064$$

whence, on addition,

$$f(6 \cdot 2) = 1483 \cdot 8336 .$$

We note that no approximations have been made in calculating this interpolated value, and so this is the exact value of $f(x)$ for $x = 6 \cdot 2$. In comparison, linear interpolation gives the approximation $1523 \cdot 2$ with a percentage error of $2 \cdot 65$ per cent. Similarly, quadratic interpolation gives the value $1476 \cdot 0$ with a percentage error of $-0 \cdot 53$ per cent. This reduction in the numerical value of the relative error is characteristic if higher differences are used in interpolation. Of course, since the tabulated function is a fourth-degree polynomial, we cannot go beyond fourth differences in interpolation.

We can also use this exact interpolation formula to determine the form of $f(x)$. We have for any a and θ

$$f(a + \theta h) = f(a) + \theta\Delta f(a) + \frac{\theta^{(2)}}{2!}\,\Delta^2 f(a) + \frac{\theta^{(3)}}{3!}\,\Delta^3 f(a) + \frac{\theta^{(4)}}{4!}\,\Delta^4 f(a).$$

Now $h = 1$, and if we set $a = 6$, then

$$f(6 + \theta) = f(6) + \theta\Delta f(6) + \frac{\theta^{(2)}}{2!}\,\Delta^2 f(6) + \frac{\theta^{(3)}}{3!}\,\Delta^3 f(6) + \frac{\theta^{(4)}}{4!}\,\Delta^4 f(6)$$

$$= 1302 + 1106\theta + 295\theta^{(2)} + 30\theta^{(3)} + \theta^{(4)}$$

If we next make the transformation $x = 6 + \theta$, then

$$
\begin{aligned}
f(x) &= 1302 + 1106(x - 6) + 295(x - 6)(x - 7) + 30(x - 6)(x - 7)(x - 8) + \\
&\quad + (x - 6)(x - 7)(x - 8)(x - 9) \\
&= 1302 + 1106(x - 6) + 295(x^2 - 13x + 42) + \\
&\quad + 30(x^3 - 21x^2 + 146x - 336) + \\
&\quad + (x^4 - 30x^3 + 335x^2 - 1650x + 3024) \\
&= x^4 + x.
\end{aligned}
$$

Thus $y = x^4 + x$ is the polynomial which gives the functional relation between x and y. It is easily verified that all given values of y in Table 3.10 are exact and so also is the value $y = 1483 \cdot 8336$, for $x = 6 \cdot 2$.

3.8 Degree of the polynomial used in interpolation

We now consider the practical point regarding the degree of the polynomial to be used for interpolation. For this purpose, we assume that we are given some function $y = f(x)$ which takes the values $y_0, y_1, y_2, \ldots, y_n$ corresponding to $x = a, a + h, a + 2h, \ldots, a + nh$, and we wish to evaluate y for $x = a + \theta h$ $(0 < \theta < 1)$. The function $f(x)$ may be

  (i)   an exact polynomial in x;

  (ii)   a mathematical non-polynomial function with a known form; or

  (iii)   an empirical function about which nothing more is known apart from the basic assumption of smoothness in the tabulated range of the argument.

The practical determination of the degree of the approximating polynomial used for interpolation in the above three instances requires slightly different considerations. However, there is one common point and this is that, with the available data, we cannot use an interpolation polynomial of degree greater than n to evaluate $f(a + \theta h)$. This is obvious since only n leading differences can be obtained from the $n + 1$ given function values.

  (i)   When $f(x)$ is, in fact, a polynomial of degree $m > n$, then the use of the first n leading differences for interpolation will necessarily involve an approximation error. Indeed, this was the case when we used linear and quadratic interpolation to obtain the interpolated values for the quartic in Example 3.5 above. However, when $m = n$, then the interpolated function value can be obtained to the same accuracy as that of the tabular y values.

  (ii)   If $f(x)$ is a non-polynomial function, then its polynomial representation is necessarily approximate whatever the degree of the polynomial used. On the other hand, the tabular values of such a function are, in general, given correct to a certain number of significant figures. The finite differences of these function values usually decrease steadily and, dependent upon the accuracy of the tabular values, differences of some order k become approximately equal. This implies that the interpolation formula need only be considered up to the kth differences. Therefore the interpolated value $f(a + \theta h)$ can be as accurate as the tabular values if $k \leqslant n$; but if a polynomial of degree less than k is used, then the interpolated value cannot be as accurate as the given y values. Such inaccuracies should be avoided in interpolation unless only a simple approximation is all that is required.

  (iii)   With an empirical function somewhat more elaborate considerations are required. The tabular values of such a function are approximate numbers, but their differences do not always decrease to zero steadily as for the values of a mathematical function. For example, it occasionally happens that the differences of an empirical function decrease numerically up to some order m and then, if differenced further, tend to increase due to the accumulation of experimental errors. Irregular behaviour of higher differences is characteristic of empirical functions. Hence, in such situations, the best interpolation is obtained by curtailing the differencing when the differences tend either to increase or to behave irregularly. Empirical functions are of frequent occurrence in statistics, and it is advisable to remember that the irregular behaviour of their higher differences restricts the degree of the polynomial used for interpolation.

We conclude this section by two examples. The first refers to interpolation of a mathematically defined non-polynomial function

$$y = \frac{1}{\sqrt{2\pi}} \int_{-\infty}^{x} e^{-\frac{1}{2}t^2} \, dt,$$

and the second example pertains to an empirical function giving the estimated age population of the U.S.S.R. in 1970 (Notestein *et al.*, 1944).

Example 3.6

The following difference table is based on the values of the function

$$y = \frac{1}{\sqrt{2\pi}} \int_{-\infty}^{x} e^{-\frac{1}{2}t^2} \, dt$$

correct to seven decimal places for $x = 0 \cdot 30(0 \cdot 02)0 \cdot 40$. Evaluate y for $x = 0 \cdot 316$.

Table 3.11   Showing the forward differences of the function
$$y = \frac{1}{\sqrt{2\pi}} \int_{-\infty}^{x} e^{-\frac{1}{2}t^2} \, dt \text{ for } x = 0 \cdot 30(0 \cdot 02)0 \cdot 40$$

x	y	Δ	Δ^2	Δ^3
0·30	0·6179114			
		76044		
0·32	0·6255158		− 485	
		75559		−27
0·34	0·6330717		− 512	
		75047		−(+) 26
0·36	0·6405764		−(−) 538	
		(−) 74509		−(+) 27
0·38	0·6480273		−(−) 565	
		(−)73944		
0·40	0·6554217			

[*Note.* The positive and negative signs in brackets are explained later, in Section 3.9.]

Here the third differences of y are almost equal except for rounding-off errors. Hence, to obtain an interpolated value of y for $x = 0 \cdot 316$ correct to seven decimal places, it is adequate to use a cubic interpolation formula.

Since $a = 0 \cdot 30$, $h = 0 \cdot 02$, we have:

$\theta = 0 \cdot 8$	$f(0 \cdot 30)$	$= 0 \cdot 6179114$
$\dfrac{\theta^{(2)}}{2!} = -0 \cdot 08$	$\theta \Delta f(0 \cdot 30)$	$= \quad 608352$
$\dfrac{\theta^{(3)}}{3!} = 0 \cdot 032$	$\dfrac{\theta^{(2)}}{2!} \Delta^2 f(0 \cdot 30) =$	388
	$\dfrac{\theta^{(3)}}{3!} \Delta^3 f(0 \cdot 30) =$	-9

Therefore, on addition, $f(0 \cdot 316) = 0 \cdot 6239987$ correct to seven decimal places.

Example 3.7

The following table gives the estimated population in millions (y) with ages below x years for the U.S.S.R. in 1970. Determine the value of y for $x = 28 \cdot 67$.

Table 3.12   Showing the forward differences of the function $y = f(x)$ representing the estimated population in millions with ages below x years for the U.S.S.R. in 1970

x	y	Δ	Δ^2	Δ^3	Δ^4
25	109·4				
		219			
30	131·3		−13		
		206		−35	
35	151·9		−48		112
		158		77	
40	167·7		29		−151
		187		−74	
45	186·4		−45		97
		142		23	
50	200·6		−22		9
		120		32	
55	212·6		10		−72
		130		−40	
60	225·6		−30		44
		100		4	
65	235·6		−26		1
		74		5	
70	243·0		−31		7
		43		12	
75	247·3		−19		−6
		24		6	
80	249·7		−13		
		11			
85	250·8				

It is clear from the above table that the higher differences become increasingly irregular, and it would certainly not be justifiable to use differences beyond the third for interpolation.

Here $a = 25$, $h = 5$, $\theta h = 3 \cdot 67$. Therefore

$$\theta = 0 \cdot 734 \qquad\qquad f(25) = 109 \cdot 4$$

$$\frac{\theta^{(2)}}{2!} = -0 \cdot 097622 \qquad\qquad \theta \Delta f(25) = 16 \cdot 07$$

$$\frac{\theta^{(3)}}{3!} = 0 \cdot 041196 \qquad\qquad \frac{\theta^{(2)}}{2!} \Delta^2 f(25) = 13$$

$$\frac{\theta^{(3)}}{3!} \Delta^3 f(25) = -14$$

Hence, correct to one decimal place,

linear interpolation gives $\qquad f(28 \cdot 67) = 125 \cdot 5,$

quadratic interpolation gives $f(28\cdot67)$ = 125$\cdot$6,

and cubic interpolation gives $f(28\cdot67)$ = 125$\cdot$5.

Thus, to the accuracy to which the y values are given, there is little advantage to be obtained by using quadratic or cubic interpolation. Indeed, adequate accuracy is achieved by linear interpolation.

3.9   Backward differences and interpolation

We have seen from the data of Table 3.11 that it is necessary to use cubic interpolation in order to obtain the accuracy of the tabular values for an interpolated y value. Hence, with the given table, it is possible to interpolate with full accuracy for the evaluation of a y value corresponding to any x in the range $0\cdot30 < x < 0\cdot36$. On the other hand, suppose it is desired to interpolate for any y value corresponding to the argument x in the interval $0\cdot36 < x < 0\cdot40$. Then it is clear that we cannot use the forward interpolation formula because of the lack of some of the required diagonal differences. Nevertheless, in such a situation, we can modify our procedure by reversing the difference table and using the differences along an ascending diagonal. We thus consider h to be negative, and the signs of the differences are suitably adjusted. The method is best explained by reference to Table 3.11 to evaluate y corresponding to $x = 0\cdot384$.

Here $a = 0\cdot40$, $h = -0\cdot02$, $\theta h = -0\cdot016$, so that $\theta = 0\cdot8$. Also, by the reversal of the differencing operations, the first differences of Table 3.11 now become negative. The second differences remain negative but the third differences become positive. The proper signs for these backward differences are indicated in brackets.

Hence the following computations:

$$f(0\cdot40) \quad = \quad 0\cdot6554217$$
$$-\theta\Delta f(0\cdot38) \quad = \quad - \quad\quad 591552$$
$$\frac{\theta^{(2)}}{2!}\,\Delta^2 f(0\cdot36) \quad = \quad\quad\quad 452$$
$$-\frac{\theta^{(3)}}{3!}\,\Delta^3 f(0\cdot34) \quad = \quad\quad\quad\quad 9$$

Therefore, on addition, $f(0\cdot384) = 0\cdot6495108$ correct to seven decimal places.

This is a useful method for interpolation in the lower end of a given difference table.

3.10   Inverse interpolation

We have so far used the finite differences of a function $y = f(x)$ to determine the value of y corresponding to any specified value of the argument x within the range of tabulation. It is also possible to use finite differences to determine the value x_0 of the argument for which $f(x_0) = c$, where c is a known quantity. This is the typical problem of *inverse interpolation*, and it is clearly equivalent to the determination of a root of the equation $f(x) = c$. In general, therefore, if this equation has more than one real root, then our first concern is to specify a relatively small interval in which the required root $x = x_0$ lies.

We may, without loss of generality, assume that $c = 0$ or, alternatively, set $f(x) - c = g(x)$, so that we are led to a search for the real roots of an equation of the type $g(x) = 0$. If initially nothing is known about these roots, then we can start from first principles and consider a range of values $\xi_1, \xi_2, \xi_3, \ldots$ of x such that $-\infty < \xi_1 < \xi_2 < \ldots < \xi_k < \ldots < \infty$. Now it is well-known that if $g(\xi_r)$ and $g(\xi_{r+1})$ are of opposite sign, then a real root of the equation $g(x) = 0$ lies in the interval (ξ_r, ξ_{r+1}). If the ξ_r are taken sufficiently close to each other, it is possible to identify separately the intervals in which all the real roots lie. Suppose that the particular root $x = x_0$ lies in the interval $(a, a + h)$, where h is relatively small. Then

$$f(x_0) = f\left[a + \left(\frac{x_0 - a}{h}\right)h\right] = f(a + \theta h), \quad \text{where } \theta \equiv \frac{x_0 - a}{h}, \quad 0 < \theta < 1.$$

Therefore, by the forward interpolation formula,

$$f(x_0) = f(a) + \theta \Delta f(a) + \frac{\theta^{(2)}}{2!} \Delta^2 f(a) + \frac{\theta^{(3)}}{3!} \Delta^3 f(a) + \ldots . \tag{1}$$

If $f(x)$ is a polynomial of degree n, then clearly the series on the right-hand side of (1) will terminate after the nth difference. More generally, suppose $f(x)$ is not a polynomial so that the series is theoretically non-terminating. However, since in this case the values of $f(x)$ can only be approximate numbers, it is clear that for any specified accuracy of the tabular values the mth (say) differences should become approximately equal. Hence, in practice, the series in (1) will have only a finite number of terms.

In the above formulation, θ is unknown, but $f(x_0) = c$ is known, and our problem becomes simply that of determining θ to any specified degree of accuracy, whence $x_0 = a + \theta h$. We can rewrite (1) as

$$\theta = \frac{1}{\Delta f(a)}\left[f(x_0) - f(a) - \frac{\theta^{(2)}}{2!}\Delta^2 f(a) - \frac{\theta^{(3)}}{3!}\Delta^3 f(a) - \ldots\right]. \tag{2}$$

Hence, to a first approximation,

$$\theta = \theta_1 = \frac{f(x_0) - f(a)}{\Delta f(a)} .$$

Substituting this first approximation in (2), we have the second approximation for θ as

$$\theta_2 = \frac{1}{\Delta f(a)}\left[f(x_0) - f(a) - \frac{\theta_1^{(2)}}{2!}\Delta^2 f(a) - \frac{\theta_1^{(3)}}{3!}\Delta^3 f(a) - \ldots\right].$$

Again, substituting θ_2 in (2), the third approximation for θ is obtained as

$$\theta_3 = \frac{1}{\Delta f(a)}\left[f(x_0) - f(a) - \frac{\theta_2^{(2)}}{2!}\Delta^2 f(a) - \frac{\theta_2^{(3)}}{3!}\Delta^3 f(a) - \ldots\right],$$

and so on till we reach stability in the value of θ. This means that for some $r \,(> 1)$, $\theta_{r-1} = \theta_r$ to a specified degree of accuracy.

Then $$x_0 = a + \theta_r h.$$

This iterative process is one way of inverse interpolation based on the forward interpolation formula. The method is simple and quite general for finding a real root of an equation, but its application requires some care to determine the root to a specified degree of accuracy. We divide our discussion of the important practical points into two parts according as to whether the function values are given, or are to be evaluated from the known functional form of $f(x)$.

(i) Firstly, if the function values are given, then their accuracy is specified and so also is h, the interval of the argument. In this case the accuracy of x_0 depends upon $\Delta f(a)$, and it is not possible to obtain the root to an accuracy greater than the significant figures in $\Delta f(a)$. Furthermore, the smaller the value of h, the smaller is the interval in which x_0 is initially known to lie. Hence, in general, the quicker will be the convergence in the determination of θ.

We shall not consider here the case when $x = x_0$ is a multiple root of the equation, but it is important to note that the method becomes difficult to apply if another root lies in close proximity to $x = x_0$. In such a situation, we need an accurate determination of the function values in the neighbourhood of the root x_0 to evaluate it precisely and without undue effort. In fact, the method works best when there is a simple root in a well-defined neighbourhood, and we shall consider only such situations in practical applications of the method. It then follows that for given h, the greater the value of $\Delta f(a)$, the faster will be the convergence for the evaluation of θ.

The form of (2) implies that the contributions arising from the second and higher differences of $f(a)$ should be considered for each approximation of θ after the first. However, in general, the required accuracy for θ will be limited, and in relation to this accuracy the contributions from the higher differences could be negligible. The following example illustrates this point clearly.

Example 3.8
Use the data of Table 3.11 to determine, correct to five decimal places, the value x_0 such that $f(x_0) = 0{\cdot}62$.

Here $f(x)$ is not a polynomial function of x but, to the accuracy of the given function values, it may be represented by a polynomial of degree three. Hence the formula for inverse interpolation gives

$$\theta = \frac{1}{\Delta f(a)} \left[f(x_0) - f(a) - \frac{\theta^{(2)}}{2!} \Delta^2 f(a) - \frac{\theta^{(3)}}{3!} \Delta^3 f(a) \right],$$

where $x_0 = a + \theta h$, $h = 0{\cdot}02$, $a = 0{\cdot}30$.

Therefore

$$\theta_1 = \frac{0{\cdot}62 - 0{\cdot}6179114}{0{\cdot}0076044}$$

$$= \frac{0{\cdot}0020886}{0{\cdot}0076044}$$

$$= 0{\cdot}2746 \text{ to four decimal places.}$$

Next,

$$\theta_1 = 0 \cdot 2746$$

$$\tfrac{1}{2}(\theta_1 - 1) = -0 \cdot 36270 \qquad\qquad \tfrac{1}{2!}\theta^{(2)}_1 = -0 \cdot 099597$$

$$\tfrac{1}{3}(\theta_1 - 2) = -0 \cdot 57513 \qquad\qquad \tfrac{1}{3!}\theta^{(3)}_1 = 0 \cdot 057281$$

Therefore

$$\begin{aligned}
\theta_2(0 \cdot 0076044) &= 0 \cdot 0020886 - (-0 \cdot 099597)(-0 \cdot 0000485) - \\
&\qquad - (0 \cdot 057281)(-0 \cdot 0000027) \\
&= 0 \cdot 0020886 - 0 \cdot 0000048 + 0 \cdot 0000002 \\
&= 0 \cdot 0020840,
\end{aligned}$$

so that $\qquad\qquad \theta_2 = 0 \cdot 2741$ correct to four decimal places.

Again,

$$\theta_2 = 0 \cdot 2741$$

$$\tfrac{1}{2}(\theta_2 - 1) = -0 \cdot 36295 \qquad\qquad \tfrac{1}{2!}\theta^{(2)}_2 = -0 \cdot 099485$$

$$\tfrac{1}{3}(\theta_2 - 2) = -0 \cdot 57530 \qquad\qquad \tfrac{1}{3!}\theta^{(3)}_2 = 0 \cdot 057233$$

Therefore

$$\begin{aligned}
\theta_3(0 \cdot 0076044) &= 0 \cdot 0020886 - 0 \cdot 0000048 + 0 \cdot 0000002 \\
&= 0 \cdot 0020840,
\end{aligned}$$

so that $\qquad\qquad \theta_3 = 0 \cdot 2741$ correct to four decimal places.

Thus, correct to four decimal places, $\theta_2 = \theta_3 = 0 \cdot 2741$, and we finally obtain

$$\begin{aligned}
x_0 &= 0 \cdot 30 + 0 \cdot 005482 \\
&= 0 \cdot 30548 \quad \text{correct to five decimal places.}
\end{aligned}$$

It is to be noted that since $h = 0 \cdot 02$, it is necessary to evaluate θ correct to three decimal places only. Accordingly, the contributions from $\Delta^3 f(a)$ in the evaluation of θ_2 and θ_3 are negligible. Indeed, by ignoring these contributions, we obtain $\theta_2 = \theta_3 = 0 \cdot 274$ correct to three decimal places, whence we are led to the same value for x_0. In this example h is rather small, and this has given a quick convergence to the required value for θ. With a larger value of h, stability in the successive determinations of θ would, in general, be attained only after several iterations.

(ii) Secondly, suppose the form of $f(x)$ is known but the function values are to be evaluated to obtain a difference table for inverse interpolation. We now have the choice of the initial accuracy of the function values; and in this case, if the value of x_0 is required correct to k decimal places, it will, in general, be appropriate to determine the function values correct to $k + 2$ decimal places. This additional accuracy in the evaluation of the function values is a useful precaution, since the accuracy of x_0 could be sensitive to the rounding-off errors in the tabular values. Apart from the accuracy of the function values, we now also have the choice of h, the interval of the argument used in the difference table. As a general rule, it is advisable to make h as small as possible such that $a < x_0 < a + h$. This implies the initial location of x_0 in a close neighbourhood, and this could involve considerable work especially if $f(x)$ is a complicated function to evaluate numerically. On the other hand, the labour required for the initial location of x_0 in a close neighbourhood has to be balanced against the fact that if h is large then the determination of θ will, in general, be slow. An appropriate choice between practical alternatives of this kind is a matter of judgement and some experience. The following example shows many of the considerations involved in the solution of a simple problem.

Example 3.9

Determine, correct to four decimal places, the positive root of the cubic

$$x^3 + \lambda(1 - \phi)x^2 + \lambda(1 - 2\phi)x + (1 - 3\phi) = 0,$$

where $\phi = \dfrac{4}{3}$, $\lambda \equiv e + 1 + e^{-1}$, and e is the Napierian constant.

It is clear that it will be inappropriate to express ϕ as a recurring decimal fraction. In fact, a simple multiplication by 3 reduces the equation to the simpler form

$$f(x) \equiv 3x^3 - \lambda(x^2 + 5x) - 9 = 0.$$

We must, of course, use a digital approximation for λ, and we obtain from *Barlow's Tables* that, correct to six decimal places, $\lambda = 4 \cdot 086161$.

Next, we know from Descartes' Rule of Signs that $f(x) = 0$ has only one positive root, and we need to determine a close interval in which this root x_0 lies.

We have

$f(0) = -9; \quad f(1) = -6 - 6\lambda; \quad f(2) = 15 - 14\lambda; \quad f(3) = 72 - 24\lambda;$
$f(3 \cdot 5) = 119 \cdot 625 - 29 \cdot 75\lambda; \qquad f(4) = 183 - 36\lambda;$
$f(4 \cdot 5) = 264 \cdot 375 - 42 \cdot 75\lambda; \qquad f(5) = 366 - 50\lambda.$

An approximate calculation indicates that $3 < x_0 < 4$ since $f(3) < 0$ and $f(4) > 0$. It is thus appropriate to start the difference table with $x = 3$ and use an argument interval $0 \cdot 5$. We need at least four function values to obtain three leading differences, and it is useful to evaluate another function value so as to use the constancy of the third differences as a check on the accuracy of the difference table. Hence Table 3.13 and the computations that follow.

Table 3.13 Showing the forward differences of $f(x)$ tabulated for $x = 3 \cdot 0(0 \cdot 5)5 \cdot 0$

x	$f(x)$	Δ	Δ^2	Δ^3
3·0	− 26·067864			
		24·129574		
3·5	− 1·938290		13·706920	
		37·836494		2·249999
4·0	35·898204		15·956919	
		53·793413		2·250001
4·5	89·691617		18·206920	
		72·000333		
5·0	161·691950			

Here $a = 3\cdot5$; $h = 0\cdot5$; $f(x_0) = 0$; and $x_0 = a + \theta h$.

Therefore
$$\theta_1 = \frac{1\cdot938290}{37\cdot836494} = 0\cdot0512.$$

$$\theta_1 = 0\cdot0512$$

$$\tfrac{1}{2}(\theta_1 - 1) = -0\cdot4744 \qquad\qquad \tfrac{1}{2!}\,\theta_1^{(2)} = -0\cdot024289$$

$$\tfrac{1}{3}(\theta_1 - 2) = -0\cdot6496 \qquad\qquad \tfrac{1}{3!}\,\theta_1^{(3)} = 0\cdot015778$$

Therefore
$$\theta_2(37\cdot836494) = 1\cdot938290 + 0\cdot387578 - 0\cdot035501$$
$$= 2\cdot290367,$$
so that
$$\theta_2 = 0\cdot06053.$$

$$\theta_2 = 0\cdot06053$$

$$\tfrac{1}{2}(\theta_2 - 1) = -0\cdot469735 \qquad\qquad \tfrac{1}{2!}\,\theta_2^{(2)} = -0\cdot028433$$

$$\tfrac{1}{3}(\theta_2 - 2) = -0\cdot646490 \qquad\qquad \tfrac{1}{3!}\,\theta_2^{(3)} = 0\cdot018382$$

Therefore
$$\theta_3(37\cdot836494) = 1\cdot938290 + 0\cdot453703 - 0\cdot041360$$
$$= 2\cdot350633,$$
so that
$$\theta_3 = 0\cdot06213.$$

$$\theta_3 = 0\cdot06213$$

$$\tfrac{1}{2}(\theta_3 - 1) = -0\cdot468935 \qquad\qquad \tfrac{1}{2!}\,\theta_3^{(2)} = -0\cdot029135$$

$$\tfrac{1}{3}(\theta_3 - 2) = -0\cdot645957 \qquad\qquad \tfrac{1}{3!}\,\theta_3^{(3)} = 0\cdot018820$$

Therefore
$$\theta_4(37\cdot836494) = 1\cdot938290 + 0\cdot464905 - 0\cdot042345$$
$$= 2\cdot360850,$$
so that
$$\theta_4 = 0\cdot06240.$$

$$\theta_4 = 0\cdot06240$$

$$\tfrac{1}{2}(\theta_4 - 1) = -0\cdot468800 \qquad\qquad \tfrac{1}{2!}\,\theta_4^{(2)} = -0\cdot029253$$

$$\tfrac{1}{3}(\theta_4 - 2) = -0\cdot645867 \qquad\qquad \tfrac{1}{3!}\,\theta_4^{(3)} = 0\cdot018894$$

Therefore
$$\theta_5(37\cdot836494) = 1\cdot938290 + 0\cdot466788 - 0\cdot042512$$
$$= 2\cdot362566,$$
so that
$$\theta_5 = 0\cdot06244.$$

$$\theta_5 = 0\cdot06244$$

$$\tfrac{1}{2}(\theta_5 - 1) = -0\cdot468780 \qquad\qquad \tfrac{1}{2!}\,\theta_5^{(2)} = -0\cdot029271$$

$$\tfrac{1}{3}(\theta_5 - 2) = -0\cdot645853 \qquad\qquad \tfrac{1}{3!}\,\theta_5^{(3)} = 0\cdot018905$$

Therefore

$$\theta_6(37 \cdot 836494) = 1 \cdot 938290 + 0 \cdot 467075 - 0 \cdot 042536$$
$$= 2 \cdot 362829,$$

so that $\qquad \theta_6 = 0 \cdot 06245.$

$$\theta_6 = 0 \cdot 06245$$

$$\frac{1}{2}(\theta_6 - 1) = -0 \cdot 468775 \qquad \qquad \frac{1}{2!} \theta_6^{(2)} = -0 \cdot 029275$$

$$\frac{1}{3}(\theta_6 - 2) = -0 \cdot 64585 \qquad \qquad \frac{1}{3!} \theta_6^{(3)} = 0 \cdot 018907$$

Therefore

$$\theta_7(37 \cdot 836494) = 1 \cdot 938290 + 0 \cdot 467139 - 0 \cdot 042541$$
$$= 2 \cdot 362888,$$

so that $\qquad \theta_7 = 0 \cdot 06245.$

Thus, correct to five decimal places, $\theta_6 = \theta_7 = 0 \cdot 06245.$ Hence

$$x_0 = 3 \cdot 5 + 0 \cdot 031225$$
$$= 3 \cdot 531225.$$

Therefore, correct to four decimal places,

$$x_0 = 3 \cdot 5312.$$

CHAPTER 4

SUMMATION OF SERIES

There is no general method for summing series, nor is it even always possible to express the sum to n terms as an elementary function of n. However, in this chapter we shall consider a variety of methods for the summation of special types of both finite and convergent infinite series. Such series are of interest in statistical theory and the techniques used for their summation have frequent application.

4.1 Arithmetical progressions

Perhaps the simplest of finite series is the standard arithmetical progression whose first term is a and common difference d. The rth term of such a series is

$$u_r = a + (r-1)d, \quad \text{for } r \geqslant 1,$$

and it is known that

$$\sum_{r=1}^{n} u_r = \frac{n}{2}[2a + (n-1)d].$$

(i) In particular, $a = d = 1$ gives $u_r = r$, and we obtain simply the sum of the first n natural numbers as

$$\sum_{r=1}^{n} r = \tfrac{1}{2}n(n+1).$$

(ii) Again, the sum of the first n odd natural numbers is obtained by putting $a = 1$, $d = 2$ so that

$$\sum_{r=1}^{n} (2r - 1) = \frac{n}{2}[2 + (n-1)2] = n^2.$$

(iii) Similarly, for the sum of the first n even natural numbers put $a = d = 2$. This gives

$$\sum_{r=1}^{n} 2r = \frac{n}{2}[4 + (n-1)2] = n(n+1).$$

4.2 The difference method

We can extend these results to sums of powers of natural numbers by using the following two theorems.

Theorem 1

If the rth term of a series $u_r \equiv f(r+1) - f(r)$, then

$$\sum_{r=1}^{n} u_r = f(n + 1) - f(1).$$

The proof is immediate by addition.

Theorem 2

For any two non-negative integers k and s such that $n + s \geqslant k + 1$

$$\sum_{r=1}^{n} (r+s)^{(k+1)} = [(n+1+s)^{(k+2)} - (1+s)^{(k+2)}]/(k+2)$$

where, in standard factorial notation,

$$x^{(m)} = x(x-1)(x-2)\ldots(x-m+1).$$

Proof

We have

$$\begin{aligned}
(r+s)^{(k+1)} &= (r+s)(r+s-1)\ldots(r+s-k+1)(r+s-k) \\
&= [(r+s+1)(r+s)\ldots(r+s-k+1)(r+s-k)- \\
&\qquad -(r+s)(r+s-1)\ldots(r+s-k)(r+s-k-1)]/(k+2) \\
&= [(r+s+1)^{(k+2)} - (r+s)^{(k+2)}]/(k+2).
\end{aligned}$$

The result now follows upon applying Theorem 1 with

$$f(r) = (r+s)^{(k+2)}/(k+2).$$

4.3 Powers of natural numbers

The general result of Theorem 2 can be used to obtain the sums of the successive integral powers of the first n natural numbers.

(i) *Sum of squares*

Since $r^2 = (r+1)r - r$, we have

$$\begin{aligned}
\sum_{r=1}^{n} r^2 &= \sum_{r=1}^{n} (r+1)^{(2)} - \sum_{r=1}^{n} r \\
&= \frac{1}{3}(n+2)^{(3)} - \frac{1}{2}(n+1)^{(2)} \\
&= \frac{1}{3}(n+2)(n+1)n - \frac{1}{2}(n+1)n = n(n+1)(2n+1)/6.
\end{aligned}$$

Therefore

$$\sum_{r=1}^{n} r^2 = n(n+1)(2n+1)/6.$$

(ii) *Sum of cubes*

Since $r(r+1)(r+2) = r^3 + 3r^2 + 2r$
$$= r^3 + 3(r+1)r - r,$$
we have $r^3 = (r+2)^{(3)} - 3(r+1)^{(2)} + r.$

Therefore

$$\begin{aligned}
\sum_{r=1}^{n} r^3 &= \sum_{r=1}^{n} (r+2)^{(3)} - 3\sum_{r=1}^{n} (r+1)^{(2)} + \sum_{r=1}^{n} r \\
&= \frac{1}{4}(n+3)^{(4)} - 3\frac{1}{3}(n+2)^{(3)} + \frac{1}{2}(n+1)^{(2)} \\
&= \frac{1}{4}n(n+1)[(n+2)(n+3) - 4(n+2) + 2] = \left[\frac{1}{2}n(n+1)\right]^2.
\end{aligned}$$

Therefore

$$\sum_{r=1}^{n} r^3 = \left[\frac{1}{2}n(n+1)\right]^2.$$

(iii) *Sum of fourth powers*

Again, $r(r+1)(r+2)(r+3) = r^4 + 6r^3 + 11r^2 + 6r$

$$= r^4 + 6r(r^2 + 3r + 2) - 7r(r+1) + r$$
$$= r^4 + 6(r+2)^{(3)} - 7(r+1)^{(2)} + r,$$

so that $r^4 = (r+3)^{(4)} - 6(r+2)^{(3)} + 7(r+1)^{(2)} - r$.

Therefore

$$\sum_{r=1}^{n} r^4 = \frac{1}{5}(n+4)^{(5)} - 6\frac{1}{4}(n+3)^{(4)} + 7\frac{1}{3}(n+2)^{(3)} - \frac{1}{2}(n+1)^{(2)}$$

$$= \frac{(n+1)^{(2)}}{30}[6(n+4)(n+3)(n+2) - 45(n+3)(n+2) + 70(n+2) - 15]$$

$$= \frac{(n+1)^{(2)}}{30}[(n+3)(n+2)(6n-21) + (70n+125)]$$

$$= \frac{(n+1)^{(2)}}{30}[6n^3 + 9n^2 + n - 1]$$

$$= n(n+1)(2n+1)(3n^2 + 3n - 1)/30.$$

Therefore

$$\sum_{r=1}^{n} r^4 = n(n+1)(2n+1)(3n^2 + 3n - 1)/30.$$

A useful check of the accuracy of a sum obtained in this way is to put $n = 1$ and verify that the sum reduces to unity. More generally, the correctness of the summation of any finite series may be tested by putting $n = 1$ and checking that the formula for the sum reduces to the value of the first term. This check should always be used.

The method used in the above summations is general, and we consider next three slightly different examples which reveal the extended power and simplicity of the technique based on Theorems 1 and 2.

Example 4.1

Sum to n terms the series whose rth term is

$$u_r = \frac{1}{r(r+1)(r+2)}, \quad \text{for } r \geqslant 1.$$

Here

$$u_r = \frac{1}{2}\left[\frac{1}{r(r+1)} - \frac{1}{(r+1)(r+2)}\right],$$

and so setting

$$f(r) = -\frac{1}{2r(r+1)},$$

we have by Theorem 1

$$\sum_{r=1}^{n} u_r = \frac{1}{4} - \frac{1}{2(n+1)(n+2)}.$$

This series is evidently convergent, as a simple limiting process for $n \to \infty$ gives

$$\sum_{r=1}^{\infty} u_r = \frac{1}{4}.$$

Example 4.2

Evaluate the sum of the double series

(i) $\sum\limits_{s=1}^{n} \sum\limits_{r<s} s(s+2)r(r+1)$ and   (ii) $\sum\limits_{s=1}^{n} \sum\limits_{r\leqslant s} s(s+2)r(r+1)$.

(i) Here $\sum\limits_{s=1}^{n} \sum\limits_{r<s} s(s+2)r(r+1) = \sum\limits_{s=1}^{n} s(s+2)\sum\limits_{r=0}^{s-1} r(r+1)$

$$= \sum\limits_{s=1}^{n} s(s+2)\sum\limits_{r=1}^{s-1} r(r+1)$$

$$= \sum\limits_{s=1}^{n} s(s+2)\frac{1}{3}(s-1)s(s+1)$$

$$= \frac{1}{3}\sum\limits_{s=1}^{n} (s+2)^{(4)}[(s+3)-3]$$

$$= \frac{1}{3}\sum\limits_{s=1}^{n} (s+3)^{(5)} - \sum\limits_{s=1}^{n} (s+2)^{(4)}$$

$$= \frac{1}{3}\times\frac{1}{6}(n+4)^{(6)} - \frac{1}{5}(n+3)^{(5)}$$

$$= (n+3)^{(5)}[5(n+4)-18]/90$$

$$= (5n+2)(n+3)^{(5)}/90.$$

Therefore

$$\sum\limits_{s=1}^{n} \sum\limits_{r<s} s(s+2)r(r+1) = (5n+2)(n+3)^{(5)}/90.$$

(ii) Again,

$$\sum\limits_{s=1}^{n} \sum\limits_{r\leqslant s} s(s+2)r(r+1) = \sum\limits_{s=1}^{n} s(s+2)\sum\limits_{r=1}^{s} r(r+1)$$

$$= \sum\limits_{s=1}^{n} s(s+2)\frac{1}{3}s(s+1)(s+2)$$

$$= \frac{1}{3}\sum\limits_{s=1}^{n} s(s+1)(s+2)[s(s+2)]$$

$$= \frac{1}{3}\sum\limits_{s=1}^{n} (s+2)^{(3)}[(s+3)(s+4)-5(s+3)+3]$$

$$= \frac{1}{3}\sum\limits_{s=1}^{n} [(s+4)^{(5)} - 5(s+3)^{(4)} + 3(s+2)^{(3)}]$$

$$= \frac{1}{3}\left[\frac{1}{6}(n+5)^{(6)} - 5\frac{1}{5}(n+4)^{(5)} + 3\frac{1}{4}(n+3)^{(4)}\right]$$

$$= (n+3)^{(4)}[2(n+4)(n+5)-12(n+4)+9]/36$$

$$= (2n^2+6n+1)(n+3)^{(4)}/36.$$

Therefore

$$\sum\limits_{s=1}^{n} \sum\limits_{r\leqslant s} s(s+2)r(r+1) = (2n^2+6n+1)(n+3)^{(4)}/36.$$

Example 4.3

Evaluate the sum of the first $m\,(< n)$ terms of the series for which

$$u_0 = n^{-a+1},$$

and $u_r = n^{-a}[(r+1)^a - 2r^a + (r-1)^a](n-r),$ for $1\leqslant r\leqslant n-1$.

The expression involving r in u_r is

$$(n-r)[(r+1)^a - 2r^a + (r-1)^a] = n[\{(r+1)^a - r^a\} - \{r^a - (r-1)^a\}] -$$
$$- (r+1)^a[(r+1)-1] + 2r^{a+1} - (r-1)^a[(r-1)+1]$$

$$
\begin{aligned}
&= n[\{(r + 1)^{\alpha} - r^{\alpha}\} - \{r^{\alpha} - (r - 1)^{\alpha}\}] - [(r + 1)^{\alpha+1} - r^{\alpha+1}] + \\
&\quad + [r^{\alpha+1} - (r - 1)^{\alpha+1}] + [(r + 1)^{\alpha} - (r - 1)^{\alpha}] \\
&= n[\{(r + 1)^{\alpha} - r^{\alpha}\} - \{r^{\alpha} - (r - 1)^{\alpha}\}] - [(r + 1)^{\alpha+1} - r^{\alpha+1}] + \\
&\quad + [r^{\alpha+1} - (r - 1)^{\alpha+1}] + [(r + 1)^{\alpha} - r^{\alpha}] + [r^{\alpha} - (r - 1)^{\alpha}] \\
&= (n + 1)[(r + 1)^{\alpha} - r^{\alpha}] - (n - 1)[r^{\alpha} - (r - 1)^{\alpha}] - \\
&\quad - [(r + 1)^{\alpha+1} - r^{\alpha+1}] + [r^{\alpha+1} - (r - 1)^{\alpha+1}].
\end{aligned}
$$

Hence

$$
\begin{aligned}
\sum_{r=0}^{m-1} u_r &= n^{-\alpha+1} + n^{-\alpha}\sum_{r=1}^{m-1}[(n + 1)\{(r + 1)^{\alpha} - r^{\alpha}\} - (n - 1)\{r^{\alpha} - (r - 1)^{\alpha}\} - \\
&\quad - \{(r + 1)^{\alpha+1} - r^{\alpha+1}\} + \{r^{\alpha+1} - (r - 1)^{\alpha+1}\}].
\end{aligned}
$$

But

$$
\sum_{r=1}^{m-1}[(r + 1)^{k} - r^{k}] = m^{k} - 1; \qquad \sum_{r=1}^{m-1}[r^{k} - (r - 1)^{k}] = (m - 1)^{k}.
$$

Therefore

$$
\begin{aligned}
\sum_{r=0}^{m-1} u_r &= n^{-\alpha+1} + n^{-\alpha}[(n + 1)\{m^{\alpha} - 1\} - (n - 1)(m - 1)^{\alpha} - \\
&\quad - \{m^{\alpha+1} - 1\} + (m - 1)^{\alpha+1}] \\
&= n^{-\alpha+1} + n^{-\alpha}[-n + (n + 1)m^{\alpha} - (n - 1)(m - 1)^{\alpha} - \\
&\quad - m^{\alpha+1} + (m - 1)^{\alpha+1}] \\
&= n^{-\alpha}[(n - m)m^{\alpha} + m^{\alpha} - (n - m)(m - 1)^{\alpha}] \\
&= n^{-\alpha}[m^{\alpha} + (n - m)\{m^{\alpha} - (m - 1)^{\alpha}\}].
\end{aligned}
$$

Hence

$$
\sum_{r=0}^{m-1} u_r = n^{-\alpha}\lfloor m^{\alpha} + (n - m)\{m^{\alpha} - (m - 1)^{\alpha}\}].
$$

For $m = n$, the sum of the series is unity.

4.4 Binomial expansions

Another class of finite series of particular importance in statistical theory is that associated with the binomial expansion of $(a + x)^{n}$ where n is an integer $\geqslant 0$.

We know that

$$
(a + x)^{n} = \sum_{r=0}^{n} \binom{n}{r} x^{r} a^{n-r},
$$

where we define the binomial coefficients as

$$
\binom{n}{r} \equiv \frac{n!}{(n - r)!\, r!} \equiv \frac{n(n - 1)(n - 2)\ldots(n - r + 1)}{r!} \equiv \frac{n^{(r)}}{r!}.
$$

In particular, for $a = x = 1$ and $a = 1, \; x = -1$ we obtain respectively

$$
\sum_{r=0}^{n} \binom{n}{r} = 2^{n}; \qquad \sum_{r=0}^{n} (-1)^{r}\binom{n}{r} = 0.
$$

These are standard results, and we next consider a few examples of how other finite series involving binomial coefficients can be summed.

Example 4.4

We have $\sum_{r=0}^{n} (r+1) \binom{n}{r} x^r = \sum_{r=0}^{n} r \binom{n}{r} x^r + (1+x)^n$

$$= nx \sum_{r=1}^{n} \binom{n-1}{r-1} x^{r-1} + (1+x)^n$$

$$= nx \sum_{t=0}^{n-1} \binom{n-1}{t} x^t + (1+x)^n$$

$$= nx(1+x)^{n-1} + (1+x)^n,$$

so that

$$\sum_{r=0}^{n} (r+1) \binom{n}{r} x^r = (1+x)^{n-1}[1 + (n+1)x].$$

In particular, for $x = 1$ and $x = -1$ we obtain respectively

$$\sum_{r=0}^{n} (r+1) \binom{n}{r} = (n+2)2^{n-1}; \qquad \sum_{r=0}^{n} (-1)^r (r+1) \binom{n}{r} = 0.$$

Example 4.5

We have $\sum_{r=0}^{n} r^2 \binom{n}{r} x^r = \sum_{r=1}^{n} r^2 \binom{n}{r} x^r$

$$= \sum_{r=1}^{n} [r(r-1) + r] \binom{n}{r} x^r$$

$$= n(n-1)x^2 \sum_{r=2}^{n} \binom{n-2}{r-2} x^{r-2} + nx \sum_{r=1}^{n} \binom{n-1}{r-1} x^{r-1}$$

$$= n(n-1)x^2 \sum_{t=0}^{n-2} \binom{n-2}{t} x^t + nx \sum_{t=0}^{n-1} \binom{n-1}{t} x^t$$

$$= n(n-1)x^2(1+x)^{n-2} + nx(1+x)^{n-1}$$

$$= nx(1+x)^{n-2}(1+nx).$$

Therefore

$$\sum_{r=0}^{n} r^2 \binom{n}{r} x^r = nx(1+x)^{n-2}(1+nx).$$

In particular, for $x = 1$ and $x = -1$ we obtain respectively

$$\sum_{r=0}^{n} r^2 \binom{n}{r} = n(n+1)2^{n-2}; \qquad \sum_{r=0}^{n} (-1)^r r^2 \binom{n}{r} = 0.$$

Example 4.6

From $\sum_{r=0}^{n} \binom{n}{r} x^r = (1+x)^n$ we obtain

$$\sum_{r=1}^{n} \binom{n}{r} x^{r-1} = [(1+x)^n - 1]/x,$$

and integration of both sides with respect to x from -1 to 0 gives

$$\int_{-1}^{0} \sum_{r=1}^{n} \binom{n}{r} x^{r-1} dx = \int_{-1}^{0} \frac{(1+x)^n - 1}{x} dx,$$

or

$$\sum_{r=1}^{n} \binom{n}{r} \left[\frac{x^r}{r}\right]_{-1}^{0} = \int_{0}^{1} \frac{y^n - 1}{y - 1} dy \quad (y = 1 + x),$$

or

$$\sum_{r=1}^{n} r^{-1} \binom{n}{r} (-1)^{r-1} = \int_{0}^{1} (y^{n-1} + y^{n-2} + \dots + y + 1) dy,$$

or

$$\sum_{r=1}^{n} r^{-1} (-1)^{r-1} \binom{n}{r} = \sum_{r=1}^{n} r^{-1}.$$

Example 4.7

We have $(1 + x)^{2n} = (1 + x)^n (1 + x)^n$.

Therefore

$$\sum_{r=0}^{2n} \binom{2n}{r} x^r = \left[\sum_{r=0}^{n} \binom{n}{r} x^r \right] \left[\sum_{s=0}^{n} \binom{n}{s} x^s \right]$$

$$= \sum_{r=0}^{n} \sum_{s=0}^{n} \binom{n}{r} \binom{n}{s} x^{r+s}.$$

Equating the coefficients of x^n on both sides, we have

$$\binom{2n}{n} = \sum_{r=0}^{n} \binom{n}{r} \binom{n}{n-r}, \text{ or } \sum_{r=0}^{n} \left[\binom{n}{r} \right]^2 = \binom{2n}{n}.$$

Again, equating the coefficients of x^{n-1}, we have

$$\binom{2n}{n-1} = \sum_{r=0}^{n} \binom{n}{r} \binom{n}{n-r-1}, \text{ or } \sum_{r=0}^{n} \binom{n}{r} \binom{n}{r+1} = \binom{2n}{n-1}.$$

These results are particular cases of Vandermonde's theorem considered in the next example.

Example 4.8 (Vandermonde's theorem)

For positive integers m,n and an integer r such that $0 \leqslant r \leqslant m + n$

$$\sum_{s=0}^{r} \binom{m}{s} \binom{n}{r-s} = \binom{m+n}{r},$$

where it is to be understood that the binomial coefficients like $\binom{m}{s}$ are zero for $s > r$.

Proof

We have $(1 + x)^m (1 + x)^n = (1 + x)^{m+n}$,

so that

$$\left[\sum_{s=0}^{m} \binom{m}{s} x^s \right] \left[\sum_{t=0}^{n} \binom{n}{t} x^t \right] = \sum_{r=0}^{m+n} \binom{m+n}{r} x^r.$$

Equating coefficients of x^r on both sides, we obtain

$$\sum_{s=0}^{r} \binom{m}{s} \binom{n}{r-s} = \binom{m+n}{r}.$$

We can derive an alternative form of this result by writing it as

$$\frac{(m+n)!}{(m+n-r)!r!} = \sum_{s=0}^{r} \frac{m!}{s!(m-s)!} \frac{n!}{(r-s)!(n-r+s)!},$$

or

$$\frac{(m+n)!}{(m+n-r)!} = \sum_{s=0}^{r} \binom{r}{s} \frac{m!n!}{(m-s)!(n-r+s)!},$$

or

$$(m+n)^{(r)} = \sum_{s=0}^{r} \binom{r}{s} m^{(s)} n^{(r-s)}.$$

This result is also known as the *binomial theorem for factorial powers*, and is due to Alexandre-Théophile Vandermonde (1735 – 1796).

4.5   Power series

The simplest series in ascending powers of a variable x is the *geometrical progression* whose rth term is

$$u_r = ax^{r-1},$$

and

$$\sum_{r=1}^{n} u_r = \frac{a(1 - x^n)}{1 - x}.$$

If $|x| < 1$, then in the limit as $n \to \infty$

$$\sum_{r=1}^{\infty} u_r = \frac{a}{1 - x}.$$

For $a = 1$, we derive the series

$$\sum_{r=1}^{\infty} x^{r-1} = (1 - x)^{-1},$$

which is a particular example of the *binomial series*. This series is an extension of the finite binomial expansion when the index is a positive integer.

Binomial series

If $|x| < 1$ and m is not a non-negative integer, then

$$\sum_{r=0}^{\infty} \binom{m}{r} x^r = (1 + x)^m$$

and

$$\sum_{r=0}^{\infty} \binom{m}{r} (-x)^r = (1 - x)^m,$$

where

$$\binom{m}{r} \equiv \frac{m(m - 1)(m - 2) \dots (m - r + 1)}{r!}.$$

In particular, if m is a negative integer and we set $-m \equiv \nu > 0$, then

$$(1 - x)^{-\nu} = \sum_{r=0}^{\infty} (-1)^r \frac{\nu(\nu + 1)(\nu + 2) \dots (\nu + r - 1)}{r!} (-x)^r$$

$$= \sum_{r=0}^{\infty} \frac{(\nu + r - 1)!}{(\nu - 1)! r!} x^r$$

$$= \sum_{r=0}^{\infty} \binom{\nu + r - 1}{r} x^r.$$

Hence

$$\sum_{r=0}^{\infty} \binom{\nu + r - 1}{r} x^r = (1 - x)^{-\nu}.$$

This is an important alternative form of the binomial expansion. We also note that this general formulation gives rise to many useful series for different values of ν. For example, for $\nu = 2$, we have

$$\sum_{r=0}^{\infty} (r + 1) x^r = (1 - x)^{-2}.$$

We observe that in this series the coefficient of x^r is $r + 1$, a linear function of r. Since these numerical coefficients form an arithmetical progression, the above series is also known as an *arithmetico-geometric series*. More generally, consider the sum

$$S_n = \sum_{r=0}^{n-1} (a + rd) x^r, \quad a \text{ and } d \text{ being constants.}$$

We now show that sums of this type can be evaluated simply by the process of differencing. Thus

$$S_n = \sum_{r=0}^{n-1} (a + rd) x^r$$

and

$$xS_n = \sum_{r=0}^{n-1}(a + rd)x^{r+1}$$

$$= \sum_{r=1}^{n}[a + (r - 1)d]x^r.$$

Therefore, on subtraction,

$$(1 - x)S_n = a + d\sum_{r=1}^{n-1} x^r - [a + (n - 1)d]x^n$$

$$= a + \frac{dx(1 - x^{n-1})}{1 - x} - [a + (n - 1)d]x^n,$$

so that

$$S_n = \frac{a - [a + (n - 1)d]x^n}{1 - x} + \frac{dx(1 - x^{n-1})}{(1 - x)^2}.$$

This method of summation is general, and if the numerical coefficient of x^r is a polynomial of the kth degree in r, then k successive differencing operations will be required for summing the series.

Also, if $|x| < 1$, we have

$$\lim_{n \to \infty} S_n = \frac{a}{1 - x} + \frac{dx}{(1 - x)^2}.$$

Example 4.9

Sum to n terms the series whose rth term is $r(r + 1)x^{r-1}$. The required sum is

$$S_n = 2 + 6x + 12x^2 + 20x^3 + \ldots + n(n + 1)x^{n-1},$$

and

$$xS_n = 2x + 6x^2 + 12x^3 + \ldots + (n - 1)nx^{n-1} + n(n + 1)x^n.$$

Therefore, on subtraction,

$$(1 - x)S_n = 2[1 + 2x + 3x^2 + \ldots + nx^{n-1}] - n(n + 1)x^n$$

and

$$x(1 - x)S_n = 2[x + 2x^2 + \ldots + (n - 1)x^{n-1} + nx^n] - n(n + 1)x^{n+1}.$$

Therefore

$$(1 - x)^2 S_n = 2[1 + x + x^2 + \ldots + x^{n-1}] - [n(n + 1) + 2n]x^n + n(n + 1)x^{n+1}$$

$$= \frac{2(1 - x^n)}{1 - x} - n(n + 3)x^n + n(n + 1)x^{n+1}.$$

Hence

$$S_n = \frac{2(1 - x^n)}{(1 - x)^3} + \frac{nx^n[(n + 1)x - (n + 3)]}{(1 - x)^2}.$$

An *alternative* method of obtaining S_n is to note that

$$S_n = \frac{d^2}{dx^2}[1 + x + x^2 + \ldots + x^{n+1}] = \frac{d^2}{dx^2}\left[\frac{1 - x^{n+2}}{1 - x}\right].$$

4.6   Summation by difference tables

It is obvious that the above process of differencing (or differentiation) if required several times, could be performed more conveniently by the formation of a difference table. The summation is then carried out by using the following theorem due to Leonhard Euler (1707 – 1783).

Theorem 3

If a_0, a_1, a_2, ... is a sequence of numbers and x is a variable, then

$$\sum_{r=0}^{\infty} a_r x^r = \sum_{r=0}^{\infty} \frac{x^r}{(1-x)^{r+1}} \Delta^r a_0 .$$

provided $|x| < 1$ and both series are convergent.

A formal proof of this theorem is obtained simply by an operational argument. This method is intuitively reasonable, but its proper justification lies outside the scope of this book.

Proof

Since $a_r = (1 + \Delta)^r a_0$, we have

$$\sum_{r=0}^{\infty} a_r x^r = \sum_{r=0}^{\infty} x^r (1 + \Delta)^r a_0$$

$$= \left[\sum_{r=0}^{\infty} \{x(1 + \Delta)\}^r \right] a_0 .$$

The series in the square brackets represents an operational function of the operator Δ, but it can be summed in the ordinary way as if it were a simple geometric series. We thus obtain

$$\sum_{r=0}^{\infty} a_r x^r = \left[\frac{1}{1 - x(1 + \Delta)} \right] a_0$$

$$= \frac{1}{1 - x} \left[1 - \left(\frac{x}{1 - x} \right) \Delta \right]^{-1} a_0 .$$

The operational function in the square brackets can be expanded in the usual way and each term of the resulting series operates on a_0. Thus

$$\sum_{r=0}^{\infty} a_r x^r = \frac{1}{1 - x} \left[\sum_{r=0}^{\infty} \left(\frac{x}{1 - x} \right)^r \Delta^r \right] a_0$$

$$= \sum_{r=0}^{\infty} \frac{x^r}{(1 - x)^{r+1}} \Delta^r a_0 ,$$

whence the theorem. It is important to note that in this method the functions of x occurring at each stage are always prefixed to the operator Δ and its powers Δ^r, which are understood to be operating on a_0.

If a_r is a kth degree polynomial in r, then all differences $\Delta^r a_0$ will be zero for $r > k$, and by the theorem the sum of the infinite series is expressed as a series involving a_0 and its first k differences.

Example 4.10

Sum the infinite series

$$S \equiv 2 + 6x + 13x^2 + 24x^3 + 40x^4 + 62x^5 + \dots$$

for suitable values of x.

Table 4.1 overleaf presents the difference table formed from the coefficients of the given powers of x.

Table 4.1 Showing the forward differences of the coefficients a_r of the series

a_r	Δ	Δ^2	Δ^3
2			
	4		
6		3	
	7		1
13		4	
	11		1
24		5	
	16		1
40		6	
	22		
62			

Thus $a_0 = 2$, $\Delta a_0 = 4$, $\Delta^2 a_0 = 3$, $\Delta^3 a_0 = 1$, and $\Delta^k a_0 = 0$, for $k \geqslant 4$.
Hence, by Theorem 3,

$$S = \frac{2}{1-x} + \frac{4x}{(1-x)^2} + \frac{3x^2}{(1-x)^3} + \frac{x^3}{(1-x)^4}$$

$$= \frac{2(1+x)}{(1-x)^2} + \frac{3x^2 - 2x^3}{(1-x)^4}$$

$$= \frac{2 - 2x + x^2}{(1-x)^4} = \frac{1 + (1-x)^2}{(1-x)^4}$$

This summation is valid for $|x| < 1$ since this is the condition for the expansion of $(1-x)^{-4}$ as a power series.

4.7   Summation by generating functions

Another important method of summing a convergent infinite series is by the use of a *generating function*. The technique depends upon the principle that the coefficients of the power series satisfy some linear relation. Consequently this method does not apply generally but only to a special class of series known as *recurring series*. The method has important statistical applications.

To indicate the method, suppose the coefficients of the infinite series

$$\sum_{r=0}^{\infty} a_r x^r$$

satisfy a recurrence relation of the form

$$a_r + a_1 a_{r-1} + a_2 a_{r-2} + \ldots + a_k a_{r-k} = 0, \text{ for } r \geqslant k,$$

and k is a positive integer $\geqslant 1$. The a's are constants whose values are not known but we merely have the knowledge that they exist.

If S denotes the required sum of the series, then

$$(1 + a_1 x + a_2 x^2 + \ldots + a_k x^k) S = (1 + a_1 x + a_2 x^2 + \ldots + a_k x^k) \sum_{r=0}^{\infty} a_r x^r$$

$$= a_0 + (a_1 + a_1 a_0)x + (a_2 + a_1 a_1 + a_2 a_0) x^2 + \ldots +$$

$$+ (a_{k-1} + a_1 a_{k-2} + \ldots + a_{k-1} a_0)x^{k-1},$$

the other terms all being zero because of the relations between the a_r for $r \geqslant k$.

Hence

$$S = \sum_{r=0}^{k-1} \sum_{j=0}^{r} a_j a_{r-j} x^r \bigg/ \sum_{r=0}^{k} a_r x^r,$$

where $a_0 \equiv 1$. This rational function is the generating function of the power series, and it represents its sum for all values of x for which the series is convergent. Alternatively, if the generating function of a power series is known, then its expansion gives the terms of the series. It is also important to remember that the given series will always be convergent for those values of x that make the power series for

$$1 \bigg/ \sum_{r=0}^{k} a_r x^r$$

converge.

We illustrate this method by the following example.

Example 4.11

Evaluate the sum of the infinite series

$$4 + 4x + 2x^2 - 8x^3 - 46x^4 - 176x^5 + \dots,$$

given that

$$a_r + a_1 a_{r-1} + a_2 a_{r-2} + a_3 a_{r-3} = 0, \text{ for } r \geqslant 3.$$

Here $a_0 = a_1 = 4$, $a_2 = 2$, $a_3 = -8$, $a_4 = -46$, $a_5 = -176$.

Therefore, for $r = 3, 4, 5$, we must have

$$
\begin{array}{lll}
-8 + 2a_1 + 4a_2 + 4a_3 = 0, & \text{or} & a_1 + 2a_2 + 2a_3 = 4 \quad (1) \\
-46 - 8a_1 + 2a_2 + 4a_3 = 0, & \text{or} & -4a_1 + a_2 + 2a_3 = 23 \quad (2) \\
-176 - 46a_1 - 8a_2 + 2a_3 = 0, & \text{or} & -23a_1 - 4a_2 + a_3 = 88 \quad (3)
\end{array}
$$

Elimination of a_3 from (1) to (3) gives

$$5a_1 + a_2 = -19; \quad 47a_1 + 10a_2 = -172.$$

A straightforward solution of these simultaneous equations gives $a_1 = -6$ and $a_2 = 11$. Finally, from (1), we obtain $a_3 = -6$.

Hence the generating function of the series is

$$S = \frac{4 + (4 - 6 \times 4)x + (2 - 6 \times 4 + 11 \times 4)x^2}{1 - 6x + 11x^2 - 6x^3}$$

$$= \frac{2(2 - 10x + 11x^2)}{(1 - x)(1 - 2x)(1 - 3x)}.$$

Thus S represents the sum of the series for $|x| < \frac{1}{3}$.

We conclude this chapter with an example illustrating the use of generating functions in the summation of a finite series.

Example 4.12

Evaluate the sum of the series whose rth term is

$$\binom{2N - r}{N} \bigg/ 2^{2N-r}, \text{ for } 0 \leqslant r \leqslant N.$$

The sum is

$$\sum_{r=0}^{N} \binom{2N-r}{N} \frac{1}{2^{2N-r}}$$

$$= \frac{1}{2^N} \sum_{r=0}^{N} \binom{2N-r}{N-r} \frac{1}{2^{N-r}}$$

$$= \frac{1}{2^N} \sum_{r=0}^{N} \left[\text{coefficient of } x^{N-r} \text{ in } \left(1 + \frac{x}{2}\right)^{2N-r} \right]$$

$$= \frac{1}{2^N} \sum_{r=0}^{N} \left[\text{coefficient of } x^{N} \text{ in } x^r \left(1 + \frac{x}{2}\right)^{2N-r} \right]$$

$$= \frac{1}{2^N} \times \text{coefficient of } x^N \text{ in } \sum_{r=0}^{N} \left(\frac{x}{1 + x/2}\right)^r \left(1 + \frac{x}{2}\right)^{2N}$$

$$= \frac{1}{2^N} \times \text{coefficient of } x^N \text{ in } \left(1 + \frac{x}{2}\right)^{2N} \left[\frac{1 - \left(\dfrac{x}{1+x/2}\right)^{N+1}}{1 - \left(\dfrac{x}{1+x/2}\right)} \right]$$

$$= \frac{1}{2^N} \times \text{coefficient of } x^N \text{ in } \left(1 + \frac{x}{2}\right)^{N} \left[\frac{(1 + x/2)^{N+1} - x^{N+1}}{(1 - x/2)} \right]$$

$$= \frac{1}{2^N} \times \text{coefficient of } x^N \text{ in } \left(1 + \frac{x}{2}\right)^{2N+1} \left(1 - \frac{x}{2}\right)^{-1}$$

$$= \frac{1}{2^N} \times \text{coefficient of } x^N \text{ in } \left[\sum_{r=0}^{N} \binom{2N+1}{r} \left(\frac{x}{2}\right)^r \times \sum_{s=0}^{N} \left(\frac{x}{2}\right)^s \right]$$

$$= \frac{1}{2^N} \sum_{r=0}^{N} \binom{2N+1}{r} \frac{1}{2^r} \times \frac{1}{2^{N-r}}$$

$$= \frac{1}{2^{2N}} \sum_{r=0}^{N} \binom{2N+1}{r} .$$

But $2^{2N+1} = \sum_{r=0}^{2N+1} \binom{2N+1}{r}$

$$= \sum_{r=0}^{N} \binom{2N+1}{r} + \sum_{r=N+1}^{2N+1} \binom{2N+1}{2N+1-r}$$

$$= \sum_{r=0}^{N} \binom{2N+1}{r} + \sum_{s=0}^{N} \binom{2N+1}{s}$$

or $\sum_{r=0}^{N} \binom{2N+1}{r} = 2^{2N}.$

Hence the sum of the series is 1.

PART II

STATISTICAL MATHEMATICS AND PROBABILITY

CHAPTER 5

INTRODUCTORY CONCEPTS OF STATISTICS

5.1   The scope of statistics

The science of statistics (or briefly "statistics") is concerned with the application of mathematics to the study of quantitative data obtained from observations; and since observation is the basis of all factual information in empirical science, it is clear that statistical principles necessarily apply to all quantitative scientific inquiry. In this sense, statistics has two main characteristics. Firstly, statistics is a branch of applied mathematics, if we agree to use this term generically to denote all applications of mathematics and not only the applications to the phenomena of physics. Secondly, as a scientific methodology, statistics is applicable to all the sciences — physical, biological, and social. These two features are the basis for the general development of this book and, more particularly, for showing how the same theoretical structure can be applied to various problems occurring in different sciences. Indeed, it is this wide applicability which gives to statistics its growing importance in modern science as a purposeful, analytical, and interpretative discipline.

In general, the subject-matter of statistics may be broadly classified under four heads:

   (i)   The Theory of Probability,
   (ii) Theoretical Statistics,
   (iii) Statistical Methods, and
   (iv) Computational Techniques.

These formal categories are convenient for a preliminary view of the scope of statistics, but they should not be regarded as rigid divisions.

All observations are subject to chance variations, and the mathematical study of chance results in the *Theory of Probability*. The idea of chance is a primitive intuitive concept, but its mathematical study began in France only in the seventeenth century with the analysis of games of chance played by tossing coins or rolling dice. Gambling was then a fashionable pastime and, as large sums were involved in stake money, attempts were made to determine mathematically the odds for winning in various complicated systems of play. These early problems still have some interest, and we shall later on study a few of them. However, the systematic development of probability theory came about under the impact of the great creativity in statistics at the beginning of this century, led by Karl Pearson (1857 – 1936) and

Ronald Aylmer Fisher (1890 – 1962). *Theoretical Statistics* is, as the name implies, the theoretical body of statistics, and it is based wholly on the mathematics of probability. This is the tool-making division of our science, and without it statistics could not be regarded as a well-structured mathematical discipline. The methods developed theoretically provide the techniques for the analysis of data and thereby also lead to a deeper understanding of the complex processes which give rise to variability in observational measurements. Not all users of statistics are mathematicians and to most, who understand the usefulness and importance of the science, an exposition of the methods is necessary. This is, for instance, what books on *Statistical Methods* usually set out to do with the help of examples of the analysis of data from different sciences. The understanding of the proper relevance of a theoretical argument to an empirical situation is a major problem of statistical education, and this book lays much stress on it. *Computational Techniques* are the devices used in actually applying statistical methods to the analysis of data. The preliminary sections of this book give some of the basic ideas of numerical analysis, but there is much else that is required in practical statistical work. It will thus be seen that the domain of statistics extends from the pure abstraction of probability theory to the rather mundane but essential level of arithmetical operations with approximate numbers; and it is worth emphasis that a coherent view of statistics emerges only when the subject is studied as an integrated combination of the four divisions. Such is the plan of this book.

To give concreteness to our conception of what this approach implies, it is perhaps best to consider some examples of problems whose solution depends upon statistics. These examples are deliberately chosen as simple illustrations of the use of statistics, but they should indicate its wide applicability and the type of reasoning which lies at the heart of statistical analysis.

(1) *Physics*

Suppose an experiment is performed to determine the specific gravity of lead, and the result 11·296 is obtained. We agree in assuming that this figure is an approximate number which has been deduced by appropriate rounding off. This implies that had we so wished, we could, at least theoretically, have used a more precise measuring technique to obtain an answer correct to four or five (say) decimal places. There is thus an inherent approximation in the value obtained, and it is therefore obvious that this number cannot be regarded as the *true* specific gravity of lead. Next suppose the experiment is repeated and another measurement made with the same care and accuracy as used in the first experiment. It is a matter of common experience that, in general, the second measurement will not be 11·296. Suppose it is found to be 11·289, and we deduce by our previous reasoning that this also is not the true specific gravity of lead. This variation from measurement to measurement is a characteristic feature of all data obtained by experimentation, and it is the main reason which necessitates the statistical approach.

Assume, for example, that ten experiments are performed under similar conditions and the following measurements obtained:

Table 5.1   Showing the measurements of the specific gravity of lead

11·296	11·289	11·302	11·305	11·299
11·294	11·287	11·306	11·297	11·295

If desired, these experiments could be continued indefinitely, but suppose we decide to stop after the tenth determination. These measurements give some idea about the specific gravity of lead, though we still cannot say what is the true specific gravity of lead. If one insists on a definite answer, it is intuitively reasonable to suggest that the average of the ten measurements, that is 11·297, is the best assessment of the specific gravity *on the available information*. Nevertheless, this average is also not the true specific gravity of lead. Yet statistical theory tells us that, under certain broad assumptions, the average obtained is a proper answer to the question "What is the specific gravity of lead?" on the basis of the available experimental evidence. Again, given the observations, statistical theory can also tell us that, *on a certain level of assurance*, the true specific gravity of lead lies between two numbers. However, it is important to realise that such a statement is not true in the absolute sense, as it is circumscribed by some degree of uncertainty. On the other hand, and this is important, we understand from our theory that no *finite* set of measurements can give the true specific gravity of lead. Indeed, in terms of measurable experience, the true value remains necessarily unknown, though it is possible to reduce the element of uncertainty by obtaining more measurements. As an old Indian proverb states, "The more sugar you add to the cup, the sweeter will be your tea!".

This is a simple but typical problem of statistics, and its solution exemplifies some of the central ideas of the science.

(2)   *Biometry*

It is known that the cuckoo does not hatch its own eggs but lays them in the nests of other birds. A biologist can identify such eggs, and it is believed that the cuckoo has the capacity to vary the size of its eggs according to the type of nest in which it lays them. We observe here that a qualitative fact — the cuckoo lays its eggs in the nests of other birds — is being augmented by a quantitative hypothesis — the cuckoo can adapt the size of its eggs according to the nest in which it lays them. This is an elementary example of how quantitative biology or biometry has been created.

Consider now the figures given in Table 5.2, which pertain to the lengths of cuckoo's eggs found in the nests of hedge-sparrows and reed-warblers.

This limited number of observations throws some light on the hypothesis about the difference of size between the eggs found in the two types of nests. The methods for the analysis and interpretation of the data require a theoretical argument which the theory of statistics provides. The interest is not in the observations *per se* but in the belief that they can give a reasonable basis for a generalisation about the totality of eggs found in the nests of

Table 5.2   Showing the lengths of cuckoo's eggs found in the nests
of hedge-sparrows and reed-warblers

Host	Length of cuckoo's eggs in mm						
Hedge-sparrow	22·0	23·9	20·9	23·8	25·0	24·0	21·7
	23·8	22·8	23·1	23·1	23·5	23·0	23·0
Reed-warbler	23·2	22·0	22·2	21·2	21·6		
	21·6	21·9	22·0	22·9	22·8		

Source: O.H.Latter (1902), *Biometrika*, Vol.1, p.173.

hedge-sparrows and reed-warblers. More specifically, we may pose the
questions:

(i) What are the true average lengths of cuckoo's eggs found in the
nests of hedge-sparrows and reed-warblers?

(ii) Are these true average lengths different and, if so, what is the extent of the true difference?

It is clear that answers to these questions would add materially to the knowledge about the behaviour of the cuckoo, which is of some interest to a biologist. However, as formulated, the questions are unanswerable for exactly
the same reasons for which it is impossible to state the true specific gravity of lead on the basis of a finite set of measurements. Yet statistical
theory again informs us that, under certain assumptions, the questions can
be answered with limited degrees of assurance. But what if the assumptions
behind the theoretical argument are not applicable to the phenomenon under
study? This is a valid and an important point. If a theoretical structure (or
"model") is inapplicable to a given empirical situation, then it is the function of statistical theory to devise another model and the method of analysis
appropriate to it. However, the appropriateness of a particular model for a
given practical situation is a valid but different question. It can be answered, but again within limits of uncertainty. The point is that, as with drinks,
it is unwise to mix questions indiscriminately. Many questions can be
formulated which may be quite relevant to a set of data, but the same statistical analysis will not necessarily answer all of them. In general, a
method of analysis is appropriate to a particular question, and the answer
is based on certain assumptions. If these assumptions are in doubt, then
another analysis and, maybe, more data are needed.

(3)   *Biochemistry*

Consider next the following data obtained from an investigation into the
absorption and accumulation of salts by living plant cells. In the experiment,
potato slices were immersed in a solution of rubidium bromide, and measurements were made at different times to determine the uptake of Rb and Br
ions by the slices.

Many questions can be asked on the basis of these data, but we shall
particularly consider the rates at which the two kinds of ions were absorbed
by the potato slices. If the amount absorbed is plotted against time on
graph paper, it is seen that the two sets of points lie approximately on two

Table 5.3   Showing the uptake of Rb and Br ions by potato slices after different times of immersion in a solution of rubidium bromide

Time of immersion in hours	Mg equivalents per 1000 g of water in the tissue	
	Rb ions	Br ions
21·7	7·2	0·7
46·0	11·4	6·4
67·0	14·2	9·9
90·2	19·1	12·8
95·5	20·0	15·8

Source: F.C.Steward and J.A.Harrison (1939), *Annals of Botany*, N.S.Vol.3, p.427.

straight lines. Accordingly, if we make the assumption that, over the time interval during which the measurements were made, the rates of uptake of the Rb and Br ions are linear, then a possible hypothesis of interest could be that the slopes of the two lines are equal. It is to be noted that even if the assumption of linearity is correct, we do not know the exact straight lines. Nevertheless, statistical theory provides a method for assessing whether the slopes of the lines could be regarded as equal. If the analysis suggested that the two rates were, in fact, not equal, we could determine the magnitude of the true difference. Of course, all such answers can only be given within specified limits of uncertainty.

Further, suppose the assumption that the uptake of both kinds of ions increases linearly with time is questioned. We know from statistical theory that it is possible to answer this, but the available data are inadequate for the purpose. Indeed, a different experiment is needed to provide the data necessary for the appropriate statistical analysis. This brings out another very important aspect of statistics. It is not correct to collect data first and then think of their analysis. The investigator should be clear about the questions he wants answered by his inquiry, and then the theory of statistics gives the best and the most economical plan for experimentation to obtain the required data. The *design* of an experiment is an essential element in the subsequent analysis of the data collected.

(4)   Medicine

The following data are on the incidence of open pulmonary tuberculosis, and refer to the number of deaths in men and women of different age-groups during the first year after the detection of infection.

This is a typical example of data which provide quantitative evidence for prognosis of diseases in medicine. It is important to note that these data pertain to *qualitative* attributes rather than *metrical* characteristics of the kind considered in the earlier illustrations. Age itself is metrical, but the age-groups are clearly defined, and within each such age-group the data give the number of men and women exposed to risk and their mortality. Thus the two qualitative attributes are the sex of individuals and their exposure to risk of death. Some of the questions which statistical analysis could answer from such data are:

(i) Is there any difference in the true mortality rates in each age-group?

Table 5.4   Showing the incidence of mortality amongst men and women of
varying ages within a year of detection of open pulmonary tuberculosis

Age-group in years	Men		Women	
	Exposed to risk	Deaths	Exposed to risk	Deaths
15–19	406	156	500	174
20–24	695	204	816	246
25–29	585	169	619	184
30–34	454	128	433	150
35–39	274	82	257	92
40–44	221	68	194	83
45–49	153	41	94	39
50–54	110	34	58	20
55–59	69	36	29	13
60–	89	43	47	28
Total	3056	961	3047	1029

Source: G. Berg (1939), *Acta Tuberculosea Scand.*, Supplement IV.

(ii) Do the age-groups consistently reveal the same sex difference in
the mortality rates?

(iii) Does the overall incidence of mortality indicate a sex difference?
If so, what quantitative assessment can be made of this difference? If
not, what is the true overall mortality due to tuberculosis irrespective of
sex?

(5)   *Genetics*

Genetics is one of the youngest of the biological sciences, and the origin of its mathematical theory is due to the work of the Austrian monk,
Gregor Johann Mendel (1822-84). We illustrate the use of statistical analysis in genetics by citing one of Mendel's original experiments. Mendel
crossed pure-bred varieties of the sweet pea, one with yellow seeds, the
other with green seeds. The hybrid progeny had yellow seeds. When these
hybrids were fertilised with their own pollen approximately one-quarter of
the offspring had green seeds and bred true when self-fertilised. The remainder were yellow. Of these, approximately one-third also bred true on
self-fertilisation, but the rest, when self-fertilised, had yellow and green
offspring in approximately the proportion 3:1. Mendel performed his experiment in 1865, and several other investigators have since then repeated
the experiment. The table below gives the results obtained.

Table 5.5   Showing the proportions of yellow and green seeded peas obtained by
seven investigators in the offspring of yellow hybrids when self-fertilised

Investigator	Date	Offspring of yellow hybrids when self-fertilised		
		Proportion yellow	Proportion green	Total
Mendel	1865	0·7505	0·2495	8023
Correns	1900	0·7547	0·2453	1847
Tschermak	1900	0·7505	0·2495	4770
Hurst	1904	0·7464	0·2536	1755
Bateson	1905	0·7530	0·2470	15806
Lock	1905	0·7367	0·2633	1952
Darbishire	1909	0·7509	0·2491	145246

An obvious question based on these data is: Are the results of the seven investigators consistent with the hypothesis that the yellow and green progenies of the self-fertilized yellow hybrids are, in fact, in the proportion 3:1? An affirmative answer provided by statistical analysis corroborates experimentally a result obtained by a simple application of Mendel's famous theory of "particulate inheritance".

(6) Education

A typical statistical problem in education is to obtain a numerical measure of agreement between the scores obtained from two different tests given to the same group of subjects. The following data pertain to the intelligence scores of 65 children born in Bath during the period 13th to 31st January 1922 The scores were determined by the administration of the Binet and Otis tests.

Table 5.6 Showing the Binet and Otis test scores obtained by 65 children

Binet score	Otis score	Binet score	Otis score	Binet score	Otis score
67	36	91	91	108	91
70	28	91	129	108	111
72	34	92	92	108	115
74	28	92	98	109	134
75	48	94	115	110	113
76	50	95	80	110	124
77	62	96	96	110	129
78	22	96	108	110	140
81	82	96	146	112	145
82	84	97	118	113	147
83	64	97	121	114	126
83	77	99	106	114	132
83	82	100	79	115	142
84	92	101	103	115	157
85	91	101	113	116	126
86	65	101	118	116	138
86	75	101	119	123	149
86	76	101	141	126	142
87	68	103	115	126	164
89	80	103	131	127	172
89	110	103	139	135	156
91	72	107	102		

Source: J.A.F.Roberts and R.Griffiths (1937), *Annals of Eugenics*, Vol.8, p.15.

The two sets of scores indicate a general association between them. Statistical analysis provides a numerical measure of this agreement and the amount of confidence to be placed in the result after making allowance for variation in the performance of the subjects due to unavoidable chance factors like previous illness, worry, excitement, etc.

(7) Economics

As a simple example of an application of statistical analysis in economics consider the relationship between the number of wage-earners and output per manshift at coal mines employing 100 or more wage-earners in Great Britain in 1945, as given in Table 5.7.

Table 5.7   Showing the distribution of 894 mines according to the number of wage-earners and the output per manshift

Number of wage-earners in a mine	Number of mines with an output per manshift of				
	Under 15 cwt	15 cwt and under 20 cwt	20 cwt and under 25 cwt	25 cwt and over	Total
100-499	103	140	76	42	361
500-999	58	131	76	39	304
1000 and over	25	73	83	48	229
Total	186	344	235	129	894

Source: Ministry of Fuel and Power *Statistical Digest*, 1945.

On the basis of these data, statistical analysis can answer the question whether output per manshift is associated with the number of wage-earners in a mine. A relationship of this kind is an obvious first consideration in the determination of the optimum number of employees in terms of output.

(8)   *Agriculture*

Modern theoretical statistics was initially largely developed to meet the needs of agricultural science, and this still provides the most extensive use of statistical methodology. We consider an interesting example from livestock production (see Table 5.8).

Table 5.8   Showing the gains in weight of five lots of ten pigs each in a comparative feeding experiment

Pig No.	Lot I	Lot II	Lot III	Lot IV	Lot V
1	165	168	164	185	201
2	156	180	156	195	189
3	159	180	189	186	173
4	167	166	138	201	193
5	170	170	153	165	164
6	146	161	190	175	160
7	130	171	160	187	200
8	151	169	172	177	142
9	164	179	142	166	184
10	158	191	155	165	149
Total	1566	1735	1619	1802	1755

Source: E.W.Crampton and J.W.Hopkins (1934), *Journal of Nutrition*, Vol.8, p.329.

For best returns, the farmer is interested in the maximum increase of weight of his stock for slaughter in a given period of time. To achieve this, producers of animal fodder offer the farmer various kinds of protein-enriched foodstuffs for livestock. In general, the prices will vary (and statistical analysis can take account of this additional variable), but suppose the price factor does not enter into the farmer's considerations, for we may, for simplicity, assume that the fodder prices are the same. To test the quality of the foodstuffs, the farmer may wish to try them out on his animals. He could then obtain data of the type recorded in Table 5.8 in a comparative feeding trial with five lots of ten pigs each. The figures give the gains in weight (in lb) recorded in a given period of time.

The individual variability of animals is an important feature in such trials. However, in an appropriately designed experiment it is possible to find out whether, on the average, the five types of foodstuffs actually produced different gains of weight.

(9) *Production engineering*

Our final example is from production engineering. The table below gives the diameters (in mm) of 200 rivet heads produced at a factory.

Table 5.9   Showing the **200** measurements of the head diameter of rivets produced at a factory

13·39	13·43	13·54	13·64	13·40	13·55	13·40	13·26
13·42	13·50	13·32	13·31	13·28	13·52	13·46	13·63
13·38	13·44	13·52	13·53	13·37	13·33	13·24	13·13
13·53	13·53	13·39	13·57	13·51	13·34	13·39	13·47
13·51	13·48	13·62	13·58	13·57	13·33	13·51	13·40
13·30	13·48	13·40	13·57	13·51	13·40	13·52	13·56
13·40	13·34	13·23	13·37	13·48	13·48	13·62	13·35
13·40	13·36	13·45	13·48	13·29	13·58	13·44	13·56
13·28	13·59	13·47	13·46	13·62	13·54	13·20	13·38
13·43	13·35	13·56	13·51	13·47	13·40	13·29	13·20
13·46	13·44	13·42	13·29	13·41	13·39	13·50	13·48
13·53	13·34	13·45	13·42	13·29	13·38	13·45	13·50
13·55	13·33	13·32	13·69	13·46	13·32	13·32	13·48
13·29	13·25	13·44	13·60	13·43	13·51	13·43	13·38
13·24	13·28	13·58	13·31	13·31	13·45	13·43	13·44
13·34	13·49	13·50	13·38	13·48	13·43	13·37	13·29
13·54	13·33	13·36	13·46	13·23	13·44	13·38	13·27
13·66	13·26	13·40	13·52	13·59	13·48	13·46	13·40
13·43	13·26	13·50	13·38	13·43	13·34	13·41	13·24
13·42	13·55	13·37	13·41	13·38	13·14	13·42	13·52
13·38	13·54	13·30	13·18	13·32	13·46	13·39	13·35
13·34	13·37	13·50	13·61	13·42	13·32	13·35	13·40
13·57	13·31	13·40	13·36	13·28	13·58	13·58	13·38
13·26	13·37	13·28	13·39	13·32	13·20	13·43	13·34
13·33	13·33	13·31	13·45	13·39	13·45	13·41	13·45

Source: A. Hald, *Statistical Theory with Engineering Applications* (John Wiley, New York: 1952), p.45. Reproduced with the kind permission of the author and publishers.

Suppose the exact diameter of a rivet head should be 13·5mm, but the rivets will be satisfactory if the diameters lie within the limits 13·5 ±0·25. Clearly, since the rivets are being mass-produced, they will not be exactly 13·5mm in head diameter (this is apart from the rounding-off error in measuring the diameter), and we expect to observe the kind of variations seen in the table above. Now, in general, it would be impracticable to measure the head diameter of each rivet and to discard those rivets found to be outside the specification limits. In fact, statistical theory provides a method by which measurements are made on a comparatively small number of rivets. This limited information is then used to assess the likely proportion of defective rivets in the factory's output. Statistical theory also gives methods for checking the outgoing product to ensure that not more than a given proportion of defective rivets will ultimately reach the customer. These

ideas, introduced by W.A.Shewhart in the 'twenties, form the basis of quality control procedures throughout industry.

5.2   Statistical variation

In the preceding section we have given a number of examples from
widely differing sciences where statistical principles are necessarily applicable for proper analysis and interpretation of the data. The illustrations could be increased indefinitely by further applications from the sciences already referred to and from others such as meteorology, anthropology,
bacteriology, sociology, etc. However, enough has been said to indicate
the wide applicability of statistics, and it is now appropriate to review
the general nature of the problems posed in our examples.

It is important to note that in each case the conclusions to be arrived
at by statistical analysis depend upon observations of some variable. Thus
in Example (1) the variable is the experimental determination of the specific gravity of lead. We have seen that each of these determinations will,
in general, be different from the others as it is affected by its own specific
error of observation. Had there been no observational error, each determination of the specific gravity would have been the same and consequently
there would have been no statistical problem. But, what is the nature of
this error? In the context of our example, we can distinguish its two constituent elements. The first is the error due to the experimental technique
used, and the second is the arithmetical error of rounding off. Clearly, it
is not possible to distinguish between the measurable effects of the two
constituent sources of error, but it is important to understand the difference
between them.

Next, consider the variable in Mendel's experiment in Example (5).
This is the number of yellow offspring in the self-fertilised progeny of the
yellow hybrids. As an experimental fact, Mendel obtained a certain number
(and this is exact) of yellow offspring out of a total of 8,023. If it is assumed that he made no mistake in the counting of the yellow offspring, the
number obtained is an integer 6,021 and there is no error of measurement
in it. If we express the fraction 6021/8023 as a digital approximation,
then we make a rounding-off error in writing it as 0·7505. On the other
hand, the fact that Mendel obtained 6,021 yellow offspring out of a total of
8,023 is a *chance* result and involves an inherent error of experimentation.
The number could, for example, have been 6,020 or 6,037 without statistically violating the theoretical Mendelian hypothesis that the true proportions of the yellow and green offspring are in the ratio of $3:1$. It is not
possible to analyse the factors which might have operated in the experimental situation to give Mendel his result of 6,021 yellow offspring out
of a total of 8,023. In fact, the number 6,021 is an observation of a *discrete
random* (or chance) *variable*, whereas the number 11·296 obtained in the
first determination of the specific gravity of lead is an observation of a
continuous random variable.

The observational error in the value of a discrete random variable depends only on the inherent and unanalysable factors affecting experimentation, but the value of a continuous random variable depends both on the experimental error (which is due to chance) and the arithmetical error of approximation. Discrete random variables arise when we are dealing with numbers (or frequencies) in a number of *qualitatively distinct* classes, and continuous random variables are associated with *metrical* characters such as specific gravity, weight, height, etc. In the numerical sense, the distinction between discrete and continuous random variables is the familiar one between integers and real numbers. However, it is useful to remember that a discrete random variable could, on occasion, be equivalently expressed as a continuous random variable. Thus, in Mendel's experiment the number 6,021 is an observation of a discrete random variable but, expressed as the proportion 0·7505, it may be thought of as an observation of a continuous random variable. Nevertheless, the distinction between discrete and continuous random variables is a fundamental one, and the methods for the analysis of data associated with them are largely different.

Apart from the nature of the error, the observations of both continuous and discrete random variables are *measurements*. Thus the fact that an experiment gives the specific gravity of lead as 11·296 is as much a measurement as the statement that Mendel obtained 6,021 yellow offspring out of a total of 8,023. All measurements are subject to chance in the sense that the values obtained are dependent upon many unanalysable factors which W.A.Shewhart has called the *unassignable causes of variation*. In so far as measurements obtained from two similar experiments are numerically different, their difference is attributable to chance or unassignable causes. On the other hand, if the experimental conditions are varied deliberately, then the observed difference is possibly affected by *identifiable* or *assignable causes of variation*. For example, the differences in the diameters of rivet heads given in Example (9) are chance differences; but if rivets produced by two different machines are measured, then differences between their products may well, in part, be due to the difference in the operational precision of the machines.

Finally, it is only because the observations can be *repeated* that statistical analysis becomes possible. In general, a solitary observation provides no answer through statistics. Indeed, the science is essentially concerned with the behaviour of numerical aggregates. For example, a single measurement of the diameter of a rivet head gives no information about the random variation in the production process. In the same way, Mendel's result is statistically meaningful because he obtained 6,021 yellow offspring out of a total of 8,023. We are thus led to the primary practical consideration in statistical analysis — the methods for studying a mass of similar measurements; and we first consider the case when these measurements are of a continuous random variable.

5.3   Frequency distribution and table

We have seen in Example (9) how observations of a continuous random variable give rise to measurements which are all subject to chance variation. In this example the random variable is the head diameter of rivets produced at a factory, and the 200 measurements made give some information about the variability in the production process. In order to assess this variability quantitatively, the first step is to summarise the data without substantially losing the relevant information about the production process that is contained in the observations. This is done by the formation of a *frequency distribution* which represents the grouping of the observations in a relatively small number of classes. A *frequency table* presents in a systematic tabular form the relationship between the observations and the "frequency" with which they occur, as implied in the concept of a frequency distribution. Indeed, a table is the usual method for the presentation of a frequency distribution based on a large number of observations.

It is to be noted that in its simplest usage in statistics the word "frequency" refers to a number of observations in a qualitative category or a linear interval and not to a rate, which is its meaning in ordinary language and the physical sciences.

The principle for making a frequency distribution is simple. The difference between the largest and smallest observations is divided into a certain number of intervals, usually of equal length, and then the frequency with which the observations fall within these intervals is determined.

The interval chosen for the groupings is called the *class-interval*, and the frequency in a particular class-interval is called the *class-frequency*.

5.4   Magnitude of the class-interval

Given the individual observations, the first consideration in forming a frequency distribution is that of the magnitude of the class-interval. In general, two conditions guide this choice. Firstly, it is desired to be able to treat all the values assigned to any one class, without serious error, as if they were equal to the midpoint of the class-interval. Secondly, for convenience and brevity, it is desired to make the class-interval as large as possible subject to the first requirement. From this it follows that, for a given total number of observations, the larger the range of variation of the observations, the greater must be the class-interval and vice versa. On the other hand, for a fixed range of variation of the observations, the larger the total number of observations the smaller will be the class-interval and vice versa. Thus to determine the magnitude of the class-interval in any particular case, it is necessary to know the range of variation of the observations and their total number.

There is no general method for the calculation of the magnitude of the class-interval, though a good working rule is that the interval should not be greater than $\frac{1}{4}\sigma$, where σ is a particular measure of variation (called "standard deviation" and to be defined later) of the population from which the observations are obtained. However, we have the following table for the values of (range of variation)$/\sigma$ for different values of N, the total number of observations. This table can be used as a reasonably safe guide for the approximate determination of the magnitude of the class-interval.

Table 5.10   Showing the ratio of range of variation and standard
deviation (σ) for samples of different size

N	(Range of variation)/σ	N	(Range of variation)/σ
20	3·7	200	5·5
30	4·1	300	5·8
50	4·5	400	5·9
75	4·8	500	6·1
100	5·0	700	6·3
150	5·3	1000	6·5

Source: L.H.C. Tippett (1925), *Biometrika*, Vol.17, p.364 (Abstract).

As an example of the use of the table, suppose $N = 500$, and the magnitude of the observations varies from 0·25 to 2·63, so that the range of variation is 2·38. Then

$$\frac{2 \cdot 38}{\sigma} = 6 \cdot 1,$$

and the magnitude of the class-interval can be taken to be approximately $\frac{1}{4}(2 \cdot 38/6 \cdot 1) = 0 \cdot 098$. In round figures, the interval may be taken to be 0·10.

It is also known from experience that in the usual practical statistical situations, the number of class-intervals lies between 15 and 25. This also gives a rough indication of the magnitude of the class-interval. In actual practice, the choice of the class-interval can be arrived at quite easily even without Tippett's table if one has had some amount of intelligent experience of handling frequency data.

5.5   Class-marks and class-limits

The second step in the formation of a frequency distribution is the clear and unambiguous definition of the successive class-intervals. For example, if a series of class-intervals is given as

(α):   0−10;   10−20;   20−30;   30−40;   ...,

then it is not clear in which classes the integer values 10,20,30,40,... are to be placed. This indefiniteness can be obviated in a number of ways. An observed value 20 may, for example, be assigned $\frac{1}{2}$ to the class-interval 10−20 and $\frac{1}{2}$ to the class-interval 20−30. However, this procedure of using fractional frequencies is not recommended and should, indeed, be avoided generally. If the observations to be classified are integers, then an appropriate method for defining the class-intervals is

(β):   0−9;   10−19;   20−29;   30−39;   ... .

Alternatively, if the observations are recorded to the nearest tenth of an integer, then the class-intervals may be defined as

(γ):   0−9·9;   10−19·9;   20−29·9;   30−39·9;   ... .

Another method for defining the class-intervals is

$$(\delta): \quad 0- \; ; \quad 10- \; ; \quad 20- \; ; \quad 30- \quad ; \quad \dots \; .$$

This means that the first class-interval includes all observations $\geqslant 0$ but < 10; the second includes all observations $\geqslant 10$ but < 20; and so on.

In methods $(\beta), (\gamma)$, and (δ) there is no ambiguity about correctly plac- ing any observation. The numbers used in defining the class-intervals in $(\beta), (\gamma)$, and (δ) are known as *class-marks*. These need to be distinguish- ed from *class-limits*, the actual terminal points of the successive class- intervals on a linear scale. On a line, a number represents a point, whereas class-intervals are lengths. Thus in (β) the first class-interval ends with the number 9 and the next begins with the number 10, and there is a break in the linear scale. This difficulty can be obviated by defining the class- limits in $(\beta), (\gamma)$, and (δ) as follows:

(β'):   0–9·5;   9·5–19·5;   19·5–29·5;   29·5–39·5;   … ;

(γ'):   0–9·95;   9·95–19·95;   19·95–29·95;   29·95–39·95;   … ;

(δ'):   0– ;   9·5– ;   19·5– ;   29·5– ;   … ;

      or, 0– ;   9·95– ;   19·95– ;   29·95– ;   … ;

depending upon the accuracy of the observations.

It is also important (not for classification but for later work) to deter- mine the correct values of the midpoints of the class-intervals. These are readily obtained as the averages of the class-limits of the class-intervals. It is to be noted that for this purpose the class-limits of the first class- interval in (β) are taken to be $-0·5$ and $9·5$, since the number 0 is strictly represented by the unit interval $(-0·5, 0·5)$. Evidently all the midpoints ob- tained for any system of grouping with equal class-intervals are equispaced and easily written down once any one midpoint is correctly determined. Corresponding to $(\alpha), (\beta), (\gamma)$, and (δ), the midpoints are:

(α''):   5, 15, 25, 35, … ;

(β''):   4·5, 14·5, 24·5, 34·5, … ;

(γ''):   4·95, 14·95, 24·95, 34·95, … ;

(δ''):   the same as in (β'') or (γ'') depending upon the accuracy of the observations.

It is usually convenient to use class-marks in defining the class- intervals in preference to class-limits for making a frequency distribution; but it is essential for some work (to be considered later) that the class- limits be known definitely.

5.6   Classification

The final step in the making of a frequency distribution is that of clas- sification. If the number of observations is not large, it will be adequate to mark the class-marks of the successive class-intervals in a column down the left-hand side of a sheet of computing paper, and transfer the entries of the original observations to this sheet by marking a 1 on the line corresponding to any class-interval for each observation assigned thereto. It is important to go through the observations systematically either according to rows or columns. Do not try to pick out from the data the observations belonging to one class-interval at a time. This takes longer and invariably leads to more mistakes. It also saves time in subsequent totalling if each fifth entry in a class-interval is marked by a diagonal across the preceding four. Each square of the computing paper should thus account for five entries. These markings are usually called *tallies* or *tally marks*.

This is a simple method but it requires a good deal of concentration for accuracy of classification. Its main disadvantage is that the process offers no facility for checking: if a repetition of the classification leads to a different result there is no means of identifying the mistakes made. If the number of observations is at all considerable and accuracy is essential, it will then be better to enter the values observed on cards, one to each observation. These are dealt out into packs according to their class-intervals. The classification can be easily checked by running through the packs corresponding to each class-interval to verify that no cards have been wrongly placed.

Example 5.1

We now apply the principles for making a frequency distribution to Hald's data in Example (9) of Section 5.1. These observations reveal an interesting feature in that the random variation in the measurements is confined to the two digits after the decimal point. This is not true of data generally, and is a special feature of these data arising from precision engineering. Indeed, the largest and the smallest observed diameters are 13·69 and 13·13 mm respectively, so that the range of random variation of the 200 observations is only 0·56 of a mm. Although it is possible to use these observations in the way indicated above, it is simpler first to make a transformation to take advantage of the fact that the integral part of all the observations is the same. If d denotes any observed diameter, the corresponding x value is obtained from the relation

$$x = 100(d - 13).$$

It is clear that the 200 x values are now two-digit integers obtained from the figures after the decimal point of the diameter measurements. We shall work with these x values and, as will be seen later, all calculations made with them can be easily transformed to corresponding results in terms of the original units of the diameter measurements.

The largest and smallest x values are 69 and 13, so that the observed range of variation is 56. The total number of observations is $N = 200$, and therefore, from Table 5.10,

$$\frac{56}{\sigma} = 5\cdot5, \quad \text{whence} \quad \frac{\sigma}{4} = \frac{28}{11} \sim 3.$$

The magnitude of the class-interval may be taken as 3. The class-marks and class-limits of the first class-interval are 12—14 and 11·5—14·5 with

Table 5.11 Showing the frequency distribution of the transformed values obtained from Hald's data of Table 5.9

Class-marks	Class-limits	Midpoint x_i	Tally Marks	Frequency n_i	$z_i^{(*)}$	$n_i z_i^{(*)}$	$n_i z_i^{2 (\dagger)}$
12 — 14	11·5 — 14·5	13	//	2	−9	−18	162
15 — 17	14·5 — 17·5	16		0	−8	− 0	0
18 — 20	17·5 — 20·5	19	////	4	−7	−28	196
21 — 23	20·5 — 23·5	22	//	2	−6	−12	72
24 — 26	23·5 — 26·5	25		8	−5	−40	200
27 — 29	26·5 — 29·5	28		12	−4	−48	192
30 — 32	29·5 — 32·5	31		14	−3	−42	126
33 — 35	32·5 — 35·5	34		17	−2	−34	68
36 — 38	35·5 — 38·5	37		19	−1	−19	19
39 — 41	38·5 — 41·5	40		23	0	0	0
42 — 44	41·5 — 44·5	43		21	1	21	21
45 — 47	44·5 — 47·5	46		17	2	34	68
48 — 50	47·5 — 50·5	49		16	3	48	144
51 — 53	50·5 — 53·5	52		15	4	60	240
54 — 56	53·5 — 56·5	55		10	5	50	250
57 — 59	56·5 — 59·5	58		11	6	66	396
60 — 62	59·5 — 62·5	61	//	5	7	35	245
63 — 65	62·5 — 65·5	64	//	2	8	16	128
66 — 68	65·5 — 68·5	67	/	1	9	9	81
69 — 71	68·5 — 71·5	70	/	1	10	10	100
Total				200		108	2708

(*) Required for the evaluation of the mean of the frequency distribution

(†) Required for the evaluation of the variance of the frequency distribution

the midpoint at 13. Hence the systematic calculations given in Table 5.11.

This frequency distribution is the first important stage in the summaris-
ation of the given set of observations. In this way, the 200 x values have
been replaced by twenty class-intervals and the frequency of observations
lying in each one of them. By doing this, a certain amount of information
about the random variability of the data has been lost but, on the other hand,
the summary reveals several aspects of this variability which are of great
importance in statistical theory.

5.7   Graphical representation: frequency-polygon and histogram

It is often useful to represent the frequency distribution by means
of a diagram which conveys to the eye the general run of the observations.
Furthermore, as we shall see later, a geometrical image also helps in intuitively grasping the sense of the limiting process when the number of observations is increased indefinitely and at the same time the length of the
class-interval becomes infinitesimal. The diagrammatic representation of
the frequency distribution is obtained as follows.

Take a sheet of computing paper and mark a suitable *horizontal* scale
for the class-intervals and a *vertical* scale for the frequencies. Mark the
midpoints of the class-intervals and then through each midpoint mark off
lengths of ordinates proportional to the frequencies in the corresponding
class-intervals. The diagram may then be completed in one of two ways.

(i) As a *frequency-polygon*, by joining up the marks representing the
upper terminal points of the ordinates by straight lines, the last points
at each end being joined to the base at the centre of the next class-
interval to give a closed figure.

(ii) As a column diagram or *histogram*, short horizontals being drawn
through the upper terminal points of the ordinates, which now form the
central axes of a series of rectangles representing the class-frequencies.

Fig.5.1 overleaf shows the frequency-polygon and the histogram of the
frequency distribution obtained in Table 5.11.

In both diagrams the whole area of the figure is proportional to the total
number of observations, but the area over every interval is not correct in the
case of the frequency-polygon, and the frequency of every fraction of any
interval is not the same, as indicated by the histogram.

The area shown by the frequency-polygon over any class-interval with a
central ordinate y_2 is only correct if the tops of the three successive ordinates y_1, y_2, y_3 lie on a straight line, that is if $y_2 = (y_1 + y_3)/2$, as then
the areas of the two little triangles shaded in Fig. 5.2 are equal. If $y_2 >
(y_1 + y_3)/2$, then the area shown by the frequency-polygon is too small.
Conversely, if $y_2 < (y_1 + y_3)/2$, then the area shown by the frequency-
polygon is too great. Accordingly, if, for this reason, the frequency-polygon
tends to become very misleading at any part of the range, it is better to use
the histogram. Indeed, for a simple representation of a frequency distribution, the histogram is generally the better method. The frequency-polygon
is useful when two or more frequency distributions are to be visually compared by superimposition.

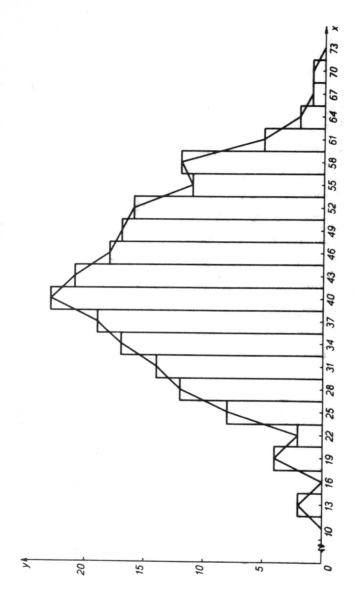

Fig. 5.1 Histogram and frequency-polygon of the frequency distribution of Table 5.11
Note The indentation in the x-axis indicates that the scale between 0 and 9 is not the same as that used for the histogram beyond $x = 9$.

The histogram can also be used when the class-intervals are unequal. Indeed, it is common practice to group the tail observations, especially if they are erratic, and this slightly modifies the histogram. The process is known as *smoothing* the tails of the frequency distribution. This is done simply by describing an area equal, on the scale adopted, to the frequency in a particular interval. The height of the new rectangle will now be the total frequency divided by the width of the interval. As an example, consider the smoothing of the lower tail of the frequency distribution of Table

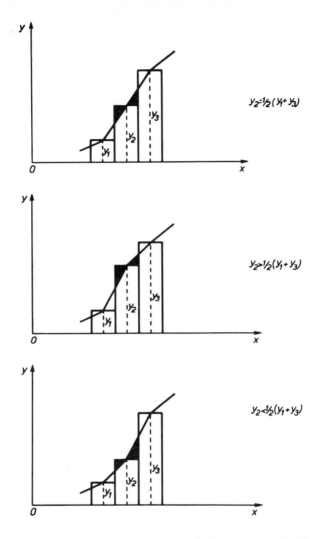

Fig. 5.2 **Differences of the areas under equivalent ranges of a histogram and a frequency-polygon**

5.11. Here the first four class-intervals have a total frequency of 8, so that the height of the rectangle with base of length equal to the four class-intervals is two units of frequency. The upper tail of the frequency distribution does not need smoothing.

The shape of the histogram depends upon the magnitude of the class-interval, and the scales adopted on the x and y axes for the graphical representation. We have already considered how the magnitude of the class-interval should be determined. Nevertheless, it is worth noting that if the class-interval is too small, then the histogram will have a rather erratic appearance. On the other hand, if the class-interval is too big, then the spread of the histogram will be too small, leading to loss of information. The graphical scales to be adopted for the histogram should ensure that in the completed diagram the height of the figure would be 60-80 per cent of its base, provided of course the class-interval has been chosen suitably.

5.8   A limiting process

Suppose a frequency distribution has k class-intervals such that the ith class-interval is of length h_i, its midpoint is x_i and its frequency is n_i, for $i = 1, 2, ..., k$. Then, in general, the statement that the area in a class-interval is proportional to the class-frequency means that

$$n_i = y_i h_i,$$

where y_i is the height of the rectangle in the histogram whose base is the ith class-interval $x_i \pm h_i/2$.

If, as is generally the case, the class-intervals are equal, that is $h_i = h$, a constant, then

$$y_i = n_i/h.$$

Now, as the number of observations increases indefinitely, it becomes possible to reduce the magnitude of the class-interval without making the histogram erratic in appearance. This implies that in the limit, as h tends to zero and the number of class-intervals increases indefinitely, the y_i tend to vary more regularly. Thus the histogram may be regarded conceptually as approaching a smooth curve. This is the *frequency curve*, and it is a concept of supreme importance in statistics. The histogram may be looked upon as an "approximation" to the "true" frequency curve realisable only as a limit. In the context of Hald's data about the head diameter of rivets, we may intuitively think of the frequency curve as representing the distribution of the variation in the head diameters of the conceptual infinity of rivets produced at the factory. Accordingly, the variation revealed by the histogram may then be regarded as a microcosmic representation of the "true" variation in the infinite universe of the frequency curve.

We shall return to these ideas later, but before that we consider certain other features of the histogram based on a finite number of observations.

5.9   Measures of location of the histogram

The histogram obtained for the frequency distribution of head diameter of rivets has a shape which is generally characteristic of histograms arising from a wide variety of data. It is seen from the shape of the histogram that the observations cluster round a central value, and for values removed from it the frequencies tend to decline more or less steadily to reach zero frequencies at the two extremes of the range of variation. In other words, the histogram has a single *point of accumulation* of frequency. It would, therefore, be intuitively reasonable to define this feature of the frequency distribution partly by some central value. Such a central value is called a *measure of location, central tendency,* or *position* of the histogram. In the main, there are three such measures of location, and we consider them one by one.

(i) *Mode*

The mode is the most common value of the random variable, that is the value which is observed with the greatest frequency. In a frequency distribution of a continuous random variable, the mode is of little use, since although the class-interval with the highest frequency is obvious enough, there is the difficulty of placing the mode accurately within that class-interval. In fact, the mode is of little practical use except perhaps in frequency distributions of discrete random variables (to be considered later).

There is no direct and accurate method of calculating the mode of a frequency distribution. However, Karl Pearson suggested an approximate empirical formula for the determination of the mode. This formula bases the calculation of the mode upon the other two measures of location, namely, the *median* and the *mean*. The formula is

$$\text{Mean} - \text{Mode} = 3\,(\text{Mean} - \text{Median}).$$

(ii) *Median*

The median is that value of the random variable which divides the frequency distribution in half. This means that half the observations are less than the median and the other half greater than it. The median is calculated from the *cumulative distribution* by inverse interpolation.

Suppose the observed frequency distribution has k class-intervals each of length h. Let the midpoint of the ith class-interval be x_i and the frequency in this interval be n_i such that $\sum_{i=1}^{k} n_i \equiv N$. With this formal specification of the frequency distribution, the lower limit of the first class-interval is $x_1 - h/2$, and the upper limits of the k class-intervals are $x_i + h/2$, for $i = 1, 2, \ldots, k$. The cumulative distribution is obtained by the summation of the observed class-frequencies up to the upper limits of the successive class-intervals. The cumulative distribution may thus be represented schematically as in Table 5.12 overleaf.

The starting point of the cumulative distribution is the lower limit of the first class-interval, and the first line in Table 5.12 means simply that there is no observed frequency below $x_1 - h/2$. Similarly, the last line of the table signifies that the total observed frequency falls below $x_k + h/2$. Thus the cumulative distribution is an empirical function, whose argument is the upper-limit of the class-intervals and the dependent variable is the total observed frequency below the argument. It is, therefore, immediately obvious that x, the median of the frequency distribution, is the value of the argument which corresponds to cumulative frequency $N/2$.

If we use linear interpolation, and the median lies in the $(r + 1)$th class-interval whose midpoint is x_{r+1}, then

$$x = x_r + \frac{h}{2} + \left[\frac{N}{2} - \sum_{i=1}^{r} n_i\right] h \Big/ n_{r+1}.$$

This simple formula is adequate for all practical purposes. However, if greater precision is wanted, inverse interpolation can be carried out by using a difference table derived from Table 5.12.

Table 5.12   Showing a schematic representation of a cumulative frequency distribution

Upper limit of class-interval	Cumulative frequency up to upper limit
$x_1 - h/2$	0
$x_1 + h/2$	n_1
$x_2 + h/2$	$n_1 + n_2$
$x_3 + h/2$	$n_1 + n_2 + n_3$
$\vdots$	$\vdots$
$x_r + h/2$	$\sum\limits_{i=1}^{r} n_i$
$\vdots$	$\vdots$
$x_k + h/2$	$\sum\limits_{i=1}^{k} n_i$

Example 5.2

As an example, we use the frequency distribution of Table 5.11 to obtain its cumulative distribution and the median. We observe from Table 5.13 that the cumulative frequency up to 38·5 is 78 and that up to 41·5 is 101.

Table 5.13   Showing the cumulative distribution of the frequency distribution of Table 5.11

Upper limit of class-interval	Cumulative frequency up to upper limit
11·5	0
14·5	2
17·5	2
20·5	6
23·5	8
26·5	16
29·5	28
32·5	42
35·5	59
38·5	78
41·5	101
44·5	122
47·5	139
50·5	155
53·5	170
56·5	180
59·5	191
62·5	196
65·5	198
68·5	199
71·5	200

Therefore the median lies between 38·5 and 41·5, and the above formula gives

$$\tilde{x} = 38\cdot5 + \frac{(100 - 78)3}{23} = 38\cdot5 + 2\cdot87 = 41\cdot37 \quad \text{correct to two}$$

decimal places.

This value of the median is in terms of the transformed x values, and we recall that the diameter values (d) are related to the x values by the relation

$$x = 100(d - 13).$$

Therefore, in terms of the original scale of measurement, the median of the frequency distribution is

$$\tilde{d} = 13 + \frac{\tilde{x}}{100} = 13\cdot4137 \quad \text{correct to four decimal places.}$$

In using the above method for evaluating $\tilde{x}$, we have assumed implicitly that the frequency of 23 observed in the class-interval whose upper limit is 41·5 is *distributed uniformly* over the length of the interval. This means that equal parts of the class-interval are proportional to equal frequency. This assumption will hardly ever be strictly true, but it is the simplest basis for the evaluation of the median.

Graphical representation of the cumulative distribution

By a technique similar to that used for the histogram, the cumulative distribution of Table 5.13 can be represented graphically. This has been done in Fig. 5.3, and the form of the graph is characteristic of such data. Generally, this discontinuous curve is known as a *step-function*. It is important to remember that the cumulative frequency is plotted against the upper limits of the class-intervals and not against their midpoints as is the case for a histogram.

Another way of representing the cumulative distribution graphically is to draw a smooth curve through the upper extremities of the ordinates representing the successive cumulative frequencies. Such a smooth curve with an elongated S shape is called a *sigmoid* or a *sigmoidal curve*. This is shown in Fig. 5.4 and is based on the cumulative distribution of Table 5.13.

Median for ungrouped data

We have introduced the median as a measure of location for a frequency distribution, but it is possible (and useful) to extend the definition to the case when the observations are not grouped. Suppose that $y_1, y_2, \dots, y_n$ are n observations which have been *ordered* so that, for example,

$$y_1 \leqslant y_2 \leqslant y_3 \leqslant \dots \leqslant y_n.$$

Then the median of these observations is $y_{\nu+1}$ when $n \equiv 2\nu + 1$, an odd integer. If $n \equiv 2\nu$, an even number, then the median is strictly undefined, but it is conventional to take its value as midway between the two central values, that is as $\frac{1}{2}(y_\nu + y_{\nu+1})$.

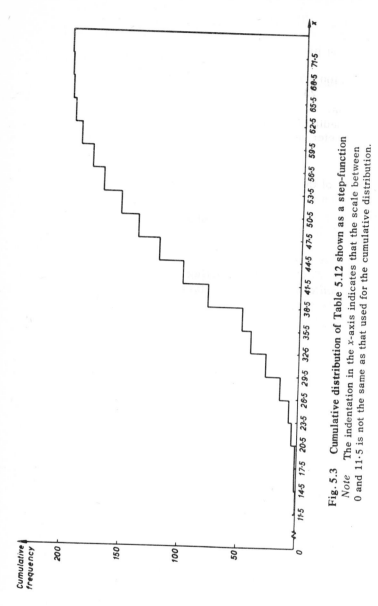

Fig. 5.3   Cumulative distribution of Table 5.12 shown as a step-function

Note   The indentation in the *x*-axis indicates that the scale between 0 and 11·5 is not the same as that used for the cumulative distribution.

(iii)   *Mean*

The mean is the most important of the three measures of location. Besides, like the median, the mean can also be defined formally for both ungrouped and grouped observations.

Thus, if $y_1, y_2, y_3, \ldots, y_n$ are any n individual observations, then their mean is

$$\bar{y} = \frac{1}{n} \sum_{i=1}^{n} y_i.$$

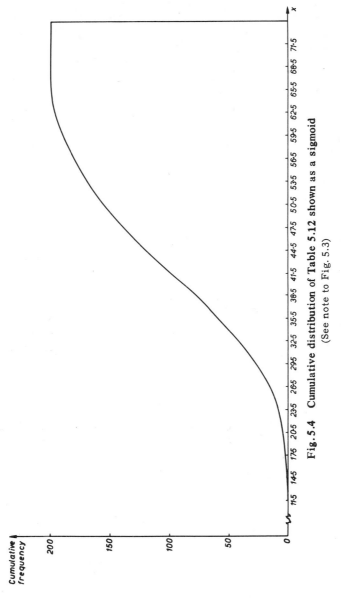

Fig. 5.4 Cumulative distribution of Table 5.12 shown as a sigmoid
(See note to Fig. 5.3)

For observations grouped into a frequency distribution, let x_i be the midpoint of the ith class-interval in which the frequency is n_i, for $i = 1, 2, \ldots, k$. Then the mean of the frequency distribution is

$$\bar{x} = \frac{1}{N} \sum_{i=1}^{k} n_i x_i, \quad \text{where } \sum_{i=1}^{k} n_i \equiv N.$$

It is to be noted that the two definitions depend upon the same principle of averaging. Indeed, if all n_i are equal to unity, then the definition of $\bar{x}$ is the same as that of $\bar{y}$, except that in one case the n observations are averaged, whereas in the other case the midpoints of the class-intervals containing a single observation each are averaged.

The general definition of $\bar{x}$ is based on the *assumption* that the mean of the observations in each class-interval is at the midpoint of the class-interval. Suppose the n_i observations in the ith class-interval are x_{i1}, x_{i2}, ..., x_{in_i}. Hence, if

$$x_i = \frac{1}{n_i} \sum_{j=1}^{n_i} x_{ij},$$

then

$$\bar{x} = \frac{1}{N} \sum_{i=1}^{k} n_i x_i = \frac{1}{N} \sum_{i=1}^{k} \sum_{j=1}^{n_i} x_{ij} = \frac{1}{N} \text{ (sum of all the observations).}$$

However, this assumption can hardly ever be strictly true, and one implicit consideration in the choice of the magnitude of the class-interval is that the error due to the above assumption should be relatively small.

The mean is also referred to as the *average* or, more specifically, as the *arithmetic mean*. Furthermore, a convenient terminological distinction is made between $\bar{y}$, which is referred to as the *simple average*, and $\bar{x}$, which is called the *weighted average*. The *weights* are the n_i, and when they are all equal, the weighted average becomes the simple average. It follows that results for the simple average may be deduced directly from the corresponding ones for the weighted average by putting the weights equal to unity.

Invariance of the mean

One of the most important properties of the mean is its *invariance* under a *linear transformation* of the variables x_i. Suppose a new variable z is defined by the relation

$$z = ax + b,$$

where a and b are known constants and $a \neq 0$. This equation represents a linear transformation from x to z, so that given any particular value of x, say x_i, it is possible to determine the corresponding value of z, that is

$$z_i = ax_i + b, \quad \text{for} \quad i = 1, 2, ..., k.$$

Hence

$$\frac{1}{N} \sum_{i=1}^{k} n_i z_i = \frac{1}{N} \sum_{i=1}^{k} n_i (ax_i + b),$$

or, in terms of the means, $\bar{z} = a\bar{x} + b$. Thus if $z = ax + b$, then $\bar{z} = a\bar{x} + b$, so that the linear relation between the means is exactly the same as that between the original observations. This property is referred to as the invariance of the mean under a linear transformation.

It is to be noted that if we consider a non-linear transformation such as, for example,

$$w = cx^2, \quad c \neq 0 \text{ a constant,}$$

then for the means, $\bar{w} \neq c\bar{x}^2$. Indeed, with such a transformation

$$\bar{w} = c\left[\bar{x}^2 + \frac{1}{N} \sum_{i=1}^{k} n_i(x_i - \bar{x})^2 \right].$$

This result is easily derived by using a simple but important identity (to be proved later)

$$\sum_{i=1}^{k} n_i x_i^2 \equiv N\bar{x}^2 + \sum_{i=1}^{k} n_i(x_i - \bar{x})^2 .$$

The invariance of the mean is a great convenience in its calculation from a frequency distribution. We illustrate this by calculating the mean of the frequency distribution given in Table 5.11.

Example 5.3

We first use the transformation

$$z = \frac{x - 40}{3}$$

to transform the x_i in Table 5.11 to the corresponding values z_i. Then, since

$$\Sigma\, n_i z_i = 108, \quad \bar{z} = 0 \cdot 54.$$

Hence $\bar{x} = 3\bar{z} + 40 = 41 \cdot 62$ correct to two decimal places.

The x to z transformation is of the type that is usually made for the calculation of the mean of a frequency distribution. In the context of the example, the transformation effectively means that we took a *working mean* of 40 and a scale factor of 3 (which is the size of the class-interval) to obtain the simpler z values. It is convenient always to try and select the working mean as close as possible to the actual mean. This can be done fairly easily in most cases by taking as the working mean the midpoint of the central class-interval with the highest frequency.

The transformation reduced the numerical calculation of $\bar{x}$ to that of $\bar{z}$. This is a simple but useful example to show that a computing formula differs from the formal algebraic expression. We can extend this argument by calculating the mean of the original diameter values on which Table 5.11 is based. We recall that the diameter values (d) are related to the x values by the relation

$$x = 100(d - 13).$$

This is also a linear transformation and so the mean of the diameter values is

$$\bar{d} = \frac{\bar{x}}{100} + 13 = 13 \cdot 4162 \text{ correct to four decimal places.}$$

It is worth emphasising the roles of the two linear transformations used. The first (from d to x values) simplified the tabulation of the frequency distribution and the second (from x to z values) materially lessened the labour of the evaluation of the mean. Furthermore, since

$$x = 100(d - 13) \quad \text{and} \quad x = 3z + 40,$$

therefore $d = 0{\cdot}03z + 13{\cdot}4.$

This is also a linear transformation so that

$$\bar{d} = 0{\cdot}03\bar{z} + 13{\cdot}4,$$

which leads to the same value for $\bar{d}$ directly without the intermediate evaluation of $\bar{x}$.

5.10   Measures of dispersion of the histogram

The second important characteristic of the histogram is the *spread* of its values, or its *dispersion*. There are four main measures of dispersion, and each of these can be defined formally either for a frequency distribution or for a set of n ungrouped observations.

(i)   *Range*

The simplest measure of dispersion is the *range*. For n ungrouped but ordered observations $y_1 \leqslant y_2 \leqslant \cdots \leqslant y_n$, the range is simply $y_n - y_1$. For grouped observations, suppose x_1 and x_k are the midpoints of the first and last class-intervals of the frequency distribution. If the length of the class-interval is h, then the range of the frequency distribution is

Upper limit of last class-interval $-$ lower limit of the first class-interval
$$= x_k + h/2 - (x_1 - h/2) = x_k - x_1 + h.$$

Example 5.4

If the individual observations of Table 5.11 are known, then the range is simply $13{\cdot}69 - 13{\cdot}13 = 0{\cdot}56$. On the other hand, if only the frequency distribution of Table 5.11 is known and not the original observations, then the range of the grouped data in terms of the x values is $71{\cdot}5 - 11{\cdot}5 = 60$. Hence, in terms of the original diameter values, the range of the frequency distribution is $0{\cdot}60$.

(ii)   *Semi-interquartile range*

For the ordered observations $y_1 \leqslant y_2 \leqslant y_3 \leqslant \cdots \leqslant y_n$, suppose $n \equiv 4\nu + 3$ where ν is an integer $\geqslant 1$. Then the observations $y_{\nu+1}$, $y_{2\nu+2}$, and $y_{3\nu+3}$ divide the n observations into four ordered sub-groups such that each sub-group contains ν observations. We define $y_{\nu+1}$, $y_{2\nu+2}$, and $y_{3\nu+3}$ as the first, second, and third *quartiles* respectively. These quartiles are usually denoted by Q_1, Q_2, and Q_3. Of course, Q_2 is the same as the median. The *semi-interquartile range* (S.I.Q.R.) is simply

$$\frac{1}{2}(Q_3 - Q_1) = \frac{1}{2}(y_{3\nu+3} - y_{\nu+1}).$$

It is possible (but not worth while) to extend the definition of the S.I.Q.R. approximately when $n \neq 4\nu + 3$ by an argument similar to that used in the conventional definition of the median of an even number of ordered observations.

The first and third quartiles of a frequency distribution are obtained from the cumulative distribution in exactly the same way as tne median. In fact, one-fourth of the total frequency lies below Q_1, and three-fourths of the total frequency lies below Q_3. The calculation of the S.I.Q.R. from a frequency distribution is illustrated best by the following example.

Example 5.5

We use the cumulative distribution of Table 5.13 to obtain Q_1 and Q_3. Since the total frequency is 200, Q_1 lies between (32·5, 35·5) and Q_3 between (47·5, 50·5). Hence, by linear interpolation,

$$Q_1 = 32\cdot5 + \frac{(50-42)3}{17} = 32\cdot5 + 1\cdot41 = 33\cdot91 \text{ correct to two decimal places,}$$

and

$$Q_3 = 47\cdot5 + \frac{(150-139)3}{16} = 47\cdot5 + 2\cdot06 = 49\cdot56 \text{ correct to two decimal places.}$$

Therefore S.I.Q.R. $= \frac{1}{2}(49\cdot56 - 33\cdot91) = 7\cdot82$ correct to two decimal places. In terms of the original diameter values the first and third quartiles are 13·3391 and 13·4956 respectively, and so the S.I.Q.R. is 0·0782 correct to four decimal places.

Percentiles

In general, the point x_p below which p per cent of the frequency distribution lies is known as the pth *percentile*. The percentiles corresponding to the p values 10, 20, 30, ... , 90 are known as the first, second, third, ... , ninth *deciles*. In particular, the median is the second quartile or the fifth decile. All percentage points are also generally known as *quantiles* or *fractiles*.

(iii) *Mean deviation*

For n individual observations $y_1, y_2, ..., y_n$ having mean $\bar{y} = \frac{1}{n}\sum_{i=1}^{n} y_i$, it is easy to verify that the sum of the deviations of the observations from their mean is zero, that is

$$\sum_{i=1}^{n} (y_i - \bar{y}) \equiv 0.$$

Since these deviations measure the extent of the variability of the observations about the mean $\bar{y}$, a non-zero average can be obtained by using the modulus of the deviations. This is the idea behind the definition of the mean deviation, which is formally expressed as

$$\text{Mean deviation} = \frac{1}{n}\sum_{i=1}^{n} |y_i - \bar{y}|.$$

For grouped data, if $x_1, x_2, ..., x_k$ are the midpoints of the k class-intervals with frequencies $n_1, n_2, ..., n_k$ respectively $\left[\sum_{i=1}^{k} n_i \equiv N\right]$, then it is again seen that

$$\sum_{i=1}^{k} n_i(x_i - \bar{x}) \equiv 0, \quad \text{where} \quad N\bar{x} = \sum_{i=1}^{k} n_i x_i.$$

Accordingly, for the frequency distribution defined,

$$\text{Mean deviation} = \frac{1}{N} \sum_{i=1}^{k} n_i |x_i - \bar{x}|.$$

The mean deviation is not a convenient measure of dispersion for analytical purposes, and it is rarely used in statistical theory.

(iv) *Standard deviation*

This is by far the most important measure of dispersion and it is formally defined as the positive square root of the *variance*. The variance is a quadratic measure of dispersion obtained by averaging the squares of the deviations of the observations from their mean.

Using the notation of (iii) above, the variance of $y_1, y_2, \ldots, y_n$ is defined as

$$s_y^2 = \frac{1}{n-1} \sum_{i=1}^{n} (y_i - \bar{y})^2.$$

The use of the divisor $n - 1$ is not arbitrary. The explanation is that the $(y_i - \bar{y})^2$ are not n independent quantities, since $\sum_{i=1}^{n} (y_i - \bar{y}) \equiv 0$, and so any one $(y_i - \bar{y})^2$ may be expressed in terms of the other $n - 1$.

In the same way, the variance of the frequency distribution stated in (iii) above is

$$s_x^2 = \frac{1}{N-1} \sum_{i=1}^{k} n_i (x_i - \bar{x})^2.$$

The divisor $N - 1$ is again introduced because $\sum_{i=1}^{k} n_i(x_i - \bar{x}) \equiv 0$.

The parallelism between the definitions of s_y^2 and s_x^2 is analogous to that between the simple and weighted averages.

Linear transformation and the variance

If we use the linear transformation $x = az + b$ defined before, then

$$\bar{x} = a\bar{z} + b \quad \text{and} \quad x_i - \bar{x} = a(z_i - \bar{z}).$$

Therefore

$$\sum_{i=1}^{k} n_i (x_i - \bar{x})^2 = a^2 \sum_{i=1}^{k} n_i (z_i - \bar{z})^2,$$

so that, if s_z^2 is the variance of the z_i, we have

$$s_x^2 = a^2 s_z^2.$$

This result is of fundamental importance. The computation of s_x^2 may thus be based on that of s_z^2. Furthermore, by definition,

$$s_z^2 = \frac{1}{N-1} \sum_{i=1}^{k} n_i (z_i - \bar{z})^2$$

$$= \frac{1}{N-1} \sum_{i=1}^{k} n_i (z_i^2 - 2\bar{z}z_i + \bar{z}^2)$$

$$= \frac{1}{N-1}\left[\sum_{i=1}^{k} n_i z_i^2 - 2\bar{z}\sum_{i=1}^{k} n_i z_i + N\bar{z}^2\right]$$

$$= \frac{1}{N-1}\left[\sum_{i=1}^{k} n_i z_i^2 - N\bar{z}^2\right], \quad \text{since} \quad N\bar{z} = \sum_{i=1}^{k} n_i z_i,$$

$$= \frac{1}{N-1}\left[\sum_{i=1}^{k} n_i z_i^2 - \left\{\sum_{i=1}^{k} n_i z_i\right\}^2/N\right]$$

$$= \frac{1}{N(N-1)}\left[N\sum_{i=1}^{k} n_i z_i^2 - \left\{\sum_{i=1}^{k} n_i z_i\right\}^2\right].$$

This is the most convenient computational formula for s_z^2. We also see that we have, in passing, established the identities

$$\sum_{i=1}^{k} n_i(z_i - \bar{z})^2 \equiv \sum_{i=1}^{k} n_i z_i^2 - N\bar{z}^2 \equiv \sum_{i=1}^{k} n_i z_i^2 - \frac{1}{N}\left[\sum_{i=1}^{k} n_i z_i\right]^2,$$

which are particularly useful in statistical theory.

Variance of the ungrouped observations

In the same way, or by putting $n_i = 1$ and $k = n$, we have

$$s_y^2 = \frac{1}{n-1}\sum_{i=1}^{n} (y_i - \bar{y})^2$$

$$= \frac{1}{n-1}\left[\sum_{i=1}^{n} y_i^2 - n\bar{y}^2\right]$$

$$= \frac{1}{n-1}\left[\sum_{i=1}^{n} y_i^2 - \frac{1}{n}\left\{\sum_{i=1}^{n} y_i\right\}^2\right]$$

$$= \frac{1}{n(n-1)}\left[n\sum_{i=1}^{n} y_i^2 - \left\{\sum_{i=1}^{n} y_i\right\}^2\right].$$

Example 5.6

We again use the data of Table 5.11 to calculate the standard deviation. The last two columns of this table give

$$\sum n_i z_i = 108 \quad \text{and} \quad \sum n_i z_i^2 = 2708.$$

Hence

$$s_z^2 = \frac{1}{199 \times 200}\left[200 \times 2708 - (108)^2\right] = \frac{529936}{39800} = 13\cdot3150.$$

Therefore

$$s_z = 3\cdot64897 \quad \text{correct to five decimal places.}$$

But $x = 3z + 40$, so that $s_x = 3s_z = 10\cdot9469$ correct to four decimal places. Also, s_d, the standard deviation of the original d values, is obtained simply since

$$d = 0\cdot03z + 13\cdot4,$$

so that $s_d = 0\cdot03s_z = 0\cdot109469$ correct to six decimal places.

Note

A convenient way of calculating $\Sigma n_i z_i$ and $\Sigma n_i z_i^2$ in the last two columns of Table 5.11 is to compute each $n_i z_i$ and $n_i z_i^2$ as a continuous machine operation. This process of successive multiplications is done especially quickly by using the automatic transfer in a "Brunsviga"-type desk calculator.

CHAPTER 6

FREQUENCY CURVES

6.1 Some limiting concepts

We have seen how the 200 measurements of head diameters of rivets produced at a factory could be grouped into a frequency distribution. Given such a frequency distribution, it is possible to represent it as a histogram and to calculate measures such as the mean and standard deviation of the histogram. These are the preliminary stages of statistical analysis, but it is not possible to go any further without making certain broad assumptions about some limiting statistical concepts. It is possible, and perhaps even usual, to define these concepts in terms of formal mathematics, but we shall approach them from an empirical point of view to establish more clearly the links between the practical situation and the formal constructs which underlie our theory.

Measurements like those of the head diameters of rivets are the basis of an important part of statistical analysis. However, the interest of the theory of statistics is not in the measurements *per se*, but in the assumption that these measurements can be analysed to reveal the characteristic features of the general production process. More specifically, the argument runs as follows. The factory is mass-producing rivets, and the machines have been set to produce rivets with a head diameter of δ mm. An essential point in the argument is that δ is unknown. The production engineers will, indeed, have a concrete target, that is, for example, they may wish to produce rivets with a head diameter of 13·5 mm. But any setting of the machines cannot for certain produce rivets of the required diameter because of the inherent uncertainty in machine calibration. This uncertainty is apart from the actual observational errors. The unknown value δ represents the "true" but unknown head diameter of the rivets produced at a given setting of the production machinery if there were no chance variations in the manufacturing process nor any errors of measurement in the assessment of the head diameters of the rivets produced.

Now, for a given setting of the production machinery, there are many factors affecting the head diameter of the rivets actually produced, and there is also the inherent approximation in the measurement of the rivets. It is, therefore, only possible to obtain measurements d_i, where

$$d_i = \delta + \epsilon_i.$$

The ϵ_i are the observational errors, which include the chance and measurement errors of each head diameter measured. It is clear that δ is unknown and so also are the ϵ_i, and we can only observe their sum. Of course, if the ϵ_i were zero or known, there would have been no difficulty; but, given that the ϵ_i are unknown and, in general, different, the statistical problem is to find an answer for the unknown quantity δ within some specified limits of uncertainty.

The 200 measurements actually obtained provide some information about the accuracy of the production process and so implicitly about δ. Furthermore, if we are prepared to measure more rivets produced by the factory, we can add to the available information about δ. Nevertheless, we will not know the exact value of δ by measuring any finite number of rivets. In formal statistical terminology the determination of the head diameter of a rivet is called an *experiment*, and the value obtained for the head diameter, an *observation*. We may thus state that, in general, we have N observations obtained from N individual experiments. We accept as a primary and intuitively reasonable axiom that we cannot obtain all the information about δ from a finite set of experiments.

To overcome this difficulty, we introduce our first limiting concept: we postulate a conceptual infinity of experiments which, if performed, would give all the possible information about δ or, rather, about the observational errors made in performing the experiments. However, since it is impossible to perform an experiment without observational error, the two statements are in essence the same. This implies that the observations arise from identical experiments, where identical is taken to mean under similar conditions. This, then, is our second limiting concept: we postulate that there is no difference or, more correctly, no difference affecting the observations, in the conditions under which the experiments are performed. Of course, this refers to the assignable causes of variation, since the unassignable or random ones will necessarily vary from experiment to experiment. In other words, the structural conditions underlying experimentation are assumed to be constant for all experiments so that the observations are measurements of the same unknown δ. This postulate obviates the logical paradox of the ancient philosopher Heraclitus that it is not possible to step into the same river twice because the river is continually changing.

We next consider that the N particular observations which we have actually obtained are just some of the conceptual infinity of observations. These N observations are said to be a *random sample* from the possible infinity of observations, which constitute what is known as an *infinite population* of observations. This statistical population is a mathematical construct defining the conceptual infinity of observations. We now introduce our third limiting concept: we postulate that the N observations are, *in some way*, representative of the infinite population from which they are drawn. Later on, we shall have occasion to explain more precisely the meaning of this indefinite expression "in some way".

Now suppose that we have actually drawn the histogram based on the N observations. Let x_i be the midpoint of the ith interval ($i = 1, 2, ..., k$), and let the magnitude of the class-intervals be a constant h, so that

$$x_{i+1} = x_i + h.$$

If n_i is the number of observations in the ith class-interval, then n_i/N is said to be the *relative frequency* in that class-interval and, of course,

$$\sum_{i=1}^{k} \frac{n_i}{N} \equiv 1.$$

Also, the size of the rectangle whose base is the ith class-interval $(x_i - h/2, x_i + h/2)$ in the histogram is a measure of n_i/N.

Let $$p_i^* \equiv \frac{n_i}{N}, \quad \text{so that} \quad \sum_{i=1}^{k} p_i^* \equiv 1.$$

Given our limited empirical knowledge, if we were asked the probability (that is the chance) that an observation would lie within the class-interval $(x_i - h/2, x_i + h/2)$, we might say that, since in the experiments performed so far we have obtained n_i out of N observations in the specified class-interval, the probability based on this experience alone is p_i^*. This implies another postulate: the probability is based wholly on the observations and on nothing extraneous to these.

With this intuitive understanding of the notion of *empirical probability*, we are now in a position to introduce our final limiting concept: we postulate that as N, the number of observations, increases, we would expect to be able to give the probability with greater assurance; and, indeed, if we had the entire infinity of observations constituting the population, we would know the probability with complete certainty. In other words, we postulate that as $N \to \infty$ (and, of course, $n_i \to \infty$), $p_i^* \to p_i$, a fixed unknown, and that within our limited experience the only assessment that can be made of p_i is p_i^*. We define p_i as the true probability of an observation lying within the class-interval $(x_i - h/2, x_i + h/2)$. This probability is the relative frequency of observations in the infinite population which lie in the specified interval.

This concept of $p_i^* \to p_i$, as $N \to \infty$, is that of *statistical convergence*. To understand, at least roughly, what this means, it is appropriate to write p_i^* as $p_i^*(N)$ to indicate that p_i^* depends upon N, the number of observations used in determining the relative frequency.

Then, for any $N = N_0$, say, we have

$$|p_i^*(N_0) - p_i| = \epsilon(N_0),$$

where $\epsilon > 0$ and is dependent upon N_0. Again, for some $N = N_1 > N_0$, we have an analogous relation

$$|p_i^*(N_1) - p_i| = \epsilon(N_1).$$

However, it does not invariably follow that since $N_1 > N_0$, therefore

$$\epsilon(N_1) < \epsilon(N_0).$$

In fact, for varying values of N, the $\epsilon(N)$ change but not systematically, and all we imply is that in the limit, as $N \to \infty$, it is certain that $\epsilon(N) \to 0$. This is the kind of uncertainty which is usually implicit in statistical inference based on a finite number of observations. We shall see later how, despite this uncertainty, statistical theory provides methods for making meaningful statements about the unknown p_i on the basis of a limited num ber of observations.

There is one more point which requires clarification. The histogram based on N observations has k class-intervals each of length h and the range of the histogram is $x_k - x_1 + h$. Now if N is sufficiently large, then it is unlikely, though not impossible, that as $N \to \infty$, observations may be obtained which are either less than $x_1 - h/2$ or greater than $x_k + h/2$. In other words, as $N \to \infty$, the range of the limiting distribution could be greater than $x_k - x_1 + h$, so that the number of class-intervals, each of length h, could be $\kappa \geqslant k$. However, the relation between p_i^* and p_i remains unaffected, though it could mean that corresponding to some p_λ, the relative frequency in N observations might well be zero. With this reasoning and assuming that there are κp_i's in all, we have the following properties of the p_i:

(i) $p_i \geqslant 0$, for $1 \leqslant i \leqslant \kappa$.

(ii) $\sum\limits_{i=1}^{\kappa} p_i \equiv 1$, for $\kappa \geqslant k$.

(iii) Given any N observations from the conceptual infinite population, the *expected frequency* of observations in the class-interval $(x_i - h/2, x_i + h/2)$ is Np_i. In general, $Np_i \neq n_i$, the number actually observed in the class-interval. Furthermore, it is possible that n_i may be zero for some i, even when $Np_i > 0$. The quantitative difference between n_i and Np_i is due to chance. Indeed, the *expected* frequencies are defined in terms of the true relative frequencies in the population, whereas the *observed* frequencies are determined from the N observations actually made. This distinction is of fundamental importance.

6.2   Limiting form of the histogram

Consider the point $x_r + h/2$ on the x axis of the histogram based on N observations. This point is the upper limit of the class-interval with midpoint x_r. Therefore the relative frequency of observations in the histogram which are less than $x_r + h/2$ is

$$\frac{1}{N} \sum_{i=1}^{r} n_i = \sum_{i=1}^{r} p_i^* \equiv P_r^*, \quad \text{say.}$$

Now, as $N \to \infty$, each $p_i^* \to p_i$, so that in the limit the relative frequency of observations less than $x_r + h/2$ becomes

$$\sum_{i=1}^{r} p_r \equiv P_r, \quad \text{say.}$$

In other words, as $N \to \infty$, $P_r^* \to P_r$. We have thus defined a limiting function $F(x_r + h/2) \equiv P_r$ at the successive upper limits of the class-intervals, that is at the points $x_r + h/2$. This is the *cumulative distribution function*.

Also, $p_r = \sum\limits_{i=1}^{r} p_i - \sum\limits_{i=1}^{r-1} p_i$

$= . P_r - P_{r-1}$

$$= F(x_r + h/2) - F(x_r - h/2),$$

or $$p_r = F(x + h) - F(x), \quad \text{where} \quad x = x_r - h/2.$$

If the ordinate of the histogram at x_r is $f^*(x_r)$, then $p_r^* = hf^*(x_r)$, since p_r^* is proportional to area. Also, since $p_r^* \to p_r$ as $N \to \infty$, we may write equivalently that, as $N \to \infty$,

$$p_r^* \equiv hf^*(x_r) \to hf(x_r) \equiv p_r.$$

Therefore, putting $x = x_r - h/2$, we have for $N \to \infty$

$$hf(x + h/2) = F(x + h) - F(x).$$

Hence, finally,

$$\lim_{h \to 0} \frac{F(x + h) - F(x)}{h} = f(x) = \frac{dF(x)}{dx}.$$

This two-fold limiting process has enabled a transition from the histogram to a smooth curve $y = f(x)$ to be made. The ordinate of the curve at any point x is $f(x)$, the area under the curve for values less than x is $F(x)$, and

$$f(x) = \frac{dF(x)}{dx}.$$

The function $f(x)$ is the *probability density function* of the continuous random variable x, and the function $F(x)$ is the *distribution function* of x. The curve $y = f(x)$ is the *frequency curve* of x, and the area below the curve gives the *probability distribution* of x.

The frequency curve may thus be regarded as giving all the information about the random variable. Of course, this curve will not be known and it exists only as a convenient conceptual model to represent mathematically the random variable under study. The existence of such a curve and the manner in which it is arrived at provide the essential link between the concrete histogram based on a finite number of actual observations and the formal mathematics which underlies statistical theory. It is more usual in introductions to statistical theory to postulate the existence of a frequency curve and then to regard the histogram as a microcosmic representation of this universe. This difference is of little consequence theoretically because what is of primary importance is the dual existence of the infinite universe of conceptual observations — the population, and the actual finite number of observations — the sample; and statistical inference may thus be regarded as inference about the population on the basis of the sample information. The logical process is that of reasoning from the particular to the general or, in statistical terminology, from the sample to the population.

6.3 Properties of the probability density function

Accordingly, for the development of statistical theory, the starting point is the study of the probability density function, which has the following properties:

(i) $f(x) \geqslant 0$, (compare $p_i \geqslant 0$)

(ii) $\int_{-\infty}^{\infty} f(x)\,dx = 1,$ (compare $\sum_{i=1}^{K} p_i = 1$)

(iii) The probability that an observation lies in the interval (a, b) is $\int_{a}^{b} f(x)\,dx$, that is, the area under the curve $y = f(x)$ for $a \leqslant x \leqslant b$.

It is to be noted that in (ii) the integration takes place over all possible values of the random variable x, which certainly lie in $(-\infty, \infty)$. If the possible values of x lie in the interval (α, β), then outside this interval the probability density function is defined to be identically zero. Hence

$$\int_{-\infty}^{\infty} f(x)\,dx = \int_{\alpha}^{\beta} f(x)\,dx = 1,$$

there being evidently no contribution to the integral outside the range (α, β).

Furthermore, $F(x)$, the distribution function of x, is the area under the curve up to the value x, that is,

$$F(x) = \int_{-\infty}^{x} f(t)\,dt, \quad \text{or} \quad f(x) = \frac{dF(x)}{dx}, \quad \text{as stated above.}$$

Hence $F(-\infty) = 0$ and $F(\infty) = 1$. Also, by using the definition of the distribution function, we can write

$$\int_{a}^{b} f(x)\,dx = \int_{-\infty}^{b} f(x)\,dx - \int_{-\infty}^{a} f(x)\,dx = F(b) - F(a).$$

In particular, if the possible range of x is (α, β), then

$$F(\alpha) = 0 \quad \text{and} \quad F(\beta) = 1.$$

Fig. 6.1 gives a geometrical picture of a probability distribution represented by a frequency curve which has a form typical of many continuous random variables.

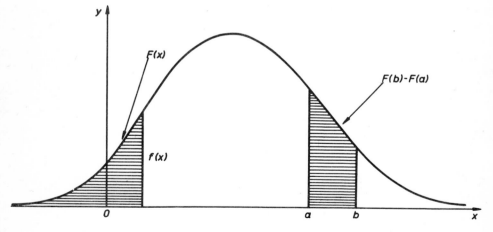

Fig. 6.1   **Typical representation of a frequency curve and the associated probability distribution**

Finally, in view of (iii) above, if of N observations a number n lie in the interval (a, b), then the *expected proportion* of observations in the same interval is $F(b) - F(a)$, and the *expected number* is $N[F(b) - F(a)]$.

6.4   Measures of location and dispersion of the frequency curve

The dual limiting process which leads from the histogram to the frequency curve also involves a corresponding transference from the measures of location and dispersion of the histogram to the measures for the frequency curve, as indicated below.

(i)   Mode

Since $y = f(x)$ is a continuous curve its mode, if any, will be given by such a stationary value of $f(x)$ for which $f(x)$ is a maximum. The stationary values of $f(x)$ are the roots of the equation $f'(x) = 0$, and so $x = m_0$ is a mode of the frequency curve if

$$f'(m_0) = 0 \quad \text{and} \quad f''(m_0) < 0.$$

It is not necessary that a frequency curve should have a mode, but in a majority of cases the curve will have one mode. Such frequency curves are called *unimodal* or *bell-shaped* curves. On the other hand, it occasionally happens that data arising from certain phenomena can only be described by a limiting frequency curve which has more than one mode. These curves are said to be *multimodal*, but we shall not deal with them in this book.

(ii)   Median

The median of the histogram $\tilde{x}$ is evidently translated into the population median $\tilde{m}$ which is determined by the equation

$$F(\tilde{m}) = \int_{-\infty}^{\tilde{m}} f(x)\,dx = \frac{1}{2}.$$

(iii)   Mean

The mean of the histogram is

$$\bar{x} = \sum_{i=1}^{k} x_i \left(\frac{n_i}{N}\right)$$

$$= \sum_{i=1}^{k} x_i p_i^*$$

$$= \sum_{i=1}^{k} x_i h f^*(x_i).$$

Thus, in the limit as $N \to \infty$ and $h \to 0$, the sample mean tends to

$$\int_{-\infty}^{\infty} x f(x)\,dx.$$

This conceptual limit is called the *population mean*, the *expected value*, or the *expectation* of x. It is usual to write the expectation as $E(x) = \mu$.

More generally, the expectation of any function $g(x)$ is defined as

$$E[g(x)] = \int_{-\infty}^{\infty} g(x) f(x)\,dx$$

(iv)   Semi-interquartile range

In general, the pth percentile of the histogram corresponds to ξ_p, the pth percentile of the population, which is determined by the equation

$$F(\xi_p) = \int_{-\infty}^{\xi_p} f(x)\,dx = \frac{p}{100}.$$

Accordingly, the first and third quartiles of the population are ξ_{25} and ξ_{75}; they are determined by the equations

$$F(\xi_{25}) = \int_{-\infty}^{\xi_{25}} f(x)\,dx = \frac{1}{4},$$

and

$$F(\xi_{75}) = \int_{-\infty}^{\xi_{75}} f(x)\,dx = \frac{3}{4}.$$

Hence the S.I.Q.R. of the population is $\frac{1}{2}(\xi_{75} - \xi_{25})$.

(v) *Mean deviation*

The mean deviation of the histogram is

$$\frac{1}{N}\sum_{i=1}^{k} n_i\,|x_i - \bar{x}| = \sum_{i=1}^{k} |x_i - \bar{x}|hf^*(x_i), \quad \text{as in (iii) above.}$$

Therefore, as $N \to \infty$ and $h \to 0$, the mean deviation tends to

$$\int_{-\infty}^{\infty} |x - \mu|f(x)\,dx = E[\,|x - \mu|\,],$$

in the extended notation for expected values. This defines the mean deviation of the population.

(vi) *Standard deviation*

The variance of the histogram is

$$s^2 = \sum_{i=1}^{k}\left(\frac{n_i}{N-1}\right)(x_i - \bar{x})^2$$

$$\sim \sum_{i=1}^{k}(x_i - \bar{x})^2\,p_i^*, \quad \text{for large } N,$$

$$= \sum_{i=1}^{k}(x_i - \bar{x})^2\,hf^*(x_i).$$

Therefore, in the limit as $N \to \infty$ and $h \to 0$, the variance of the histogram tends to

$$\int_{-\infty}^{\infty}(x - \mu)^2 f(x)\,dx = E[(x - \mu)^2\,].$$

This defines the variance of the population, and it is usual to write it as var(x). It is also conventionally frequently denoted by σ^2. Then

$$\sigma = \sqrt{\text{var}(x)} \text{ is the population standard deviation.}$$

We also observe that

$$\text{var}(x) = E[(x - \mu)^2\,]$$

$$= \int_{-\infty}^{\infty}(x - \mu)^2 f(x)\,dx$$

$$= \int_{-\infty}^{\infty}(x^2 - 2\mu x + \mu^2)f(x)\,dx$$

$$= E(x^2) - E^2(x).$$

Therefore

$$\operatorname{var}(x) = E(x^2) - E^2(x).$$

This is a fundamental relation and it is of great use in statistical theory. It is to be noted that $E^2(x) \equiv E(x) \times E(x) \equiv \mu^2$.

6.5 Expectation and variance of a linear function of a random variable

If $g(x) = ax + b$, where a and b are constants and $a \neq 0$, then

$$E(ax + b) = \int_{-\infty}^{\infty} (ax + b) f(x) \, dx$$

$$= aE(x) + b.$$

In particular, if $a = 1$ and $b = -\mu$, then $E(x-\mu) \equiv 0$. This result for the population corresponds with the sample identity

$$\sum_{i=1}^{k} n_i (x_i - \bar{x}) \equiv 0.$$

Again,

$$ax + b - E(ax + b) = a(x - \mu)$$

so that

$$\operatorname{var}(ax + b) = E[\{ax + b - E(ax + b)\}^2]$$
$$= a^2 E[(x - \mu)^2] = a^2 \operatorname{var}(x).$$

Therefore

$$\operatorname{var}(ax + b) = a^2 \operatorname{var}(x).$$

This is also an important result and it corresponds with the one obtained earlier, namely, that for the histogram

$$s_x^2 = a^2 s_z^2, \quad \text{where} \quad x = az + b.$$

6.6 Minimal property of the variance

We have defined $\operatorname{var}(x)$ as $E[(x - \mu)^2]$. If $a \neq \mu$ is any arbitrary constant, then

$$E[(x - a)^2] = E[\{(x - \mu) + (\mu - a)\}^2]$$
$$= E[(x - \mu)^2 + 2(\mu - a)(x - \mu) + (\mu - a)^2]$$
$$= E[(x - \mu)^2] + (\mu - a)^2,$$

or

$$\operatorname{var}(x) = E[(x - a)^2] - (\mu - a)^2.$$

Hence $\operatorname{var}(x) \leqslant E[(x-a)^2]$, the equality holding only when $a = \mu$. Thus $\operatorname{var}(x)$ may be regarded as the minimum value of $E[(x - a)^2]$, which is attained when $a = \mu$, the mean of x.

6.7 Moments of a random variable about the origin

We have seen that in terms of expectations, the population mean is simply $E(x)$ and the population variance is

$$E[(x - \mu)^2] = E(x^2) - E^2(x).$$

It is useful to generalise this notion of expectation of a power of x by defining

$$E(x^r) = \int_{-\infty}^{\infty} x^r f(x) \, dx,$$

for all integers $r \geqslant 0$. It is usual to call $E(x^r)$ the *rth moment* or *the moment of order r of x about the origin*, and to write

$$\mu'_r = E(x^r).$$

According to this moment notation,

$$\mu'_0 = 1; \quad \mu = E(x) = \mu'_1; \quad \text{and}$$

$$\text{var}(x) = E[(x - \mu)^2] = E(x^2) - E^2(x) = \mu'_2 - (\mu'_1)^2.$$

Such definitions are formal in the sense that for any specified form of $f(x)$ all the moments for $r \geqslant \nu$, a constant $\geqslant 1$, can become infinite. We then say that the corresponding moments of x *do not exist* or *diverge*.

6.8   Central moments of a random variable

We extend the moment notation a step further by defining what are known as *central moments* or *moments about the mean*. As the name suggests, the rth central moment of x is

$$\mu_r = \int_{-\infty}^{\infty} (x - \mu)^r f(x)\, dx = E[(x - \mu)^r], \quad \text{for} \quad r \geqslant 0.$$

In particular,

$$\mu_0 = 1; \quad \mu_1 = E(x - \mu) \equiv 0;$$

and

$$\mu_2 = \text{var}(x) = \mu'_2 - (\mu'_1)^2.$$

6.9   Skewness and kurtosis

It is clear that for any symmetrical population with a finite mean

$$E[(x - \mu)^{2r+1}] = \int_{-\infty}^{\infty} (x - \mu)^{2r+1} f(x)\, dx \equiv 0, \quad \text{for} \quad r \geqslant 0,$$

since the integrand is an odd function of x. Accordingly any odd central moment which is not zero may be considered as a measure of *asymmetry* or *skewness* of the population. The simplest of these is μ_3, which is of the third dimension in units of x. We can thus obtain a measure of asymmetry which is of zero dimension (that is a pure number) by dividing μ_3 by $\mu_3^{3/2}$. Therefore we define a measure of skewness as

$$\gamma_1 = \mu_3 / \mu_2^{3/2}.$$

To understand the geometrical meaning of skewness, consider a unimodal curve of the type that usually occurs in statistical theory. Then, in relation to the mode, the right-hand side of the curve is called the *upper tail* and the left-hand side the *lower tail*. Now, if a unimodal frequency curve has a long upper and a short lower tail, then for such a curve $\gamma_1 > 0$, and the curve is said to have *positive skewness* or is *positively skew*. In the opposite event, $\gamma_1 < 0$ and the curve has *negative skewness* or is *negatively skew*. If $\gamma_1 = 0$, the curve is defined to be *symmetrical*. It is to be noted that this definition of symmetry is not in conformity with its usual mathematical meaning. For example, suppose all the moments of x exist. Then the frequency curve of x will be symmetrical in the strict mathematical sense only if all the odd central moments of x vanish. However, in statistical usage, the curve is

said to be symmetrical if $\mu_3 = 0$, so that it is possible, at least theoretically, that the higher odd central moments may not be zero. This difference should be noted, though, in the commoner frequency curves, the
vanishing of μ_3 generally implies that all the higher odd central moments
are also zero.

Fig.6.2 below gives two typical illustrations of curves with positive
and negative skewness.

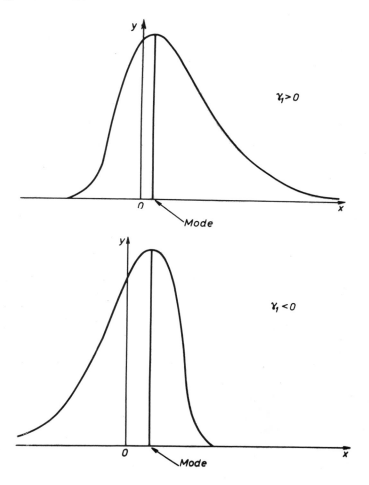

Fig. 6.2   Two typical frequency curves with positive and negative skewness

As will be seen later, the mathematical models for representing the
populations of many random variables can be regarded as approximately
symmetrical, and this assumption is of considerable help in the development of the appropriate theory. However, lack of symmetry is a real empirical phenomenon, and the statistical analysis of data arising from a skew
population presents many theoretical difficulties. We shall not be concerned
with such problems in this book, but it is important to realise that skewness

in many statistical populations is a fundamental characteristic, whence the need for a convenient measure of asymmetry.

The fourth central moment μ_4 can also be reduced to zero dimension by dividing it by μ_2^2. We then define the coefficient of *kurtosis* or *excess* as

$$\gamma_2 = \frac{\mu_4}{\mu_2^2} - 3.$$

This is sometimes used as a measure of the degree of flattening of a frequency curve near its mode. The reason for this apparently arbitrary definition will become clear a little later when we study the "normal distribution" in Chapter 7. Positive values of γ_2 are generally supposed to indicate that the curve is taller and slimmer than the normal curve in the neighbourhood of the mode, and conversely for negative values. However, this is not strictly true, as it is known that values of γ_2 depend considerably upon the tails of the frequency curve. Nevertheless, in the simpler cases, the modal behaviour of the curve is correctly described by the values of γ_2. When $\gamma_2 = 0$, the curve is said to be *mesokurtic*, when $\gamma_2 > 0$, *leptokurtic*, and when $\gamma_2 < 0$, *platykurtic*. A convenient mnemonic is suggested by the two animals — the kangaroo and the platypus: the kangaroo leaps and the platypus has a flat bill.

The coefficients γ_1 and γ_2 were introduced by R.A. Fisher, though Karl Pearson had earlier used the coefficients β_1 and β_2, where

$$\beta_1 = \gamma_1^2 \quad \text{and} \quad \beta_2 = \gamma_2 + 3.$$

Theoretically, it is somewhat more appropriate to use Fisher's coefficients, though in the type of applications with which we shall be concerned, there is little to choose between the two alternative sets of coefficients.

Invariance of the coefficients γ_1 *and* γ_2 *under a linear transformation*

The fact that γ_1 and γ_2 are pure numbers implies that they should be invariant under a linear transformation of x. A transformation of the type

$$x = az + b \quad (a, b \text{ constants and } a \neq 0, b \neq 0)$$

means a change both of origin and of scale of the variable x. With such a transformation,

$$E(x) = aE(z) + b,$$

and
$$[x - E(x)]^r = a^r [z - E(z)]^r.$$

Hence, taking expectations, we have for the central moments of x and z

$$\mu_r(x) = a^r \mu_r(z),$$

that is, the rth central moment of x is a^r times the rth central moment of z. Therefore

$$\gamma_1(x) = \frac{\mu_3(x)}{\mu_2^{3/2}(x)} = \frac{a^3 \mu_3(z)}{a^3 \mu_2^{3/2}(z)} = \gamma_1(z).$$

Similarly,

$$\gamma_2(x) = \frac{\mu_4(x) - 3\mu_2^2(x)}{\mu_2^2(x)} = \frac{a^4 \mu_4(z) - 3a^4 \mu_2^2(z)}{a^4 \mu_2^2(z)} = \gamma_2(z).$$

Thus the coefficients of skewness and kurtosis are invariant under a linear transformation.

We conclude by emphasising that the mean and the central moments μ_2, μ_3, and μ_4 are associated with important and geometrically meaningful characteristics of a frequency curve. In the mathematical study of curves which are important as models of statistical variation, it is useful to be able to evaluate the above four moments conveniently. This can, of course, be done by using the defining integrals, but there are simpler methods of evaluation. We next consider a general exposition of these methods before taking up the study of some particularly useful frequency curves.

6.10 Moment-generating functions

If for every value of t in a certain finite interval the function

$$M_0(t) \equiv E(e^{tx})$$

exists, then for each value of t in the interval $M_0(t)$ is defined to be the *moment-generating function of x about the origin*. The reason for this definition becomes clear from the following argument. We have

$$M_0(t) = \int_{-\infty}^{\infty} e^{tx} f(x)\, dx$$

$$= \int_{-\infty}^{\infty} \left[\sum_{r=0}^{\infty} \frac{(tx)^r}{r!} \right] f(x)\, dx,$$

or, interchanging the order of summation and integration,

$$= \sum_{r=0}^{\infty} \frac{t^r}{r!} \int_{-\infty}^{\infty} x^r f(x)\, dx = \sum_{r=0}^{\infty} \frac{t^r}{r!} \mu'_r.$$

Thus the function $M_0(t)$ is such that, on its expansion as a power series in t, the coefficient of $t^r/r!$ is μ'_r, when it exists. In other words, $M_0(t)$ generates the moments.

In the same way, it is possible to define a generating function which, on expansion, gives the central moments μ_r. In fact, we define the *moment-generating function of x about the mean* as

$$M(t) = E[e^{t(x-\mu'_1)}]$$

$$= \int_{-\infty}^{\infty} e^{t(x-\mu'_1)} f(x)\, dx$$

$$= \int_{-\infty}^{\infty} \sum_{r=0}^{\infty} \frac{t^r}{r!} (x - \mu'_1)^r f(x)\, dx$$

$$= \sum_{r=0}^{\infty} \frac{t^r}{r!} \int_{-\infty}^{\infty} (x - \mu'_1)^r f(x)\, dx = \sum_{r=0}^{\infty} \frac{t^r}{r!} \mu_r.$$

We also observe that

$$M(t) = e^{-\mu'_1 t} M_0(t), \tag{1}$$

which is a fundamental relation. It permits the determination of the relations between moments about the origin and the central moments. By definition

$$M(t) = \sum_{r=0}^{\infty} \frac{t^r}{r!} \mu_r \qquad (\mu_0 = 1, \quad \mu_1 = 0)$$

and

$$M_0(t) \;=\; \sum_{r=0}^{\infty} \frac{t^r}{r!}\, \mu_r' \qquad (\mu_0' = 1, \quad \mu_1' = \mu).$$

Hence, from (1),

$$\sum_{r=0}^{\infty} \frac{t^r}{r!}\, \mu_r \;=\; e^{-\mu_1' t} \sum_{r=0}^{\infty} \frac{t^r}{r!}\, \mu_r'$$

$$=\; \sum_{\nu=0}^{\infty} \frac{(-\mu_1' t)^{\nu}}{\nu!} \times \sum_{r=0}^{\infty} \frac{t^r}{r!} \mu_r'.$$

Equating coefficients of $t^r/r!$ on both sides, we have

$$\mu_r \;=\; r! \sum_{j=0}^{r} \frac{\mu_{r-j}'(-\mu_1')^j}{(r-j)!\, j!}$$

$$=\; \sum_{j=0}^{r} \binom{r}{j} \mu_{r-j}'(-\mu_1')^j.$$

Therefore, in particular,

$$\mu_0 \;=\; \mu_0' \;=\; 1$$
$$\mu_1 \;=\; 0$$
$$\mu_2 \;=\; \mu_2' - (\mu_1')^2$$
$$\mu_3 \;=\; \mu_3' - 3\mu_2'\mu_1' + 2(\mu_1')^3$$
$$\mu_4 \;=\; \mu_4' - 4\mu_3'\mu_1' + 6\mu_2'(\mu_1')^2 - 3(\mu_1')^4,$$

and so on for higher moments. These expressions for the central moments in terms of moments about the origin become increasingly complicated for higher values of r. However, our main interest is in the first four moments, for which the expressions are quite simple. Thus if the first four moments about the origin are known, it is possible to evaluate the central moments by using the above formulae. This indirect method for obtaining the central moments of x is useful when, as happens in some cases, the form of $f(x)$ does not permit an easy evaluation of $M(t)$, though the moments about the origin can be evaluated readily.

6.11   Cumulant-generating function

The *cumulants* are another class of coefficients associated with a frequency curve and, like the moments, they are also simply generated, by the *cumulant-generating function*. This function is formally defined as

$$\kappa(t) \;\equiv\; \log M_0(t)$$

$$=\; \sum_{r=1}^{\infty} \frac{t^r}{r!}\, \kappa_r, \quad \text{where } \kappa_r \text{ is the } r\text{th cumulant.}$$

Now, by definition,

$$\log M_0(t) \;=\; \log\left[1 + \sum_{r=1}^{\infty} \frac{t^r}{r!}\, \mu_r'\right]$$

$$=\; \left[\frac{t}{1!}\mu_1' + \frac{t^2}{2!}\mu_2' + \frac{t^3}{3!}\mu_3' + \frac{t^4}{4!}\mu_4' + \cdots\right] -$$

$$-\; \frac{1}{2}\left[\frac{t}{1!}\mu_1' + \frac{t^2}{2!}\mu_2' + \frac{t^3}{3!}\mu_3' + \cdots\right]^2 + \frac{1}{3}\left[\frac{t}{1!}\mu_1' + \frac{t^2}{2!}\mu_2' + \cdots\right]^3 -$$

$$- \frac{1}{4} \left[\frac{t}{1!} \mu'_1 + \cdots \right]^4 + \cdots$$

$$= \frac{t}{1!} \mu'_1 + \frac{t^2}{2!} [\mu'_2 - (\mu'_1)^2] + \frac{t^3}{3!} [\mu'_3 - 3\mu'_1\mu'_2 + 2(\mu'_1)^3] +$$

$$+ \frac{t^4}{4!} [\mu'_4 - 3(\mu'_2)^2 - 4\mu'_1\mu'_3 + 12\mu'_2(\mu'_1)^2 - 6(\mu'_1)^4] + \cdots$$

$$\equiv \kappa(t).$$

Hence, equating coefficients of $t^r/r!$ in the two expressions for $\kappa(t)$, we have

$$\kappa_1 = \mu'_1$$
$$\kappa_2 = \mu'_2 - (\mu'_1)^2$$
$$\kappa_3 = \mu'_3 - 3\mu'_1\mu'_2 + 2(\mu'_1)^3$$
$$\kappa_4 = \mu'_4 - 3(\mu'_2)^2 - 4\mu'_3\mu'_1 + 12\mu'_2(\mu'_1)^2 - 6(\mu'_1)^4.$$

Furthermore, substituting for μ'_r in terms of the central moments, we obtain

$$\kappa_1 = \mu'_1; \quad \kappa_2 = \mu_2; \quad \kappa_3 = \mu_3; \quad \kappa_4 = \mu_4 - 3\mu_2^2.$$

Therefore, in terms of the cumulants,

$$\gamma_1 = \kappa_3/\kappa_2^{3/2} \quad \text{and} \quad \gamma_2 = \kappa_4/\kappa_2^2.$$

It was, in fact, in this way that Fisher was led to his coefficients γ_1 and γ_2 as dimensionless equivalents of the third and fourth cumulants.

The relations between the cumulants and the central moments can also be obtained directly by using the defining equations

$$M_0(t) = e^{\mu'_1 t} M(t)$$

and

$$\log M_0(t) = \kappa(t).$$

Thus

$$\kappa(t) = \mu'_1 t + \log M(t), \tag{1}$$

so that

$$\sum_{r=1}^{\infty} \frac{t^r}{r!} \kappa_r = \mu'_1 t + \log \left[1 + \sum_{r=2}^{\infty} \frac{t^r}{r!} \mu_r \right],$$

whence the stated results connecting the cumulants and central moments. It is also seen from (1) that a change in the origin of measurement of the random variable x only affects the first cumulant. This invariance of the κ_r $(r \geqslant 2)$ under a change of origin is the reason why in older literature the cumulants were referred to as *semi-invariants*.

If we make a general linear transformation

$$w = ax + b, \quad \text{where } a \neq 0 \text{ and } b \text{ are constants,}$$

then

$$E(e^{tw}) = E[e^{t(ax+b)}] = e^{tb} E(e^{tax}) = e^{tb} M_0(at), \tag{2}$$

whence the moment-generating function of w about its mean is

$$e^{-t(\mu'_1 a+b)} E(e^{tw}) = e^{-\mu'_1 at} M_0(at) = M(at), \tag{3}$$

where $M_0(t)$ and $M(t)$ are as defined in Section 6.10. Hence the rth central moments of w and x satisfy the relation

$$\mu_r(w) = a^r \mu_r(x), \quad \text{for } r \geqslant 2.$$

Furthermore, for the cumulants,

$$K_1(w) = aK_1(x) + b$$

and
$$K_r(w) = a^r K_r(x), \quad \text{for } r \geqslant 2,$$

since the cumulant-generating function of w is from (2)

$$bt + \log M_0(at) = bt + K(at),$$

where $K(t)$ is the cumulant-generating function of x.

For our purpose, the convenience of the cumulants lies in the fact that $\log M_0(t)$ is occasionally easier to expand in a power series than $M_0(t)$. In such cases, the cumulants provide the simpler analytical approach to the main geometrical features of the frequency curve.

6.12 Skewness and kurtosis of the histogram

We have seen that the histogram may be regarded as a microcosmic representation of the infinite universe of the frequency curve. Furthermore, there is also a correspondence between the mean and variance of the histogram and the measures of the population. We now extend this argument to show how the moments of the histogram may be used to define its coefficients of skewness and kurtosis.

We use the notation that the histogram has k class-intervals, each of length h, with midpoints x_i, and class-frequencies n_i, for $1, 2, ..., k$. Also, $N \equiv \sum_{i=1}^{k} n_i$. Using this formal specification of the histogram, we define its rth moment about the origin as

$$m_r' = \frac{1}{N} \sum_{i=1}^{k} n_i x_i^r, \quad \text{for } r \geqslant 0.$$

In this notation, the mean of the histogram is denoted by m_1'. We now define the rth central moment of the histogram as

$$m_r = \frac{1}{N} \sum_{i=1}^{k} n_i (x_i - m_1')^r$$

$$= \frac{1}{N} \sum_{i=1}^{k} n_i \sum_{j=0}^{r} \binom{r}{j} x_i^{r-j} (-m_1')^j$$

$$= \sum_{j=0}^{r} \binom{r}{j} (-m_1')^j m_{r-j}' .$$

This relation between the central moments and moments about the origin is the same as that for the moments of the frequency curve. Hence, in particular,

$$m_2 = m_2' - (m_1')^2$$
$$m_3 = m_3' - 3m_1' m_2' + 2(m_1')^3$$
$$m_4 = m_4' - 4m_1' m_3' + 6(m_1')^2 m_2' - 3(m_1')^4.$$

It is important to note that according to the above definition of m_r

$$s_x^2 = \frac{N}{N-1} m_2,$$

where s_x^2 is the variance of the histogram as defined earlier.

Given the central moments of the histogram, the coefficients of skewness and kurtosis of the histogram are defined as

$$g_1 = m_3/m_2^{3/2}; \quad g_2 = (m_4 - 3m_2^2)/m_2^2.$$

It is also clear that g_1 and g_2 are invariant under a general linear transformation of the type

$$x = az + b, \quad a \neq 0 \text{ and } b \text{ being constants,}$$

since

$$m_r(x) = a^r m_r(z), \quad \text{for } r \geqslant 2,$$

are the relations between the central moments of x and z. Accordingly, it is only necessary to evaluate the central moments of the transformed variable z to determine g_1 and g_2 for the random variable x.

Example 6.1

We conclude this section with an example of the calculation of g_1 and g_2 for the frequency distribution obtained for the stature (in inches) of 6,999 males aged 40·5 years. The data and the calculations are given in the following table.

Table 6.1 Showing the frequency distribution of the stature of 6,999 males aged 40·5 years, and the calculations for determining the coefficients of skewness and kurtosis

Midpoint of interval x_i	Frequency n_i	$z_i = \dfrac{x_i - 67}{2}$	$n_i z_i$	$n_i z_i^2$	$n_i z_i^3$	$n_i z_i^4$
57	4	−5	−20	100	−500	2500
59	18	−4	−72	288	−1152	4608
61	86	−3	−258	774	−2322	6966
63	454	−2	−908	1816	−3632	7264
65	1277	−1	−1277	1277	−1277	1277
67	2001	0	0	0	0	0
69	1886	1	1886	1886	1886	1886
71	912	2	1824	3648	7296	14592
73	301	3	903	2709	8127	24381
75	52	4	208	832	3328	13312
77	6	5	30	150	750	3750
79	2	6	12	72	432	2592
Total	6999		2328	13552	12936	83128

Source: H.A. Ruger (1932 − 33), *Annals of Eugenics*, Vol. 5, p. 95.

$m_1'(z) = 0\cdot332619;$

$m_2'(z) = 1\cdot936277; \quad m_2(z) = 1\cdot936277 - 0\cdot110635 = 1\cdot825642;$

$m_3'(z) = 1\cdot848264; \quad m_3(z) = 1\cdot848264 - 1\cdot932128 + 0\cdot073599 = -0\cdot010265;$

$m_4'(z) = 11\cdot877125; \quad m_4(z) = 11\cdot877125 - 2\cdot459071 + 1\cdot285325 - 0\cdot036720 =$
$$= 10\cdot666659.$$

$\sqrt{m_2(z)} = 1\cdot351163; \quad m_2^2(z) = 3\cdot332969.$

Therefore the coefficients of skewness and kurtosis are

$$g_1 = -\frac{0\cdot010265}{2\cdot466740} = -0\cdot004161;$$

$$g_2 = \frac{0\cdot667752}{3\cdot332969} = 0\cdot200347.$$

CHAPTER 7

CONTINUOUS PROBABILITY DISTRIBUTIONS

Some mathematical preliminaries

7.1 Introduction

Our preceding study of frequency curves has been formal in the sense that we did not specify the functional form of $f(x)$ in the equation of the frequency curve $y = f(x)$. However, from the applicational point of view, we are more interested in the probability distributions which arise when $f(x)$ has some specified form. Evidently, it is possible to define an unlimited number of forms for $f(x)$ satisfying the two essential conditions for it to be a probability density function, namely, that $f(x)$ is non-negative and its integral over the specified range of variation of the random variable x is unity. In this chapter we shall study only a few such forms of $f(x)$ and the associated probability distributions which are of fundamental importance in statistical theory. It is clear that such analysis will be necessarily of a mathematical character, the link with probability distributions being established by interpreting the area under the specified curve $y = f(x)$ as probability. In essence, therefore, the present study is largely an investigation of the properties of certain special types of curves. A convenient unifying feature of our study is that the properties of several of these curves are seen to depend mathematically upon two closely related functions which are well known in the calculus. These are the *gamma* and *beta functions*, and they have the interesting feature that they are functions defined by integrals. Their properties were first studied by Euler and because of this the functions are also occasionally referred to as *Eulerian integrals*.

The idea of defining a function by an integral is novel for a beginner, and it is therefore important to understand what such a definition means. Expressions like

(i) $3x^2 + 5x + 9$;

(ii) $(4x^3 - 8x + 7)/(2x^4 - 5x - 12)$;

(iii) $x^4 \log x + x^2 - 7x$; and

(iv) $x^2 \tan x - 3x \sin^2 x + 4 \cos x$

are all functions of a variable x in the ordinary pure mathematical sense. As an extension of this concept, we may define a function $g(x)$ as

(v) $g(x) = \int_0^x (\cosh t + 3t + 4) \, dt$.

The variable of integration t is of no consequence here, and the integral, when evaluated, gives a function of x which is defined to be $g(x)$. In fact, on integration, we obtain

$$g(x) = \left[\sinh t + \frac{3}{2}t^2 + 4t\right]_0^x$$

$$= \sinh x + \frac{3}{2}x^2 + 4x.$$

The definition of $g(x)$ by the above integral implies that $g(x)$ is a function of the upper limit of the integral, that is of x. It was in this sense that we defined the distribution function $F(x)$ as

$$F(x) = \int_{-\infty}^{x} f(t)dt,$$

the integrand being a probability density function. Finally, there is another way in which a function can be defined by an integral. Thus if α is some non-zero constant, then

(vi) $g(\alpha) = \int_1^2 (e^{-\alpha t} + 4t)dt$

is a function of α. It is to be observed that α does not occur in the limits of integration but only in the integrand. On performing the integration, we have

$$g(\alpha) = \left[-\frac{e^{-\alpha t}}{\alpha} + 2t^2\right]_1^2$$

$$= 6 + e^{-\alpha}(1 - e^{-\alpha})/\alpha.$$

In this case we say that $g(\alpha)$ is a function of α, which is a parameter in the integrand of the defining integral. Nevertheless, $g(\alpha)$ is a function of α in the usual sense. The point is that α is a parameter, that is a constant, so far as the evaluation of the integral defining $g(\alpha)$ is concerned, but once the integral is evaluated, the argument α of $g(\alpha)$ may also be interpreted as a variable in the conventional manner.

It is in this last sense that the gamma and beta functions are functions defined by integrals. With this understanding, we now proceed to investigate the simplest properties of these functions. We shall then use this mathematical knowledge to study some important continuous probability distributions.

7.2   The gamma function

For any given number $n > 0$, the gamma function is defined by the equation

$$\Gamma(n) = \int_0^{\infty} e^{-x} x^{n-1} dx. \tag{1}$$

The expression $\Gamma(n)$ is read as "gamma n", and it is a function of n, which is a positive parameter of the defining integral. The range of integration is infinite, but the integral is finite for finite n. We now evaluate this integral to give an alternative meaning to $\Gamma(n)$.

To begin with, if $n = 1$, we have

$$\Gamma(1) = \int_0^\infty e^{-x} dx = 1.$$

Next, if $n - 1$ is positive, we have, on integration by parts,

$$\Gamma(n) = [- e^{-x} x^{n-1}]_0^\infty + \int_0^\infty e^{-x}(n - 1)x^{n-2} dx,$$

so that

$$\Gamma(n) = (n - 1)\Gamma(n - 1).$$

This is the fundamental recurrence relation of the gamma function. We next show how this relation can be used to evaluate $\Gamma(n)$. Here two cases arise.

(i) Suppose n is an integer. Then, by repeated applications of the recurrence relation, we have

$$\begin{aligned}
\Gamma(n) &= (n - 1)\Gamma(n - 1) \\
&= (n - 1)(n - 2)\Gamma(n - 2) \\
&= (n - 1)(n - 2)(n - 3)\Gamma(n - 3),
\end{aligned}$$

and so on till

$$\Gamma(n) = (n - 1)(n - 2)(n - 3) \ldots 3.\,2.\,1.\,\Gamma(1) = (n - 1)!$$

Hence, if n is an integer,

$$\Gamma(n) = (n - 1)!$$

Because of this property $\Gamma(n)$ is also sometimes denoted by $(n - 1)!$ whether n is an integer or not.

(ii) Next, suppose n is not an integer. We now write

$$n = \nu + \delta,$$

where ν is a positive integer and $0 < \delta < 1$. This means that ν is the largest integer contained in n. For example, if $n = 14 \cdot 63$, then $\nu = 14$ and $\delta = 0 \cdot 63$. Again, using the recurrence relation, we have

$$\begin{aligned}
\Gamma(n) &= \Gamma(\nu + \delta) \\
&= (\nu + \delta - 1)\Gamma(\nu + \delta - 1) = \ldots \\
&= (\nu + \delta - 1)(\nu + \delta - 2) \ldots (\delta)\Gamma(\delta),
\end{aligned}$$

or $\Gamma(n) = (\nu + \delta - 1)^{(\nu)} \Gamma(\delta).$

The integral defining $\Gamma(\delta)$ is known to be finite but, in general, it can only be evaluated numerically for specified values of δ. However, this is no problem since extensive tables exist for determining $\Gamma(\delta)$. For our present purpose, it is enough to note the difference in the methods for evaluating $\Gamma(n)$ when n is or is not an integer. Some other important properties of the gamma function are established in conjunction with the beta function.

7.3   The beta function

For any two positive numbers m and n, the beta function is defined by the equation

$$B(m,n) = \int_0^1 x^{m-1}(1 - x)^{n-1} dx. \tag{2}$$

The expression $B(m, n)$ is read as "beta m, n", and it is a function of the two positive parameters of the defining integral.

If we put $x = 1 - z$, then (2) gives

$$B(m, n) = -\int_1^0 (1 - z)^{m-1} z^{n-1} dz$$

$$= \int_0^1 z^{n-1} (1 - z)^{m-1} dz = B(n, m).$$

Thus the beta function is a symmetrical function of m and n.

Also, in particular,

$$B(1, 1) = 1.$$

Again, if we put $x = \sin^2\theta$, then $dx = 2\sin\theta \cos\theta\, d\theta$, so that (2) gives

$$B(m, n) = 2 \int_0^{\pi/2} \sin^{2m-1}\theta \cos^{2n-1}\theta\, d\theta.$$

Hence, in particulai,

$$B\left(\frac{1}{2}, \frac{1}{2}\right) = 2 \int_0^{\pi/2} d\theta = \pi.$$

An important *alternative* definition of the beta function is obtained by making the substitution $x = (1 + y)^{-1}$ in (2).
Then

$$dx = -\frac{dy}{(1 + y)^2} \quad \text{and} \quad 1 - x = \frac{y}{1 + y}, \quad \text{so that (2) gives}$$

$$B(m, n) = -\int_\infty^0 \frac{y^{n-1} dy}{(1 + y)^{m+n}},$$

$$\text{or } B(m, n) = \int_0^\infty \frac{y^{n-1} dy}{(1 + y)^{m+n}} = \int_0^\infty \frac{y^{m-1} dy}{(1 + y)^{m+n}}, \quad \text{by symmetry.} \tag{3}$$

This is the alternative definition of the beta function.

7.4 Relation between the gamma and beta functions

The gamma and beta functions are related by the equation

$$B(m, n) = \Gamma(m)\Gamma(n)/\Gamma(m + n). \tag{4}$$

We do not prove this standard result here since the proof requires mathematics beyond the level assumed for this book.

7.5 Further properties of the gamma function

Putting $m = n = \frac{1}{2}$ in (4), we have

$$B\left(\frac{1}{2}, \frac{1}{2}\right) = \Gamma\left(\frac{1}{2}\right)\Gamma\left(\frac{1}{2}\right)\Big/\Gamma(1) = \pi.$$

Hence

$$\Gamma\left(\frac{1}{2}\right) = \sqrt{\pi}, \quad \text{or} \quad \int_0^\infty e^{-x} x^{-\frac{1}{2}} dx = \sqrt{\pi}.$$

This result is of great importance in our theory.

Again, putting $m = n$ in (2) and using (4), we obtain

$$\frac{\Gamma^2(n)}{\Gamma(2n)} = \int_0^1 x^{n-1} (1 - x)^{n-1} dx.$$

If we make the substitution $y = 2x - 1$, then

$$\frac{\Gamma^2(n)}{\Gamma(2n)} = 2^{1-2n} \int_{-1}^{1} (1 - y^2)^{n-1} \, dy$$

$$= 2^{1-2n} \left[\int_{-1}^{0} (1 - y^2)^{n-1} dy + \int_{0}^{1} (1 - y^2)^{n-1} dy \right]$$

$$= 2^{2-2n} \int_{0}^{1} (1 - y^2)^{n-1} dy,$$

because the integrand is an even function. Hence, putting $y^2 = z$, we obtain

$$\frac{\Gamma^2(n)}{\Gamma(2n)} = 2^{1-2n} \int_{0}^{1} (1 - z)^{n-1} z^{-\frac{1}{2}} dz$$

$$= \frac{B\left(n, \frac{1}{2}\right)}{2^{2n-1}} = \frac{\Gamma(n)\,\Gamma\left(\frac{1}{2}\right)}{2^{2n-1}\,\Gamma\left(n + \frac{1}{2}\right)} ,$$

so that

$$\Gamma(2n) = \frac{2^{2n-1}}{\sqrt{\pi}} \, \Gamma(n)\,\Gamma\left(n + \frac{1}{2}\right) .$$

This is the important *duplication formula* for the gamma function.

7.6 Stirling's approximation for the gamma function

It is clear that as $n \to \infty$, $\Gamma(n) \to \infty$, but for large n

$$n! \equiv \Gamma(n + 1) \sim \sqrt{2\pi n} \left(\frac{n}{e}\right)^n \left[1 + \frac{1}{12n} + \frac{1}{288n^2} + \cdots\right].$$

This approximation is due to James Stirling (1692 – 1770), who gave it in his *Methodus Differentialis* (1730). We do not prove this result here but only mention that the formula is of great importance in statistics. Furthermore, although the result is true for large n, the first approximation

$$n! \sim \sqrt{2\pi n} \, (n/e)^n$$

is effectively accurate for values of n as small as 9. Indeed, a simple calculation shows that

$$9! = 362880 \quad \text{and} \quad \sqrt{18\pi} \, (9/e)^9 = 359533 \cdot 22 .$$

The percentage error in the approximation is less than one per cent. In many statistical applications such a small approximation error can be readily ignored. However, if greater accuracy is desired, then the second approximation

$$n! \sim \sqrt{2\pi n} \left(\frac{n}{e}\right)^n \left[1 + \frac{1}{12n}\right]$$

is almost wholly accurate. For $n = 9$, this gives the value $362862 \cdot 23$ with an approximation error of less than $0 \cdot 005$ per cent. We will rarely need greater accuracy in statistical work. As a matter of fact, Stirling's formula is the basis of many important approximate results in statistical theory as, in the type of questions that our theory seeks to answer, small errors of approximation are necessarily ignored to derive results whose simplicity rather than "complete" mathematical accuracy is the best pragmatic criterion of their value.

Another useful result can be deduced simply from Stirling's formula. Thus, if n is large and h finite,

$$\frac{\Gamma(n+h)}{\Gamma(n)} \sim \frac{\sqrt{2\pi}\,(n+h-1)^{n+h-\frac{1}{2}}\,e^{n-1}}{\sqrt{2\pi}\,(n-1)^{n-\frac{1}{2}}\,e^{n+h-1}} = \frac{n^h}{e^h}\left[1+\frac{h}{n-1}\right]^{n-\frac{1}{2}}\left[1+\frac{h-1}{n}\right]^h.$$

Therefore

$$\log\left[\frac{\Gamma(n+h)}{\Gamma(n)}\right] \sim h\log n - h + \left(n-\frac{1}{2}\right)\log\left[1+\frac{h}{n-1}\right] + h\log\left[1+\frac{h-1}{n}\right]$$

$$= h\log n - h + \left(n-\frac{1}{2}\right)\left[\frac{h}{n-1} - \frac{h^2}{2(n-1)^2} + \cdots\right] +$$

$$+ h\left[\frac{h-1}{n} - \frac{(h-1)^2}{2n^2} + \cdots\right]$$

$$= h\log n - h + n\left[1-\frac{1}{2n}\right]\left[\frac{h}{n}\left(1-\frac{1}{n}\right)^{-1} - \frac{h^2}{2n^2}\left(1-\frac{1}{n}\right)^{-2} + \cdots\right] +$$

$$+ \frac{h(h-1)}{n}\left[1-\frac{h-1}{2n} + \cdots\right],$$

or, neglecting terms involving n^{-2},

$$\sim h\log n - h + \left[1-\frac{1}{2n}\right]\left[h\left(1+\frac{1}{n}\right) - \frac{h^2}{2n}\right] + \frac{h(h-1)}{n}$$

$$\sim h\log n - h + h\left[1+\frac{1}{2n}\right] - \frac{h^2}{2n} + \frac{h(h-1)}{n}$$

$$= h\log n + \frac{h(h-1)}{2n}.$$

Hence

$$\frac{\Gamma(n+h)}{\Gamma(n)} \sim n^h e^{h(h-1)/2n} \sim n^h\left[1+\frac{h(h-1)}{2n}\right]. \tag{1}$$

This approximation is also very good for relatively small n. For example, if $n = 10$ and $h = 3$, then

$$\Gamma(13)/\Gamma(10) = 1320,$$

and the approximation gives the value 1300, with a percentage error less than 1·6 per cent. Again, for $n = 20$ and $h = 3$,

$$\Gamma(23)/\Gamma(20) = 9240,$$

and the approximate value is 9200 with a percentage error of less than 0·5 per cent.

7.7 Bernoulli numbers

We indicated in Chapter 6 that, in general, the simplest method for determining the moments (cumulants) of a random variable is by a power-series expansion of its moment- (cumulant-)generating function. Quite frequently such an expansion is straightforward and its general term can be written without any difficulty. However, in some cases, the expansion is not simple, though it can be expressed in terms of a power series whose coefficients are the *Bernoulli numbers,* so named after their discoverer James Bernoulli (1654 – 1705).

Formally, the Bernoulli number of order j – denoted by B_j – is defined as the coefficient of $z^j/j!$ in the power series representation of the function $z/(e^z - 1)$. Thus

$$\frac{z}{e^z - 1} \equiv \sum_{j=0}^{\infty} B_j \frac{z^j}{j!} \,. \tag{1}$$

Direct expansion of the left-hand side of (1) gives

$$\frac{z}{e^z - 1} = \frac{1}{\sum_{r=1}^{\infty} z^{r-1}/r!}$$

$$= \left[1 + \sum_{r=2}^{\infty} z^{r-1}/r! \right]^{-1}.$$

Hence, comparing coefficients of $z^j/j!$ in

$$\left[1 + \sum_{r=2}^{\infty} z^{r-1}/r! \right]^{-1} = \sum_{j=0}^{\infty} B_j \frac{z^j}{j!} \,,$$

we obtain

$$B_0 = 1, \; B_1 = -\frac{1}{2}, \; B_{2j+1} = 0 \text{ for all } j \geqslant 1;$$

and $B_2 = \dfrac{1}{6}, \; B_4 = -\dfrac{1}{30}, \; B_6 = \dfrac{1}{42}, \; B_8 = -\dfrac{1}{30}, \; B_{10} = \dfrac{5}{66}, \; B_{12} = -\dfrac{691}{2730},$

$B_{14} = \dfrac{7}{6}, \;$ and so on.

A useful expression is obtained by writing (1) in the form

$$\frac{z}{e^z - 1} - 1 + \frac{z}{2} = \sum_{j=2}^{\infty} B_j \frac{z^j}{j!} \,,$$

or

$$\frac{1}{e^z - 1} - \frac{1}{z} + \frac{1}{2} = \sum_{j=2}^{\infty} B_j \frac{z^{j-1}}{j!} \,,$$

or

$$\frac{1}{2} \frac{\cosh z/2}{\sinh z/2} - \frac{1}{z} = \sum_{j=2}^{\infty} B_j \frac{z^{j-1}}{j!} \,,$$

or

$$\frac{d}{dz} \left[\log \left\{ \frac{\sinh z/2}{z/2} \right\} \right] = \sum_{j=2}^{\infty} B_j \frac{z^{j-1}}{j!} \,.$$

Hence, integrating with respect to z from 0 to z, we have

$$\log \left[\frac{\sinh z/2}{z/2} \right] = \sum_{j=2}^{\infty} B_j \frac{z^j}{j \, j!} \,,$$

which is the required result.

Some continuous probability distributions

7.8   The uniform distribution

This simple distribution is defined by the probability density function

$$f(x) = \frac{1}{\alpha}, \text{ for } 0 \leqslant x \leqslant \alpha,$$

and

$$f(x) = 0, \text{ otherwise.}$$

Here α is a positive constant and it is known as a parameter of the distribution. Also, it is easily seen that the integral of $f(x)$ over the specified range of x is unity, and so $f(x)$ does represent a proper probability density function. It is an important feature of the distribution that the range of variation of the random variable x depends upon α. Furthermore, it is evident that the probability density function remains constant throughout the permissible range of x, and the total probability is confined to a rectangular region with base length α and height $1/\alpha$. Accordingly, this distribution is also known as the *rectangular distribution*. With the transformation $z = x/\alpha$, any rectangular distribution can be transformed into a uniform distribution defined in the unit square. Such a transformed distribution is known as the *standardised* form of the uniform distribution.

Cumulants

The moment-generating function of x about the origin is

$$M_0(t) \equiv E(e^{tx}) = \frac{1}{\alpha} \int_0^{\alpha} e^{tx} \, dx$$

$$= \frac{e^{t\alpha} - 1}{t\alpha} = \frac{e^{t\alpha/2}(e^{t\alpha/2} - e^{-t\alpha/2})}{t\alpha} = e^{t\alpha/2} \frac{\sinh t\alpha/2}{t\alpha/2}.$$

Hence the cumulant-generating function of x is

$$\kappa(t) = \frac{t\alpha}{2} + \log \left[\frac{\sinh t\alpha/2}{t\alpha/2} \right]$$

$$= \frac{t\alpha}{2} + \sum_{j=2}^{\infty} B_j \frac{(t\alpha)^j}{j \, j!}, \quad \text{where } B_j \text{ are the Bernoulli numbers,}$$

$$= t\frac{\alpha}{2} + \sum_{r=1}^{\infty} B_{2r} \frac{\alpha^{2r}}{2r} \frac{t^{2r}}{(2r)!}, \quad \text{since } B_{2r+1} = 0 \text{ for } r \geqslant 1.$$

Therefore the cumulants of x are

$$\kappa_1 = \frac{\alpha}{2}; \quad \kappa_{2r} = \frac{B_{2r}}{2r} \alpha^{2r}; \quad \kappa_{2r+1} = 0, \text{ for } r \geqslant 1.$$

Hence, in particular,

$$\kappa_1 = \frac{\alpha}{2}; \quad \kappa_2 = \frac{\alpha^2}{12}; \quad \kappa_3 = 0; \quad \kappa_4 = -\frac{\alpha^4}{120}.$$

Therefore the coefficients of skewness and kurtosis of x are

$$\gamma_1 = 0 \text{ and } \gamma_2 = -\frac{\alpha^4}{120} \times \frac{144}{\alpha^4} = -\frac{6}{5}.$$

We thus see that the distribution is symmetrical and platykurtic.

For this distribution, we could alternatively obtain the central moments by first evaluating

$$\mu_r' \equiv E(x^r) = \alpha^r/(r+1), \text{ for } r \geqslant 0,$$

and then transforming from the moments about the origin to the central moments. However, the evaluation of the moment-generating function is generally the more convenient way of determining the moment coefficients of a distribution. Of course, in the present case, the use of Bernoulli numbers has enabled compact evaluation of the cumulants of x.

Mean deviation

Since $E(x) = \frac{1}{2}a$, the mean deviation of x is

$$E\left[\left|x - \frac{1}{2}a\right|\right] = \frac{1}{a}\int_0^a \left|x - \frac{1}{2}a\right|dx$$

$$= a\int_0^1 \left|z - \frac{1}{2}\right|dz, \text{ where } z = x/a,$$

$$= a\left[\int_0^{\frac{1}{2}}\left(\frac{1}{2} - z\right)dz + \int_{\frac{1}{2}}^1\left(z - \frac{1}{2}\right)dz\right]$$

$$= a\left\{\left[\frac{1}{2}z - \frac{1}{2}z^2\right]_0^{\frac{1}{2}} + \left[\frac{1}{2}z^2 - \frac{1}{2}z\right]_{\frac{1}{2}}^1\right\} = \frac{1}{4}a.$$

Percentiles

The pth percentile of the distribution is ξ_p, and it is determined by the equation

$$\frac{1}{a}\int_0^{\xi_p}dx = \frac{p}{100}, \text{ so that } \xi_p = \frac{ap}{100}.$$

Therefore, in particular, the median of the distribution is $\xi_{50} = \frac{1}{2}a$, which is the same as the mean — a result intuitively obvious for a symmetrical distribution. Again, the S.I.Q.R. $= \frac{1}{2}\left(\xi_{75} - \xi_{25}\right) = \frac{1}{2}\left(\frac{3}{4}a - \frac{1}{4}a\right) = \frac{1}{4}a$. Thus the mean deviation and the S.I.Q.R. are both equal to $\frac{1}{4}a$ but the standard deviation of x is $a/2\sqrt{3}$.

Clearly, the distribution has no mode since the curve representing the distribution is simply the straight line $y = 1/a$. A diagrammatic representation of the distribution is given in Fig. 7.1 below.

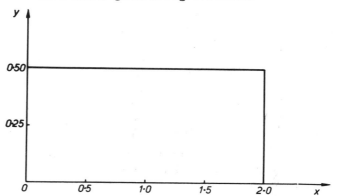

Fig.7.1   Typical form of the rectangular distribution $(a = 2)$

7.9   The negative exponential distribution

As suggested by the name of this distribution, its probability density function is defined by

$$f(x) = \frac{1}{\lambda} e^{-x/\lambda}, \quad \text{for } 0 \leqslant x < \infty,$$

and

$$f(x) = 0, \quad \text{otherwise.}$$

Here λ is a parameter of the distribution, but in this case the range of variation of x is independent of λ. In general, such distributions are more convenient to work with theoretically. Also, since $f(x)$ is essentially non-negative, it is clear that $\lambda > 0$.

It is readily verified that the integral of $f(x)$ over the specified range of x is unity so that $f(x)$ does define a proper probability density function. The parameter λ is a scale parameter, and if we put $z = x/\lambda$, then the probability density function of z is

$$f(z) = e^{-z}, \quad \text{for } 0 \leqslant z < \infty,$$

and

$$f(z) = 0, \quad \text{otherwise.}$$

The distribution of z is the standardised form of the negative exponential distribution, and any such distribution can be reduced to this standard form.

Cumulants

The simplest approach to the moments of x is again through the moment-generating function. We have

$$M_0(t) \equiv E(e^{tx}) = \frac{1}{\lambda} \int_0^\infty e^{tx - x/\lambda} \, dx$$

$$= \frac{1}{\lambda} \int_0^\infty e^{-x(1-\lambda t)/\lambda} \, dx = (1 - \lambda t)^{-1}, \quad \text{if } 1 - \lambda t > 0.$$

Hence the cumulant-generating function is

$$K(t) = -\log(1 - \lambda t)$$

$$= \sum_{r=1}^\infty \frac{(\lambda t)^r}{r} = \sum_{r=1}^\infty \frac{t^r}{r!} \lambda^r (r - 1)!$$

Therefore the rth cumulant $(r \geqslant 1)$ of the distribution is given by

$$\kappa_r = \lambda^r (r - 1)!,$$

and, in particular,

$$\kappa_1 = \lambda; \quad \kappa_2 = \lambda^2; \quad \kappa_3 = 2\lambda^3; \quad \kappa_4 = 6\lambda^4.$$

Hence the coefficients of skewness and kurtosis of the distribution are

$$\gamma_1 = 2; \quad \gamma_2 = 6.$$

This is a positively skew and leptokurtic distribution. The parameter λ is seen to be the mean of the distribution, and the mean is equal to the standard deviation. These are important characteristics of the distribution. There is no mode of the distribution since, starting from the value $1/\lambda$ at the origin, the ordinates approach zero asymptotically. The distribution may be represented as in Fig.7.2 facing. A curve of this general shape is called *J-shaped*.

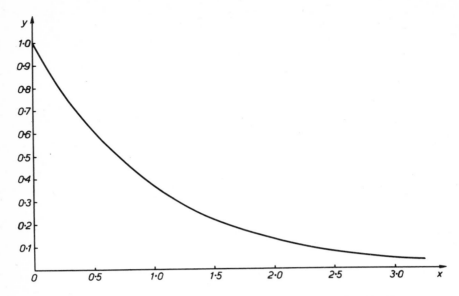

Fig.7.2   Form of the negative exponential distribution ($\lambda = 1$)

Mean deviation

The mean deviation of the distribution is

$$E\left[|x - \lambda|\right] = \frac{1}{\lambda} \int_0^\infty |x - \lambda| e^{-x/\lambda} dx$$

$$= \lambda \int_0^\infty |z - 1| e^{-z} dz, \quad \text{where } z = x/\lambda,$$

$$= \lambda \left[\int_0^1 (1 - z) e^{-z} dz + \int_1^\infty (z - 1) e^{-z} dz \right].$$

But $\displaystyle \int_0^1 (1 - z) e^{-z} dz = \left[-e^{-z}(1 - z) \right]_0^1 - \int_0^1 e^{-z} dz = e^{-1}$,

and $\displaystyle \int_1^\infty (z - 1) e^{-z} dz = \int_0^\infty u e^{-(u+1)} du = e^{-1} \Gamma(2) = e^{-1}$,

Therefore $E[|x - \lambda|] = 2\lambda/e.$

Percentiles

The equation for ξ_p, the pth percentile of the distribution, is

$$\frac{1}{\lambda} \int_0^{\xi_p} e^{-x/\lambda} dx = \frac{p}{100},$$

or $\displaystyle 1 - e^{-\xi_p/\lambda} = \frac{p}{100},$

so that $\displaystyle \xi_p = -\lambda \log \left[1 - \frac{p}{100} \right].$

Hence the median of the distribution is

$$\xi_{50} = -\lambda \log \frac{1}{2} = \lambda \log 2 < \lambda.$$

Also, the first and third quartiles of the distribution are $\lambda \log \frac{4}{3}$ and

$\lambda \log 4$. Therefore

$$S.I.Q.R. = \frac{1}{2}\lambda\left[\log 4 - \log\frac{4}{3}\right] = \frac{1}{2}\lambda \log 3.$$

Now $e^{1 \cdot 10}$ is approximately 3, so that to the same accuracy the S.I.Q.R. is $0 \cdot 55\lambda$.

7.10 The gamma distribution

The negative exponential distribution is, in fact, a particular case of the important gamma distribution. This is a two-parameter distribution which is defined by the probability density function

$$f(x) = \frac{1}{\lambda^{\alpha}\Gamma(\alpha)} e^{-x/\lambda} x^{\alpha-1}, \text{ for } 0 \leqslant x < \infty,$$

and $\qquad f(x) = 0$, otherwise.

The parameters λ and α are both positive; and for $\alpha = 1$, the distribution degenerates into the negative exponential distribution. We also observe that

$$\frac{1}{\lambda^{\alpha}\Gamma(\alpha)} \int\limits_{0}^{\infty} e^{-x/\lambda} x^{\alpha-1} dx = \frac{1}{\Gamma(\alpha)} \int\limits_{0}^{\infty} e^{-z} z^{\alpha-1} dz = 1.$$

Therefore $f(x)$, as defined, does represent a proper probability distribution.

Cumulants

The moment-generating function of x about the origin is

$$M_0(t) \equiv E(e^{tx}) = \frac{1}{\lambda^{\alpha}\Gamma(\alpha)} \int\limits_{0}^{\infty} e^{-x(1-\lambda t)/\lambda} x^{\alpha-1} dx.$$

This integral converges if $1 - \lambda t > 0$; and if we now make the substitution

$$z = (1 - \lambda t)x/\lambda,$$

we have

$$M_0(t) = \frac{1}{(1-\lambda t)^{\alpha}\Gamma(\alpha)} \int\limits_{0}^{\infty} e^{-z} z^{\alpha-1} dz = (1 - \lambda t)^{-\alpha}.$$

Therefore the cumulant-generating function of x is

$$\kappa(t) = -\alpha \log(1 - \lambda t)$$
$$= \alpha \sum_{r=1}^{\infty} \frac{(\lambda t)^r}{r} = \alpha \sum_{r=1}^{\infty} \frac{t^r}{r!} \lambda^r (r-1)!,$$

so that the rth cumulant of x is $\kappa_r = \alpha\lambda^r(r-1)!$ In particular,

$$\kappa_1 = \alpha\lambda; \quad \kappa_2 = \alpha\lambda^2; \quad \kappa_3 = 2\alpha\lambda^3; \quad \kappa_4 = 6\alpha\lambda^4.$$

Hence the coefficients of skewness and kurtosis of the distribution are

$$\gamma_1 = \frac{2\alpha\lambda^3}{\alpha^{3/2}\lambda^3} = \frac{2}{\sqrt{\alpha}}; \quad \gamma_2 = \frac{6\alpha\lambda^4}{\alpha^2\lambda^4} = \frac{6}{\alpha}.$$

It is important to note that as $\alpha \to \infty$, both γ_1 and $\gamma_2 \to 0$, though for any finite α the distribution is, irrespective of λ, positively skew and lepto-kurtic.

Percentiles

The *p*th percentile of the distribution is determined from the equation

$$\frac{1}{\lambda^{a}\,\Gamma(a)}\int_{0}^{\xi_{p}} e^{-x/\lambda}\,x^{a-1}\,dx \;=\; \frac{p}{100}.$$

If we set $z = x/\lambda$ and $z_{p} \equiv \xi_{p}/\lambda$, then

$$\frac{1}{\Gamma(a)}\int_{0}^{z_{p}} e^{-z}\,z^{a-1}\,dz \;=\; \frac{p}{100}. \tag{1}$$

The above integral defines what is known as the *incomplete gamma function*, and we write the general defining relation for any $y > 0$ as

$$\Gamma_{y}(a) \;\equiv\; \frac{1}{\Gamma(a)}\int_{0}^{y} e^{-u}\,u^{a-1}\,du.$$

According to this definition, $\Gamma_{y}(a) \to 1$ as $y \to \infty$, and for all finite values of y $0 < \Gamma_{y}(a) < 1$. By an integration by parts we can derive a useful recurrence relation for this function as follows:

$$\Gamma_{y}(a) \;=\; \frac{1}{\Gamma(a)}\Big[-e^{-u}u^{a-1}\Big]_{0}^{y} + \frac{a-1}{\Gamma(a)}\int_{0}^{y} e^{-u}\,u^{a-2}\,du$$

$$=\; -e^{-y}\,y^{a-1}/\Gamma(a) + \Gamma_{y}(a-1), \text{ if } a-1 > 0,$$

whence

$$\Gamma_{y}(a) - \Gamma_{y}(a-1) \;=\; -e^{-y}\,y^{a-1}/\Gamma(a). \tag{2}$$

With the help of this formula we can evaluate $\Gamma_{y}(a)$ exactly when a is an integer $\geqslant 2$. Thus, if we replace a by r in (2), and then sum, we obtain

$$\sum_{r=2}^{a}\Big[\Gamma_{y}(r) - \Gamma_{y}(r-1)\Big] \;=\; -e^{-y}\sum_{r=2}^{a} y^{r-1}/\Gamma(r),$$

or

$$\Gamma_{y}(a) - \Gamma_{y}(1) \;=\; -e^{-y}\sum_{r=2}^{a} y^{r-1}/\Gamma(r).$$

But

$$\Gamma_{y}(1) \;=\; \int_{0}^{y} e^{-u}\,du = 1 - e^{-y}.$$

Hence

$$\Gamma_{y}(a) \;=\; 1 - e^{-y}\sum_{r=1}^{a} y^{r-1}/\Gamma(r)$$

$$=\; 1 - e^{-y}\sum_{t=0}^{a-1} y^{t}/\Gamma(t+1).$$

We note that this formula also retains its validity for $a = 1$.

For non-integral values of a, an elementary evaluation of $\Gamma_{y}(a)$ is not possible. However, we are not primarily concerned with the evaluation of $\Gamma_{y}(a)$ for known values of y and a, but in determining z_{p} from equation (1) for given p and a. The solution of such an equation can only be carried out numerically. In statistical terminology, $\Gamma_{y}(a)$ is the distribution function of the standardised gamma distribution with parameter a. Extensive tables of this function entitled *Tables of the Incomplete Gamma Function* were published in 1922 by the office of *Biometrika* under the editorship of Karl Pearson. In these tables, the variable of integration used is $v = u/\sqrt{a}$ in order to facilitate tabulation.

By the use of Pearson's tables, it is possible either to read off directly or to interpolate for z_p, whence $\xi_p = \lambda z_p$. Hence the median of the two-parameter distribution of x is

$$\xi_{50} = \lambda z_{50}, \quad \text{where} \quad \Gamma_{z_{50}}(\alpha) = 0\cdot 5.$$

Similarly, the S.I.Q.R. of the distribution is

$$\frac{1}{2}\lambda(z_{75} - z_{25}), \quad \text{where} \quad \Gamma_{z_{75}}(\alpha) = 0\cdot 75 \quad \text{and} \quad \Gamma_{z_{25}}(\alpha) = 0\cdot 25.$$

Mean deviation

Since $E(x) = \alpha\lambda$, the mean deviation of x is

$$E[|x - \alpha\lambda|] = \frac{1}{\lambda^\alpha\Gamma(\alpha)} \int_0^\infty |x - \alpha\lambda| e^{-x/\lambda} x^{\alpha-1} dx$$

$$= \frac{\lambda}{\Gamma(\alpha)} \int_0^\infty |z - \alpha| e^{-z} z^{\alpha-1} dz$$

$$= \frac{\lambda}{\Gamma(\alpha)} \left[\int_0^\alpha (\alpha - z) e^{-z} z^{\alpha-1} dz + \int_\alpha^\infty (z - \alpha) e^{-z} z^{\alpha-1} dz \right]$$

$$= \alpha\lambda \left[\frac{1}{\Gamma(\alpha)} \int_0^\alpha e^{-z} z^{\alpha-1} dz - \frac{1}{\Gamma(\alpha+1)} \int_0^\alpha e^{-z} z^\alpha dz + \right.$$

$$\left. + \frac{1}{\Gamma(\alpha+1)} \int_\alpha^\infty e^{-z} z^\alpha dz - \frac{1}{\Gamma(\alpha)} \int_\alpha^\infty e^{-z} z^{\alpha-1} dz \right]$$

$$= \alpha\lambda[\Gamma_\alpha(\alpha) - \Gamma_\alpha(\alpha+1) + \{1 - \Gamma_\alpha(\alpha+1)\} - \{1 - \Gamma_\alpha(\alpha)\}]$$

$$= 2\alpha\lambda[\Gamma_\alpha(\alpha) - \Gamma_\alpha(\alpha+1)],$$

so that

$$E[|x - \alpha\lambda|] = 2\alpha\lambda[\Gamma_\alpha(\alpha) - \Gamma_\alpha(\alpha+1)]$$

$$= 2\lambda(\alpha/e)^\alpha/\Gamma(\alpha), \quad \text{by (2) above.}$$

As a check, we observe that, for $\alpha = 1$, this gives

$$E[|x - \lambda|] = 2\lambda/e,$$

which is the mean deviation of the negative exponential distribution, as shown earlier.

Mode

To obtain the mode and the points of inflexion of the distribution curve, it is convenient to use the following lemma.

Lemma

The stationary values of $f(x)$ (assumed positive) occur at the same points as those of $\log f(x)$, and are of the same nature. The points of inflexion of $f(x)$ occur where

$$\frac{d^2 \log f(x)}{dx^2} + \left[\frac{d \log f(x)}{dx} \right]^2 = 0.$$

Proof

If $y = \log f(x)$, then $f(x) = e^y$, $f'(x) = e^y y'$, and $f''(x) = e^y(y'' + y'^2)$. Hence the condition $f'(x) = 0$ for a stationary value of $f(x)$ is equivalent to $y' = 0$, and the condition $f''(x) < 0$, $f''(x) > 0$ for a maximum or minimum of

$f(x)$ is then equivalent to $y'' < 0$, $y'' > 0$. The condition $f''(x) = 0$ for a point of inflexion is equivalent to $y'' + y'^2 = 0$.

This lemma is useful when $f(x)$ is a product of functions of x so that logarithmic differentiation is simpler than straightforward determination of the derivatives of $f(x)$.

For the gamma distribution defined above,

$$\log f(x) = \text{constant} - \frac{x}{\lambda} + (\alpha - 1)\log x.$$

Hence

$$\frac{d \log f(x)}{dx} = -\frac{1}{\lambda} + \frac{\alpha - 1}{x},$$

and

$$\frac{\cdot d^2 \log f(x)}{dx^2} = -\frac{\alpha - 1}{x^2}.$$

Therefore $f(x)$ has a single stationary value given by $x = \lambda(\alpha - 1)$ provided $\alpha > 1$. Evidently, $\dfrac{d^2 \log f(x)}{dx^2} < 0$ and so the mode of the distribution is at $x = \lambda(\alpha - 1)$.

Points of inflexion

For the points of inflexion of $f(x)$, we have from the lemma

$$-\frac{\alpha - 1}{x^2} + \left[\frac{\alpha - 1}{x} - \frac{1}{\lambda}\right]^2 = 0,$$

so that

$$[(\alpha - 1)\lambda - x]^2 = \lambda^2(\alpha - 1).$$

Therefore the points of inflexion are at

$$x = \lambda(\alpha - 1)[1 \pm (\alpha - 1)^{-\frac{1}{2}}], \quad \text{if } \alpha > 2.$$

However, if $1 < \alpha \leqslant 2$, then there is a single point of inflexion at

$$x = \lambda(\alpha - 1)[1 + (\alpha - 1)^{-\frac{1}{2}}].$$

Forms of the distribution

We may therefore conclude that for variation in α, the gamma distribution can take the following forms:

(i) For $0 < \alpha < 1$, the distribution is J-shaped with both the x and y axes as asymptotes. The ordinates of the curve steadily decrease as x varies from 0 to ∞.

(ii) For $\alpha = 1$, the distribution degenerates into the negative exponential distribution. This is also a J-shaped curve but with a finite ordinate for $x = 0$.

(iii) For $1 < \alpha \leqslant 2$, the distribution is unimodal and has a single point of inflexion at a distance $\lambda(\alpha - 1)^{\frac{1}{2}}$ to the right of the mode at $x = \lambda(\alpha - 1)$.

(iv) For $\alpha > 2$, the distribution is unimodal and has two points of inflexion symmetrically placed on either side of the mode at $x = \lambda(\alpha - 1)$ and at a distance $\lambda(\alpha - 1)^{\frac{1}{2}}$ from it.

(v) As $\alpha \to \infty$, the distribution becomes symmetrical and mesokurtic.

The parameter λ is only a scale factor and of not much direct statistical interest, since it is always possible to standardise a gamma distribution with $\lambda \neq 1$ by the transformation $z = x/\lambda$.

Fig. 7.3 below shows some of the variety of forms that this distribution curve can take for $\lambda = 1$ and varying α.

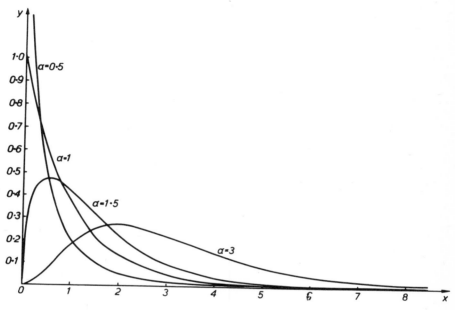

Fig.7.3   Typical forms of the gamma distribution $(\lambda = 1)$

7.11   The beta distribution of the first kind

This two-parameter distribution is based on the first definition of the beta function, and its probability density function is

$$f(x) = \frac{1}{B(\alpha,\beta)} x^{\alpha-1}(1 - x)^{\beta-1}, \quad \text{for} \quad 0 \leqslant x \leqslant 1,$$

and $\qquad f(x) = 0$, otherwise.

Evidently, the parameters α and β must both be positive, and the total area under the curve $y = f(x)$ is unity. Thus $f(x)$ does define a proper probability distribution in the $(0, 1)$ interval.

If we make the transformation $y = 1 - x$, then the probability density function of the distribution of y is

$$\frac{1}{B(\beta,\alpha)} (1 - y)^{\alpha-1} y^{\beta-1}, \quad \text{for} \quad 0 \leqslant y \leqslant 1,$$

and zero otherwise. The distribution of y is thus the mirror image of the distribution of x. It is an important characteristic of the beta distribution that its mirror image is again a beta distribution with the parameters interchanged.

Cumulants

Another important property of this distribution is that its moments can be more easily derived directly than by attempting to evaluate its moment-generating function. Thus the rth moment about the origin of x is

$$\mu'_r \equiv E(x^r) = \frac{1}{B(\alpha,\beta)} \int_0^1 x^{\alpha+r-1}(1-x)^{\beta-1}dx$$

$$= \frac{B(\alpha+r,\beta)}{B(\alpha,\beta)} = \frac{\Gamma(\alpha+r)\Gamma(\alpha+\beta)}{\Gamma(\alpha)\Gamma(\alpha+\beta+r)},$$

or $\qquad \mu'_r = \dfrac{(\alpha+r-1)^{(r)}}{(\alpha+\beta+r-1)^{(r)}},\qquad$ in standard factorial notation.

Hence the first four cumulants of x are obtained as follows:

$$\kappa_1 \equiv \mu'_1 = \frac{\alpha}{\alpha+\beta}.$$

$$\kappa_2 \equiv \mu_2 = \mu'_2 - (\mu'_1)^2$$

$$= \frac{\alpha(\alpha+1)}{(\alpha+\beta+1)^{(2)}} - \frac{\alpha^2}{(\alpha+\beta)^2}$$

$$= \frac{\alpha\beta}{(\alpha+\beta)^2(\alpha+\beta+1)}.$$

$$\kappa_3 \equiv \mu_3 = \mu'_3 - 3\mu'_1\mu'_2 + 2(\mu'_1)^3$$

$$= \frac{(\alpha+2)^{(3)}}{(\alpha+\beta+2)^{(3)}} - \frac{3\alpha(\alpha+1)^{(2)}}{(\alpha+\beta)(\alpha+\beta+1)^{(2)}} + \frac{2\alpha^3}{(\alpha+\beta)^3}$$

$$= \frac{\alpha(\alpha+1)}{(\alpha+\beta+1)^{(2)}}\left[\frac{\alpha+2}{\alpha+\beta+2} - \frac{\alpha}{\alpha+\beta}\right] - \frac{2\alpha^2}{(\alpha+\beta)^2}\left[\frac{\alpha+1}{\alpha+\beta+1} - \frac{\alpha}{\alpha+\beta}\right]$$

$$= \frac{2\alpha\beta(\alpha+1)}{(\alpha+\beta)^2(\alpha+\beta+2)^{(2)}} - \frac{2\alpha^2\beta}{(\alpha+\beta)^3(\alpha+\beta+1)}$$

$$= \frac{2\alpha\beta(\beta-\alpha)}{(\alpha+\beta)^3(\alpha+\beta+2)^{(2)}}$$

Finally,

$$\mu_4 \equiv \mu'_4 - 4\mu'_1\mu'_3 + 6\mu'_2(\mu'_1)^2 - 3(\mu'_1)^4$$

$$= \frac{(\alpha+3)^{(4)}}{(\alpha+\beta+3)^{(4)}} - \frac{4\alpha(\alpha+2)^{(3)}}{(\alpha+\beta)(\alpha+\beta+2)^{(3)}} + \frac{6\alpha^2(\alpha+1)^{(2)}}{(\alpha+\beta)^2(\alpha+\beta+1)^{(2)}} - $$

$$- \frac{3\alpha^4}{(\alpha+\beta)^4}$$

$$= \frac{(\alpha+2)^{(3)}[(\alpha+3)(\alpha+\beta) - \alpha(\alpha+\beta+3)]}{(\alpha+\beta)^2(\alpha+\beta+3)^{(3)}} - $$

$$- \frac{3\alpha(\alpha+1)^{(2)}[(\alpha+2)(\alpha+\beta) - \alpha(\alpha+\beta+2)]}{(\alpha+\beta)^3(\alpha+\beta+2)^{(2)}} + $$

$$+ \frac{3\alpha^3[(\alpha+1)(\alpha+\beta) - \alpha(\alpha+\beta+1)]}{(\alpha+\beta)^4(\alpha+\beta+1)}$$

$$= \frac{3(\alpha+2)^{(3)}\beta}{(\alpha+\beta)^2(\alpha+\beta+3)^{(3)}} - \frac{6\alpha(\alpha+1)^{(2)}\beta}{(\alpha+\beta)^3(\alpha+\beta+2)^{(2)}} + \frac{3\alpha^3\beta}{(\alpha+\beta)^4(\alpha+\beta+1)}$$

$$= \frac{3(\alpha+1)^{(2)}\beta[(\alpha+2)(\alpha+\beta) - \alpha(\alpha+\beta+3)]}{(\alpha+\beta)^3(\alpha+\beta+3)^{(3)}} -$$

$$- \frac{3\alpha^2\beta[(\alpha+1)(\alpha+\beta) - \alpha(\alpha+\beta+2)]}{(\alpha+\beta)^4(\alpha+\beta+2)^{(2)}}$$

$$= \frac{3\alpha\beta[\alpha\beta(\alpha+\beta+2) + 2(\beta-\alpha)^2]}{(\alpha+\beta+3)^{(3)}(\alpha+\beta)^4}, \quad \text{on some simplification.}$$

Therefore

$$\kappa_4 \equiv \mu_4 - 3\mu_2^2$$

$$= \frac{3\alpha\beta[\alpha\beta(\alpha+\beta+2) + 2(\beta-\alpha)^2]}{(\alpha+\beta+3)^{(3)}(\alpha+\beta)^4} - \frac{3\alpha^2\beta^2}{(\alpha+\beta)^4(\alpha+\beta+1)^2}$$

$$= \frac{6\alpha\beta[(\alpha+\beta+1)(\alpha-\beta)^2 - \alpha\beta(\alpha+\beta+2)]}{(\alpha+\beta)^4(\alpha+\beta+1)^2(\alpha+\beta+3)^{(2)}}, \quad \text{on reduction.}$$

Hence the coefficients of skewness and kurtosis of the distribution are

$$\gamma_1 \equiv \frac{\kappa_3}{\kappa_2^{3/2}} = \frac{2(\beta-\alpha)}{\alpha+\beta+2}\left[\frac{\alpha+\beta+1}{\alpha\beta}\right]^{\frac{1}{2}};$$

and

$$\gamma_2 \equiv \frac{\kappa_4}{\kappa_2^2} = \frac{6[(\alpha-\beta)^2(\alpha+\beta+1) - \alpha\beta(\alpha+\beta+2)]}{\alpha\beta(\alpha+\beta+3)^{(2)}}.$$

As checks, we use the following two:

(i) When $\alpha = \beta = 1$, then the beta distribution degenerates into the uniform distribution, and we correctly have $\gamma_1 = 0$, $\gamma_2 = -6/5$.

(ii) When α is finite and $\beta \to \infty$, then $\gamma_1 \to 2/\sqrt{\alpha}$ and $\gamma_2 \to 6/\alpha$.

These are the values of the corresponding coefficients obtained for the gamma distribution. Indeed, we can verify simply that the gamma distribution is the appropriate limiting form of the beta distribution when $\beta \to \infty$ and α remains finite. To prove this, we first make the transformation $\beta x = z$. Then the distribution of z becomes

$$\frac{\Gamma(\alpha+\beta)}{\beta^\alpha \Gamma(\beta)} \frac{1}{\Gamma(\alpha)} z^{\alpha-1}\left(1 - \frac{z}{\beta}\right)^{\beta-1} dz, \quad \text{for} \quad 0 \leqslant z \leqslant \beta.$$

Now, as $\beta \to \infty$ and α remains finite, we have from (1) of Section 7.6 that

$$\frac{\Gamma(\alpha+\beta)}{\beta^\alpha \Gamma(\beta)} \to 1, \quad \text{and} \quad \left(1 - \frac{z}{\beta}\right)^{\beta-1} \to e^{-z}.$$

Therefore the limiting distribution of z becomes

$$\frac{1}{\Gamma(\alpha)} e^{-z} z^{\alpha-1} dz, \quad \text{for} \quad 0 \leqslant z < \infty.$$

Percentiles

The pth percentile, ξ_p, of the distribution of x is obtained from the equation

$$\frac{1}{B(\alpha,\beta)} \int_0^{\xi_p} x^{a-1}(1-x)^{\beta-1}\,dx \;=\; \frac{p}{100}\;.$$

In general, for any y in the interval $0 < y < 1$, the integral

$$B_y(\alpha,\beta) \;\equiv\; \frac{1}{B(\alpha,\beta)} \int_0^y x^{a-1}(1-x)^{\beta-1}\,dx$$

is defined as the *incomplete beta function*. If either α or β is an integer, then the integral can be evaluated simply.

(i) First, suppose β is an integer. Then

$$B_y(\alpha,\beta) \;=\; \frac{1}{B(\alpha,\beta)} \int_0^y x^{a-1} \sum_{r=0}^{\beta-1} \binom{\beta-1}{r}(-x)^r dx.$$

Now, by reversing the order of summation and integration, we have

$$B_y(\alpha,\beta) \;=\; \frac{1}{B(\alpha,\beta)} \sum_{r=0}^{\beta-1} \binom{\beta-1}{r}(-1)^r \int_0^y x^{a+r-1}dx$$

$$=\; \frac{1}{B(\alpha,\beta)} \sum_{r=0}^{\beta-1} \binom{\beta-1}{r}(-1)^r\, y^{a+r}/(\alpha+r).$$

(ii) Next, suppose α is an integer. Then

$$B_y(\alpha,\beta) \;=\; 1 - \frac{1}{B(\alpha,\beta)} \int_y^1 x^{a-1}(1-x)^{\beta-1}dx$$

$$=\; 1 - \frac{1}{B(\beta,\alpha)} \int_0^{1-y} u^{\beta-1}(1-u)^{a-1}du, \quad \text{where} \quad u = 1-x,$$

$$=\; 1 - B_{1-y}(\beta,\alpha)$$

$$=\; 1 - \frac{1}{B(\beta,\alpha)} \sum_{r=0}^{a-1} \binom{\alpha-1}{r}(-1)^r(1-y)^{\beta+r}/(\beta+r),$$

by using (i).

(iii) Finally, if both α and β are fractional, then

$$B_y(\alpha,\beta) \;=\; \frac{1}{B(\alpha,\beta)} \int_0^y x^{a-1} \sum_{r=0}^{\infty} \binom{\beta-1}{r}(-x)^r dx$$

$$=\; \frac{1}{B(\alpha,\beta)} \sum_{r=0}^{\infty} \binom{\beta-1}{r}(-1)^r\, y^{a+r}/(\alpha+r).$$

Although this is an infinite series, it converges rapidly, particularly for small values of y. In fact, for most numerical computations of incomplete beta functions just the first few terms would give the necessary accuracy.

The function $B_y(\alpha,\beta)$ is the distribution function of the beta distribution with parameters α and β. Karl Pearson also edited the *Tables of the Incomplete Beta Function*, but since two parameters are involved the tables are not particularly convenient for interpolation. Direct evaluation of $B_y(\alpha,\beta)$ is in many cases a more reliable and simpler method.

The median of the beta distribution is now simply obtained as ξ_{50}, where

$$B_{\xi_{50}}(\alpha,\beta) \;=\; 0\cdot5.$$

In the same way, the S.I.Q.R. of the distribution of x is $\frac{1}{2}(\xi_{75} - \xi_{25})$, where

$$B_{\xi_{75}}(\alpha,\beta) = 0\cdot75 \quad \text{and} \quad B_{\xi_{25}}(\alpha,\beta) = 0\cdot25.$$

Mode

To obtain the mode of the distribution of x, we have

$$\log f(x) = \text{constant} + (\alpha - 1)\log x + (\beta - 1)\log(1 - x),$$

so that

$$\frac{d\log f(x)}{dx} = \frac{\alpha - 1}{x} - \frac{\beta - 1}{1 - x},$$

and

$$\frac{d^2\log f(x)}{dx^2} = -\frac{\alpha - 1}{x^2} - \frac{\beta - 1}{(1 - x)^2}.$$

Hence for the stationary values of $f(x)$

$$(\alpha - 1)(1 - x) - (\beta - 1)x = 0, \quad \text{or} \quad x = (\alpha - 1)/(\alpha + \beta - 2).$$

Thus there is a stationary value within the permissible range of x if both the parameters are either > 1 or < 1. At the stationary value

$$\frac{d^2\log f(x)}{dx^2} = -\frac{(\alpha + \beta - 2)^2}{\alpha - 1} - \frac{(\alpha + \beta - 2)^2}{\beta - 1} = -\frac{(\alpha + \beta - 2)^3}{(\alpha - 1)(\beta - 1)}.$$

Therefore, if both α and β are < 1, then $\dfrac{d^2\log f(x)}{dx^2} > 0$ and the point $x = (\alpha - 1)/(\alpha + \beta - 2)$ gives a minimum or an *anti-mode*. The distribution is then known as *U-shaped*. If α and β are both > 1, then $\dfrac{d^2\log f(x)}{dx^2} < 0$, and so $x = (\alpha - 1)/(\alpha + \beta - 2)$ is the mode of the distribution.

Points of inflexion

Furthermore, $f''(x) = 0$ if

$$-\frac{\alpha - 1}{x^2} - \frac{\beta - 1}{(1 - x)^2} + \left[\frac{\alpha - 1}{x} - \frac{\beta - 1}{1 - x}\right]^2 = 0,$$

or $(\alpha - 1)(\alpha - 2) - 2(\alpha - 1)(\alpha + \beta - 3)x + (\alpha + \beta - 2)(\alpha + \beta - 3)x^2 = 0$.
The roots of this quadratic are

$$x = \frac{(\alpha - 1)(\alpha + \beta - 3) \pm [(\alpha - 1)^2(\alpha + \beta - 3)^2 - (\alpha - 1)(\alpha - 2)(\alpha + \beta - 2)(\alpha + \beta - 3)]}{(\alpha + \beta - 2)(\alpha + \beta - 3)}.$$

$$= \frac{\alpha - 1}{\alpha + \beta - 2} \pm \frac{[(\alpha - 1)(\beta - 1)(\alpha + \beta - 3)]^{\frac{1}{2}}}{(\alpha + \beta - 2)(\alpha + \beta - 3)}.$$

There are thus two points of inflexion symmetrically placed on either side of of the mode at a distance $[(\alpha - 1)(\beta - 1)(\alpha + \beta - 3)]^{\frac{1}{2}}/(\alpha + \beta - 2)^{(2)}$, provided both parameters are greater than 2. If either of the parameters is equal to 2 and the other greater than 2, then there is only one point of inflexion. Thus, if $\alpha = 2,\ \beta > 2$, the point of inflexion is at $x = 2/\beta$, whereas if $\alpha > 2, \beta = 2$, then the point of inflexion is at $(\alpha - 2)/\alpha$.

Forms of the distribution

We thus conclude that, for finite values of α and β, the beta distribution can take the following variety of forms:

(i) If both parameters are less than 1, then the distribution is U-shaped. This means that the distribution curve has infinite ordinates for both $x = 0$ and $x = 1$ with an anti-mode at $x = (\alpha - 1)/(\alpha + \beta - 2)$.

(ii) If one of the parameters is less than 1 and the other greater than 1, then the distribution is J-shaped. The curve has an infinite ordinate at $x = 0$ or $x = 1$ according as $\alpha < 1$ or $\beta < 1$.

(iii) If $\alpha = \beta = 1$, then the distribution becomes rectangular, and is defined in the unit square of the positive quadrant of the (x,y) plane.

(iv) If one of the parameters is equal to 1 and the other equal to 2, then the distribution is *semi-triangular*, that is, the area is confined to a right-angled triangle with the base as the x axis. Thus, if $\alpha = 1$, $\beta = 2$, the probability density function is $2(1 - x)$, and the triangular region is bounded by $x = 0$, $y = 0$, and $y = 2(1 - x)$. Conversely, if $\alpha = 2$, $\beta = 1$, then the probability density function is $2x$ and the triangular region is now bounded by $y = 0$, $x = 1$, and $y = 2x$.

(v) If one of the parameters is equal to 1 and the other greater than 2, then the curve is monotonic. The ordinates increase or decrease steadily according as $\beta = 1$ or $\alpha = 1$. For $\beta = 1$, the probability density function is $\alpha x^{\alpha - 1}$, and for $\alpha = 1$ it is $\beta(1 - x)^{\beta - 1}$.

(vi) The distribution is unimodal for $\alpha > 1$ and $\beta > 1$. However, if one of the parameters is equal to 2 and the other greater than 2, then there is only one point of inflexion, there being two such points when both parameters are greater than 2. The mode is at $x = (\alpha - 1)/(\alpha + \beta - 2)$, and the points of inflexion are symmetrically placed on either side of the mode at a distance $[(\alpha - 1)(\beta - 1)(\alpha + \beta - 3)]^{\frac{1}{2}}/(\alpha + \beta - 2)^{(2)}$. For $\alpha = 2$, the point of inflexion is at $x = 2/\beta$, whereas for $\beta = 2$, the point is at $x = (\alpha - 2)/\alpha$.

Fig. 7.4 overleaf shows a few of the forms that this distribution curve can take for different values of the parameters α and β.

7.12 The beta distribution of the second kind

As its name suggests, this is another two-parameter distribution which is based on the alternative definition of the beta function. The probability density function of this distribution is

$$f(x) = \frac{1}{B(\alpha,\beta)} \frac{x^{\alpha-1}}{(1 + x)^{\alpha+\beta}}, \quad \text{for} \quad 0 \leqslant x < \infty,$$

and $\qquad f(x) = 0$, otherwise.

Clearly α and β are both positive. Also, it is readily seen that the total area under the curve $y = f(x)$ is unity, so that $f(x)$ does define a proper probability distribution. Furthermore, if we make the transformation $x = 1/y$, then it follows that the probability density function of y is

$$\frac{1}{B(\beta,\alpha)} \frac{y^{\beta-1}}{(1 + y)^{\alpha+\beta}}, \quad \text{for} \quad 0 \leqslant y < \infty,$$

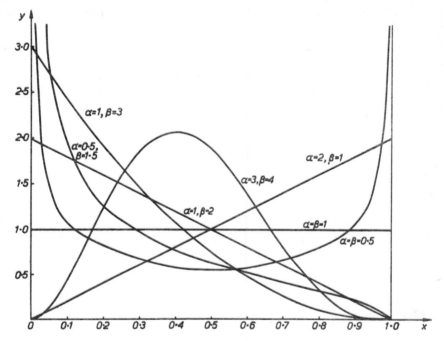

Fig. 7.4 Typical forms of the beta distribution of the first kind

and zero otherwise. This is another beta distribution of the second kind but with the roles of the parameters α and β interchanged as compared with those in the distribution of x.

Cumulants

The moments of the distribution of x are obtained directly and not from the expansion of the moment-generating function. Thus the rth moment of x about the origin is

$$\mu'_r \equiv E(x^r) = \frac{1}{B(\alpha,\beta)} \int_0^\infty \frac{x^{\alpha+r-1}\, dx}{(1+x)^{\alpha+\beta}}$$

$$= \frac{1}{B(\alpha,\beta)} \int_0^\infty \frac{x^{\alpha+r-1}\, dx}{(1+x)^{\alpha+r+\beta-r}}$$

$$= \frac{B(\alpha+r,\beta-r)}{B(\alpha,\beta)}, \quad \text{if} \quad \beta - r > 0.$$

It is clear that although the distribution is well-defined for $\beta > 0$, its moments can diverge. Indeed, its rth moment exists only if $\beta > r$. Assuming that $\beta > r$, we can write

$$\mu'_r = \frac{\Gamma(\alpha+r)\,\Gamma(\beta-r)}{\Gamma(\alpha)\,\Gamma(\beta)}$$

$$= \frac{(\alpha+r-1)^{(r)}}{(\beta-1)^{(r)}}, \quad \text{in standard factorial notation.}$$

Hence the first four cumulants of the distribution of x are obtained as follows:

$$\kappa_1 \equiv \mu_1' = \frac{\alpha}{\beta - 1}.$$

$$\kappa_2 \equiv \mu_2 = \mu_2' - (\mu_1')^2$$

$$= \frac{(\alpha + 1)^{(2)}}{(\beta - 1)^{(2)}} - \frac{\alpha^2}{(\beta - 1)^2}$$

$$= \frac{\alpha(\alpha + \beta - 1)}{(\beta - 1)^2(\beta - 2)}.$$

$$\kappa_3 \equiv \mu_3 = \mu_3' - 3\mu_1'\mu_2' + 2(\mu_1')^3$$

$$= \frac{(\alpha + 2)^{(3)}}{(\beta - 1)^{(3)}} - \frac{3\alpha(\alpha + 1)^{(2)}}{(\beta - 1)(\beta - 1)^{(2)}} + \frac{2\alpha^3}{(\beta - 1)^3}$$

$$= (\alpha + 1)^{(2)}\left[\frac{(\alpha + 2)(\beta - 1) - \alpha(\beta - 3)}{(\beta - 1)^2(\beta - 2)^{(2)}}\right] - 2\alpha^2\left[\frac{(\alpha + 1)(\beta - 1) - \alpha(\beta - 2)}{(\beta - 1)^3(\beta - 2)}\right]$$

$$= \frac{2(\alpha + 1)^{(2)}(\alpha + \beta - 1)}{(\beta - 1)^2(\beta - 2)^{(2)}} - \frac{2\alpha^2(\alpha + \beta - 1)}{(\beta - 1)^3(\beta - 2)}$$

$$= \frac{2\alpha(\alpha + \beta - 1)(2\alpha + \beta - 1)}{(\beta - 1)^3(\beta - 2)^{(2)}}.$$

Lastly, $\mu_4 = \mu_4' - 4\mu_1'\mu_3' + 6(\mu_1')^2\mu_2' + 3(\mu_1')^4$

$$= \frac{(\alpha + 3)^{(4)}}{(\beta - 1)^{(4)}} - \frac{4\alpha(\alpha + 2)^{(3)}}{(\beta - 1)(\beta - 1)^{(3)}} + \frac{6\alpha^2(\alpha + 1)^{(2)}}{(\beta - 1)^2(\beta - 1)^{(2)}} - \frac{3\alpha^4}{(\beta - 1)^4}$$

$$= (\alpha + 2)^{(3)}\left[\frac{(\beta - 1)(\alpha + 3) - \alpha(\beta - 4)}{(\beta - 1)^2(\beta - 2)^{(3)}}\right] -$$

$$- 3\alpha(\alpha + 1)^{(2)}\left[\frac{(\beta - 1)(\alpha + 2) - \alpha(\beta - 3)}{(\beta - 1)^3(\beta - 2)^{(2)}}\right] +$$

$$+ 3\alpha^3\left[\frac{(\beta - 1)(\alpha + 1) - \alpha(\beta - 2)}{(\beta - 1)^4(\beta - 2)}\right]$$

$$= \frac{3(\alpha + \beta - 1)(\alpha + 2)^{(3)}}{(\beta - 1)^2(\beta - 2)^{(3)}} - \frac{6(\alpha + \beta - 1)\alpha(\alpha + 1)^{(2)}}{(\beta - 1)^3(\beta - 2)^{(2)}} + \frac{3(\alpha + \beta - 1)\alpha^3}{(\beta - 1)^4(\beta - 2)}$$

$$= 3(\alpha + \beta - 1)(\alpha + 1)^{(2)}\left[\frac{(\beta - 1)(\alpha + 2) - \alpha(\beta - 4)}{(\beta - 1)^3(\beta - 2)^{(3)}}\right] -$$

$$- 3(\alpha + \beta - 1)\alpha^2\left[\frac{(\beta - 1)(\alpha + 1) - \alpha(\beta - 3)}{(\beta - 1)^4(\beta - 2)^{(2)}}\right]$$

$$= \frac{3\alpha(\alpha + \beta - 1)[2(\beta - 1)^2 + \alpha(\alpha + \beta - 1)(\beta + 5)]}{(\beta - 1)^4(\beta - 2)^{(3)}}, \text{ on reduction.}$$

Therefore

$$\kappa_4 \equiv \mu_4 - 3\mu_2^2$$

$$= \frac{3\alpha(\alpha + \beta - 1)[2(\beta - 1)^2 + \alpha(\alpha + \beta - 1)(\beta + 5)]}{(\beta - 1)^4(\beta - 2)^{(3)}} - \frac{3\alpha^2(\alpha + \beta - 1)^2}{(\beta - 1)^4(\beta - 2)^2}$$

$$= \frac{6\alpha(\alpha + \beta - 1)[(\beta - 2)\{(\beta - 1)^2 + 5\alpha(\alpha + \beta - 1)\} - \alpha(\alpha + \beta - 1)]}{(\beta - 1)^4(\beta - 2)^2(\beta - 3)^{(2)}},$$

on simplification. Hence the coefficients of skewness and kurtosis of the distribution of x are

$$\gamma_1 \equiv \frac{\kappa_3}{\kappa_2^{3/2}} = \frac{2(2\alpha + \beta - 1)}{\beta - 3}\left[\frac{\beta - 2}{\alpha(\alpha + \beta - 1)}\right]^{\frac{1}{2}};$$

and

$$\gamma_2 \equiv \frac{\kappa_4}{\kappa_2^2} = \frac{6[(\beta - 2)\{(\beta - 1)^2 + 5\alpha(\alpha + \beta - 1)\} - \alpha(\alpha + \beta - 1)]}{\alpha(\alpha + \beta - 1)(\beta - 3)^{(2)}}.$$

As an instructive check, we observe that if $\beta \to \infty$ and α remains finite then

$$\gamma_1 \to 2/\sqrt{\alpha} \quad \text{and} \quad \gamma_2 \to 6/\alpha.$$

As we have seen, these are the values of γ_1 and γ_2 for the gamma distribution, and we can verify directly that as $\beta \to \infty$, the beta distribution of the second kind does indeed tend to the gamma distribution. Thus, if we make the transformation $x = z/\beta$, then the distribution of z is

$$\frac{\Gamma(\alpha + \beta)}{\beta^\alpha \Gamma(\beta)} \frac{1}{\Gamma(\alpha)} z^{\alpha - 1}\left(1 + \frac{z}{\beta}\right)^{-\alpha-\beta}dz, \quad \text{for} \quad 0 \leqslant z < \infty.$$

Now for $\beta \to \infty$ and α finite, we have from (1) of Section 7.6 that

$$\frac{\Gamma(\alpha + \beta)}{\beta^\alpha \Gamma(\beta)} \to 1, \quad \text{and} \quad \left(1 + \frac{z}{\beta}\right)^{-\alpha-\beta} \to e^{-z}.$$

Hence the limiting form of the distribution of z is

$$\frac{1}{\Gamma(\alpha)} e^{-z}z^{\alpha - 1}dz, \quad \text{for} \quad 0 \leqslant z < \infty,$$

which is a gamma distribution.

Percentiles

The pth percentile of the distribution of x is ξ_p, and it is determined from the equation

$$\frac{1}{B(\alpha,\beta)} \int_0^{\xi_p} \frac{x^{\alpha - 1}dx}{(1 + x)^{\alpha+\beta}} = \frac{p}{100}.$$

If we now make the substitution $y = 1/(1 + x)$, then the above equation may be rewritten as

$$\frac{1}{B(\alpha,\beta)} \int_{y_p}^1 y^{\beta-1}(1 - y)^{\alpha-1}dy = \frac{p}{100}, \quad \text{where} \quad y_p \equiv 1/(1 + \xi_p),$$

or

$$1 - \frac{1}{B(\beta,\alpha)} \int_0^{y_p} y^{\beta-1}(1 - y)^{\alpha-1}dy = \frac{p}{100},$$

or, in the notation of the incomplete beta function,

$$B_{y_p}(\beta,\alpha) = 1 - \frac{p}{100}.$$

This equation shows that the percentiles of the distribution of x can be determined from incomplete beta functions, whose values may be either obtained from published tables or determined by numerical methods.

In particular, the median of the distribution of x is

$$\xi_{50} = (1 - y_{50})/y_{50}, \quad \text{where} \quad B_{y_{50}}(\beta, \alpha) = 0.5.$$

Similarly, the S.I.Q.R. of the distribution of x is $\frac{1}{2}(\xi_{75} - \xi_{25})$, where

$$\xi_{75} = (1 - y_{75})/y_{75}; \quad \xi_{25} = (1 - y_{25})/y_{25};$$

and

$$B_{y_{75}}(\beta, \alpha) = 0.25; \quad B_{y_{25}}(\beta, \alpha) = 0.75.$$

Hence S.I.Q.R. $= \frac{1}{2}\left[(1 - y_{75})/y_{75} - (1 - y_{25})/y_{25}\right] = \frac{1}{2}\left[\dfrac{1}{y_{75}} - \dfrac{1}{y_{25}}\right].$

Mode

To determine the mode of the distribution of x, we have

$$\log f(x) = \text{constant} + (\alpha - 1) \log x - (\alpha + \beta) \log(1 + x)$$

Therefore

$$\frac{d \log f(x)}{dx} = \frac{\alpha - 1}{x} - \frac{\alpha + \beta}{1 + x},$$

and

$$\frac{d^2 \log f(x)}{dx^2} = -\frac{\alpha - 1}{x^2} + \frac{\alpha + \beta}{(1 + x)^2}.$$

Thus $f(x)$ has one stationary value given by the equation

$$(\alpha - 1)(1 + x) - (\alpha + \beta) x = 0,$$

or

$$x = (\alpha - 1)/(\beta + 1), \quad \text{if} \quad \alpha > 1.$$

Also, at this stationary value,

$$\frac{d^2 \log f(x)}{dx^2} = \frac{(\beta + 1)^2}{\alpha + \beta} - \frac{(\beta + 1)^2}{\alpha - 1} = -\frac{(\beta + 1)^3}{(\alpha - 1)(\alpha + \beta)},$$

which is < 0 for $\alpha > 1$. Hence there is a mode at $x = (\alpha - 1)/(\beta + 1)$, for $\alpha > 1$.

Points of inflexion

For the points of inflexion of the distribution, the equation $f''(x) = 0$ gives

$$-\frac{\alpha - 1}{x^2} + \frac{\alpha + \beta}{(1 + x)^2} + \left[\frac{\alpha - 1}{x} - \frac{\alpha + \beta}{1 + x}\right]^2 = 0,$$

or

$$(\alpha - 1)(\alpha - 2) - 2(\alpha - 1)(\beta + 2) x + (\beta + 1)(\beta + 2) x^2 = 0.$$

The roots of this equation are

$$x = \frac{(\alpha - 1)(\beta + 2) \pm [(\alpha - 1)^2 (\beta + 2)^2 - (\alpha - 1)(\alpha - 2)(\beta + 1)(\beta + 2)]^{\frac{1}{2}}}{(\beta + 1)(\beta + 2)}$$

$$= \frac{\alpha - 1}{\beta + 1} \pm \frac{[(\alpha - 1)(\beta + 2)(\alpha + \beta)]^{\frac{1}{2}}}{(\beta + 1)(\beta + 2)}.$$

Therefore, for $\alpha > 1$, the points of inflexion are symmetrically placed on either side of the mode at a distance $[(\alpha - 1)(\beta + 2)(\alpha + \beta)]^{\frac{1}{2}}/(\beta + 1)(\beta + 2)$ from it.

Forms of the distribution

Thus for variation in α, the distribution of x can take the following forms:

(i) If $\alpha < 1$, the distribution is J-shaped and with an infinite ordinate at $x = 0$.

(ii) If $\alpha = 1$, the distribution is J-shaped and with a finite ordinate at $x = 0$.

(iii) If $\alpha > 1$, the distribution is unimodal with a mode at $x = (\alpha - 1)/(\beta + 1)$.

There are now two points of inflexion symmetrically placed on either side of the mode at a distance $[(\alpha - 1)(\beta + 2)(\alpha + \beta)]^{\frac{1}{2}}/(\beta + 1)(\beta + 2)$ from it.

(iv) For variation in $\beta > 0$, the distribution approaches the gamma distribution with parameter α. $E(x^r)$ exists only if $\beta - r > 0$, so that all moments of the distribution diverge if $\beta < 1$.

Fig. 7.5 overleaf shows some of the forms of this distribution curve for different values of the parameters α and β.

7.13   The normal distribution

The normal distribution is undoubtedly the most important one in statistical theory. It was discovered by Abraham de Moivre (1667 – 1754), whose published works in 1733 gave a derivation of this distribution as a limiting form of the binomial distribution – a distribution of a discrete random variable which we shall study later. Pierre Simon Laplace (1749 – 1827) was also aware of the normal distribution no later than 1774, but due to an historical error, the normal distribution has come to be associated with the name of Karl Friedrich Gauss (1777 – 1855), whose earliest published reference to it appeared in 1809. Nevertheless, usage has now established the term "Gaussian distribution" as a synonym for the "normal distribution" in statistical theory.

The initial importance of the normal distribution can be traced to the many attempts made in the eighteenth and nineteenth centuries to justify this distribution as the one unifying model for the representation of the random variation of measurable natural phenomena. This explains the name "normal" as applicable to the distribution. However, all these attempts failed, and for the simple reason that random variables which arise from metrical characteristics do not invariably show the kind of symmetry which the mathematical form of the normal curve reveals. We shall see some of these differences when we have studied the properties of the normal distribution. Nevertheless, despite this initial failure, the importance of the distribution in our theory is well justified for at least four main reasons. Firstly, there are many observable phenomena whose random variation can be represented by the normal model to a high degree of accuracy. Secondly, the distribution has exceptionally convenient properties which have made this model the basis for much of statistical theory. Thirdly, many non-normal distributions approach normality under certain limiting conditions which can

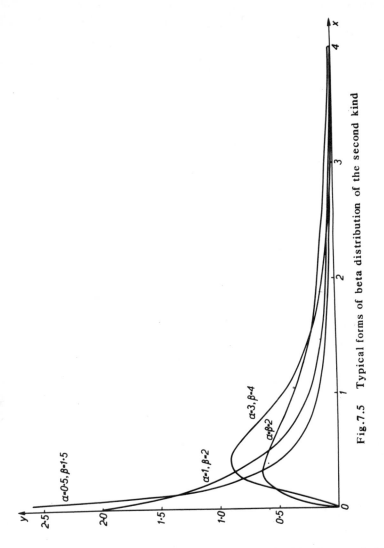

Fig.7.5   Typical forms of beta distribution of the second kind

be frequently simulated in practical situations. Fourthly, by a suitable transformation of a random variable, it is often possible to transform a non-normal distribution into an equivalent distribution which is approximately normal. For example, suppose x is a random variable having a non-normal distribution. Then $u = g(x)$ is also a random variable. If now u has a distribution which has, at least approximately, zero skewness and kurtosis, then it will be usually possible to represent the distribution of u by a normal distribution to a satisfactory degree of approximation. The transformation $u = g(x)$ is known as a *normalising* transformation, and the existence of such practically useful transformations is an important reason for the pre-eminent place of the normal distribution in statistical theory.

The normal distribution is defined by the probability density function

$$f(x) = \frac{1}{\sigma\sqrt{2\pi}} e^{-(x-\mu)^2/2\sigma^2}, \quad \text{for} \quad -\infty < x < \infty,$$

where $\sigma > 0$ and μ are the two parameters of the distribution. We shall see their physical meaning a little later. It is clear that $f(x)$ is always non-negative, and we next verify that the total area under the normal curve is unity.

We have

$$I \equiv \frac{1}{\sigma\sqrt{2\pi}} \int_{-\infty}^{\infty} e^{-(x-\mu)^2/2\sigma^2} dx = \frac{1}{\sqrt{2\pi}} \int_{-\infty}^{\infty} e^{-\frac{1}{2}z^2} dz, \quad \text{where} \quad z = \frac{x-\mu}{\sigma},$$

$$= \frac{1}{\sqrt{2\pi}} \left[\int_{-\infty}^{0} e^{-\frac{1}{2}z^2} dz + \int_{0}^{\infty} e^{-\frac{1}{2}z^2} dz \right].$$

Now since the integrand of the first integral is an even function of z, we can make the transformation $w = -z$ to obtain

$$\int_{-\infty}^{0} e^{-\frac{1}{2}z^2} dz = \int_{0}^{\infty} e^{-\frac{1}{2}w^2} dw.$$

Therefore

$$I = 2\frac{1}{\sqrt{2\pi}} \int_{0}^{\infty} e^{-\frac{1}{2}z^2} dz$$

$$= \left(\frac{2}{\pi}\right)^{\frac{1}{2}} \int_{0}^{\infty} e^{-v} \frac{dv}{\sqrt{2v}}, \quad \text{where} \quad v = \frac{1}{2}z^2,$$

$$= \frac{1}{\sqrt{\pi}} \int_{0}^{\infty} e^{-v} v^{-\frac{1}{2}} dv = \frac{1}{\sqrt{\pi}} \Gamma\left(\frac{1}{2}\right) = 1.$$

Thus

$$\frac{1}{\sigma\sqrt{2\pi}} \int_{-\infty}^{\infty} e^{-(x-\mu)^2/2\sigma^2} dx = \frac{1}{\sqrt{2\pi}} \int_{-\infty}^{\infty} e^{-\frac{1}{2}z^2} dz = \left(\frac{2}{\pi}\right)^{\frac{1}{2}} \int_{0}^{\infty} e^{-\frac{1}{2}z^2} dz = 1. \quad (1)$$

Hence the total area under the normal curve is unity and the probability density function $f(x)$ defines a proper distribution of x.

Cumulants

Before evaluating the moment-generating function of x, it is convenient to determine the expected value of $x - \mu$. We have

$$E(x - \mu) = \frac{1}{\sigma\sqrt{2\pi}} \int_{-\infty}^{\infty} (x - \mu) e^{-(x-\mu)^2/2\sigma^2} dx$$

$$= \frac{\sigma}{\sqrt{2\pi}} \int_{-\infty}^{\infty} z e^{-\frac{1}{2}z^2} dz, \quad \text{where} \quad z = \frac{x-\mu}{\sigma},$$

$$= \frac{\sigma}{\sqrt{2\pi}} [-e^{-\frac{1}{2}z^2}]_{-\infty}^{\infty} = 0.$$

Thus $E(x - \mu) = 0$, or $E(x) = \mu$, so that μ is the mean of the distribution. It is now theoretically simpler to evaluate the moment-generating function of x about its mean μ. We therefore have

$$M(t) \equiv E[e^{t(x-\mu)}] = \frac{1}{\sigma\sqrt{2\pi}} \int_{-\infty}^{\infty} e^{t(x-\mu)-(x-\mu)^2/2\sigma^2} dx,$$

or, again using the transformation $z = (x - \mu)/\sigma$,

$$M(t) = \frac{1}{\sqrt{2\pi}} \int_{-\infty}^{\infty} e^{\sigma tz - \frac{1}{2}z^2} dz$$

$$= \frac{1}{\sqrt{2\pi}} \int_{-\infty}^{\infty} e^{-\frac{1}{2}(z - \sigma t)^2 + \frac{1}{2}\sigma^2 t^2} dz$$

$$= \frac{e^{\frac{1}{2}\sigma^2 t^2}}{\sqrt{2\pi}} \int_{-\infty}^{\infty} e^{-\frac{1}{2}u^2} du, \quad \text{where} \quad u = z - \sigma t,$$

$$= e^{\frac{1}{2}\sigma^2 t^2} .$$

On expansion,

$$M(t) = \sum_{r=0}^{\infty} \left(\frac{1}{2}\sigma^2\right)^r \frac{t^{2r}}{r!}$$

$$= \sum_{r=0}^{\infty} \frac{\sigma^{2r}(2r)!}{2^r r!} \frac{t^{2r}}{(2r)!} .$$

Hence $\qquad \mu_{2r} = \dfrac{\sigma^{2r}(2r)!}{2^r r!} ; \qquad \mu_{2r+1} = 0, \text{ for } r \geqslant 0.$

We thus see that all the odd central moments of the distribution of x are zero. Furthermore, the cumulant-generating function of x is

$$\kappa(t) = \log M_0(t)$$

$$= \log[e^{\mu t} M(t)]$$

$$= \mu t + \frac{1}{2}\sigma^2 t^2.$$

Hence $\qquad \kappa_1 = \mu, \qquad \kappa_2 = \sigma^2, \quad \text{and} \quad \kappa_r = 0 \quad \text{for} \quad r \geqslant 3.$

It is therefore clear that $\text{var}(x) = \sigma^2$, and σ is the standard deviation of x. Also, $\kappa_3 = \kappa_4 = 0$ so that the coefficients of skewness and kurtosis of the distribution are both zero. The distribution is symmetrical and mesokurtic. In fact, we now see the essential meaning of the coefficients γ_1 and γ_2, which may be regarded as measuring departures from normality of any non-normal distribution in terms of the standardised third and fourth cumulants. As remarked earlier, a unimodal distribution with small γ_1 and γ_2 could, in general, be approximately represented by a normal distribution. This is the basis of many approximations in theoretical statistics.

Mean deviation

The mean deviation of x is

$$E[|x - \mu|] = \frac{1}{\sigma\sqrt{2\pi}} \int_{-\infty}^{\infty} |x - \mu| e^{-(x-\mu)^2/2\sigma^2} dx$$

$$= \frac{\sigma}{\sqrt{2\pi}} \int_{-\infty}^{\infty} |z| e^{-\frac{1}{2}z^2} dz, \quad \text{where} \quad z = \frac{x - \mu}{\sigma},$$

$$= \frac{\sigma}{\sqrt{2\pi}} \left[\int_{-\infty}^{0} z e^{-\frac{1}{2}z^2} dz + \int_{0}^{\infty} z e^{-\frac{1}{2}z^2} dz \right]$$

$$= \frac{2\sigma}{\sqrt{2\pi}} \int_{0}^{\infty} z e^{-\frac{1}{2}z^2} dz$$

$$= \sigma \left(\frac{2}{\pi}\right)^{\frac{1}{2}} .$$

Standardised normal distribution

The random variable $z = (x - \mu)/\sigma$ has zero mean and unit variance, and by using such a transformation we can reduce any normal distribution to standard form, that is, a form in which the new random variable has zero mean and unit variance. Thus the probability density function of z is

$$f(z) = \frac{1}{\sqrt{2\pi}} e^{-\frac{1}{2}z^2}, \quad \text{for} \quad -\infty < z < \infty.$$

The normal distribution like that of z is known as the *standardised normal distribution* or the *unit normal distribution*. It is conventional to denote briefly a normal distribution with mean μ and variance σ^2 by $N(\mu, \sigma^2)$, and in this convenient notation the unit normal distribution is denoted by $N(0, 1)$. It is clear from the symmetry of the unit normal distribution that

$$\frac{1}{\sqrt{2\pi}} \int_{-\infty}^{0} e^{-\frac{1}{2}z^2} dz = \frac{1}{\sqrt{2\pi}} \int_{0}^{\infty} e^{-\frac{1}{2}z^2} dz = \frac{1}{2}. \tag{2}$$

Median

We can use this result to determine the median ξ_{50} of the normal distribution $N(\mu, \sigma^2)$. The required equation for ξ_{50} is

$$\frac{1}{\sigma\sqrt{2\pi}} \int_{-\infty}^{\xi_{50}} e^{-(x-\mu)^2/2\sigma^2} dx = \frac{1}{2}.$$

Therefore, using the standardising transformation $z = (x - \mu)/\sigma$, we obtain

$$\frac{1}{\sqrt{2\pi}} \int_{-\infty}^{(\xi_{50}-\mu)/\sigma} e^{-\frac{1}{2}z^2} dz = \frac{1}{2},$$

whence, by (2), $(\xi_{50} - \mu)/\sigma = 0$, or $\xi_{50} = \mu$.

Thus the median of x is the same as its mean, a result intuitively obvious because of the symmetry of the distribution.

Mode

To determine the mode of the distribution of x, we have

$$\log f(x) = \text{constant} - \frac{1}{2\sigma^2}(x - \mu)^2,$$

$$\frac{d \log f(x)}{dx} = -\frac{x - \mu}{\sigma^2},$$

and
$$\frac{d^2 \log f(x)}{dx^2} = -\frac{1}{\sigma^2}.$$

Therefore $x = \mu$ is a stationary value of $f(x)$, and evidently this point is a mode. The height of the maximum ordinate of the distribution of x is $1/\sigma\sqrt{2\pi}$.

Points of inflexion

For the points of inflexion of the curve, we have

$$-\frac{1}{\sigma^2} + \frac{(x - \mu)^2}{\sigma^4} = 0, \quad \text{or} \quad x = \mu \pm \sigma.$$

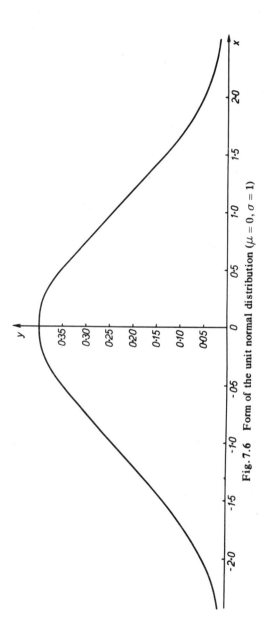

Fig. 7.6   Form of the unit normal distribution ($\mu = 0$, $\sigma = 1$)

Thus there are two points of inflexion at a distance σ on either side of the mode. Fig. 7.6 gives a diagrammatic representation of a typical normal curve.

Percentiles

We have seen that the transformation $z = (x - \mu)/\sigma$ standardises a normal distribution, and so the distribution function of x can be expressed in terms of the distribution function of z. We define the distribution function of z as

$$\Phi(z) = \frac{1}{\sqrt{2\pi}} \int_{-\infty}^{z} e^{-\frac{1}{2}t^2} dt,$$

and the distribution function of x is

$$F(x) = \frac{1}{\sigma\sqrt{2\pi}} \int_{-\infty}^{x} e^{-(u-\mu)^2/2\sigma^2} du = \frac{1}{\sqrt{2\pi}} \int_{-\infty}^{(x-\mu)/\sigma} e^{-\frac{1}{2}t^2} dt = \Phi\left(\frac{x-\mu}{\sigma}\right).$$

In this notation ξ_p, the pth percentile of the distribution of x, is given by the equation

$$\Phi\left(\frac{\xi_p - \mu}{\sigma}\right) = \frac{p}{100}. \tag{3}$$

The function $\Phi(z)$ is not expressible in terms of simple functions but extensive tables exist. Also, if $z > 0$, then

$$\Phi(-z) = \frac{1}{\sqrt{2\pi}} \int_{-\infty}^{-z} e^{-\frac{1}{2}t^2} dt$$

$$= \frac{1}{\sqrt{2\pi}} \int_{z}^{\infty} e^{-\frac{1}{2}u^2} du, \quad \text{where} \quad u = -t,$$

$$= 1 - \Phi(z).$$

Hence $\Phi(z)$ is only tabulated for the range of values $0 \leqslant z < \infty$.

The third and fourth quartiles of the distribution of x are obtained from (3) as solutions of the equations

$$\Phi\left(\frac{\xi_{75} - \mu}{\sigma}\right) = 0.75; \quad \Phi\left(\frac{\xi_{25} - \mu}{\sigma}\right) = 0.25.$$

We find from the tables of the function $\Phi(z)$ that

$$(\xi_{75} - \mu)/\sigma = 0.675 \quad \text{and} \quad (\xi_{25} - \mu)/\sigma = -0.675.$$

Hence the S.I.Q.R. of the distribution of x is

$$\frac{1}{2}(\xi_{75} - \xi_{25}) = 0.675\sigma.$$

The S.I.Q.R. of a normal distribution is also known as the *probable error*, but this expression is now used rather infrequently.

7.14   Evaluation of the distribution function $\Phi(z)$

It is of some theoretical interest to see how the function $\Phi(z)$ can be evaluated. This is usually done numerically by the use of infinite series expansions, and the forms of these series depend upon whether z is small or large.

(i) If z is small and positive, then a computationally convenient series for z is obtained as follows.

$$\Phi(z) = \frac{1}{\sqrt{2\pi}} \int_{-\infty}^{z} e^{-\frac{1}{2}u^2} du$$

$$= \frac{1}{\sqrt{2\pi}} \left[\int_{-\infty}^{0} e^{-\frac{1}{2}u^2} du + \int_{0}^{z} e^{-\frac{1}{2}u^2} du \right]$$

$$= \frac{1}{2} + \frac{1}{\sqrt{2\pi}} \int_{0}^{z} \sum_{r=0}^{\infty} \frac{\left(-\frac{1}{2}u^2\right)^r}{r!} du$$

$$= \frac{1}{2} + \frac{1}{\sqrt{2\pi}} \sum_{r=0}^{\infty} \frac{(-1)^r}{2^r r!} \int_{0}^{z} u^{2r} du,$$

by changing the order of summation and integration,

$$= \frac{1}{2} + \frac{1}{\sqrt{2\pi}} \sum_{r=0}^{\infty} \frac{(-1)^r z^{2r+1}}{(2r + 1)2^r r!} .$$

If the first n terms of this series are used to evaluate $\Phi(z)$, then the remainder after the nth term is

$$R_n(z) \equiv \frac{1}{\sqrt{2\pi}} \sum_{r=n}^{\infty} \frac{(-1)^r z^{2r+1}}{(2r + 1)2^r r!}$$

$$= \frac{1}{\sqrt{2\pi}} \sum_{\nu=0}^{\infty} \frac{(-1)^{\nu+n} z^{2(\nu+n)+1}}{(2\nu + 2n + 1)2^{\nu+n}(\nu + n)!}$$

$$= \frac{(-1)^n z^{2n+1}}{\sqrt{2\pi}\, 2^n} \sum_{\nu=0}^{\infty} \frac{\left(-\frac{1}{2}z^2\right)^\nu}{(2\nu + 2n + 1)(\nu + n)!}$$

Therefore

$$|R_n(z)| < \frac{z^{2n+1}}{\sqrt{2\pi}\, 2^n} \sum_{\nu=0}^{\infty} \frac{\left(\frac{1}{2}z^2\right)^\nu}{(2\nu + 2n + 1)(\nu + n)!}$$

$$< \frac{z^{2n+1}}{\sqrt{2\pi}\, 2^n (2n + 1)} \sum_{\nu=0}^{\infty} \frac{\left(\frac{1}{2}z^2\right)^\nu}{\nu!}$$

$$= \frac{z^{2n+1} e^{\frac{1}{2}z^2}}{\sqrt{2\pi}\, 2^n (2n + 1)} .$$

The series for $\Phi(z)$, though valid for all $z > 0$, converges rapidly for values of $z \leqslant 1$, and the first few terms suffice to determine $\Phi(z)$ to sufficient accuracy for practical purposes.

(ii) On the other hand, if z is large and positive, then a different computationally useful series for $\Phi(z)$ is derived in the following way.

$$\Phi(z) = \frac{1}{\sqrt{2\pi}} \int_{-\infty}^{z} e^{-\frac{1}{2}u^2} du$$

$$= 1 - \frac{1}{\sqrt{2\pi}} \int_{z}^{\infty} e^{-\frac{1}{2}u^2} du$$

$$= 1 - \frac{1}{\sqrt{2\pi}} \int_z^\infty (-ue^{-\frac{1}{2}u^2}) \frac{du}{(-u)}.$$

Hence integration by parts gives

$$\Phi(z) = 1 - \frac{1}{\sqrt{2\pi}} \left\{ \left[-\frac{e^{-\frac{1}{2}u^2}}{u} \right]_z^\infty - \int_z^\infty e^{-\frac{1}{2}u^2} \frac{du}{u^2} \right\}$$

$$= 1 - \frac{1}{\sqrt{2\pi}} \left[\frac{e^{-\frac{1}{2}z^2}}{z} - \int_z^\infty (-ue^{-\frac{1}{2}u^2}) \frac{du}{(-u^3)} \right].$$

Integration by parts again gives

$$\Phi(z) = 1 - \frac{1}{\sqrt{2\pi}} \left\{ \frac{e^{-\frac{1}{2}z^2}}{z} - \left[-\frac{e^{-\frac{1}{2}u^2}}{u^3} \right]_z^\infty + 1.3 \int_z^\infty e^{-\frac{1}{2}u^2} \frac{du}{u^4} \right\}$$

$$= 1 - \frac{1}{\sqrt{2\pi}} \left[\frac{e^{-\frac{1}{2}z^2}}{z} - \frac{e^{-\frac{1}{2}z^2}}{z^3} + 1.3 \int_z^\infty (-ue^{-\frac{1}{2}u^2}) \frac{du}{(-u^5)} \right].$$

After n successive integrations in this way, we have

$$\Phi(z) = 1 - \frac{1}{\sqrt{2\pi}} \left[\frac{e^{-\frac{1}{2}z^2}}{z} \left\{ 1 + \sum_{r=1}^n \frac{1.3.5...(2r-1)(-1)^r}{z^{2r}} \right\} + \right.$$

$$\left. + 1.3.5...(2n+1)(-1)^{n+1} \int_z^\infty \frac{e^{-\frac{1}{2}u^2} du}{u^{2n+2}} \right]$$

$$= 1 - \frac{e^{-\frac{1}{2}z^2}}{z\sqrt{2\pi}} \left[1 + \sum_{r=1}^n \frac{(-1)^r (2r)!}{2^r z^{2r} r!} \right] + R_n(z),$$

where

$$|R_n(z)| = \frac{(2n+1)!}{\sqrt{2\pi}\, 2^n n!} \int_z^\infty \frac{e^{-\frac{1}{2}u^2} du}{u^{2n+2}}$$

$$= \frac{(2n+1)!}{\sqrt{2\pi}\, 2^n n!} \int_0^\infty \frac{e^{-\frac{1}{2}(t+z)^2} dt}{(t+z)^{2n+2}}$$

$$< \frac{(2n+1)! e^{-\frac{1}{2}z^2}}{\sqrt{2\pi}\, 2^n n!} \int_0^\infty \frac{dt}{(t+z)^{2n+2}}$$

$$= \frac{(2n+1)! e^{-\frac{1}{2}z^2}}{\sqrt{2\pi}\, 2^n n!} \left[-\frac{1}{(2n+1)(t+z)^{2n+1}} \right]_0^\infty,$$

or

$$|R_n(z)| < \frac{(2n)! e^{-\frac{1}{2}z^2}}{\sqrt{2\pi}\, 2^n z^{2n+1} n!}.$$

Hence, for large z, we can write approximately

$$\Phi(z) \sim 1 - \frac{e^{-\frac{1}{2}z^2}}{z\sqrt{2\pi}} \left[1 + \sum_{r=1}^\infty \frac{(-1)^r (2r)!}{2^r z^{2r} r!} \right],$$

with the understanding that the absolute error made in using this series is less than the last term taken into account. It is to be noted that this series for $\Phi(z)$ is divergent, but for large z the first few terms of the series give a good evaluation of $\Phi(z)$.

7.15   The problem of specification

A statistical population is a conceptual representation of all the possible values that a random variable can take with their associated probabilities, whereas a histogram based on a frequency distribution of a finite number of observations is a concrete realisation of the variation actually
obtained in the sample. The shape of the histogram is a chance event, as
it depends considerably on the particular set of observations constituting
the sample; but we postulate that all histograms, which could be obtained
on the basis of different finite samples from the population, reflect, at least
in some degree, the main characteristics of the universe sampled. The differences between such histograms would be attributable to chance, and so
also the differences between the histogram actually obtained and the population from which the sample was derived. It is possible to obtain different
samples from the same population and then to determine the differences between the corresponding histograms. On the other hand, it is not at all possible to determine the differences between the sample histogram and the
population from which the sample is obtained. This is a fundamental difficulty because the population is, in essence, unknowable. And the starting
point of statistical theory lies in being able to infer from the finite sample
the *possible* nature of the population with *some degree of assurance*. In
other words, given the limited information contained in the sample, we wish
to infer a "reasonable" model for the mathematical representation of the
population. This is the *problem of specification* in statistical theory and,
like the answers to all problems of statistical inference, the specification
of the population can never be a complete certainty. Nevertheless, this link
between empirical information and formal mathematics is essential for the
purposeful development of statistical theory.

An intuitive approach to this problem rests on the assumption that if
the sample is reasonably large (say of the order of a few hundred observations or more), then the histogram will, in general, reflect rather closely the
basic characteristics of the population sampled such as, for example, its
unimodality, symmetry, kurtosis, etc. Accordingly, on the basis of such
sample information, it should be possible to suggest a mathematical form of
the curve which might be reasonably regarded as a suitable representation
of the population. A choice of this kind is evidently subject to much uncertainty, and a guiding element in the selection is the simplicity of the
mathematical form of the curve. However, a choice of the form of the curve
alone is not adequate because, as we have seen, a mathematical equation
of a curve involves one or more parameters. Hence a complete specification
of the curve necessarily implies that it should also be possible to indicate
the plausible values for the parameters of the curve. Thus, for example, it
is not enough that the population sampled be regarded as normal, but it is
also necessary to determine the appropriate values of the parameters μ and
σ on the basis of the sample information. We are thus led to the conclusion
that the specification of a population has two related aspects:

(i)   an indication of the mathematical form of the curve;   and

(ii) given the form of the curve, an evaluation of the parameters of the curve.

The solution of this two-fold problem is called *fitting a curve to a frequency distribution*. We do not consider this problem in full generality, but we illustrate its solution in the particular case when we have reason to believe that the normal curve might be a suitable one "to fit".

7.16 Fitting a normal curve to a frequency distribution

Suppose we have a frequency distribution based on a sample of N observations, there being in all k class-intervals with midpoints x_i and class-frequencies $n_i(i = 1, 2, ..., k)$. Let the length of the class-intervals be a constant h so that $x_{i+1} = x_i + h$. Schematically, the frequency distribution may be represented as follows:

Midpoint: $x_1, x_2, x_3, ..., x_k$;

Frequency: $n_1, n_2, n_3, ..., n_k$,   where $\sum\limits_{i=1}^{k} n_i \equiv N$.

It usually happens that when such a frequency distribution is represented as a histogram, the form of the latter suggests that the sample observations might have arisen from some normal population. Obviously, the histogram will not have an ideal shape, and we consider that the differences between the histogram and the normal curve might be due just to the chance errors of sampling. More concretely, we evaluate g_1 and g_2, the coefficients of skewness and kurtosis of the histogram, and if these turn out to be small numerically, then it would be a fairly safe conclusion that a normal curve might give an adequate representation of the variation in the conceptual population from which the N observations were obtained. Consequently, our problem reduces to that of determining the "best" normal curve that might be "fitted" to the data, and of then evaluating the discrepancies between the histogram and the fitted curve. We shall consider later the important problem of assessing whether such discrepancies can be reasonably attributed to chance or whether they indicate that the initial assumption about the normality of the population was possibly incorrect.

The equation of any normal curve is

$$y = \frac{1}{\sigma\sqrt{2\pi}} e^{-(x-\mu)^2/2\sigma^2}, \quad \text{for} \quad -\infty < x < \infty,$$

where μ and σ are unknown parameters. As these are varied, we obtain a family of normal curves, and we wish to determine the parameters so that the particular normal curve will have the "closest" agreement with the observed frequency distribution. The range of this distribution is $x_k - x_1 + h$ and that of the normal curve infinite. This difference is of no material consequence since, for all practical purposes, the range of the normal curve is 6σ, as 99·7 per cent of the total area lies between $\mu - 3\sigma$ and $\mu + 3\sigma$. The agreement desired is that over this "effective" range of the normal curve and the observed range of the histogram. Indeed, as we shall see, if this agreement is good, then the area under the fitted normal curve outside the limits $(x_1 - h/2, x_k + h/2)$ must be necessarily negligibly small.

It follows from intuitive considerations that the best fitted normal curve should have the same mean and variance as the observed histogram. Accordingly, if $\bar{x}$ and s^2 are the values for the mean and variance of the observed frequency distribution, then the equation of the best fitted normal curve may be written as

$$y = \frac{1}{s\sqrt{2\pi}} e^{-(x-\bar{x})^2/2s^2}$$

$$= \frac{1}{s}\left[\frac{1}{\sqrt{2\pi}} e^{-\frac{1}{2}z^2}\right], \quad \text{where} \quad (x-\bar{x})/s = z.$$

Thus to each x_i there corresponds a z_i such that $z_i = (x_i - \bar{x})/s$; and corresponding to the values z_i, the ordinates of the standardised curve are

$$\phi(z_i) \equiv \frac{1}{\sqrt{2\pi}} e^{-\frac{1}{2}z_i^2}, \quad \text{for} \quad i = 1, 2, \dots, k.$$

These $\phi(z_i)$ values can be calculated from the published tables of the negative exponential or obtained more readily from the tables of the unit normal curve, which are tabulated according to the convention that the total area under the curve is unity. And since the total frequency for the histogram is N, therefore in the same units the ordinates of the fitted curve with mean $\bar{x}$ and variance s^2 are

$$\frac{N}{s}\phi(z_i), \quad \text{for} \quad i = 1, 2, \dots, k.$$

The corresponding heights of the columns of the histogram are n_i/h. It is however simpler to express these heights in units of h, and then, equivalently, the fitted curve can be plotted in terms of the points (x_i, y_i), where

$$y_i = \frac{Nh}{s}\phi(z_i), \quad \text{for} \quad i = 1, 2, \dots, k.$$

Our main interest is not in this pictorial super-positioning of the normal curve over the histogram, but to find out how far the frequency obtained from the fitted curve for each of the k class-intervals differs from the corresponding observed frequency. In other words, we want the areas under the fitted normal curve for the class-intervals

$$x_i - h/2 \leqslant x \leqslant x_i + h/2, \quad \text{for} \quad i = 1, 2, \dots, k,$$

or, in terms of the standardised fitted curve, for the intervals

$$(x_i - h/2 - \bar{x})/s \leqslant z \leqslant (x_i + h/2 - \bar{x})/s.$$

The required areas are

$$\Phi[(x_i + h/2 - \bar{x})/s] - \Phi[(x_i - h/2 - \bar{x})/s] = \Delta\Phi[(x_i - h/2 - \bar{x})/s].$$

These are obtained from the published tables of the Φ function without a preliminary calculation of the y_i. Of course, the ordinates are necessary for a pictorial representation of the fitted curve.

As pointed out earlier, in the histogram, the lower limit of the first class-interval is $x_1 - h/2$ and the upper limit of the last class-interval is $x_k + h/2$, whereas for the normal curve the range is $-\infty < x < \infty$. The area under the fitted curve below $x_1 - h/2$ is $\Phi[(x_1 - h/2 - \bar{x})/s]$ and that beyond

$x_k + h/2$ is $1 - \Phi[(x_k + h/2 - x)/s]$. However, if the curve is really a good fit, then these areas should be negligible. We could therefore extend the first class-interval backward to $-\infty$ and the last class-interval forward to ∞ and then obtain the areas $\Delta\Phi[(x_i - h/2 - \bar{x})/s]$.

The class-frequencies $N \times \Delta\Phi[(x_i - h/2 - \bar{x})/s]$ are the *expected frequencies* corresponding to the *observed frequencies* n_i. Evidently,

$$\sum_{i=1}^{k} N \times \Delta\Phi[(x_i - h/2 - \bar{x})/s] \equiv N,$$

and this is an essential check for assessing the accuracy of the computations. The overall agreement between the observed and expected frequencies is the real indication of the "goodness of fit". The observed frequencies are random variables and the expected frequencies are also subject to chance since they are determined by using the sample quantities $\bar{x}$ and s^2. Any assessment of the overall agreement must take account of chance variations, and we shall see later how statistical theory provides a solution of this problem. For the present, the important point is to understand how the information contained in the sample can lead to a possible mathematical visualisation of the conceptual population sampled. We have dealt here only with a normal model, but the principles are the same whatever the nature of the curve fitted. This process is, in fact, a concrete manifestation of what is implied in the statement that a random variable is normally distributed or that the population sampled is normal. In the development of statistical theory this or a similar assumption is usually made. This is a fundamental link between the finite data of samples and the structural formalism underlying statistical theory which makes inference possible.

Example 7.1

To conclude, we consider an illustration of the numerical fitting of a normal curve to the frequency distribution of the weight in grams of 327 ears of maize obtained randomly from a field. The data and the calculation of g_1 and g_2 for the frequency distribution are given in Table 7.1. The further calculations of the ordinates of the fitted curve and the expected frequencies are given separately in Table 7.2. A pictorial representation of the histogram and the fitted normal curve is shown in Fig. 7.7 (page 152).

For the moments of u about the origin, we have

$$m'_1 = 51/327 = 0 \cdot 15596; \qquad m'_2 = 1821/327 = 5 \cdot 56881;$$
$$m'_3 = -537/327 = -1 \cdot 64220; \qquad m'_4 = 31245/327 = 95 \cdot 55046.$$

Therefore the moments of u about the mean are as follows:

$$m_2 = 5 \cdot 56881 - 0 \cdot 02432 = 5 \cdot 54449;$$
$$m_3 = -1 \cdot 64220 - 2 \cdot 60553 + 0 \cdot 00759 = -4 \cdot 24014;$$
$$m_4 = 95 \cdot 55046 + 1 \cdot 02447 + 0 \cdot 81272 - 0 \cdot 00177 = 97 \cdot 38588.$$

Hence for the distribution of x,

$$g_1^2 = 17 \cdot 97879/170 \cdot 44522 = 0 \cdot 105481, \quad \text{so that} \quad g_1 = -0 \cdot 3248;$$

and

$$g_2 = 97 \cdot 38588/30 \cdot 74137 - 3 = 0 \cdot 1679.$$

Table 7.1 Showing the frequency distribution of the weights of 327 ears of maize, and the calculation of the coefficients of skewness and kurtosis

Weight in grams x	Frequency n	$u = \dfrac{x-194 \cdot 5}{20}$	nu	nu^2	nu^3	nu^4
45–	2	−7	−14	98	−686	4802
65–	4	−6	−24	144	−864	5184
85–	6	−5	−30	150	−750	3750
105–	12	−4	−48	192	−768	3072
125–	14	−3	−42	126	−378	1134
145–	33	−2	−66	132	−264	528
165–	49	−1	−49	49	− 49	49
185–	61	0	0	0	0	0
205–	51	1	51	51	51	51
225–	43	2	86	172	344	688
245–	30	3	90	270	810	2430
265–	15	4	60	240	960	3840
285–	5	5	25	125	625	3125
305–	2	6	12	72	432	2592
Total	327		51	1821	−537	31245

Source: E.W. Lindstrom (1935), *The American Naturalist*, Vol.39, p.311.

The mean of x is $\bar{x} = 194 \cdot 5 + 3 \cdot 12 = 197 \cdot 62$.

The variance of u is $s_u^2 = 5 \cdot 5615$.

Therefore the variance of x is $s_x^2 = 2224 \cdot 60$,

whence $\qquad s_x = 47 \cdot 1657 \quad$ and $\quad 1/s_x = 0 \cdot 021202$.

Computational notes

The calculations of Tables 7.1 and 7.2 are largely self-explanatory, but there are several important practical points which need specific mention.

(i) The values obtained for g_1 and g_2 are −0·3248 and 0·1679 respectively. They are both small numerically, though g_1 could perhaps be regarded as showing a certain amount of negative skewness. This obviously means that the data suggest a greater frequency of bigger ears of maize. We could therefore have fitted a negatively skew curve to these data, but the normal curve can certainly be regarded as a possible model. As it turns out, the latter curve gives a remarkably good fit.

(ii) In the first column of Table 7.2, it is not essential to consider the points 44·5 and 324·5 if a simple comparison of the observed and expected frequencies is all that is required. We have included these two extra points to evaluate separately the areas under the fitted normal curve for $x \leqslant 44 \cdot 5$ and $x \geqslant 324 \cdot 5$. These are found to be 0·0006 and 0·0036, which are sufficiently small to be negligible. Thus the expected proportion of values $\leqslant 64 \cdot 5$ is 0·0024 as compared with the observed frequency of 2 in the first class-interval (44·5 – 64·5). In the same way, the expected proportion of values $\geqslant 304 \cdot 5$ is 0·0117 which corresponds with the observed frequency of 2 in the last class-interval (304·5–324·5).

Table 7.2 Showing the calculation of the ordinates of the fitted normal curve and the expected frequencies in the class-intervals

$x-\dfrac{h}{2}$	x	$\dfrac{x-h/2-\bar{x}}{s_x}$	$\Phi\left(\dfrac{x-h/2-\bar{x}}{s_x}\right)$	$\Delta\Phi$	$N\times\Delta\Phi$	n	$z=\dfrac{x-\bar{x}}{s_x}$	$\phi(z)$	$\dfrac{Nh}{s_x}\,\phi(z)$
$-\infty$		$-\infty$	0						
44·5		$-3\cdot246$	0·0006	$\lceil\,0\cdot0006$					
	54·5			18	0·785	2	$-3\cdot034$	0·0015	0·21
64·5		$-2\cdot822$	24						
	74·5			59	1·929	4	$-2\cdot610$	132	1·83
84·5		$-2\cdot398$	83						
	94·5			159	5·199	6	$-2\cdot186$	366	5·07
104·5		$-1\cdot974$	242						
	114·5			364	11·903	12	$-1\cdot762$	845	11·72
124·5		$-1\cdot550$	606						
	134·5			695	22·726	14	$-1\cdot338$	1630	22·60
144·5		$-1\cdot126$	1301						
	154·5			1113	36·395	33	$-0\cdot914$	2627	36·43
164·5		$-0\cdot702$	2414						
	174·5			1491	48·756	49	$-0\cdot490$	3538	49·06
184·5		$-0\cdot278$	3905						
	194·5			1675	54·772	61	$-0\cdot066$	3981	55·20
204·5		0·146	5580						
	214·5			1577	51·568	51	0·358	3742	51·89
224·5		0·570	7157						
	234·5			1244	40·679	43	0·782	2938	40·74
244·5		0·994	8399						
	254·5			820	26·814	30	1·206	1928	26·73
264·5		1·418	9219						
	274·5			453	14·813	15	1·630	1057	14·66
284·5		1·842	9672						
	294·5			211	6·900	5	2·054	484	6·72
304·5		2·266	9883						
	314·5			81	3·825	2	2·478	185	2·57
324·5		2·690	9964	$\lfloor\,36$					
∞		∞	1·0000						
Total					327·065	327			

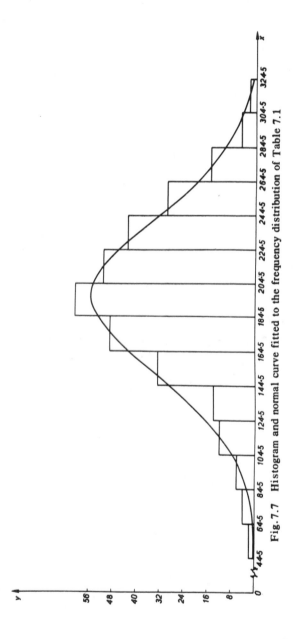

Fig. 7.7 Histogram and normal curve fitted to the frequency distribution of Table 7.1

(iii) The values $(x - h/2 - \bar{x})/s_x$ are obtained by a continuous machine operation using the reciprocal $1/s_x$ as a multiplier put on the setting levers of the desk calculator. The first value of $x - h/2 - \bar{x}$ is $-153 \cdot 12$, and the answer $-3 \cdot 246$ is obtained in the products register when the revolutions counter shows $153 \cdot 12$. The remaining negative values $(x - h/2 - \bar{x})/s_x$ are then obtained by successively *subtracting* $20 \cdot 00$ from the revolutions counter at each operation. The smallest positive value $0 \cdot 146$ is obtained when the revolutions counter shows $6 \cdot 68$, which is $204 \cdot 5 - 197 \cdot 62$, and then the remaining positive values are obtained by successively *adding* $20 \cdot 00$ to the number present in the revolutions counter.

For doing the calculations for this column, it is important to determine $1/s_x$ at least to five significant figures. The values $(x - h/2 - x)/s_x$ are recorded correct to only three decimal places, since this accuracy is adequate for obtaining the areas under the normal curve. In fact, in the shorter tables of the Φ function the argument is given to two decimal places and linear interpolation is sufficient for the third decimal place. Furthermore, some tables give the Φ values to a large number of decimal places, but four decimal places are usually sufficiently accurate for most applications.

(iv) The expected frequencies $N \times \Delta\Phi$ are recorded to three decimal places, though they could well be rounded off to two decimal places. The expected frequencies add up to $327 \cdot 065$, with error in the fifth significant figure.

(v) The values of $z = (x - \bar{x})/s_x$ are also obtained by continuous multiplication, and three decimal places are retained for the same reasons as in (iii). The values of $\phi(z)$ are easily read off from the tables of ordinates of a unit normal curve using linear interpolation for the third decimal place of the argument. Of course, because of the symmetry of a normal curve, $\phi(z)$ is the same for $\pm z$.

(vi) The calculation of the y values is adequately correct to two decimal places.

7.17 Some applications of the normal distribution

We have seen how a normal distribution may be regarded as a mathematical representation of a random variable. If the mean and variance of the distribution are also specified, then we have a theoretically complete description of the behaviour of the random variable. We now illustrate how such a formulation can be used to draw probabilistic conclusions in a variety of simple situations.

Example 7.2

It is known that the height of adult males in a certain country is a random variable x with mean $68 \cdot 5$ inches and standard deviation $3 \cdot 8$ inches. If it can be assumed that x is normally distributed, then find the probability that a randomly selected adult male has (i) height $\leqslant 60$ inches; and (ii) height $\geqslant 72$ inches.

Hence determine the expected number of adult males in a random sample of 80 who have heights within the range $60 - 72$ inches.

The probability of an adult male having a height $\leqslant 60$ inches is

$$P(x \leqslant 60) = \Phi[(60 - 68 \cdot 5)/3 \cdot 8]$$
$$= \Phi(-2 \cdot 237) = 1 - \Phi(2 \cdot 237) = 1 - 0 \cdot 9874 = 0 \cdot 0126.$$

Similarly,

$$P(x \geqslant 72) = 1 - P(x \leqslant 72)$$
$$= 1 - \Phi[(72 - 68 \cdot 5)/3 \cdot 8]$$
$$= 1 - \Phi(0 \cdot 921) = 1 - 0 \cdot 8215 = 0 \cdot 1785.$$

Therefore the probability that an adult male has a height between 60 and 72 inches is

$$P(60 \leqslant x \leqslant 72) = 1 - 0 \cdot 0126 - 0 \cdot 1785 = 0 \cdot 8089.$$

Hence the expected number of adult males in the sample of 80 with heights between 60 and 72 inches is

$$80 \, P(60 \leqslant x \leqslant 72) = 64 \cdot 7, \quad \text{or 65 persons.}$$

Example 7.3

The use of intelligence tests to determine the I.Q. (intelligence quotient) of children is a common procedure in educational research. It is known that although, on the average, boys tend to score a little higher than girls, the latter are the more consistent performers. In a particular region, the mean I.Q.'s of the boys and girls are 104·6 and 102·8 respectively with standard deviations 3·15 and 2·63 points. If the distributions of I.Q.'s of boys and girls may be assumed to be normal, determine the expected percentage of

(i) the boys who have a score greater than the mean I.Q. of the girls, and

(ii) the girls who have a score less than the mean I.Q. of the boys.

Let x and y denote the I.Q. of a boy and a girl respectively. Then

$$P(x \geqslant 102 \cdot 8) = 1 - \Phi[(102 \cdot 8 - 104 \cdot 6)/3 \cdot 15]$$
$$= 1 - \Phi(-0 \cdot 571) = \Phi(0 \cdot 571) = 0 \cdot 7160.$$

Therefore the expected percentage of boys having I.Q. greater than the mean I.Q. of the girls is 71·60.

Similarly,

$$P(y \leqslant 104 \cdot 6) = \Phi[(104 \cdot 6 - 102 \cdot 8)/2 \cdot 63]$$
$$= \Phi(0 \cdot 684) = 0 \cdot 7530.$$

Therefore the expected percentage of girls having I.Q. less than the mean I.Q. of the boys is 75·30.

Example 7.4

A factory produces light bulbs by two different processes A and B. The bulbs produced by method A have an average life-time of 1,800 hours with a standard deviation of 200 hours, whereas those produced by method B have a mean life of 2,000 hours with a standard deviation of 350 hours. According to a customer's specifications, bulbs with a mean life of less than 1,500 hours are to be regarded as definitely bad. In the light of this, determine which of the two processes would be the better one to use, as-suming that the life-time distributions of bulbs produced by A and B are both normal.

The better process is the one which gives a smaller expected percent-age of bulbs with a life-time less than 1,500 hours. Let x and y be random variables denoting life-times of bulbs produced by A and B respectively.
Then

$$P(x \leqslant 1500) = \Phi[(1500 - 1800)/200]$$
$$= \Phi(-1 \cdot 5) = 1 - \Phi(1 \cdot 5) = 1 - 0 \cdot 9332 = 0 \cdot 0668.$$

Similarly,

$$P(y \leqslant 1500) = \Phi[(1500 - 2000)/350]$$
$$= \Phi(-1 \cdot 429) = 1 - \Phi(1 \cdot 429) = 1 - 0 \cdot 9235 = 0 \cdot 0765.$$

The expected percentages of bad bulbs produced by A and B are 6·68 and 7·65 respectively so that A is the better process despite its lower mean life-time.

Example 7.5

Of the telephone calls through an exchange over a long period of time, 10 per cent were of less than three minutes' and 6 per cent of more than fifteen minutes' duration. If the length of a call may be assumed to be normally distributed find the mean and variance of the distribution. Hence determine the expected number of calls out of a total of 400 which will be of length between five and ten minutes.

Let μ and σ be the mean and standard deviation of the distribution of the length of calls. Then

$$\Phi[(3 - \mu)/\sigma] = 0 \cdot 10 \quad \text{and} \quad 1 - \Phi[(15 - \mu)/\sigma] = 0 \cdot 06,$$

or $\quad \Phi[(\mu - 3)/\sigma] = 0 \cdot 90 \quad \text{and} \quad \Phi[(15 - \mu)/\sigma] = 0 \cdot 94.$

Hence $\quad (\mu - 3)/\sigma = 1 \cdot 282 \quad \text{and} \quad (15 - \mu)/\sigma = 1 \cdot 555.$

Therefore $\quad -3 = -\mu + 1 \cdot 282\sigma \quad \text{and} \quad 15 = \mu + 1 \cdot 555\sigma.$

The solution of these equations gives $\mu = 8 \cdot 42$ and $\sigma = 4 \cdot 23$. Hence

$$P(x \leqslant 10) = \Phi[(10 - 8 \cdot 42)/4 \cdot 23] = \Phi(0 \cdot 374) = 0 \cdot 6458 ; \text{ and}$$
$$P(x \leqslant 5) = \Phi[(5 - 8 \cdot 42)/4 \cdot 23] = \Phi(-0 \cdot 809) = 1 - \Phi(0 \cdot 809)$$
$$= 1 - 0 \cdot 7907 = 0 \cdot 2093.$$

Therefore the probability of the length of a random call being between five and ten minutes is

$$P(5 \leqslant x \leqslant 10) = P(x \leqslant 10) - P(x \leqslant 5) = 0 \cdot 4365.$$

Hence the expected number of calls of this length out of a total of 400 is

$$400 \, P(5 \leqslant x \leqslant 10) = 174 \cdot 6 = 175 \quad \text{to the nearest integer.}$$

Example 7.6

A particular kind of electronic tube produced for television sets has a mean life-time of 1,600 hours with a standard deviation of 275 hours. If the life-time x is assumed to be normally distributed, find x_1 and x_2 such that

$$P(x \leqslant x_1) = P(x \geqslant x_2) = 0 \cdot 10.$$

If the mean of x remains unchanged, determine the reduced standard deviation of the distribution which would ensure that

$$P(x \leqslant x_1) = P(x \geqslant x_2) = 0 \cdot 05.$$

We have

$$P(x \geqslant x_2) = 1 - \Phi[(x_2 - 1600)/275] = 0 \cdot 10,$$

or

$$\Phi[(x_2 - 1600)/275] = 0 \cdot 90.$$

Therefore

$$(x_2 - 1600)/275 = 1 \cdot 282, \quad \text{so that} \quad x_2 = 1600 + 352 \cdot 55 = 1952 \cdot 55.$$

Hence, by symmetry,

$$x_1 = 1600 - 352 \cdot 55 = 1247 \cdot 45.$$

For the second part, if σ is the reduced standard deviation, we have

$$P(x \geqslant x_2) = 1 - \Phi[(1952 \cdot 55 - 1600)/\sigma] = 0 \cdot 05,$$

or

$$\Phi(352 \cdot 55/\sigma) = 0 \cdot 05.$$

Therefore $352 \cdot 55/\sigma = 1 \cdot 645, \quad \text{or} \quad \sigma = 214 \cdot 32.$

Example 7.7

At a flour-mill, nominally 5 lb bags of flour are filled by an automatic process. The machine is so set that the average weight of flour in the bags is 5·05 lb with a standard deviation of 0·02 lb. Find the expected percentage of underweight bags produced by this process, assuming that the weight of flour in the bags is normally distributed.

If the government permits the sale of not more than 5 per cent underweight bags, determine the lowest value of the mean weight of flour which would, on the average, just satisfy the government's stipulation for underweight bags. Hence determine the expected saving of.flour to the mill in 1,000 bags filled by the new method.

The proportion of underweight bags produced by the old method is $P(x \leqslant 5)$, where x is the weight of flour in any bag. Therefore we have

$$P(x \leqslant 5) = \Phi[(5 - 5 \cdot 05)/0 \cdot 02]$$
$$= \Phi(-2 \cdot 5) = 1 - \Phi(2 \cdot 5) = 1 - 0 \cdot 9938 = 0 \cdot 0062.$$

Therefore the expected percentage of underweight bags is 0·62.

Let μ be the new mean for the distribution of weights. Then

$$\Phi[(5 \cdot 00 - \mu)/0 \cdot 02] = 0 \cdot 05, \quad \text{or} \quad \Phi[(\mu - 5 \cdot 00)/0 \cdot 02] = 0 \cdot 95,$$

so that $(\mu - 5 \cdot 00)/0 \cdot 02 = 1 \cdot 645.$

Therefore $\mu = 5 \cdot 00 + 0 \cdot 0329 = 5 \cdot 0329.$

Thus the average saving in flour per bag is $5 \cdot 05 - 5 \cdot 0329 = 0 \cdot 0171$. Hence the average saving of flour in 1,000 bags is 17·1 lb.

Example 7.8

A man goes by car to his office, and the route through the city centre takes him, on the average, 27 minutes with a standard deviation of 5 minutes. With the opening of a new ring road, the man can bypass the congestion of the city centre, but the journey now takes, on the average, 29 minutes with a standard deviation of 2 minutes. Assuming that both journey times are normally distributed, determine which route is the better one if the man has (i) 28 minutes; and (ii) 32 minutes to reach his office for an appointment.

Obviously in each case the better route must give the smaller probability of the man's being late for the appointment.

For the old route, let x be the time taken for the journey. Then the probability of being late for (i) is

$$P(x \geqslant 28) = 1 - P(x \leqslant 28)$$
$$= 1 - \Phi[(28 - 27)/5] = 1 - \Phi(0 \cdot 2) = 1 - 0 \cdot 5793$$
$$= 0 \cdot 4207 ;$$

and for (ii) it is

$$P(x \geqslant 32) = 1 - P(x \leqslant 32)$$
$$= 1 - \Phi[(32 - 27)/5] = 1 - \Phi(1) = 1 - 0 \cdot 8413 = 0 \cdot 1587.$$

Similarly, if y is the time taken for the journey by the new route, then the probability of being late for (i) is

$$P(y \geqslant 28) = 1 - P(y \leqslant 28)$$
$$= 1 - \Phi[(28 - 29)/2] = 1 - \Phi(-0 \cdot 5) = \Phi(0 \cdot 5) = 0 \cdot 6915 ;$$

and that for (ii) is

$$P(y \geqslant 32) = 1 - P(y \leqslant 32)$$
$$= 1 - \Phi[(32 - 29)/2] = 1 - \Phi(1 \cdot 5) = 1 - 0 \cdot 9332$$
$$= 0 \cdot 0668.$$

Hence the old route is better for (i) and the new route for (ii).

The examples given above show how the assumption of the normality of the probability distribution of a random variable can be used to answer conveniently many kinds of questions. However, it is important to note that the answers provided by statistical analysis refer *either* to results which might be *expected*, that is happen in the long run, *or* to the determination of probabilities of chance events. These are typical characteristics of statistical inference, although the methods for arriving at such inference become deeper and more sophisticated as the complexities of the phenomena being studied, and of the questions being asked, increase.

7.18   Moment-generating function of a function of a random variable

If x is a random variable having a probability density function $f(x)$ in the range $-\infty < x < \infty$, then we have defined the moment-generating function of x about the origin as

$$M_0(t) \equiv E(e^{tx}) = \int_{-\infty}^{\infty} e^{tx} f(x) \, dx.$$

Further, if the moments of x exist, then on expanding $M_0(t)$ as a power series in t, the coefficient of $t^r/r!$ gives the rth moment of x about the origin.

It is useful to extend this idea to that of a moment-generating function of any function $g(x)$ of the random variable x. Formally, the moment-generating function of $g(x)$ is

$$M_g(t) \equiv E[e^{tg(x)}] = \int_{-\infty}^{\infty} e^{tg(x)} f(x) \, dx.$$

Depending upon the forms of $f(x)$ and $g(x)$, it is frequently possible to evaluate the above integral to obtain $M_g(t)$. If the moments of $g(x)$ exist, then on expanding $M_g(t)$ as a power series in t, the coefficient of $t^r/r!$ gives the rth moment of $g(x)$ about the origin. In the same way, the expansion of $\log M_g(t)$ gives the cumulants of $g(x)$.

We consider two simple examples.

Example 7.9

Suppose x is a unit normal variable, and let $g(x) = x^2$. Then the moment-generating function of x^2 is

$$M_g(t) \;\equiv\; E(e^{tx^2}) \;=\; \frac{1}{\sqrt{2\pi}} \int_{-\infty}^{\infty} \exp\left(tx^2 - \frac{1}{2}x^2\right) dx$$

$$=\; \frac{1}{\sqrt{2\pi}} \int_{-\infty}^{\infty} \exp -\frac{1}{2}(1 - 2t)\, x^2 dx,$$

where $1 - 2t > 0$. To evaluate this integral, we make the substitution $y = x(1 - 2t)^{\frac{1}{2}}$. Then

$$M_g(t) \;=\; \frac{1}{\sqrt{2\pi(1 - 2t)}} \int_{-\infty}^{\infty} e^{-\frac{1}{2}y^2} dy \;=\; (1 - 2t)^{-\frac{1}{2}},$$

so that

$$\log M_g(t) \;=\; -\frac{1}{2}\log(1 - 2t)$$

$$=\; \frac{1}{2}\sum_{r=1}^{\infty} \frac{(2t)^r}{r} \;=\; \sum_{r=1}^{\infty} \frac{t^r}{r!}\, 2^{r-1}(r-1)!$$

Hence the rth cumulant of x^2 is

$$\kappa_r(x^2) \;=\; 2^{r-1}(r-1)!\,, \quad \text{for } r \geqslant 1.$$

In particular,

$$\kappa_1(x^2) = 1; \quad \kappa_2(x^2) = 2; \quad \kappa_3(x^2) = 8; \quad \kappa_4(x^2) = 48;$$

and the coefficients of skewness and kurtosis of the distribution of x^2 are

$$\gamma_1(x^2) = 2\sqrt{2}; \quad \gamma_2(x^2) = 12.$$

Example 7.10

Again, suppose x is a unit normal variable and let $g(x) = |x|$.

Then
$$M_g(t) \equiv E[e^{t|x|}] \;=\; \frac{1}{\sqrt{2\pi}} \int_{-\infty}^{\infty} e^{(t|x| - \frac{1}{2}x^2)} dx$$

$$=\; \frac{1}{\sqrt{2\pi}} \int_{-\infty}^{0} e^{(t|x| - \frac{1}{2}x^2)} dx \;+\; \frac{1}{\sqrt{2\pi}} \int_{0}^{\infty} e^{(t|x| - \frac{1}{2}x^2)} dx$$

$$=\; (2/\pi)^{\frac{1}{2}} \int_{0}^{\infty} e^{(tx - \frac{1}{2}x^2)} dx$$

$$=\; (2/\pi)^{\frac{1}{2}} e^{\frac{1}{2}t^2} \int_{0}^{\infty} e^{-\frac{1}{2}(x-t)^2} dx$$

$$=\; (2/\pi)^{\frac{1}{2}} e^{\frac{1}{2}t^2} \int_{-t}^{\infty} e^{-\frac{1}{2}u^2} du$$

$$=\; (2/\pi)^{\frac{1}{2}} e^{\frac{1}{2}t^2} \int_{-\infty}^{t} e^{-\frac{1}{2}u^2} du \;=\; 2e^{\frac{1}{2}t^2}\Phi(t),$$

where $\Phi(t)$ is the distribution function of a unit normal variable. But we have proved in Section 7.14 that for $t > 0$

$$\Phi(t) = \frac{1}{2} + \frac{1}{\sqrt{2\pi}} \sum_{r=0}^{\infty} \frac{(-1)^r t^{2r+1}}{(2r+1) 2^r r!},$$

whence the cumulant-generating function of $|x|$ is

$$\log M_g(t) = \frac{1}{2} t^2 + \log \left[1 + \lambda \sum_{r=0}^{\infty} \frac{(-1)^r t^{2r+1}}{(2r+1) 2^r r!} \right],$$

where we have set $\lambda = (2/\pi)^{\frac{1}{2}}$,

$$= \frac{1}{2} t^2 + \log \left[1 + \lambda \left(t - \frac{t^3}{3!} + \frac{3t^5}{5!} - \cdots \right) \right]$$

$$= \frac{1}{2} t^2 + \lambda \left(t - \frac{t^3}{3!} + \cdots \right) - \frac{\lambda^2}{2} \left(t^2 - \frac{t^4}{3} + \cdots \right) + \frac{\lambda^3}{3} \left(t^3 - \cdots \right) - \frac{\lambda^4 t^4}{4} + \cdots$$

$$= t(2/\pi)^{\frac{1}{2}} + \frac{t^2}{2!} (\pi - 2)/\pi + \frac{t^3}{3!} (4-\pi)\sqrt{2}/\pi^{3/2} + \frac{t^4}{4!} 8(\pi - 3)/\pi^2 + \cdots .$$

Therefore the first four cumulants of $|x|$ are

$$\kappa_1(|x|) = (2/\pi)^{\frac{1}{2}}; \qquad \kappa_2(|x|) = (\pi - 2)/\pi;$$

$$\kappa_3(|x|) = (4-\pi)\sqrt{2}/\pi^{3/2}; \qquad \kappa_4(|x|) = 8(\pi - 3)/\pi^2.$$

Hence the coefficients of skewness and kurtosis of the distribution of $|x|$ are

$$\gamma_1(|x|) = (4-\pi)\sqrt{2}/(\pi - 2)^{3/2};$$

$$\gamma_2(|x|) = 8(\pi - 3)/(\pi - 2)^2.$$

7.19 Approximations for the mean and variance of a function of a random variable

In studying the distributional behaviour of a function $g(x)$ of a random variable, it is often of interest to evaluate the moments of the function $g(x)$. As we have seen, this can be done quite simply when it is possible to evaluate the moment-generating function of $g(x)$. However, there are frequent cases where the form of $g(x)$ is such that no convenient method exists for evaluating its moment-generating function. Situations of this kind arise quite often when x is a non-normal random variable and $g(x)$ is a function which is approximately normally distributed. We are then obliged to resort to approximations to obtain the moments of $g(x)$. The basis of these approximations rests on two assumptions, namely, that the moments of x exist and that a valid Taylor's expansion of $g(x)$ can be obtained about the point $E(x)$. Subject to these conditions, the method is quite general for determining the moments of $g(x)$, but we shall be particularly concerned with the approximate evaluation of the mean and variance of $g(x)$. The following lemma is useful in deriving the variance of $g(x)$.

Lemma

If a random variable z has expectation $\alpha + \beta$, where α and β are constants, then

$$\text{var}(z) = E(z-\alpha)^2 - \beta^2.$$

The proof of this lemma is simple. Clearly, $E(z-a) = \beta$, so that

$$
\begin{aligned}
\mathrm{var}(z) &= E[z-(a+\beta)]^2 \\
&= E[(z-a)^2 - 2\beta(z-a) + \beta^2] \\
&= E(z-a)^2 - \beta^2.
\end{aligned}
$$

To derive the main results, we write

$$
E(x) = \mu \quad \text{and} \quad E(x-\mu)^r = \mu_r, \quad \text{for } r \geqslant 2.
$$

Then, by Taylor's theorem, we have

$$
g(x) = g[\mu + (x-\mu)] = g(\mu) + \sum_{r=1}^{\infty} \frac{(x-\mu)^r}{r!} g^{(r)}(\mu), \tag{1}
$$

where $g^{(r)}(\mu)$ denotes the rth derivative of $g(x)$ at the point $x = \mu$. Therefore, taking expectations,

$$
E[g(x)] = g(\mu) + \sum_{r=2}^{\infty} g^{(r)}(\mu)\, \mu_r / r!
$$

If the successive terms of the series are decreasing then, as a first approximation, we have

$$
E[g(x)] = g(\mu),
$$

and a second approximation is that

$$
E[g(x)] = g(\mu) + \frac{1}{2}\mu_2 g''(\mu). \tag{2}
$$

In many applications the first approximation is adequate, though for greater accuracy the second approximation is necessary. We note that the second approximation is of the form $a + \beta$, and so the lemma is applicable for the derivation of the variance of $g(x)$. Thus, retaining terms to the fourth power of $x - \mu$, we have from (1)

$$
[g(x) - g(\mu)]^2 = (x-\mu)^2[g'(\mu)]^2 + (x-\mu)^3[g'(\mu)g''(\mu)] + \\
+ (x-\mu)^4[\{g''(\mu)\}^2/4 + \{g'(\mu)g'''(\mu)\}/3].
$$

Therefore

$$
E[g(x) - g(\mu)]^2 = \mu_2[g'(\mu)]^2 + \mu_3[g'(\mu)g''(\mu)] + \\
+ \mu_4[\{g''(\mu)\}^2/4 + \{g'(\mu)g'''(\mu)\}/3],
$$

whence, by the lemma,

$$
\mathrm{var}[g(x)] = \mu_2[g'(\mu)]^2 + \mu_3[g'(\mu)g''(\mu)] + \frac{1}{4}(\mu_4 - \mu_2^2)[g''(\mu)]^2 + \\
+ \frac{1}{3}\mu_4[g'(\mu)g'''(\mu)]. \tag{3}
$$

This result is correct to second-order terms, the first approximation being simply

$$
\mathrm{var}[g(x)] = \mu_2[g'(\mu)]^2.
$$

In particular, if x is normally distributed with mean μ and variance σ^2, then $\mu_2 = \sigma^2$, $\mu_3 = 0$, and $\mu_4 = 3\sigma^4$. Therefore

$$
E[g(x)] = g(\mu) + \frac{1}{2}\sigma^2 g''(\mu); \quad \text{and}
$$

$$
\mathrm{var}[g(x)] = \sigma^2[g'(\mu)]^2 + \sigma^4[\{g'(\mu)g'''(\mu)\} + \frac{1}{2}\{g''(\mu)\}^2].
$$

The first approximations for the mean and variance of $g(x)$ are of considerable practical value, but we shall derive the second-order approximations in the following illustrations of the theory.

Order notation

It is convenient to introduce here a useful order notation. Suppose a function $\psi(x)$ can be written in the form of a series

$$\psi(x) \;=\; a_0(x) + \frac{a_1(x)}{n} + \frac{a_2(x)}{n^2} + \frac{a_3(x)}{n^3} + \dots,$$

where the functions $a_i(x)$ are finite and independent of the parameter n. If n is large, then it is usual to write as a first approximation

$$\psi(x) \;=\; a_0(x) + O(n^{-1}).$$

The expression $O(n^{-1})$ is read as "terms of order n to the minus one", and it means that the terms after $a_0(x)$ all contain n^{-1} as a factor. Similarly, as a second approximation,

$$\psi(x) \;=\; a_0(x) + \frac{a_1(x)}{n} + O(n^{-2});$$

and $O(n^{-2})$ here indicates that the terms after $a_1(x)/n$ all contain n^{-2} as a factor. This notation can be extended in an obvious manner to indicate higher approximations of $\psi(x)$. It is to be noted that this usage is not in complete conformity with the conventional meaning of the order symbol O in pure mathematics, though its present use is well-established in statistical literature.

Example 7.11

Suppose x is a "Poisson variable" with mean μ (large) so that

$$E(x) \;=\; \text{var}(x) \;=\; \mu_3(x) \;=\; \mu \quad \text{and} \quad \mu_4(x) \;=\; \mu + 3\mu^2.$$

If $g(x) = \sqrt{x}$, then

$$g(\mu) = \sqrt{\mu}; \quad g'(\mu) = \frac{1}{2\sqrt{\mu}}; \quad g''(\mu) = -\frac{1}{4}\mu^{-3/2}; \quad g'''(\mu) = \frac{3}{8}\mu^{-5/2}.$$

Hence we have

$$E(\sqrt{x}) \;=\; \sqrt{\mu} + \frac{1}{2}\mu \frac{(-1)}{4}\mu^{-3/2} \;=\; \sqrt{\mu}\left[1 - \frac{1}{8\mu}\right];$$

and

$$\text{var}(\sqrt{x}) \;=\; \mu\frac{1}{4\mu} + \left[\mu \frac{1}{2\sqrt{\mu}}\frac{(-1)}{4}\mu^{-3/2} + \frac{1}{4}(\mu+3\mu^2-\mu^2)\frac{1}{16}\mu^{-3} \right.$$
$$\left. + \frac{1}{3}(\mu+3\mu^2)\frac{1}{2\sqrt{\mu}}\frac{3}{8}\mu^{-5/2}\right]$$

$$=\; \frac{1}{4}\left[1 + \frac{3}{8\mu} + O(\mu^{-2})\right].$$

Thus the second approximations are

$$E(\sqrt{x}) \;=\; \sqrt{\mu}\left[1 - \frac{1}{8\mu}\right]; \quad \text{var}(\sqrt{x}) \;=\; \frac{1}{4}\left[1 + \frac{3}{8\mu}\right].$$

The first approximations $E(\sqrt{x}) = \sqrt{\mu}$ and $\text{var}(\sqrt{x}) = 1/4$ are of considerable use for large values of μ.

Example 7.12

Suppose x is a "χ^2 variable" having the first four moments

$$E(x) = \nu; \quad \mu_2(x) = 2\nu; \quad \mu_3(x) = 8\nu; \quad \mu_4(x) = 12\nu(\nu+4),$$

where ν is a large positive integer.

If $g(x) = (x/\nu)^{1/3}$, then

$$g(\nu) = 1; \quad g'(\nu) = \frac{1}{3\nu}; \quad g''(\nu) = -\frac{2}{9\nu^2}; \quad g'''(\nu) = \frac{10}{27\nu^3}.$$

Therefore we have

$$E[(x/\nu)^{1/3}] \;=\; 1 + \frac{1}{2}2\nu\frac{(-2)}{9\nu^2} \;=\; \left[1 - \frac{2}{9\nu}\right];$$

and

$$\mathrm{var}[(x/\nu)^{1/3}] \;=\; 2\nu\frac{1}{9\nu^2} + \left[8\nu\frac{1}{3\nu}\frac{(-2)}{9\nu^2} + \frac{1}{4}\frac{4}{81\nu^4} \times \right.$$

$$\left. \times \{12\nu(\nu+4)-4\nu^2\} + \frac{1}{3}\frac{1}{3\nu}\frac{10}{27\nu^3}12\nu(\nu+4)\right]$$

$$=\; \frac{2}{9\nu} + O(\nu^{-3}).$$

It is to be noted that the fact that the terms in ν^{-2} on the right-hand side cancel out is not enough to infer that the approximation for the variance is correct to $O(\nu^{-3})$. However, it is known that this is true for the variable considered here. Indeed, it was also shown by Wilson and Hilferty that $(x/\nu)^{1/3}$ is approximately normally distributed with mean $1 - 2/9\nu$ and variance $2/9\nu$. This is a famous example of a normalising transformation for a χ^2 variable.

Example 7.13

If x is again assumed to be a χ^2 variable, as in Example 7.12, and

$$g(x) \;=\; \log(x/\nu),$$

then $\quad g(\nu) = 0; \quad g'(\nu) = 1/\nu; \quad g''(\nu) = -1/\nu^2; \quad g'''(\nu) = 2/\nu^3.$

Therefore

$$E[\log(x/\nu)] \;=\; \frac{1}{2}2\nu\frac{(-1)}{\nu^2} \;=\; -\frac{1}{\nu};$$

and $\quad \mathrm{var}[\log(x/\nu)] \;=\; 2\nu\frac{1}{\nu^2} + \left[8\nu\frac{1}{\nu}\frac{(-1)}{\nu^2} + \frac{1}{4}\frac{1}{\nu^4}\{12\nu(\nu+4)-4\nu^2\} + \right.$

$$\left. + \frac{1}{3}\frac{1}{\nu}\frac{2}{\nu^3}12\nu(\nu+4)\right]$$

$$=\; \frac{2(\nu+1)}{\nu^2} + O(\nu^{-3}).$$

Thus $\log(x/\nu)$ has mean $-1/\nu$ and variance $2(\nu+1)/\nu^2$.

CHAPTER 8

DISCRETE PROBABILITY

8.1   Statistical probability

We have so far used the concept of probability in an intuitive way by defining it in terms of the area under a frequency curve. Thus, if the random
variable x has the probability density function $f(x)$, then the probability
that any realisation of x will lie in the interval (a, b) is

$$\int_a^b f(t)\,dt.$$

We now propose to extend and deepen an understanding of the ideas of probability by a more formal approach to the subject. Although, in principle,
there is no difference whether we think in terms of continuous or discrete
variation, it is simpler to study probability formally in the discrete case.
With this restriction, the basic formal structure of probability theory can be
presented within the framework of elementary mathematics and consequently
we shall adopt this course here. Apart from this mathematical convenience,
there is one more fundamental limitation. We shall be here concerned only
with what is termed *statistical probability*. This is a more specialised
study of chance than that implied by the everyday use of the words
"probable" and "probability". For example, statements like "John is prob-
ably telling the truth" or "There is little probability of finding human beings
on Mars" are concerned with the subjective assessment of chance. Such
ideas are of interest to the logician or the philosopher, but they do not come
within the scope of the definition of statistical probability. We define this
as the study of chance in relation to the possible outcomes of a conceptual
experiment.

Suppose an experiment can have any one of N possible outcomes
$E_1, E_2, ..., E_N$. The E_i are called *simple events*. For example,

(i) the experiment of tossing a coin once has two possible outcomes –
head (H) or tail (T); and

(ii) the experiment of rolling a die once has six possible outcomes –
the points 1,2,3,4,5,6.

It is important to note that the notion of an experiment is, in essence, an
abstraction, but the appeal to empirical ideas like tossing a coin or rolling
a die is helpful in providing an initial intuitive background to our theoretical development. Besides, as we shall see later, such experiments can be
also identified with statistical situations directly relevant to the study of
the observational data of science.

163

The totality of outcomes E_i constitute what is termed a *sample space*, denoted by Ω, and each E_i is represented by a *unique* point in Ω. We may thus think indifferently in terms of either the simple events which are the possible outcomes of the conceptual experiment, or the points of the sample space Ω. There is a one-to-one correspondence between the simple events and the points of Ω.

From the abstract point of view, the sample space and its points are the primitive, undefined notions of our theory, and statistical probability is defined only in terms of these concepts.

Definition 1

Given the sample space Ω and its points E_i, we attach to each E_i a non-negative number $p_i = P(E_i)$ such that

$$\sum_{i=1}^{N} p_i = 1.$$

These p_i are the probabilities of the N possible outcomes.

It is to be noted that there is nothing in this definition of probability to imply that the p_i are equal. But when there is no reason to distinguish between the possible outcomes E_i their probabilities are *assumed* to be equal. We then say that the E_i are all *equally likely*, and $p_i = 1/N$. Thus, for example,

(i) in tossing an unbiased coin once,

$$p_1 = P(H) = \frac{1}{2}, \quad p_2 = P(T) = \frac{1}{2}, \quad \text{and}$$

(ii) in rolling an unbiased die once,

$$p_1 = P(1) = \frac{1}{6}, \quad p_2 = P(2) = \frac{1}{6}, \quad \ldots, \quad p_6 = P(6) = \frac{1}{6}.$$

If E_j is an impossible event, then $P(E_j) = p_j = 0$. For example, with a single throw of an ordinary die, $P(7) = 0$.

It is important to remember that whether the probabilities of the E_i are equal or not depends on the phenomenon which the conceptual experiment is expected to simulate. Thus, our intuitive experience with coin tossing suggests that heads and tails should be equally frequent, and we therefore agree to write

$$P(H) = P(T) = \frac{1}{2}.$$

On the other hand, if we consider a model to represent the sex of a new-born baby, then empirical evidence strongly suggests that the two possibilities are not equally frequent. Indeed, it is known that in present-day Britain

$$P(\text{boy}) \sim 0.51 \quad \text{and} \quad P(\text{girl}) \sim 0.49.$$

Finally, since one of the events E_i must occur, we write

$$\sum_{i=1}^{N} P(E_i) = 1, \quad \text{or} \quad P(\Omega) = 1.$$

We illustrate the consequences of our definition by two elementary examples.

Example 8.1

A bag contains two red, three green, and four black balls, which are assumed to be identical except for the differences in colour. Three balls are drawn at random. Find the probability that

(i) the three balls have different colours;

(ii) two balls have the same colour, the third different; and

(iii) all three balls have the same colour.

There are $\binom{9}{3}$ = 84 points in the sample space, and each point E_i may be denoted by a triplet (x_i, y_i, z_i) according to the colours of the three balls drawn. To each E_i we attach a probability $1\big/\binom{9}{3}$. The solution of our problem depends upon identifying the sample points which correspond to (i), (ii), and (iii) respectively.

(i) There are $\binom{2}{1}\binom{3}{1}\binom{4}{1}$ sample points which correspond to balls of three different colours.

Hence
$$P(\text{i}) = \frac{2.3.4}{84} = \frac{2}{7}.$$

(ii) Denote the red, green, and black balls by R, G, and B respectively. Then for two balls of the same colour and the third different, the six possibilities are

$$R^2B, \quad R^2G, \quad G^2R, \quad G^2B, \quad B^2R, \quad \text{and } B^2G.$$

The corresponding number of sample points is

$$\binom{2}{2}\binom{4}{1} + \binom{2}{2}\binom{3}{1} + \binom{3}{2}\binom{2}{1} + \binom{3}{2}\binom{4}{1} + \binom{4}{2}\binom{2}{1} + \binom{4}{2}\binom{3}{1}$$

$$= 4 + 3 + 6 + 12 + 12 + 18 = 55.$$

Hence
$$P(\text{ii}) = \frac{55}{84}.$$

(iii) For three balls of the same colour, the possibilities are B^3 and G^3. The corresponding number of sample points is

$$\binom{3}{3} + \binom{4}{3} = 5.$$

Hence
$$P(\text{iii}) = \frac{5}{84}.$$

Therefore $P(\text{i}) + P(\text{ii}) + P(\text{iii}) = 1$, a result which is otherwise obvious.

In this example, we have divided the 84 sample points of the sample space into three distinct subsets containing 24, 55, and 5 points respectively. This partitioning is meaningful because it is associated with the events (i), (ii), and (iii), which are mutually exclusive.

Example 8.2

A pack (deck) of 52 playing cards contains four aces and 48 other cards. Find the probability that a random hand of 13 cards contains one ace.

There are $\binom{52}{13}$ sample points, each point being represented by 13 elements denoting the cards in the hand. Thus to each point we attach a probability $1/\binom{52}{13}$. The number of sample points containing exactly one ace is $\binom{4}{1}\binom{48}{12}$. Therefore the required probability is

$$\binom{4}{1}\binom{48}{12}/\binom{52}{13} = \frac{27417}{62475} = 0\cdot4388.$$

8.2   Compound events

In both the above examples the calculation of the required probabilities was based on the elementary idea of enumerating the sample points associated with any event whose probability was to be determined. In this context, we have extended the idea of an event beyond the notion of a simple event, which is merely a possible realisation of the experiment. We consider next this extended concept of events.

Consider an experiment — the rolling of two dice once — which results in the simple event $(3, 3)$. Then the same trial also resulted in the following events: "two odd faces", "sum of points six", "no ace", etc. These events are not mutually exclusive, so they can occur simultaneously. They are *compound events* in the sense that they can be decomposed into simple events. Accordingly, the events whose probabilities were determined in the previous two examples are also compound events. In general, the probability of a compound event is the sum of the probabilities of the simple events constituting it.

For the above experiment of rolling two dice once, the simple events are the 36 ordered doublets (i, j), for $i, j = 1, 2, ..., 6$. Therefore the compound event "sum of points six" means the aggregate of sample points

$$(1, 5), (2, 4), (3, 3), (4, 2), (5, 1).$$

Similarly, the sample points of the compound event "two odd faces" are

$$(1, 1), (1, 3), (1, 5), (3, 1), (3, 3), (3, 5), (5, 1), (5, 3), (5, 5).$$

Hence $P(\text{sum of points } 6) = \frac{5}{36};$ $P(\text{two odd faces}) = \frac{9}{36} = \frac{1}{4}.$

In general, two compound events A_1 and A_2 are not necessarily mutually exclusive, that is, the occurrence of one does not imply the non-occurrence of the other. For two mutually exclusive events A_1 and A_2 we have the following theorem.

Theorem 1

If A_1 and A_2 are two mutually exclusive events with probabilities $P(A_1)$ and $P(A_2)$, then the probability of either A_1 or A_2 occurring is

$$P(A_1 + A_2) = P(A_1) + P(A_2).$$

The proof of this theorem is immediate on addition of the probabilities of the aggregates of points corresponding to the events A_1 and A_2.

Example 8.3
Two unbiased dice are rolled once. Suppose A_1 is the event "sum of points obtained is five", and A_2 is the event "sum of points obtained is nine". Therefore

A_1 consists of the sample points: $(4,1)$, $(3,2)$, $(2,3)$, $(1,4)$;   and
A_2 consists of the sample points: $(6,3)$, $(5,4)$, $(4,5)$, $(3,6)$.

Hence $P(A_1) = \dfrac{4}{36} = \dfrac{1}{9}$, $P(A_2) = \dfrac{4}{36} = \dfrac{1}{9}$,   and $P(A_1 + A_2) = \dfrac{2}{9}$.

Theorem 1 has an obvious generalisation.

Theorem 1a
If A_1, A_2, ..., A_k are k mutually exclusive events with probabilities $P(A_1)$, $P(A_2)$, ..., $P(A_k)$, then the probability of any one of them happening is

$$P(A_1 + A_2 + ... + A_k) = \sum_{i=1}^{k} P(A_i).$$

The proof is again immediate because the aggregates of sample points for the A_i are distinct.

Theorem 2
If A_1 and A_2 are not mutually exclusive events, then the probability of either A_1 or A_2 or both occurring is

$$P(A_1 + A_2) = P(A_1) + P(A_2) - P(A_1 A_2),$$

where $P(A_1 A_2)$ denotes the probability for both A_1 and A_2 to occur.

Proof
Let $E_1, E_2, ..., E_m$ be the sample points of A_1, and $E_\nu, E_{\nu+1}, ..., E_n$ the sample points of A_2, where $1 < \nu \leqslant m < n$.

There are n sample points and m of these correspond to A_1, $n - \nu + 1$ to A_2, and $m - \nu + 1$ to $A_1 A_2$. Now the event $A_1 + A_2$ corresponds to the sample points $E_1, E_2, ..., E_n$, each E_i being considered once only.

But
$$\sum_{i=1}^{n} E_i = \sum_{i=1}^{m} E_i + \sum_{i=\nu}^{n} E_i - \sum_{i=\nu}^{m} E_i,$$

whence, on addition of the corresponding probabilities, we have
$$P(A_1 + A_2) = P(A_1) + P(A_2) - P(A_1 A_2).$$

Example 8.4
Two unbiased dice are rolled once. Let A_1 be the event "sum of points obtained is an odd number", and A_2 the event "one of the faces shown is a six".

The 36 sample points are:

$(1,1)$	$(1,2)$	$(1,3)$	$(1,4)$	$(1,5)$	$(1,6)$
$(2,1)$	$(2,2)$	$(2,3)$	$(2,4)$	$(2,5)$	$(2,6)$
$(3,1)$	$(3,2)$	$(3,3)$	$(3,4)$	$(3,5)$	$(3,6)$
$(4,1)$	$(4,2)$	$(4,3)$	$(4,4)$	$(4,5)$	$(4,6)$
$(5,1)$	$(5,2)$	$(5,3)$	$(5,4)$	$(5,5)$	$(5,6)$
$(6,1)$	$(6,2)$	$(6,3)$	$(6,4)$	$(6,5)$	$(6,6)$

Hence $\qquad P(A_1) = \frac{1}{2}; \quad P(A_2) = \frac{11}{36}; \quad P(A_1A_2) = \frac{1}{6}.$

Therefore $\qquad P(A_1 + A_2) = \frac{1}{2} + \frac{11}{36} - \frac{1}{6} = \frac{23}{36}.$

Definition 2

If there are no sample points corresponding to an arbitrary event A, then A is defined to be an impossible event, and we write

$$A = 0; \quad P(A) = 0.$$

Accordingly, if A_1 and A_2 are mutually exclusive events, then

$$A_1A_2 = 0; \quad P(A_1A_2) = 0.$$

Definition 3

If A is an arbitrary event, then the *negation* of A is the *complementary event* denoted by $\overline{A}$. In other words, the event $\overline{A}$ consists of all sample points of Ω which are not sample points of A. We then write

$$\overline{A} = \Omega - A.$$

Accordingly, $A + \overline{A} = \Omega$, and since A and $\overline{A}$ are mutually exclusive and exhaustive events, we have from Theorem 1

$$P(A) + P(\overline{A}) = P(\Omega) = 1,$$

so that $\qquad P(\overline{A}) = 1 - P(A).$

Also, $\qquad A\overline{A} = 0; \quad P(A\overline{A}) = 0.$

8.3   Statistical independence

Definition 4

Two events A_1 and A_2 are said to be *statistically independent* (or briefly *independent*) if the probability of their joint occurrence is given by

$$P(A_1A_2) = P(A_1) P(A_2).$$

This equation defines independence, and is of fundamental importance in our theory.

Example 8.5

Suppose a card is drawn at random from an ordinary pack of 52 playing cards. We denote by A_1 the event that the card is a heart, and by A_2 the event that it is a court card (a jack, queen or king). There are 52 sample points, of which 13 correspond to the event A_1, 12 to A_2, and 3 to A_1A_2. Since all sample points are equally probable,

$$P(A_1) = \frac{13}{52} = \frac{1}{4}, \quad P(A_2) = \frac{12}{52} = \frac{3}{13},$$

and

$$P(A_1A_2) = \frac{3}{52} = P(A_1) P(A_2).$$

Therefore the events A_1 and A_2 are independent. This result is intuitively reasonable since the proportion of court cards among the hearts is the same as among the pack as a whole.

On the other hand, suppose a card is drawn at random from a pack from which the king of spades is missing. If A_1 and A_2 denote the same events as before, then the number of sample points corresponding to $A_1, A_2, A_1 A_2$ become 13, 11, 3 respectively.

Hence
$$P(A_1) = \frac{13}{51}, \quad P(A_2) = \frac{11}{51},$$

and
$$P(A_1 A_2) = \frac{3}{51} \neq P(A_1) P(A_2).$$

Therefore the events A_1 and A_2 are no longer independent. In fact, knowledge that the selected card is a heart now slightly increases the probability of its being a court card. In general, we note that Definition 4 may be interpreted as expressing the assertion that the knowledge that one of the events A_1 and A_2 has occurred does not affect the probability that the other has occurred. This will be further clarified when we discuss "conditional probability".

Repeated trials

The concept of statistical independence is of particular importance since it enables us to give now an analytical formulation of our earlier intuitive notion of *identical experiments* (that is experiments repeated under similar conditions) referred to in Chapter 6.

Suppose Ω is the sample space representing all the possible outcomes of a conceptual experiment, and let $E_1, E_2, \ldots, E_N$ be the points of Ω with
$$P(E_i) = p_i, \quad \text{for} \quad 1 \leqslant i \leqslant N.$$

Consider now a succession of two identical experiments. Then, quite clearly the possible outcomes must be pairs (E_i, E_j), and these constitute a new sample space. Now the statement that the experiment is repeated under similar conditions has the statistical implication that the outcomes on the two occasions are independent. In other words, the first outcome should in no way affect the second outcome. This means that the two events "first outcome E_i" and "second outcome E_j" should be statistically independent so that
$$P(E_i, E_j) = p_i p_j.$$

This equation assigns a probability to every pair (E_i, E_j) of possible outcomes. Clearly, this definition of the probabilities to be attached to the points of the new sample space is valid since
$$\sum_{i=1}^{N} \sum_{j=1}^{N} p_i p_j = \left[\sum_{i=1}^{N} p_i \right] \left[\sum_{j=1}^{N} p_j \right] = 1.$$

Next, suppose A and B are two arbitrary events in the original sample space Ω. We denote by (A, B) the event "A occurred at first trial and B at second". If the sample points $E_{a_1}, E_{a_2}, \ldots, E_{a_n}$ correspond to the event

A, and the sample points $E_{b_1}, E_{b_2}, ..., E_{b_m}$ to the event B, then the event (A, B) is the aggregate of the pairs (E_{a_i}, E_{b_j}). Accordingly,

$$P(A, B) = \sum_{i=1}^{n} \sum_{j=1}^{m} p_{a_i} p_{b_j}$$
$$= \left[\sum_{i=1}^{n} p_{a_i} \right] \left[\sum_{j=1}^{m} p_{b_j} \right]$$
$$= P(A) \, P(B).$$

Hence A and B are independent events. We therefore observe that the independence of (E_i, E_j) entails that the outcomes of the second trial are independent of the outcomes of the first. This is the proper description of identical experiments in probability theory.

Example 8.6

The two possible outcomes of tossing a coin once are a head (H) and a tail (T). These are the two sample points of Ω, and if the coin is unbiased, then

$$P(H) = P(T) = \frac{1}{2}.$$

If we consider next the experiment of tossing the coin twice, then the four possible outcomes are two heads (H, H); a head and then a tail (H, T); a tail and then a head (T, H); and two tails (T, T). These are the points of the new sample space, and since the coin is unbiased, we have

$$P(H, H) = P(H, T) = P(T, H) = P(T, T) = \frac{1}{4}.$$

Besides, the independence of the two tosses constituting the experiment means that

$$P(H, H) = P(H) \times P(H) = \frac{1}{4},$$

and so also for the probabilities of the other three possible outcomes.

It is to be noted that the same probabilities are obtained if two unbiased coins are tossed once.

Example 8.7

An urn contains two white balls W_1, W_2 and three black balls B_1, B_2, B_3. Consider an experiment which consists of selecting a ball at random from the urn, replacing it, and then again selecting a ball at random. Let A_1 and A_2 denote the events "first ball white" and "second ball black" respectively. The sample space consists of the 25 sample points (X, Y) representing all possible outcomes $(W_1, W_1), (W_1, W_2), (W_1, B_1), ..., (B_3, B_3)$ of the experiment. Of these, the 10 sample points (W_i, Y) correspond to A_1, the 15 sample points (X, B_j) to A_2, and the 6 sample points (W_i, B_j) to $A_1 A_2$. We therefore have

$$P(A_1) = \frac{10}{25} = \frac{2}{5}, \quad P(A_2) = \frac{15}{25} = \frac{3}{5},$$

and
$$P(A_1 A_2) = \frac{6}{25} = P(A_1) P(A_2).$$

Thus the events A_1 and A_2 are statistically independent. This is intuitively acceptable since the two drawings were made from an urn with the same constitution.

In the preceding discussion, we have considered the repetition of the same experiment, but the argument can be readily extended to two unlike experiments. For example, let the first experiment be "rolling two unbiased dice once", and the second experiment "tossing an unbiased coin once". Intuitively, we assume the two experiments to be independent. Hence, for instance, we have

$$P(\text{sum of points 10, a head}) = \frac{3}{36} \times \frac{1}{2} = \frac{1}{24}.$$

This is equivalent to assigning equal probabilities to the 72 points of the new sample space.

More generally, suppose $E_1, E_2, ..., E_N$ are the points of the sample space Ω with respective probabilities $p_1, p_2, ..., p_N$. Similarly, let $E_1', E_2', ..., E_M'$ be the points of another sample space Ω' with probabilities $p_1', p_2', ..., p_M'$. Then the statistical independence of the two experiments implies that

$$P(E_i, E_j') = p_i p_j'.$$

Furthermore, suppose A and B are arbitrary events defined over Ω and Ω' respectively. Then the independence of A and B means that

$$P(A, B) = P(A) P(B).$$

Definition 4a

A group of k events is said to be statistically independent if whenever $A_1, A_2, ..., A_r (1 \leqslant r \leqslant k)$ are distinct events of the group, the probability of their joint occurrence is given by

$$P(A_1 A_2 ... A_r) = \prod_{i=1}^{r} P(A_i).$$

This defining relationship provides the basis for an analytical definition of the probability of k repetitions of an experiment under similar conditions, which is analogous to that obtained from Definition 4 for two repetitions of an experiment. Thus, if A_i corresponds to the ith repetition of an experiment performed successively under similar conditions, then the independence of the k repetitions implies that the probability of the successive occurrence of $A_1, A_2, ..., A_k$ is

$$P(A_1 A_2 ... A_k) = \prod_{i=1}^{k} P(A_i).$$

This product rule also holds if the A_i are not repetitions of the same experiment but are defined over sample spaces $\Omega_1, \Omega_2, ..., \Omega_k$ respectively corresponding to k different successive experiments.

8.4 Conditional probability

A population of N persons consists of W females and $N - W$ males. Consider a conceptual experiment of taking one person at random from this population and noting the sex of the individual. If A is the event that the person selected is a female and $\bar{A}$ the event complementary to A, then the probabilities of the two possible outcomes are

$$P(A) \;=\; \frac{W}{N} \quad \text{and} \quad P(\overline{A}) \;=\; \frac{N-W}{N} \,.$$

Next, suppose we wish to determine the probability that a person chosen at random is a female who smokes. Then, if there are w female smokers in the population, the probability that a randomly selected person is a female smoker is

$$\frac{w}{N} \;=\; \frac{W}{N} \times \frac{w}{W} \tag{1}$$

$\qquad =$ Probability that a randomly selected person is a female

$\qquad \times$ Probability that the person is a smoker knowing that the person is a female.

The second probability refers to the *subpopulation* of females, and we need a new notation to express this fact. Accordingly, suppose B is the event that a randomly selected person is a smoker.

Then
$$P(AB) \;=\; \frac{w}{N}.$$

Hence, from (1),

$$P(AB) \;=\; P(A)\,P(B|A), \tag{2}$$

where $P(B|A) = w/W$. In this notation, $P(B|A)$ is the *conditional probability* of the event B, that is, conditional upon the occurence of A. It is usual to read $P(B|A)$ as "the probability of B given A". This probability is defined in terms of the sub-population or the subspace of the event A. The probability $P(B|A)$ is not defined if $P(A) = 0$.

Expressed as $P(B|A) = P(AB)/P(A)$, equation (2) gives a formal definition of a conditional probability. Furthermore, equation (2) is a symmetrical relation, and we have for $P(A) \neq 0, P(B) \neq 0$

$$P(AB) \;=\; P(A)\,P(B|A) \;=\; P(B)\,P(A|B). \tag{3}$$

In the context of our example, $P(A|B)$ is the conditional probability that a randomly selected person is a female given that the person is a smoker.

If A and B are independent events, then

$$P(A|B) \;=\; P(A) \quad \text{and} \quad P(B|A) \;=\; P(B),$$

so that
$$P(AB) \;=\; P(A)\,P(B), \tag{4}$$

which is in agreement with our earlier definition of independent events.

A generalisation

The formal definition (2) involving two events can be extended to a number of events $A_1, A_2, ..., A_k$. Thus, if we define the event

$$B_1 \;=\; A_2 A_3 \ldots A_k,$$

then
$$P(A_1 A_2 \ldots A_k) \;=\; P(A_1 B_1)$$
$$=\; P(A_1|B_1)\,P(B_1).$$

Again, if we define the event

$$B_2 = A_3 A_4 \dots A_k,$$

then
$$P(B_1) = P(A_2 B_2)$$
$$= P(A_2 | B_2) P(B_2),$$

whence
$$P(A_1 A_2 \dots A_k) = P(A_1 | B_1) P(A_2 | B_2) P(B_2).$$

This procedure can be applied $k-1$ times to give finally

$$P(A_1 A_2 \dots A_k) = P(A_1 | B_1) P(A_2 | B_2) \dots P(A_{k-1} | B_{k-1}) P(B_{k-1}),$$

where $B_i = A_{i+1} A_{i+2} \dots A_k$ for $i = 1, 2, \dots, k-1$. It is, of course, assumed here that the events A_i are such that $P(B_i) \neq 0$ for all i.

If the A_i are independent, then $P(A_i | B_i) = P(A_i)$, and we have

$$P(A_1 A_2 \dots A_k) = \prod_{i=1}^{k} P(A_i). \tag{5}$$

Bayes' theorem

Suppose $H_1, H_2, \dots, H_k$ are mutually exclusive and exhaustive events so that $H_1 + H_2 + \dots + H_k = \Omega$, the sample space. If A is some other event, then

$$A = A\Omega = \sum_{i=1}^{k} AH_i.$$

Hence
$$P(A) = \sum_{i=1}^{k} P(AH_i).$$

since the events AH_i are also mutually exclusive,

$$= \sum_{i=1}^{k} P(A|H_i) P(H_i). \tag{6}$$

This is a convenient formula for evaluating $P(A)$ when it is simple to determine the probabilities $P(A|H_i)$.

Again, for any given j,

$$P(AH_j) = P(A) P(H_j | A) = P(H_j) P(A | H_j),$$

so that
$$P(H_j | A) = \frac{P(A | H_j) P(H_j)}{P(A)}, \text{ or using (6)},$$

$$= \frac{P(A | H_j) P(H_j)}{\sum_{i=1}^{k} P(A | H_i) P(H_i)}, \tag{7}$$

which is the celebrated *Bayes' theorem* due to the English philosopher Thomas Bayes (1702-1761). This result was presented in his posthumous paper "An essay towards solving a problem in the doctrine of chances", which appeared in the *Philosophical Transactions of the Royal Society* in 1763. In Bayesian terminology, the $P(H_j)$ are known as *a priori* probabilities and the $P(H_j|A)$ as *a posteriori* probabilities. The difference between them is simply that $P(H_j)$ is the *absolute* or *unconditional* probability of H_j, and $P(H_j|A)$ is the *conditional* probability of H_j given that A has occurred. There are many deep and controversial problems of Bayesian inference, but they are beyond the scope of this book. It is sufficient for our purpose to realise that (7) is a convenient formula for the evaluation of a conditional probability.

We now consider two simple examples to illustrate the use of (6) and (7).

Example 8.8
Suppose there are n urns such that the rth urn contains b black and rw white balls ($r = 1, 2, ..., n$). An urn is selected at random and a ball drawn from it. Find the probability that the ball drawn is white.

Let H_r be the event of selecting the rth urn. Then $P(H_r) = 1/n$. Also, let A be the event of drawing a white ball. Therefore we have

$$P(A|H_r) = \frac{rw}{rw + b},$$

whence, by (6),

$$P(A) = \frac{1}{n} \sum_{r=1}^{n} \frac{rw}{rw + b}.$$

Example 8.9
The probability that a family has r children is π_r ($1 \leqslant r \leqslant k$), and the probabilities for a family to have a boy or a girl are p and q respectively ($p + q = 1$). A family is known to include exactly one boy. Find the probability that the boy is an only child, assuming that p is constant for all families.

Let H_r be the event that a family has r children.

Then $P(H_r) = \pi_r$ and $\sum_{r=1}^{k} H_r = \Omega$, the sample space, since the H_r are mutually exclusive and exhaustive events. Next, let A be the event that a family has one boy. We wish to determine $P(H_1|A)$.

We have
$$P(AH_1) = P(A|H_1) P(H_1) = \pi_1 p,$$

and
$$P(H_1|A) = \frac{P(AH_1)}{P(A)}$$

$$= \frac{P(AH_1)}{\sum\limits_{r=1}^{k} P(A|H_r) P(H_r)}.$$

Now $P(A|H_r)$ is the probability that a family with r children has just one boy. It is easily seen that

$$P(A|H_r) = \binom{r}{1} pq^{r-1},$$

since the one boy in the family could have any one of r positions in order of birth. Therefore

$$P(H_1|A) = \frac{\pi_1 p}{\sum\limits_{r=1}^{k} \pi_r rpq^{r-1}} = \frac{\pi_1}{\sum\limits_{r=1}^{k} r\pi_r q^{r-1}}.$$

8.5   Some further applications

In the preceding sections we have surveyed briefly the important features of the basic theorems of probability theory. These results are simple and they follow from principles which are intuitively acceptable. The real difficulty of the subject lies in trying to visualise a problem in terms of formally defined events which would make possible the direct application of the theorems. In principle, all problems of discrete probability could be solved by enumeration of the sample points; but, in general, this method would be unnecessarily laborious. The theorems provide short-cuts to direct enumeration, and their correct and consistent use can only be achieved through experience and a clear analysis of the problems posed. As a possibly useful aid in this matter, we consider in some detail the solution of a few more examples.

Example 8.10

An unbiased die is rolled n times. Determine the probability that at least one six is observed in the n trials. Hence prove that n must be 4 if this probability is to be approximately $1/2$.

Also, prove that with two unbiased dice approximately 25 trials are required to ensure that the probability of observing at least one double-six is $1/2$.

With one die, it is theoretically possible to obtain $0, 1, 2, ..., n$ sixes in n trials. If $P(r)$ is the probability of observing exactly r sixes in n trials, then

$$\sum_{r=0}^{n} P(r) = 1.$$

Hence the required probability is

$$\sum_{r=1}^{n} P(r) = 1 - P(0).$$

It is thus simpler to determine the probability of the complementary event. In any one trial there are six possible outcomes, and so there are 6^n sample points in the n trials. Each sample point has n elements corresponding to the points that could possibly be obtained in n trials. Of these 6^n sample points, there are 5^n sample points which contain no six corresponding to any trial. Hence

$$P(0) = \left(\frac{5}{6}\right)^n,$$

and so the probability for at least one six in n trials is

$$1 - \left(\frac{5}{6}\right)^n.$$

If this is to be equal to $1/2$, then the equation for n is

$$1 - \left(\frac{5}{6}\right)^n = \frac{1}{2},$$

or
$$n = \frac{\log 2}{\log 6 - \log 5} = 3 \cdot 8.$$

But since n is necessarily an integer, the appropriate value is 4. However, with $n = 4$, the probability for at least one six would be a little greater than 1/2.

For two dice, a similar argument shows that the probability for at least one double-six in n trials is

$$1 - \left(\frac{35}{36}\right)^n.$$

Hence the equation for n is

$$1 - \left(\frac{35}{36}\right)^n = \frac{1}{2},$$

or

$$n = \frac{\log 2}{\log 36 - \log 35} = 24 \cdot 7.$$

For integral n, the least value of n is 25, which necessarily gives a probability greater than 1/2.

This problem is the historic one of the Chevalier de Méré, a passionate French gambler. He had observed that it seemed safer to bet on a six in four throws of a die than to bet on a double-six in 24 throws of two dice. He posed the problem to Pascal, and thus unwittingly initiated the correspondence between Pascal and his friends which led to the beginnings of probability theory.

Example 8.11

Suppose that it is equally likely for a person's birthday to fall on any one of the 365 days (excluding leap years) of the year. Find the probability that of a group of n persons, all have their birthdays on different days. Hence deduce that if $n = 23$ then the probability is approximately 1/2 that at least two persons of the group will have their birthdays on the same day.

The total number of sample points is 365^n, and $365^{(n)}$ of these sample points correspond to the event that all n persons have birthdays on different days of the year. Hence the probability that at least one pair will have their birthdays on the same day is

$$1 - \frac{365^{(n)}}{365^n}.$$

This probability will be 1/2 if n satisfies the equation

$$\frac{365^{(n)}}{365^n} = \frac{1}{2},$$

or

$$\sum_{r=1}^{n-1} \log\left(1 - \frac{r}{365}\right) = -\log 2.$$

Expanding the logarithms and summing over r, we have

$$\frac{n(n-1)}{730} = 0 \cdot 693 \text{ approximately,}$$

or

$$n^2 - n = 505 \cdot 89,$$

or

$$n^2 - n - 506 = 0 \text{ approximately,}$$

or $$(n + 22) (n - 23) = 0.$$

Therefore approximately $n = 23$. This surprising result is due to H. Davenport.

Example 8.12

An experiment consists of rolling two unbiased dice once and noting the total of points obtained. If this experiment is performed twice, find the probability that the total Y on the second occasion is less than the total X on the first. From this derive also the respective probabilities that

$$Y \leqslant X, \quad Y = X, \quad Y \geqslant X, \quad Y > X.$$

The two totals are independent, and $P(X = r) = P(Y = r)$ for $r = 2, 3,$ ..., 12. Therefore

$$P(Y < X) = \sum_{r=2}^{12} P(X = r) P(Y < X) = \sum_{r=2}^{12} P(X = r) P(X < r);$$

$$P(X < Y) = P(Y < X);$$
$$P(Y \leqslant X) = P(Y \geqslant X) = 1 - P(Y < X);$$
$$P(Y = X) = P(Y \leqslant X) - P(Y < X).$$

Direct enumeration gives the following sets of values:

r	$P(X = r)$	$P(X < r)$
2	1/36	0
3	2/36	1/36
4	3/36	3/36
5	4/36	6/36
6	5/36	10/36
7	6/36	15/36
8	5/36	21/36
9	4/36	26/36
10	3/36	30/36
11	2/36	33/36
12	1/36	35/36

Hence $\quad P(Y < X) = \sum_{r=2}^{12} P(X = r) P(X < r) = \dfrac{575}{1296} = 0\cdot4437;$

$$P(Y \leqslant X) = 1 - P(Y < X) = \frac{721}{1296} = 0\cdot5563;$$

$$P(Y = X) = P(Y \leqslant X) - P(Y < X) = \frac{146}{1296} = 0\cdot1127;$$

$$P(Y \geqslant X) = P(Y \leqslant X) = 0\cdot5563;$$
$$P(Y > X) = P(Y < X) = 0\cdot4437.$$

Example 8.13

A set of n cards, of which one is a joker, is distributed randomly between two players A and B so that A has r (fixed) cards and B the rest. Find the probability that A receives the joker. Each of the players discards ν cards, where ν is an integer less than both r and $n-r$. If this discard is done randomly, determine the probability that A has the joker. How is this probability altered if the discard is not random in the sense that the player who has the joker retains it?

Let X be the event that A has the joker initially.

Then
$$P(X) = \binom{n-1}{r-1}\bigg/\binom{n}{r} = \frac{r}{n},$$

and $P(\overline{X}) = 1 - \frac{r}{n}$, where $\overline{X}$ is the event complementary to X.

Next, let Y be the event that A has the joker finally if the discard is done randomly. Now, since the sample space $\Omega = X + \overline{X}$, we have
$$Y = YX + Y\overline{X}.$$
Therefore
$$\begin{aligned} P(Y) &= P(YX) + P(Y\overline{X}) \\ &= P(Y|X)\,P(X) + P(Y|\overline{X})\,P(\overline{X}). \end{aligned}$$

If the discard is random, and A had the joker initially, then
$$P(Y|X) = \binom{r-1}{\nu}\bigg/\binom{r}{\nu} = \frac{r-\nu}{r}.$$
Also, clearly, $P(Y|\overline{X}) = 0,$

whence
$$P(Y) = \frac{r-\nu}{r} \times \frac{r}{n} = \frac{r-\nu}{n}.$$

Again, let Z be the event that A has the joker finally if the discard is not random.

Then $P(Z) = P(Z|X)\,P(X) + P(Z|\overline{X})\,P(\overline{X}),$

where now $P(Z|X) = 1$ and $P(Z|\overline{X}) = 0.$

Therefore $P(Z) = r/n.$

Example 8.14

According to a customer's specifications, screws of lengths between α and β are of acceptable quality. The production at the factory is so arranged that the probability of a screw of length less than α being produced is p_1 and that for a screw of length greater than β is p_2. If two screws are selected randomly and tested, find the probability that

(i)   both screws are of length less than α;

(ii) one screw is of length less than α and the other greater than β;

(iii) one screw is of acceptable length and the other is not;

(iv) at least one of the screws is of unacceptable length;

(v)   both screws are of unacceptable length; and

(vi) one screw is of length greater than α and the other is of length less than β.

We define three events as follows: A is the event that the length of a screw is less than α; B is the event that the length of a screw is greater than β; and C is the event that the screw is of acceptable length.

Then $P(A) + P(B) + P(C) = 1,$ so that $P(C) = 1 - p_1 - p_2.$

Hence
$$P(\text{i}) = P^2(A) = p_1^2;$$
$$P(\text{ii}) = 2P(A)\,P(B) = 2p_1 p_2,$$

since the first or the second screw may be less than α;

P(iii) $= 2P(C) [P(A) + P(B)] = 2(p_1 + p_2)(1 - p_1 - p_2),$

since the first or the second screw may be acceptable;

$$P\text{(iv)} = 1 - P^2(C)$$
$$= 1 - (1 - p_1 - p_2)^2 = (p_1 + p_2)(2 - p_1 - p_2);$$
$$P\text{(v)} = P^2(A) + P^2(B) + 2P(A) P(B)$$
$$= p_1^2 + p_2^2 + 2p_1 p_2 = (p_1 + p_2)^2;$$

P(vi) = Probability of one event being B or C, and the other event being A or C

$$= P^2(C) + 2P(A) P(C) + 2P(B) P(C) + 2P(A) P(B)$$
$$= 1 - P^2(A) - P^2(B)$$
$$= 1 - p_1^2 - p_2^2.$$

Example 8.15

In the game of poker, five cards are dealt out at random to a player from a standard pack of playing cards. Find the probability that the player has a trio, that is, exactly three cards of the same denomination. Also, find the probability that the player has a full house, that is, a trio and a pair of a different denomination. Hence deduce the probability that a player has a trio but not a full house.

A pack has 52 cards and there are four cards of each of the 13 denominations. The total number of sample points is $\binom{52}{5}$. To determine the number of sample points which correspond to a trio we argue as follows.

A denomination can be chosen in $\binom{13}{1}$ ways and the three cards out of four in $\binom{4}{3}$ ways. This leaves 48 cards from which two cards can be selected in $\binom{48}{2}$ ways. Hence the number of sample points containing a trio is

$$\binom{13}{1} \binom{4}{3} \binom{48}{2}.$$

These include sample points in which the two odd cards may be of the same denomination, thus giving a full house.

To obtain the number of sample points corresponding to a full house, the number of sample points for a trio and a pair is, by an extension of the above argument,

$$\binom{13}{1} \binom{4}{3} \binom{12}{1} \binom{4}{2}.$$

Hence the number of sample points giving only a trio is

$$\binom{13}{1} \binom{4}{3} \binom{48}{2} - \binom{13}{1} \binom{4}{3} \binom{12}{1} \binom{4}{2}.$$

Therefore the probability for a trio (the other two cards unrestricted out of 12 denominations) is

$$\binom{13}{1} \binom{4}{3} \binom{48}{2} \bigg/ \binom{52}{5} = \frac{94}{4165} = 0\cdot0226.$$

The probability for a full house is

$$\binom{13}{1}\binom{4}{3}\binom{12}{1}\binom{4}{2}\Big/\binom{52}{5} = \frac{6}{4165} = 0\cdot0014.$$

Hence the probability for only a trio (the other two cards unpaired) is

$$\frac{88}{4165} = 0\cdot0211.$$

Example 8.16

From a standard pack of 52 playing cards, a hand of 13 cards is dealt out randomly. Find the probability that the hand contains

   (i) exactly four aces;   (ii) no aces;

  (iii) at least one ace;   (iv) at least two aces;

   (v) exactly two aces;   (vi) exactly two specified aces; and

  (vii) at least two specified aces.

There are $\binom{52}{13}$ sample points for a random distribution of 13 cards. Of these, the sample points corresponding to the hand containing four aces are $\binom{48}{9}$. Therefore the probability of the hand containing four aces is

$$P(\text{i}) = \binom{48}{9}\Big/\binom{52}{13} = \frac{11}{4165} = 0\cdot0026.$$

The sample points corresponding to no aces in the hand are those in which 13 cards are obtained from the 48 non-aces, that is $\binom{48}{13}$ in number.

Therefore $P(\text{ii}) = \binom{48}{13}\Big/\binom{52}{13} = \frac{6327}{20825} = 0\cdot3038.$

The probability of at least one ace in the hand

$$= 1 - \text{Probability of no aces,}$$

or $P(\text{iii}) = 1 - P(\text{ii}) = \frac{14498}{20825} = 0\cdot6962.$

The probability of at least two aces in the hand

$$= \text{Probability of at least one ace in the hand}$$
$$- \text{Probability of exactly one ace in the hand.}$$

or $P(\text{iv}) = P(\text{iii}) - \text{Probability of exactly one ace in the hand.}$

The sample points corresponding to the hands containing just one ace are those which contain one of the four aces and 12 cards out of the remaining 48 non-ace cards. Therefore the probability of exactly one ace in the hand of 13 cards is

$$\binom{4}{1}\binom{48}{12}\Big/\binom{52}{13} = \frac{9139}{20825}.$$

Hence $P(\text{iv}) = \frac{14498}{20825} - \frac{9139}{20825} = \frac{5359}{20825} = 0\cdot2573.$

The sample points corresponding to the hand containing exactly two aces are those which contain two out of the four aces and 11 cards out of the other 48 cards.

Therefore $\quad P(v) = \binom{4}{2}\binom{48}{11}\Big/\binom{52}{13} = \dfrac{741}{3400} = 0\cdot2179.$

If the hand contains exactly two specified aces, then the number of sample points corresponds to the selection of 11 cards out of 48 non-aces.

Therefore $\quad P(vi) = \binom{48}{11}\Big/\binom{52}{13} = \dfrac{741}{20825} = 0\cdot0356.$

Finally, if the hand contains at least two specified aces, then the number of sample points is simply due to the selection of 11 cards from the other 50 cards.

Therefore $\quad P(vii) = \binom{50}{11}\Big/\binom{52}{13} = \dfrac{1}{17} = 0\cdot0588.$

Example 8.17

In a game of chance, four and five cards respectively are dealt out randomly from a standard pack of playing cards to two players A and B. A wins the game if

(i) A and B have no hearts; or

(ii) A has one more heart than B.

Determine the probability of A winning the game. Hence verify that the probability of A winning two out of three such games played is approximately $0\cdot1444$.

Let E_i be the event that A has i hearts, for $0 \leqslant i \leqslant 4$, and let X be the event that A wins the game.

Then $\qquad\qquad\qquad\qquad \displaystyle\sum_{i=0}^{4} E_i = \Omega,$

the sample space, and

$$X = \sum_{i=0}^{4} XE_i.$$

Therefore

$$P(X) = \sum_{i=0}^{4} P(XE_i) = \sum_{i=0}^{4} P(X|E_i)\, P(E_i).$$

But $\qquad\qquad P(E_i) = \binom{13}{i}\binom{39}{4-i}\Big/\binom{52}{4},$

$$P(X|E_0) = \binom{35}{5}\Big/\binom{48}{5},$$

and $\qquad\quad P(X|E_j) = \binom{13-j}{j-1}\binom{35+j}{6-j}\Big/\binom{48}{5} \quad \text{for } 1 \leqslant j \leqslant 4.$

Hence
$$P(X) = \left[\binom{39}{4}\binom{35}{5} + \sum_{j=1}^{4}\binom{13}{j}\binom{39}{4-j}\binom{13-j}{j-1}\binom{35+j}{6-j}\right]\Big/\binom{52}{4}\binom{48}{5}$$

$$= (26701306632 + 44789288544 + 41989958010 + 4234281480 +$$
$$+ 44504460)/463563500400$$

$$= 117759339126/463563500400 = 0\cdot2540.$$

Therefore the probability of A winning two out of three such games played is

$$3(0 \cdot 2540)^2 \times 0 \cdot 7460 = 0 \cdot 1444,$$

which is correct to four decimal places.

The factor 3 occurs because A could lose any one of the three games played.

Example 8.18

There are six unbiased dice $D_1, D_2, \ldots, D_6$, the six faces of D_i being numbered $(i, i, i, 4, 5, 6)$ or $(i, i, i, 1, 2, 3)$ according as $i \leqslant 3$ or $i > 3$. If one of the dice is selected randomly and rolled twice, find the probability that a double-six is obtained.

Further, given that a double-six has been observed, prove that the probability of the selected die being D_6 is 3/4. How is this probability altered if it is known that D_3 is a biased die such that, though it is equally likely to show a three or a number greater than three, the probability of obtaining a six with it is twice that of obtaining a four or a five?

Let X be the event that the two successive throws of the die selected show a double-six.

Then

$$P(X) = \sum_{i=1}^{6} P(X|D_i) \, P(D_i), \tag{1}$$

where $P(D_i) = \dfrac{1}{6}$ is the probability of selecting the ith die. Now

$$P(X|D_i) = (1/6)^2, \text{ for } i = 1, 2, 3;$$
$$= 0, \qquad \text{ for } i = 4, 5;$$

and

$$= (1/2)^2, \text{ for } i = 6.$$

Therefore

$$P(X) = \frac{1}{6}\left[3 \times \frac{1}{36} + \frac{1}{4}\right] = \frac{1}{18}.$$

Again,

$$P(XD_6) = P(X|D_6) \, P(D_6) = P(D_6|X) \, P(X),$$

so that

$$P(D_6|X) = P(X|D_6) \, P(D_6)/P(X) = \frac{1}{4} \times \frac{1}{6} \Big/ \frac{1}{18} = \frac{3}{4} = 0 \cdot 7500.$$

If D_3 is a biased die, and if x is the probability of obtaining a four or a five with it, then the probability of a six is $2x$. Therefore

$$4x = 1/2, \quad \text{or} \quad x = 1/8; \quad \text{and} \quad P(X|D_3) = (1/4)^2.$$

Hence, using (1),

$$P(X) = \frac{1}{6}\left[2 \times \frac{1}{36} + \frac{1}{16} + \frac{1}{4}\right] = \frac{53}{864} = 0 \cdot 0613.$$

Therefore

$$P(D_6|X) = \frac{1}{4} \times \frac{1}{6} \times \frac{864}{53} = \frac{36}{53} = 0 \cdot 6792.$$

Example 8.19

In a large population, the probability of a person suffering from tuberculosis is α, and the usual method of detecting infected persons is by radiographic examination. However, this method is not completely reliable as there is a small but not negligible probability that an examination will not lead to the correct diagnosis of a person. It may therefore be assumed that the probability of a radiographic examination giving the correct result is β, where $1-\beta > 0$ is small. A randomly selected person had three examinations and of these one showed him to be positive and the other two negative (non-tubercular). Find the probability that the person is, in fact, tubercular.

Let A be the event that the person is tubercular and B the event that of three radiographic examinations one shows him to be positive. Therefore $P(B) = P(B|A) P(A) + P(B|\bar{A}) P(\bar{A})$, where $\bar{A}$ is the event complementary to A;

$$P(A) = a; \quad P(\bar{A}) = 1 - a;$$
$$P(B|A) = 3\beta(1-\beta)^2; \quad P(B|\bar{A}) = 3\beta^2(1-\beta).$$

Hence $P(B) = 3\beta(1-\beta)^2 a + 3\beta^2(1-\beta)(1-a)$
$$= 3\beta(1-\beta)[a(1-\beta) + \beta(1-a)].$$

Now $P(AB) = P(A|B) P(B) = P(B|A) P(A)$,
so that
$$P(A|B) = P(B|A) P(A)/P(B) = a(1-\beta)/[a(1-\beta) + \beta(1-a)],$$

which is the required probability for the person to be, in fact, tubercular.

Example 8.20

A number X is selected at random from the first n integers, and then another integer Y is chosen randomly from the integers 1 to X. Determine the probability that $Y = 1$. Also, if it is known that $Y = 1$, evaluate the probability that $X = k$, where $1 \leqslant k \leqslant n$. Hence show that, for large n, this probability is approximately $(k \log n)^{-1}$.

Verify that for $n = 10$,
$$P(Y = 1) = 7381/25200 = 0\cdot 2929.$$
and
$$P(X = 8|Y = 1) = 315/7381 = 0\cdot 0427.$$

Let A_r be the event that $X = r$, so that $P(A_r) = 1/n$, and
$$\sum_{r=1}^{n} A_r = \Omega, \text{ the sample space.}$$

If B is the event that $Y = 1$, then
$$P(B) = \sum_{r=1}^{n} P(BA_r) = \sum_{r=1}^{n} P(B|A_r) P(A_r).$$

Now $P(B|A_r) = \dfrac{1}{r}$, so that $P(B) = \dfrac{1}{n}\sum_{r=1}^{n} \dfrac{1}{r}$.

Hence $P(A_k|B) = P(B|A_k) P(A_k)/P(B)$
$$= \frac{(1/k)(1/n)}{(1/n)\sum_{r=1}^{n}(1/r)} = \left[k\sum_{r=1}^{n}(1/r)\right]^{-1} \sim (k \log n)^{-1}$$

for large n, which is the required expression for $P(X = k|Y = 1)$.

For $n = 10$,
$$P(Y = 1) = P(B) = \frac{1}{10}\left[1 + \frac{1}{2} + \frac{1}{3} + \frac{1}{4} + \frac{1}{5} + \frac{1}{6} + \frac{1}{7} + \frac{1}{8} + \frac{1}{9} + \frac{1}{10}\right]$$

$$= \frac{7381}{25200} = 0 \cdot 2929.$$

Finally,
$$
\begin{aligned}
P(X = 8 \mid Y = 1) &= P(A_8 \mid B) \\
&= P(B \mid A_8)\, P(A_8)/P(B) \\
&= \frac{1}{8} \times \frac{1}{10} \times \frac{25200}{7381} = \frac{315}{7381} = 0 \cdot 0427.
\end{aligned}
$$

Example 8.21

In a deal of cards at bridge, North and South hold in all nine trumps. Find the probabilities of the possible distributions of the remaining four trumps between East and West.

Suppose North holds in his own hand both the ace and king of trumps, but the queen is with the opponents. Find the probability that if the ace and king of trumps are played in succession then

   (i) the queen of trumps will fall; and

   (ii) North will exhaust the opponents' trumps.

Hence determine the probability that the queen of trumps will not fall and North will not exhaust the opponents' trumps after the ace and king of trumps have been played.

The possible distributions of the four trumps between East (E) and West (W) are:

$(E = 4, W = 0)$, $(E = 3, W = 1)$, $(E = 2, W = 2)$, $(E = 1, W = 3)$, $(E = 0, W = 4)$,

where the notation $(E = a, W = b)$ denotes that East and West have a and b trumps respectively.

There are in all $\binom{26}{13}$ sample points corresponding to the distribution of the 26 cards between E and W.

Hence
$$
\begin{aligned}
P(E = 4, W = 0) &= P(E = 0, W = 4) \\
&= \binom{4}{4}\binom{22}{9} \Big/ \binom{26}{13} = \frac{11}{230}.
\end{aligned}
$$

Similarly,
$$
P(E = 2, W = 2) = \binom{4}{2}\binom{22}{11} \Big/ \binom{26}{13} = \frac{234}{575},
$$

and
$$
\begin{aligned}
P(E = 1, W = 3) &= P(E = 3, W = 1) \\
&= \binom{4}{1}\binom{22}{12} \Big/ \binom{26}{13} = \frac{143}{575}.
\end{aligned}
$$

Now the queen will certainly fall if the distribution of trumps between E and W is $(E = 2, W = 2)$. It will also fall if either E or W holds it as a singleton trump. Denoting the queen of trumps by Q, we have
$$
\begin{aligned}
P(E = Q, W = 3) &= P(E = 3, W = Q) \\
&= \binom{1}{1}\binom{22}{12} \Big/ \binom{26}{13} = \frac{143}{2300}.
\end{aligned}
$$

Hence the probability that the queen of trumps will fall
$$
= \frac{234}{575} + 2\frac{143}{2300} = \frac{1222}{2300} = 0 \cdot 5313.
$$

On the other hand, North will exhaust the opponents' trumps only if $(E = 2, W = 2)$, so that the probability of this is simply $234/575 = 0\cdot4070$. Finally, if A is the event that the queen will fall, and B the event that the opponents' trumps are exhausted, then

$$P(\bar{A}\,\bar{B}) = P[\Omega - (A + B)], \quad \bar{A}, \bar{B} \text{ being events complementary to } A, B$$

respectively and Ω the sample space,

$$= 1 - P(A + B)$$
$$= 1 - [P(A) + P(B) - P(AB)].$$

But B implies A, and so $P(B) = P(AB)$.

Hence $P(\bar{A}\,\bar{B}) = 1 - P(A) = 1 - 1222/2300 = 1078/2300 = 0\cdot4687$. This result is, of course, otherwise obvious.

Example 8.22

From a standard pack (deck) of 52 playing cards, a hand of 13 cards is dealt out at random to a bridge player. Find the probability that, of the four aces in the pack, the player has

  (i) only one ace;   (ii) at least one ace;
  (iii) the ace of hearts;   (iv) only the ace of hearts.
  (v) Suppose the player declares that he has an ace. Find the probability that the player has
  (a) another ace;   (b) exactly one other ace.

How are these probabilities altered if, alternatively, the player initially declares that he has the ace of hearts?

  (i) There are $\binom{52}{13}$ sample points representing all the possible hands

of 13 cards. Of these the number containing only one ace is $\binom{4}{1}\binom{48}{12}$, so that

$$P(\text{i}) = \binom{4}{1}\binom{48}{12} \bigg/ \binom{52}{13} = \frac{9139}{20825} = 0\cdot4388.$$

  (ii) Again, $P(\text{ii})$, the probability of the player having at least one ace

$$= 1 - \text{Probability of the player having no aces}$$
$$= 1 - \binom{48}{13} \bigg/ \binom{52}{13} = \frac{14498}{20825} = 0\cdot6962.$$

  (iii) For the probability of the player having the ace of hearts, we need only consider the sample points corresponding to the selection of 12 cards out of the remaining 51 cards.

Hence $$P(\text{iii}) = \binom{51}{12} \bigg/ \binom{52}{13} = \frac{1}{4} = 0\cdot2500.$$

  (iv) The sample points corresponding to the player having only the ace of hearts are $\binom{48}{12}$ in number, so that

$$P(\text{iv}) = \binom{48}{12} \bigg/ \binom{52}{13} = \frac{9139}{83300} = 0\cdot1097.$$

(v) (a) Let A be the event that the player has an ace and B the event that he has another ace, that is, at least another ace. The required probability is

$$P(B|A) \; = \; P(AB)/P(A).$$

Now $P(A)$ = Probability of the player having at least an ace

$$= \; 1 - \binom{48}{13} \Big/ \binom{52}{13},$$

and $P(AB)$ = Probability of two or more aces

= 1 − Probability of no aces − Probability of **exactly one ace**

$$= \; 1 - \left[\binom{48}{13} + \binom{4}{1} \binom{48}{12} \right] \Big/ \binom{52}{13}.$$

Hence

$$P(B|A) \; = \; \left[\binom{52}{13} - \binom{48}{13} - 4\binom{48}{12} \right] \Big/ \left[\binom{52}{13} - \binom{48}{13} \right]$$

$$= \; 5359/14498 \; = \; 0 \cdot 3696.$$

(v) (b) Let B' be the event that the player has only one more ace. The required probability is $P(B'|A) = P(AB')/P(A)$, where

$P(AB')$ = Probability of the player having exactly two aces

$$= \; \binom{4}{2} \binom{48}{11} \Big/ \binom{52}{13}.$$

Hence

$$P(B'|A) \; = \; 6\binom{48}{11} \Big/ \left[\binom{52}{13} - \binom{48}{13} \right] \; = \; \frac{4446}{14498} \; = \; 0 \cdot 3067.$$

Alternatively, suppose C is the event that the player has the ace of hearts and D the event that he has another ace. The required probability is

$$P(D|C) \; = \; P(CD)/P(C), \quad \text{where} \quad P(C) \; = \; \binom{51}{12} \Big/ \binom{52}{13},$$

and $\quad P(CD)$ = Probability of the player having the ace of hearts and one or more of the other three aces

$$= \; \left[\binom{51}{12} - \binom{48}{12} \right] \Big/ \binom{52}{13}.$$

Therefore

$$P(D|C) \; = \; 1 - \binom{48}{12} \Big/ \binom{51}{12} \; = \; \frac{11686}{20825} \; = \; 0 \cdot 5612.$$

Finally, let D' be the event that the player has only one ace other than the ace of hearts. The required probability is

$$P(D'|C) \; = \; P(CD')/P(C),$$

where $P(CD')$ = Probability of the player having the ace of hearts and one of the other three aces

$$= \; \binom{1}{1} \binom{3}{1} \binom{48}{11} \Big/ \binom{52}{13}.$$

Therefore

$$P(D'|C) \; = \; 3\binom{48}{11} \Big/ \binom{51}{12} \; = \; \frac{8892}{20825} \; = \; 0 \cdot 4270.$$

CHAPTER 9

DISCRETE PROBABILITY DISTRIBUTIONS

9.1 Discrete random variables

In the study of discrete probability in the preceding chapter, we have implicitly introduced the idea of a discrete random variable. For example, in a random distribution of a bridge hand from a standard pack of 52 cards, the hand could include 0,1,2,3 or 4 aces. There are in all $\binom{52}{13}$ sample points corresponding to the possible selections of the 13 cards constituting a bridge hand. Of these there are

$$\binom{48}{13}, \qquad \binom{4}{1}\binom{48}{12}, \qquad \binom{4}{2}\binom{48}{11}, \qquad \binom{4}{3}\binom{48}{10}, \qquad \binom{4}{4}\binom{48}{9}$$

sample points corresponding to hands containing exactly

$$0, \qquad\qquad 1, \qquad\qquad 2, \qquad\qquad 3, \qquad\qquad 4$$

aces respectively.

If A_ν is the event that the hand contains exactly ν aces, then for $0 \leqslant \nu \leqslant 4$ we simply have

$$P(A_0) = \binom{48}{13} \Big/ \binom{52}{13} = \frac{6327}{20825} = 0\cdot3038;$$

$$P(A_1) = \binom{4}{1}\binom{48}{12} \Big/ \binom{52}{13} = \frac{9139}{20825} = 0\cdot4388;$$

$$P(A_2) = \binom{4}{2}\binom{48}{11} \Big/ \binom{52}{13} = \frac{4446}{20825} = 0\cdot2135;$$

$$P(A_3) = \binom{4}{3}\binom{48}{10} \Big/ \binom{52}{13} = \frac{858}{20825} = 0\cdot0412;$$

$$P(A_4) = \binom{4}{4}\binom{48}{9} \Big/ \binom{52}{13} = \frac{55}{20825} = 0\cdot0026.$$

The sample points corresponding to the five events A_ν are distinct since the events are mutually exclusive. Besides, since the A_ν form an exhaustive set of events,

$$\sum_{\nu=0}^{4} P(A_\nu) \equiv 1.$$

In other words, the events A_ν represent a unique partitioning of the sample space into five distinct subspaces such that all the points in any one sub-space are associated with the realisation of one and only one of the A_ν.

It is possible to look at this partitioning of the sample space in a slightly different way. Thus, suppose x represents the number of aces in the bridge hand, so that x can take any one of the integral values 0,1,2,3,4. It is clear that

$$P(x = \nu) \equiv P(A_\nu) = \binom{4}{\nu} \binom{48}{13-\nu} \Big/ \binom{52}{13}, \quad \text{for } 0 \leqslant \nu \leqslant 4.$$

In fact, these equations define the probability distribution of a discrete random variable which can take any one of the five non-negative integral values 0,1,2,3,4. Such a probability distribution is the discrete analogue of the probability distribution of a continuous random variable.

More formally, we specify a discrete random variable x as a *function* defined over a sample space. The argument of x is the random experiment, and the value of x depends upon the chance outcome of the experiment. This definition of a discrete random variable implies a unique partitioning of the points of the sample space so that corresponding to each point one value of the random variable can be assigned unambiguously. However, it is not essential to the use of discrete probability distributions in statistical theory that the underlying experimental basis and the associated sample space be brought out explicitly, though this background is always present implicitly. We also observe that in the above example the sample space contains a very large number of sample points since

$$\binom{52}{13} \sim 63{\cdot}5 \times 10^{10},$$

and each of these sample points represents a possible hand of 13 cards. In contrast, the possible values of the random variable x (the number of aces in the hand) are only five in number. In principle, it is also possible to define a discrete random variable having any finite number n of values by a suitable partitioning of the sample space. Furthermore, a limiting process then leads us to define a discrete probability distribution in which the random variable x can take any one of the non-negative integral values. Then, in general, the set of equations

$$P(x = \nu) = p_\nu, \quad \text{for } 0 \leqslant \nu < \infty,$$

where $p_\nu \geqslant 0$ and $\sum_{\nu=0}^{\infty} p_\nu \equiv 1$, defines the probability distribution of the random variable x. The numbers p_ν are known as the *point-probabilities* corresponding to the values x can take. We shall use this definition of the probability distribution of a discrete random variable with the understanding that in particular cases the upper limit of variation of x may be finite.

9.2   Expectations

The process of evaluating the expectation of a function $g(x)$ of x is carried out simply as a summation instead of an integration as in the case of continuous random variables. We thus have

$$E[g(x)] = \sum_{\nu=0}^{\infty} g(\nu) p_\nu,$$

provided the series converges. Hence, in particular,

$$E(x) = \sum_{\nu=0}^{\infty} \nu p_\nu ,$$

and $\quad \mathrm{var}(x) = \sum_{\nu=0}^{\infty} [\nu - E(x)]^2 p_\nu = \sum_{\nu=0}^{\infty} \nu^2 p_\nu - E^2(x) = E(x^2) - E^2(x).$

In the same way, we can define the higher moments of x, and thus also the coefficients of skewness and kurtosis of the distribution of x. All the moments may also be obtained, as before, by evaluating the moment-generating function of x about the origin

$$M_0(t) \equiv E(e^{tx}) = \sum_{\nu=0}^{\infty} e^{t\nu} p_\nu .$$

9.3   Probability of an event

It would thus appear that, in principle, there is no difference in the evaluation of generating functions associated with discrete and continuous random variables, provided the processes of summation of series and integration are properly taken into account. It is therefore possible to study the main properties of discrete probability distributions in much the same way as is usual for probability models of continuous variation. There is, however, one fundamental difference between the probability distributions of discrete and continuous random variables. A discrete random variable can take distinct values (such as the non-negative integers) with non-zero probabilities. On the other hand, a continuous random variable can, in theory, take all the possible real number values in any given finite or infinite interval over which the probability distribution is defined. This fact leads to a simple but important result.

Suppose y is a continuous random variable with the distribution function $F(y)$. If $\alpha < \beta$ are two values within the range of variation of y, then the event $y \leqslant \beta$ can be decomposed into two mutually exclusive events $y \leqslant \alpha$ and $\alpha < y \leqslant \beta$. Hence the probability

$$P(y \leqslant \beta) = P(y \leqslant \alpha) + P(\alpha < y \leqslant \beta),$$

whence $\quad\quad P(\alpha < y \leqslant \beta) = P(y \leqslant \beta) - P(y \leqslant \alpha)$

or, in terms of the distribution function of y,

$$P(\alpha < y \leqslant \beta) = F(\beta) - F(\alpha).$$

In particular, if η is any possible value of y and $\epsilon > 0$ is arbitrarily small,

$$P(\eta - \epsilon < y \leqslant \eta + \epsilon) = F(\eta + \epsilon) - F(\eta - \epsilon). \tag{1}$$

In the limit, as $\epsilon \to 0$, the interval $\eta - \epsilon < y \leqslant \eta + \epsilon$ tends to the single point $y = \eta$. Therefore the left-hand side of (1) will tend to $P(y = \eta)$, while if $F(y)$ is assumed to be continuous the right-hand side will tend to zero. Hence

$$P(y = \eta) = 0, \quad \text{for all } \eta.$$

This surprising result means that, in contrast to the point-probabilities of a discrete random variable, the probability that a continuous random variable y with a continuous distribution function $F(y)$ will take any preassigned value is "zero". This does not mean that $y = \eta$ is an impossible event but that, given the infinity of values that y could take in a given range, it is exceedingly unlikely to take any specified value η.

We can look at this conclusion more concretely in the light of the earlier example pertaining to the number of aces in a hand of 13 cards. The possible number of different hands is, as we have seen, $63 \cdot 5 \times 10^{10}$, a number effectively infinite for all practical purposes. Therefore the probability of obtaining a hand containing any specified 13 cards, say all spades, is *almost* zero. Nevertheless, it is true that on each deal *some* one of the possible hands will be realised, and it could well be a hand of 13 spades. The effective impossibility refers to a realisation of a hand of 13 cards which has been *initially* stipulated.

9.4   Distribution function of a discrete random variable

The distribution function of a discrete random variable x is obtained by the cumulative sum of the point-probabilities. Thus the distribution function is

$$P(x \leqslant k) = \sum_{\nu=0}^{k} p_{\nu}, \quad \text{for } 0 \leqslant k \leqslant \infty.$$

This is a step-function with finite jumps at each of the integral values of x for which $p_{\nu} > 0$. A typical form of such a function is similar to that of the cumulative distribution based on an observed frequency distribution if the ordinates are taken to represent cumulative point-probabilities instead of cumulative frequencies.

We now turn to a consideration of some of the important discrete probability distributions which are of use in statistical theory.

9.5   The uniform distribution

This is the simplest of discrete probability distributions and it corresponds to the uniform distribution of a continuous random variable. Formally, a random variable x has the discrete uniform distribution if the point-probabilities are

$$P(x = \nu) = 1/(n+1), \quad \text{for } \nu = 0, 1, 2, ..., n,$$

where n is a positive integer.

Cumulants

Therefore the moment-generating function of x about the origin is

$$M_0(t) \equiv E(e^{tx}) = \sum_{\nu=0}^{n} e^{t\nu/(n+1)}$$

$$= \frac{e^{(n+1)t} - 1}{(n+1)(e^t - 1)}$$

$$= \frac{e^{(n+1)t/2}[e^{(n+1)t/2} - e^{-(n+1)t/2}]}{(n+1)e^{t/2}(e^{t/2} - e^{-t/2})}$$

$$= \frac{e^{nt/2}\sinh(n+1)t/2}{(n+1)\sinh t/2}$$

$$= e^{nt/2} \frac{\sinh(n+1)t/2}{(n+1)t/2} \bigg/ \frac{\sinh t/2}{t/2}.$$

Hence the cumulant-generating function of x is

$$\kappa(t) \equiv \log M_o(t) = \frac{1}{2}nt + \log\left[\frac{\sinh(n+1)t/2}{(n+1)t/2}\right] - \log\left[\frac{\sinh t/2}{t/2}\right]$$

$$= \frac{1}{2}nt + \sum_{j=2}^{\infty} B_j \frac{[(n+1)t]^j}{jj!} - \sum_{j=2}^{\infty} B_j \frac{t^j}{jj!},$$

where B_j are the Bernoulli numbers,

$$= \frac{1}{2}nt + \sum_{j=2}^{\infty} B_j \frac{[(n+1)^j - 1]t^j}{jj!}$$

$$= \frac{1}{2}nt + \sum_{r=1}^{\infty} B_{2r} \frac{[(n+1)^{2r} - 1]t^{2r}}{2r(2r)!},$$

since $B_{2r+1} = 0$ for $r \geqslant 1$. Thus for the cumulants of x

$$\kappa_1 = \tfrac{1}{2}n; \quad \kappa_{2r+1} = 0; \quad \kappa_{2r} = [(n+1)^{2r} - 1]B_{2r}/2r, \quad \text{for } r \geqslant 1.$$

In particular,

$$\kappa_2 = \tfrac{1}{12}[(n+1)^2 - 1] = n(n+2)/12; \quad \kappa_3 = 0;$$

$$\kappa_4 = -\tfrac{1}{120}[(n+1)^4 - 1] = -n(n+2)(n^2+2n+2)/120.$$

Therefore the coefficients of skewness and kurtosis of the distribution of x are

$$\gamma_1 = 0;$$

$$\gamma_2 = -\frac{n(n+2)(n^2+2n+2)}{120} \times \frac{144}{n^2(n+2)^2} = -\frac{6}{5}\left[1 + \frac{2}{n(n+2)}\right].$$

9.6 The hypergeometric distribution

This distribution provides a basic approach to all the standard discrete probability distributions in the sense that other distributions can be derived from it by elementary and physically meaningful limiting processes. We have already come across a simple particular case of the hypergeometric distribution in the example regarding the number of aces in a random distribution of a bridge hand of 13 cards from a standard pack of playing cards. We shall now consider the general form of this distribution and some of its important properties. The distribution is best deduced by associating its derivation with a sampling experiment from a finite collection of two kinds of objects which are similar to each other except in one respect. Thus, for example, the objects may be likened to identical balls of two different colours.

Suppose, then, that the collection consists of N identical balls of which W are white and B black in colour ($W+B = N$). If with these balls we associate symbols $X_1, X_2, ..., X_N$ such that $X_i = 1$ if the ith ball in the collection is white, and $X_i = 0$ if the ith ball in the collection is black, then an alternative way of representing the collection is in terms of a collection of W 1's and B 0's. Furthermore, we recall from our previous definitions that if all the X_i are equally likely, then the rth moment about the origin of the X_i is simply

$$m_r' = \sum_{i=1}^{N} X_i^r/N, \quad \text{for } r \geqslant 0.$$

Now, although it is not certain whether any particular X_j is scored 1 or 0 since the colour of the jth ball in the collection is unknown, nevertheless it is definite that of all the X_i exactly W are scored 1 and the rest 0. Accordingly,

$$m'_r = (W.1^r + B.0^r)/N = W/N \equiv p, \quad \text{say.}$$

Hence, in particular,

$$m'_1 = p,$$

so that the mean represents the proportion of white balls in the collection. The second, third, and fourth central moments are readily obtained as

$$m_2 \equiv m'_2 - (m'_1)^2 = pq, \qquad \text{where } p+q = 1,$$
$$m_3 \equiv m'_3 - 3m'_2 m'_1 + 2(m'_1)^3 = pq(q-p),$$
$$m_4 \equiv m'_4 - 4m'_3 m'_1 + 6m'_2(m'_1)^2 - 3(m'_1)^4 = pq(1-3pq).$$

These are the moments of a finite population of W 1's and B 0's which correspond to the white and black balls in the collection. However, we can also look upon these moments as those of an infinite population of 1's and 0's in which the proportion of 1's is p. This infinite population is generated simply by repeated random selection of a ball from the collection. The random variable x represents the colour of the ball drawn, and $x = 1$ if the ball drawn is white and $x = 0$ if the ball turns out to be black. The ball drawn is replaced each time before the next drawing is made so that the probability of drawing a white ball $(x = 1)$ remains constant. If this experiment is carried out a large number of times, then a sequence of values of x, say $x_1, x_2, x_3, \ldots$, is generated in which each x_i represents the score (1 or 0) corresponding to whether the ball drawn on the ith occasion is white or black. The x_i are the possible realisations of the random variable x and, in their conceptual totality, constitute an infinite population. The essential point is that though the finite collection of W 1's and B 0's and the infinite population have the same moments, the latter is a *derived* population generated by a simple sampling procedure. Accordingly the random variable x of the infinite population is a *derived random variable*. To put it in a slightly different way, we may also state that the probability distribution of x is the *sampling distribution* of the colour of a ball drawn from a collection in which the proportion of white balls is p. In fact, with an unambiguous definition of the sampling experiment, it is possible to generate different kinds of sampling distributions. The one we have considered here is perhaps the simplest one possible.

We next extend the sampling procedure to define a more elaborate experiment in which a random sample of $n\,(1 \leqslant n < N)$ balls is taken from the collection. The sampling is done without replacement so that there are in all $\binom{N}{n}$ distinct samples, each of n balls, of which one could be drawn from the collection on any one occasion. However, if the sampling experiment is to be repeated then the n balls taken at a drawing have to be replaced before the next drawing is made. We are not at the moment concerned with a repetition of the experiment, but only with the constitution of the possible samples of n (fixed) balls which could be drawn. If we denote the white and black balls of the collection by the symbols $a_1, a_2, \ldots, a_W$ and $b_1, b_2, \ldots, b_B$, then the n balls drawn in the sample will be a group of some a_i and b_j. The particular interest of this sampling experiment lies in finding out the number of white balls obtained in the sample. Clearly, this is a discrete random variable w which could *formally* take any one of the possible integral values $0, 1, 2, \ldots, n$. If the sample contains w white and b black balls ($w + b = n$), then the number of sample points corresponding to this event is $\binom{W}{w}\binom{B}{b}$. Hence, assuming that the $\binom{N}{n}$ distinguishable samples are equally likely, the probability of a sample of n balls containing w white ones is

$$
\begin{aligned}
P(w) &= \binom{W}{w}\binom{B}{b} \Big/ \binom{N}{n}, \quad \text{for } 0 \leqslant w \leqslant n, \\
&= \binom{n}{w} \frac{W!\, B!\, (N-n)!}{(W-w)!\, (B-b)!\, N!} \\
&= \binom{n}{w} \frac{W^{(w)} B^{(b)}}{N^{(n)}},
\end{aligned}
$$

in standard factorial notation. It then follows immediately from Vandermonde's theorem that

$$
\sum_{w=0}^{n} P(w) = \sum_{w=0}^{n} \binom{W}{w}\binom{B}{b} \Big/ \binom{N}{n} = \sum_{w=0}^{n} \binom{n}{w} \frac{W^{(w)} B^{(b)}}{N^{(n)}} = 1.
$$

Therefore the probabilities $P(w)$ define a proper discrete probability distribution. The number of aces drawn in a bridge hand is an example of this distribution in which the aces are the white and the non-aces the black balls. Thus $N = 52$, $W = 4$, $B = 48$, $n = 13$, and w, the number of aces in the hand, has *formally* an integral value in the range $(0, n)$. However, since $W < n$, we have $P(w) \equiv 0$ for $w > W$. Accordingly, non-zero values for $P(w)$ are obtained only for $0 \leqslant w \leqslant W$.

The distribution of w obtained in the general case is known as the *hypergeometric distribution*. It is so called because its probabilities can be shown to be expressible in terms of the *hypergeometric function* in mathematical analysis. We shall not derive the moment-generating function of w, but we next show that the central moments of w can be obtained from its *factorial moments*.

Moments

For any integer $r \geqslant 0$, the rth-order factorial moment of w about the origin is

$$
\begin{aligned}
\mu'_{(r)} \equiv E[w^{(r)}] &= \sum_{w=0}^{n} w^{(r)} \binom{n}{w} \frac{W^{(w)} B^{(b)}}{N^{(n)}} \\
&= \sum_{w=r}^{n} \frac{n!\, W^{(w)} B^{(b)}}{(n-w)!\,(w-r)!\, N^{(n)}}
\end{aligned}
$$

$$= \sum_{w=r}^{n} \frac{n^{(r)}(n-r)!\, W^{(r)}(W-r)^{(w-r)}B^{(b)}}{(n-w)!\,(w-r)!\,N^{(r)}(N-r)^{(n-r)}}$$

$$= \frac{n^{(r)}W^{(r)}}{N^{(r)}} \sum_{w=r}^{n} \binom{n-r}{w-r}\frac{(W-r)^{(w-r)}B^{(b)}}{(N-r)^{(n-r)}}.$$

If we now put $w-r = c$, $n-r = \nu$, $N-r = M$, and $W-r = C$, then

$$\mu'_{(r)} = \frac{n^{(r)}W^{(r)}}{N^{(r)}} \sum_{c=0}^{\nu} \binom{\nu}{c}\frac{C^{(c)}B^{(b)}}{M^{(\nu)}},$$

where $B+C = M$, $b+c = \nu$, whence, by Vandermonde's theorem, the summation is unity and we obtain

$$\mu'_{(r)} \equiv E[w^{(r)}] = n^{(r)}W^{(r)}/N^{(r)}.$$

This is an important result, and it can be used to obtain the central moments of w. Thus, for $r = 1$,

$$\mu'_{(1)} \equiv E(w) = nW/N = np,$$

so that the expected number of white balls in the sample of n balls is np. Again,

$$\begin{aligned}
\mathrm{var}(w) &= E(w^2) - E^2(w) \\
&= E[w^{(2)} + w] - E^2(w) \\
&= E[w^{(2)}] + E(w) - E^2(w) \\
&= \frac{n^{(2)}W^{(2)}}{N^{(2)}} + np - n^2p^2,
\end{aligned}$$

whence, on some simplification,

$$\mathrm{var}(w) = npq(N-n)/(N-1).$$

Similarly, the third central moment of w is

$$\begin{aligned}
\mu_3(w) &= E(w-np)^3 \\
&= E[w^3 - 3npw^2 + 3n^2p^2w - n^3p^3] \\
&= E[w^{(3)} + 3w^{(2)} + w - 3np\{w^{(2)}+w\} + 3n^2p^2w - n^3p^3] \\
&= E[w^{(3)} + 3(1-np)w^{(2)} + (1-3np+3n^2p^2)w - n^3p^3] \\
&= \frac{n^{(3)}(Np)^{(3)}}{N^{(3)}} + 3(1-np)\frac{n^{(2)}(Np)^{(2)}}{N^{(2)}} + (1-3np+3n^2p^2)np - n^3p^3,
\end{aligned}$$

which, on some reduction, gives

$$\mu_3(w) = npq(N-n)(N-2n)/(N-1)^{(2)}.$$

Finally, the fourth central moment of w is

$$\begin{aligned}
\mu_4(w) &= E(w-np)^4 \\
&= E[w^4 - 4npw^3 + 6n^2p^2w^2 - 4n^3p^3w + n^4p^4] \\
&= E[\{w^{(4)} + 6w^{(3)} + 7w^{(2)} + w\} - 4np\{w^{(3)} + 3w^{(2)} + w\} + \\
&\qquad\qquad + 6n^2p^2\{w^{(2)} + w\} - 4n^3p^3w + n^4p^4] \\
&= E[w^{(4)} + (6 - 4np)w^{(3)} + (7 - 12np + 6n^2p^2)w^{(2)} + \\
&\qquad\qquad + (1 - 4np + 6n^2p^2 - 4n^3p^3)w + n^4p^4]
\end{aligned}$$

$$= \frac{n^{(4)}(Np)^{(4)}}{N^{(4)}} + (6-4np)\frac{n^{(3)}(Np)^{(3)}}{N^{(3)}} + (7-12np+6n^2p^2) \times$$

$$\times \frac{n^{(2)}(Np)^{(2)}}{N^{(2)}} + np(1-4np+6n^2p^2-3n^3p^3)$$

which, on reduction, gives

$$\mu_4(w) = \frac{n(N-n)pq}{(N-1)^{(3)}} [N(N+1) - 6n(N-n) + 3pq\{(n-2)N^2 - n(n-6)N - 6n^2\}].$$

As a check, we observe that for $n = 1$ the moments of w reduce to the moments of x, the random variable denoting the colour of a single ball drawn from the collection. Hence the distribution of w is an appropriate generalisation when n balls are drawn at a time from the collection.

Sampling distributions

We have related the derivation of the hypergeometric distribution to a sampling experiment from a finite collection of balls of two colours. This gives the derived distribution a direct physical significance since the balls of the model can be identified with different kinds of objects in meaningful statistical problems. However, we could quite legitimately define the hypergeometric distribution as the representation of an infinite population of a discrete random variable w such that

$$P(w) = \binom{n}{w}\frac{W^{(w)}B^{(b)}}{N^{(n)}}, \quad \text{for } 0 \leqslant w \leqslant n,$$

where the symbols satisfy the earlier relations of magnitude. Indeed, it was in this formal way that we introduced the probability distributions of continuous random variables. The interest of our present derivation of the distribution lies in its natural association with a sampling experiment. And in the context of this experiment, $P(w)$ may be interpreted as the *expected proportion* of samples of n balls containing exactly w white balls, if the sampling experiment were to be repeated a large number of times under similar conditions. Of course, in any finite number of repetitions the expectation will not be realised. Thus, if the experiment is performed M times we expect that $M \times P(w)$ samples would each contain exactly w white balls. The number actually observed will be some $f_w \neq M \times P(w)$.

The concept of a sampling distribution is the basis of statistical inference. This is a representation of the behaviour of a random variable whose independent realisations are obtained from a conceptually unending sequence of random samples of the same size and drawn from the same universe or collection. This also applies to continuous random variables, but the mathematics of deriving a sampling distribution is beyond the scope of this work.

9.7 The probability- and factorial moment-generating functions

The factorial moments of the hypergeometric distribution were evaluated by direct summation, but it is often simpler to obtain such moments by an expansion of a *factorial moment-generating function* in exactly the same way as ordinary moments are obtained from a moment-generating function. The simplest method of evaluating the factorial moment-generating function of a random variable is through the *probability-generating function* which, as its name suggests, generates the probabilities of the distribution.

Suppose x is a discrete random variable which can take all the non-negative integral values with the point-probabilities

$$P(x = \nu) = p_\nu, \quad \text{for } \nu \geqslant 0, \quad \text{where } \sum_{\nu=0}^{\infty} p_\nu = 1.$$

The probability-generating function of x is the function of a real variable θ defined by

$$G(\theta) \equiv E(\theta^x) = \sum_{\nu=0}^{\infty} \theta^\nu p_\nu.$$

The fundamental property of $G(\theta)$ is that if it is expanded as a power series in θ, then the coefficient of θ^ν is $P(x = \nu)$. Clearly, $G(1) \equiv 1$. Furthermore, differentiation with respect to θ gives

$$\frac{dG(\theta)}{d\theta} = \sum_{\nu=1}^{\infty} \theta^{\nu-1} \nu p_\nu,$$

so that

$$\left[\frac{dG(\theta)}{d\theta} \right]_{\theta=1} \equiv G'(1) = E(x).$$

Similarly,

$$\left[\frac{d^2 G(\theta)}{d\theta^2} \right]_{\theta=1} \equiv G''(1) = E[x^{(2)}].$$

Hence

$$\text{var}(x) = G''(1) + G'(1)[1 - G'(1)].$$

Thus $G(\theta)$ can also be used to generate the moments of x. Furthermore, if we make the substitution $\theta = e^t$ in $G(\theta)$, then

$$G(e^t) = E(e^{tx}) \equiv M_0(t),$$

the moment-generating function of x about the origin.

Finally, if we make the transformation $\theta = 1 + \alpha$ in $G(\theta)$, then

$$G(1+\alpha) \equiv E[(1+\alpha)^x]$$

$$= \sum_{\nu=0}^{\infty} (1+\alpha)^\nu p_\nu$$

$$= \sum_{\nu=0}^{\infty} p_\nu \sum_{r=0}^{\nu} \binom{\nu}{r} \alpha^r$$

$$= \sum_{\nu=0}^{\infty} p_\nu \sum_{r=0}^{\nu} \nu^{(r)} \frac{\alpha^r}{r!}.$$

Hence, reversing the order of summation in the double series, we have

$$G(1+\alpha) = \sum_{r=0}^{\infty} \frac{\alpha^r}{r!} \sum_{\nu=r}^{\infty} p_\nu \nu^{(r)}$$

$$= \sum_{r=0}^{\infty} \mu'_{(r)} \frac{\alpha^r}{r!} \equiv H(\alpha), \quad \text{say}.$$

Then $H(\alpha)$ is the factorial moment-generating function of x. This function has the basic property that when it is expanded as a power series in α, the coefficient of $\alpha^r/r!$ is $\mu'_{(r)}$, the rth factorial moment of x about the origin.

We shall use these generating functions in studying the properties of the next important discrete probability distribution.

9.8   The binomial distribution

This classical distribution was discovered by James Bernoulli (1654 – 1705), and first published in his posthumous work *Ars Conjectandi* in 1713. Accordingly, the distribution is also commonly known as the Bernoulli distribution. There are many elegant and physically meaningful ways of deriving this distribution directly, but we shall here obtain it as a simple limiting form of the hypergeometric distribution. The binomial distribution is deduced by assuming that N, the number of balls in the collection, is infinite, with the proviso that

$$\lim_{N\to\infty} W/N = p \quad \text{and} \quad \lim_{N\to\infty} B/N = q,$$

where p and q are fixed numbers and $p+q=1$. This means that in the infinite collection the proportion of white balls is p and that of black balls q. As in the case of the finite collection, we are interested in the colour of the balls drawn from the infinite collection of balls. We may, therefore, equivalently represent the infinite collection to be made up of 1's and 0's corresponding to white and black balls respectively. If x represents the colour of a randomly selected ball from the collection, then

$$E(x^r) = 1^r.p + 0^r.q = p.$$

Hence $E(x) = p;$ $\text{var}(x) = pq;$ $\mu_3(x) = pq(q-p);$ $\mu_4(x) = pq(1-3pq).$

These are the same as the moments obtained for x when the sampling was from a finite collection.

The binomial distribution is the sampling distribution of the number of white balls found in samples of n balls drawn from the infinite collection. We may thus simply obtain the point-probabilities of the binomial distribution by taking the limit as $N\to\infty$ of the hypergeometric probabilities. If a random sample of n balls contains w white and b black balls, then for finite N

$$P(w) = \binom{n}{w} \frac{W^{(w)}B^{(b)}}{N^{(n)}},$$

where we may take $0 \leqslant w \leqslant n$ for $n < W$, since as $N\to\infty$ both W and $B\to\infty$. Now, setting $W = Np$ and $B = Nq$, we have

$$P(w) = \binom{n}{w}\frac{Np(Np-1)(Np-2)...(Np-w+1)Nq(Nq-1)(Nq-2)...(Nq-b+1)}{N(N-1)(N-2)...(N-n+1)}$$

$$= \binom{n}{w}\frac{p(p-1/N)(p-2/N)...[p-(w-1)/N]q(q-1/N)(q-2/N)...[q-(b-1)/N]}{1(1-1/N)(1-2/N)...[1-(n-1)/N]}.$$

Hence, for fixed w, b, and n,

$$\lim_{N \to \infty} P(w) \;=\; \binom{n}{w} p^w q^{n-w} \;=\; B(w), \quad \text{say.}$$

These limiting probabilities define the binomial distribution, and we readily verify that

$$\sum_{w=0}^{n} B(w) \;=\; \sum_{w=0}^{n} \binom{n}{w} p^w q^{n-w} \;=\; (p+q)^n \;\equiv\; 1.$$

The reason for the name "binomial distribution" is clear: the point-probabilities are obtained by the binomial expansion of $(p+q)^n$, where n is a positive integer.

Cumulants

The probability-generating function of w is

$$G(\theta) \;\equiv\; E(\theta^w) \;=\; \sum_{w=0}^{n} \binom{n}{w} p^w q^{n-w} \theta^w \;=\; (q + p\theta)^n.$$

Hence the factorial moment-generating function of w is

$$H(a) \;=\; [q + p(1+a)]^n$$

$$=\; (1 + pa)^n \;=\; \sum_{r=0}^{n} \binom{n}{r} p^r a^r \;=\; \sum_{r=0}^{n} n^{(r)} p^r \frac{a^r}{r!}.$$

Thus
$$E[w^{(r)}] \;\equiv\; \mu'_{(r)} \;=\; n^{(r)} p^r, \quad \text{for } r \geqslant 0.$$

This result could also be deduced from the limit as $N \to \infty$ of the rth factorial moment of the hypergeometric distribution. In fact,

$$\lim_{N \to \infty} \frac{n^{(r)} W^{(r)}}{N^{(r)}} \;=\; \lim_{N \to \infty} n^{(r)}\, \frac{W(W-1)(W-2)...(W-r+1)}{N(N-1)(N-2)...(N-r+1)}$$

$$=\; \lim_{N \to \infty} n^{(r)}\, \frac{p(p-1/N)(p-2/N)...[p-(r-1)/N]}{1(1-1/N)(1-2/N)...[1-(r-1)/N]}$$

$$=\; n^{(r)} p^r, \quad \text{for fixed } r.$$

Next, using the factorial moments of the binomial distribution, we have

$$E(w) \;=\; np; \quad \text{and}$$

$$\mathrm{var}(w) \;=\; E(w^2) - n^2 p^2$$

$$=\; E[w^{(2)} + w] - n^2 p^2 \;=\; n^{(2)} p^2 + np - n^2 p^2 \;=\; npq, \quad \text{on reduction.}$$

Similarly, for the third central moment of w, we obtain

$$\mu_3(w) \;=\; E(w-np)^3$$

$$=\; n^{(3)} p^3 + 3(1-np) n^{(2)} p^2 + (1-3np+2n^2 p^2) np,$$

which, on some straightforward reduction, gives

$$\mu_3(w) \;=\; npq(q-p).$$

Finally, the fourth central moment of w is

$$\mu_4(w) \;=\; E(w-np)^4$$

$$=\; n^{(4)} p^4 + (6-4np) n^{(3)} p^3 + (7-12np+6n^2 p^2) n^{(2)} p^2 +$$

$$+\; (1-4np+6n^2 p^2 - 3n^3 p^3) np,$$

whence, on simplification,

$$\mu_4(w) \;=\; npq[1 + 3(n-2)pq].$$

Therefore the first four cumulants of the binomial distribution are

$$\kappa_1 = np; \quad \kappa_2 = npq; \quad \kappa_3 = npq(q-p); \quad \kappa_4 = npq(1-6pq).$$

It is to be noted that for the binomial distribution $\kappa_2 < \kappa_1$. This is an important feature of the distribution. Furthermore, γ_1 and γ_2, the coefficients of skewness and kurtosis of the binomial distribution, are

$$\gamma_1 = (q-p)/(npq)^{\frac{1}{2}}; \quad \gamma_2 = (1-6pq)/npq.$$

It thus follows that the distribution is symmetrical when $p = q = \frac{1}{2}$, that is, the proportions of white and black balls are the same in the infinite collection sampled. The distribution is negatively skew for $q < p$, and the skewness is positive for $q > p$. In the same way, the distribution is mesokurtic if $1 - 6pq = 0$. This is a quadratic in p with roots $p = \frac{1}{2}(1 \pm 1/\sqrt{3})$, that is, $p = 0 \cdot 7887$ or $0 \cdot 2113$.

Recurrence relation in $B(w)$

The calculation of the binomial point-probabilities is found to be necessary for some statistical applications of the distribution. For given n and p, these probabilities can be conveniently evaluated by the use ot a simple recurrence relation in $B(w)$. The probability of exactly w white balls in a sample of n balls is

$$B(w) = \binom{n}{w} p^w q^{n-w} = \frac{p}{q}\left(\frac{n-w+1}{w}\right)\binom{n}{w-1} p^{w-1} q^{n-w+1}.$$

Thus $\qquad B(w) = \frac{p}{q}\left(\frac{n-w+1}{w}\right) B(w-1), \quad$ for $w \geqslant 1,$

which is the required recurrence relation. Now, for $w = 0$, $B(0) = q^n$. Hence we have successively

$$B(1) = \frac{p}{q}\binom{n}{1} B(0),$$

$$B(2) = \frac{p}{q}\left(\frac{n-1}{2}\right) B(1),$$

$$B(3) = \frac{p}{q}\left(\frac{n-2}{3}\right) B(2), \quad \text{and so on till}$$

$$B(n) = \frac{p}{q}\left(\frac{1}{n}\right) B(n-1).$$

We also observe from the general recurrence relation that

$$\frac{B(w)}{B(w-1)} = \frac{p(n-w+1)}{(1-p)w}.$$

Therefore the left-hand ratio of probabilities is $>$ or <1 according as

$$p(n-w+1) > \quad \text{or} \quad < (1-p)w.$$

Hence the values of $B(w)$ increase with w as long as $w < (n+1)p$, and decrease with w when $w > (n+1)p$. In general, $B(w)$ is maximum when w is equal to the integer next below $(n+1)p$. Thus the binomial distribution is unimodal.

We next consider two simple examples to illustrate how the binomial distribution can be used in statistical analysis. In this context, it is important to realise that an essential element in the practical use of a probability model is the understanding of the limitations under which the model is correctly applicable.

Example 9.1

The following data pertain to the number of occurrences of $0, 1, 2, \ldots, 8$ boys observed in 53,680 German families each having eight children.

Table 9.1   Showing the number of boys in 53,680 German
families each having eight children

Number of boys in a family	Observed number of families	Expected frequency of families
0	215	165·22
1	1485	1401·69
2	5331	5202·65
3	10649	11034·65
4	14959	14627·60
5	11929	12409·87
6	6678	6580·24
7	2092	1993·78
8	342	264·30
Total	53680	53680·00

Source: Geissler (1889), Beiträge zur Frage des Geschlechtsverhältnisses der Geborenen. *Zeitschrift des K. Sächsischen Statistischen Bureaus.*

The total number of boys observed in the sampled families is 221,023, and the total number of children $8 \times 53680 = 429440$. Hence the observed proportion p of boys in the families sampled is $221023/429440 = 0·51468$. This value of p is in close agreement with the proportion of male births observed in other large populations.

Consider any one of the sampled families. This has eight children of whom some number w $(0 \leqslant w \leqslant 8)$ are boys. We may regard the eight children as a sample of $n = 8$ balls of which w are white (boys) and the rest black (girls). On the basis of all the sampled families, the probability of a child in a family being a boy is $p = 0·51468$. Hence the probability that a family has w boys is

$$\binom{8}{w} (0·51468)^{w} (0·48532)^{8-w}.$$

Next, if we assume that the 53,680 families correspond to independent samples of size 8, then the expected number of families with w boys is

$$53680 \binom{8}{w} (0·51468)^{w} (0·48532)^{8-w}.$$

These expected frequencies are only valid if the families are assumed to be independent, and each of the eight children in any family has a constant probability of being a boy or a girl. However, it is clear that both these assumptions could not be strictly true for at least two reasons. Firstly, in such a survey, inter-familial connections could hardly be eliminated completely. Secondly, it is known that large families usually include twins which often turn out to be monozygotic so that they are necessarily of the same sex. Nevertheless, despite these reservations, the binomial model is still useful in so far as it could reveal the differences between the observed and expected frequencies. The calculated expected frequencies given in Table 9.1 show consistently fewer families with 0, 1, 2, 6, 7 or 8 boys than the observed frequencies. At the moment, this conclusion is based on a numerical comparison only, but we shall see later how the occurrence of such differences can be given a probabilistic basis. We naturally anticipate some differences between observed data and the expectations based on a probability model. The interesting question is whether such differences may reasonably be described as due to chance, or do they indicate that the model is invalid? Thus, in the present example, it may well be that parents of some families have a propensity for producing children of the same sex.

Example 9.2

In production engineering, large numbers of small components such as screws or ball-bearings are made on automatic machines. The components are produced subject to clearly defined specifications, and inspection of their quality may involve either measurement of characteristics or simply qualitative discrimination between a component satisfying the specifications or not. In any case, the assessment of the quality of the components may be regarded as a dichotomous classification into defective and non-defective items. The components produced are grouped into large batches which are then despatched to customers. Suppose it is known that over a long period of time, the proportion of defective components produced has been 0·05.

Batches are checked by the inspection of a random sample of 40 components.

    (i) Find the probability that such a sample contains at least four defectives.

Also, find the probability that of three such successive samples

    (ii) each contains at least four defectives;

    (iii) each contains exactly two defectives;

    (iv) some two contain not more than two defectives each, and the third sample has at least three defectives; and

    (v) some two contain less than four defectives each, and the third sample has at least two defectives.

The questions raised in this problem can be answered simply by an application of the binomial distribution, but it is worth while first to state explicitly the major assumptions under which this application is valid. Firstly, it is assumed that the production process is stable, so that the past proportion of defectives can be regarded as correct for the components produced later. Secondly, the defective and non-defective components are identified with the white and black balls of the model. Thirdly, it is assumed that the sample size of 40 components is small compared with the number of components in a batch, so that the binomial model may be used instead of the hypergeometric distribution. Fourthly, the different samples are assumed to be independent, so that the multiplicative law of probabilities is applicable.

In terms of the binomial model, $p = 0.05$ and $q = 0.95$. Hence the probability of w defectives in a sample of 40 components is

$$B(w) \; = \; \binom{40}{w} (0.05)^{w} (0.95)^{40-w}.$$

Therefore $B(0) \; = \; (0.95)^{40} \; = \; 0.12851,$

whence, using the recurrence relations, we have

$$B(1) \; = \; 40 \, B(0)/19 \; = \; 0.27055,$$
$$B(2) \; = \; 19.5 \, B(1)/19 \; = \; 0.27767,$$
$$B(3) \; = \; 38 \, B(2)/57 \; = \; 0.18511,$$
$$B(4) \; = \; 37 \, B(3)/76 \; = \; 0.09012, \quad \text{and so on.}$$

(i) The probability of a sample containing at least four defectives is

$$P(\text{i}) \; = \; P(w \geqslant 4) \; = \; 1 - P(w \leqslant 3)$$
$$= \; 1 - [B(0) + B(1) + B(2) + B(3)] \; = \; 0.13816.$$

(ii) Therefore the probability of four or more defectives in each of three samples is

$$P(\text{ii}) \; = \; [P(w \geqslant 4)]^{3} \; = \; (0.13816)^{3} \; = \; 0.00264.$$

(iii) Similarly, the probability of each of the three samples containing exactly two defectives is

$$P(\text{iii}) \; = \; [B(2)]^{3} \; = \; (0.27767)^{3} \; = \; 0.02141.$$

(iv) The probability of a sample containing not more than two defectives is

$$P(w \leqslant 2) \; = \; B(0) + B(1) + B(2) \; = \; 0.67673,$$

and the probability of its having at least three defectives is

$$P(w \geqslant 3) \; = \; 1 - P(w \leqslant 2) \; = \; 0.32327.$$

If we now regard the samples of 40 components each as constituting an infinite population of samples, then the probability that of three samples two contain not more than two defectives each, and the third at least three defectives is

$$P(\text{iv}) \; = \; \binom{3}{2} P^{2}(w \leqslant 2) P(w \geqslant 3)$$
$$= \; 3 (0.67673)^{2} \times 0.32327 \; = \; 0.44414.$$

(v) If X be the event $w \leqslant 3$ and Y the event $w \geqslant 2$, then X and Y are not mutually exclusive. However, if we define the mutually exclusive events

$$A: w = 0, 1; \quad B: w = 2, 3; \quad C: w \geqslant 4,$$

then $X = A + B$ and $Y = B + C$.

Also, $P(A) \; = \; 0.39906, \quad P(B) \; = \; 0.46278,$

and $P(C) \; = \; 1 - P(A) - P(B) \; = \; 0.13816.$

We require the probability that in three samples two X's and one Y will be realised. Now, in terms of A, B, and C, the events *not* consistent with the realisation of two X's and one Y are

$$A^{3}, \quad C^{3}, \quad C^{2}A, \quad C^{2}B.$$

Hence, allowing for possible permutations of A, B, and C in the three independent samples, the total probability for X^2Y is

$$
\begin{aligned}
P(v) &= 1 - P^3(A) - P^3(C) - 3P^2(C)\,P(A) - 3P^2(C)\,P(B) \\
&= 1 - P^3(A) - P^3(C) - 3P^2(C)[1 - P(C)] \\
&= 1 + 2P^3(C) - P^3(A) - 3P^2(C) \\
&= 1 + 0.00527 - 0.06355 - 0.05726 = 0.88446.
\end{aligned}
$$

9.9 The Poisson distribution

This distribution was first presented by Siméon Denis Poisson (1781 – 1840) in his book on probability published in 1837. Together with the normal and binomial distributions, the Poisson distribution constitutes the famous trio of probability models which have a central place in statistical theory. There are many instructive ways of arriving at the Poisson distribution, but we shall consider here the method used by Poisson himself. This derivation has the merit of simplicity and also reveals very clearly the link between the Poisson and binomial distributions.

We have seen that if a random sample of n balls is taken from an infinite population of white and black balls in proportions p and q respectively $(p + q = 1)$, then the probability that there are w white balls in the sample is

$$
B(w) = \binom{n}{w} p^w q^{n-w}, \quad \text{for } 0 \leqslant w \leqslant n.
$$

Here p is a finite number between 0 and 1, and n is a positive integer. There are many practical applications of the binomial distribution where, comparatively speaking, n is large and p is small, whereas $\mu \equiv np$ is of moderate magnitude. In such cases, the binomial probabilities $B(w)$ can be approximated well by the equivalent probabilities of the Poisson distribution. Following Poisson's derivation, we therefore consider the limit of $B(w)$ when $p \to 0$, $n \to \infty$, but $np \to \mu$, a finite parameter. Thus

$$
\begin{aligned}
B(w) &= \binom{n}{w} [p/(1-p)]^w (1-p)^n \\
&= n^{(w)} [\mu/(n-\mu)]^w (1-\mu/n)^n/w! \\
&= \frac{\mu^w}{w!} (1-\mu/n)^n \frac{1(1-1/n)(1-2/n)\ldots[1-(w-1)/n]}{(1-\mu/n)^w}.
\end{aligned}
$$

Now $\lim_{n\to\infty}(1-\mu/n)^n = e^{-\mu}$, and for fixed μ and finite w

$$
\lim_{n\to\infty} \frac{1(1-1/n)(1-2/n)\ldots[1-(w-1)/n]}{(1-\mu/n)^w} = 1.
$$

Hence, for fixed μ and w,

$$
\lim_{n\to\infty} B(w) = e^{-\mu}\mu^w/w! = Q(w), \quad \text{say.}
$$

The quantities $Q(w)$ define the probability distribution of a random variable w which can take all the integral values for $w \geqslant 0$. Since $Q(w) > 0$ and

$$
\sum_{w=0}^{\infty} Q(w) = e^{-\mu} \sum_{w=0}^{\infty} \mu^w/w! \equiv 1,
$$

it is clear that the $Q(w)$ do define a proper probability distribution. This is the Poisson distribution. The above derivation gives the exact limiting relationship between the Poisson and binomial distributions, but we shall also show that even for moderately large n and reasonably small p, the Poisson distribution provides a close and convenient approximation for the binomial distribution.

An alternative derivation of the Poisson distribution

There is another instructive and mathematically powerful method of establishing the Poisson distribution as a limit of the binomial distribution. This method rests on the limiting behaviour of the moment-generating function of the binomial distribution, and the equivalence of the two distributions is established by invoking a theorem that if two random variables have identical moments then they have the same probability distribution. We know that the moment-generating function of the binomial distribution about the origin is

$$E(e^{tw}) = \sum_{w=0}^{n} \binom{n}{w} p^w q^{n-w} e^{tw}$$

$$= (q + pe^t)^n = [1 + (e^t - 1)p]^n = [1 + (e^t - 1)\mu/n]^n, \quad \text{for } p = \mu/n.$$

Hence $\qquad \lim_{n \to \infty} E(e^{tw}) = \lim_{n \to \infty} [1 + (e^t - 1)\mu/n]^n = \exp[\mu(e^t - 1)].$

This limit is, in fact, the moment-generating function about the origin of the Poisson distribution since

$$M_0(t) \equiv \sum_{w=0}^{\infty} e^{tw} Q(w)$$

$$= \sum_{w=0}^{\infty} e^{tw} e^{-\mu} \mu^w / w! = e^{-\mu} \sum_{w=0}^{\infty} (\mu e^t)^w / w! = \exp[\mu(e^t - 1)],$$

as before. We may also work back from the moment-generating function of the Poisson distribution to obtain its probability-generating function by putting $\theta = e^t$, so that

$$G(\theta) \equiv E(\theta^w) = \exp \mu(\theta - 1)$$

whence, on expansion, we again reach the Poisson probabilities $Q(w)$.

Cumulants

The cumulant-generating function of the Poisson distribution is

$$\kappa(t) \equiv \log M_0(t) = \mu(e^t - 1) = \mu \sum_{r=1}^{\infty} t^r / r!$$

Thus all the cumulants κ_r of the distribution are equal to μ. In particular, the mean and variance are equal, which is a fundamental characteristic of the distribution. Also, the coefficients of skewness and kurtosis are

$$\gamma_1 = 1/\sqrt{\mu}; \quad \gamma_2 = 1/\mu.$$

Hence, for large μ, both γ_1 and γ_2 approach zero and the distribution becomes both symmetrical and mesokurtic. In fact, we shall see that for moderately large μ the distribution approaches normality.

Recurrence relation in $Q(w)$

The numerical evaluation of the Poisson point-probabilities is considerably eased by the use of a simple recurrence relation. We have

$$Q(w) = e^{-\mu}\mu^w/w!, \qquad \text{for } w \geqslant 0$$
$$= e^{-\mu}\mu^{w-1}(\mu/w)/(w-1)!,$$
$$\text{or} \qquad Q(w) = (\mu/w)Q(w-1), \qquad \text{for } w \geqslant 1,$$

which is the required recurrence relation. Thus

$$Q(0) = e^{-\mu},$$
$$Q(1) = \mu Q(0),$$
$$Q(2) = (\mu/2)Q(1),$$
$$Q(3) = (\mu/3)Q(2), \qquad \text{and so on.}$$

Since $w \geqslant 0$, it is often useful to evaluate the probability of $w \geqslant k$ as

$$P(w \geqslant k) = 1 - P(w < k) = 1 - \sum_{w=0}^{k-1} Q(w).$$

This formula is particularly convenient when k is a small integer.

9.10   A comparison of the binomial and Poisson distributions

We next consider an example to illustrate the numerical agreement of the Poisson approximation with the binomial distribution for moderately large n and small p.

Example 9.3

A college has 500 students who may be regarded as a random sample from an effectively infinite population of students. Excluding leap years, a year has 365 days, and we assume, as a first approximation, that the probability of a student having his birthday on the first of January is $p = 1/365$. This assumption is not strictly true since births are not uniformly distributed over the year. However, this assumption is reasonable for our present illustrative purpose. Then the probability of any w students having their birthdays on the first of January is

$$B(w) = \binom{500}{w}\left(\frac{1}{365}\right)^w\left(\frac{364}{365}\right)^{500-w}$$

Since the sample size $n = 500$ and $p = 1/365$, the Poisson parameter

$$\mu = np = 500/365 = 1\cdot3699.$$

Hence the Poisson approximation gives the probability of w students having birthdays on the first of January as

$$Q(w) = e^{-1\cdot3699}(1\cdot3699)^w/w!$$

Table 9.2 gives a numerical comparison of the probabilities $B(w)$ and $Q(w)$, for $0 \leqslant w \leqslant 6$ and $w \geqslant 7$. The general agreement shown in the table is remarkably good since the approximation error does not extend beyond the fourth place of decimals. This approximating role of the Poisson distribution is a useful practical convenience, but it would be wrong to infer this as the only reason for the importance of the Poisson distribution.

9.11   Observations fitting the Poisson distribution

The physical significance of the approximation used in the above

Table 9.2   Showing a comparison of the binomial point-probabilities and
their Poisson approximations for $n = 500$ and $p = 1/365$.

w	$B(w)$	$Q(w)$
0	0·2537	0·2541
1	0·3484	0·3481
2	0·2388	0·2385
3	0·1089	0·1089
4	0·0372	0·0373
5	0·0101	0·0102
6	0·0023	0·0023
$w \geqslant 7$	0·0006	0·0006

example is that the binomial probability for an event (birthday on first of
January) to occur is small, but since a large number of individuals are exposed to the risk of its occurrence, therefore the expectation np is finite.
As we have seen, these conditions lead to the Poisson approximation. In
the past, the smallness of p led to the Poisson distribution being described
as a model for the satisfactory representation of the variation in the occurrence of "rare events". This is a complete misnomer since random events
whose frequency of occurrence is correctly described by the Poisson distribution are very common indeed. This distribution has a wide and important role in statistical theory as the following examples indicate.

Example 9.4

An historic illustration of the Poisson distribution pertains to the number
of deaths due to horse-kicks in the former Prussian army corps. The probability of a man being killed by a horse-kick on any one day is exceedingly
small, but if an army corps of men are exposed to this risk for a year, it is
possible that one or more of them will be killed in this way. Table 9.3
gives the distribution of deaths obtained from the records of ten army corps
for twenty years.

Table 9.3   Showing the frequency distribution of deaths by horse-kicks
in ten Prussian army corps over twenty years

Deaths in a corps in a year (w)	Observed frequency	Expected frequency
0	109	108·67
1	65	66·29
2	22	20·22
3	3	4·11
4	1	0·63
$w \geqslant 5$	0	0·09
Total	200	200·00

Source: L.v. Bortkiewicz, *Das Gesetz der kleinen Zahlen*, Leipzig, 1898.

An army corps consists of a large number of men and it is assumed
that over a period of a year the probability that w of them will be killed by
horse-kicks is

$$Q(w) = e^{-\mu}\mu^{w}/w! , \quad \text{for } w \geqslant 0.$$

The ten corps are assumed to be independent and so also are the years as time-intervals over which the observations are made. There are thus 200 independent observations which are spread spatially (that is over the corps) and temporally (that is over twenty years). The total number of deaths is 122, and so the mean of the observed frequency distribution is $122/200 = 0.61$. Therefore the expected frequencies are

$$200\,e^{-0.61}\,(0.61)^{w}/w! \,, \quad \text{for } w \geqslant 0.$$

However, with two decimal place accuracy, the expected frequencies for $w \geqslant 6$ are negligible. The agreement between the observed and expected frequencies in Table 9.3 is excellent, and we shall show later how a probabilistic measure of assurance can be assigned to this excellence.

Example 9.5

Deaths by horse-kicks are somewhat of a genuine rarity in a mechanised society, and the importance of the Poisson distribution as a model for the representation of the commoner biological phenomena can be indicated more appropriately by another famous example pertaining to the counting of yeast cells with a special instrument known as a haemocytometer.

Table 9.4 Showing the frequency distribution of yeast cells obtained by a count with a haemocytometer

Number of cells in a square (w)	Observed frequency	Expected frequency
0	0	3·71
1	20	17·37
2	43	40·65
3	53	63·41
4	86	74·19
5	70	69·44
6	54	54·16
7	37	36·21
8	18	21·18
9	10	11·02
10	5	5·16
11	2	2·19
12	2	0·86
$w \geqslant 13$	0	0·45
Total	400	400·00

Source: "Student" (1907), *Biometrika*, Vol. 5, p. 351.

In "Student's" experiment the liquid containing the yeast cells was first well mixed and then spread out in a thin layer over the 400 units of area of the haemocytometer. The thickness of the layer was 0·01 mm and the unit of area 1/400 sq. mm. The above table presents the distribution of the number of yeast cells in the 400 squares of the haemocytometer. The assumption is now made that the probability of occurrence of the yeast cells in each square can be represented by a Poisson distribution. Thus, if μ is the mean number of cells per square, then the probability that a square contains w cells is

$$Q(w) = e^{-\mu}\mu^w/w! , \quad \text{for } w \geqslant 0.$$

The total number of cells observed in the 400 squares is 1,872, so that the observed average per square is $1872/400 = 4\cdot68$. Therefore, using this value of μ, we obtain the expected frequency of squares each containing w cells as

$$400\,e^{-4\cdot68}(4\cdot68)^w/w! , \quad \text{for } w \geqslant 0.$$

These expected frequencies agree well with the observations made by "Student". Indeed, as we shall see, the differences between observation and expectation can be reasonably ascribed to sampling variation.

This illustration is not an isolated one, for the distributions obtained from bacterial and blood counts all tend to show the same general form of spatial distribution of organisms which is adequately described by a Poisson model. Besides these, other biological phenomena such as the density of plants in the wild have spatial distributions which are often of the Poisson form. This is illustrated by the next example.

Example 9.6

The following table gives the distribution of density of *Eryngium maritimum* obtained by the method of *quadrat sampling*.

Table 9.5   Showing the frequency distribution of *Eryngium maritimum* in 147 quadrats sampled

Number of plants in a quadrat (w)	Observed frequency	Expected frequency
0	16	20·03
1	41	39·93
2	49	39·79
3	20	26·44
4	14	13·17
5	5	5·25
6	1	1·74
7	1	0·50
$w \geqslant 8$	0	0·15
Total	147	147·00

Source: G.E. Blackman (1935), *Annals of Botany*, Vol. 49, p. 749.

The quadrat is a small square frame which is located at random points in the region in which the plants of the species are found. After location, the number of plants in each quadrat is determined. Blackman's frequency distribution is based on the count of 147 quadrats. As before, the assumption is made that the probability of finding w plants in a quadrat is

$$Q(w) = e^{-\mu}\mu^w/w! , \quad \text{for } w \geqslant 0,$$

where μ is the mean number of plants per quadrat. The total number of plants observed is 293 so that the average number of plants per quadrat is $293/147 = 1\cdot9931$. Hence the expected frequency of quadrats containing w plants is

$$147\,e^{-1\cdot9931}(1\cdot9931)^w/w! , \quad \text{for } w \geqslant 0.$$

The agreement between observation and expectation is excellent.

Example 9.7

There is not much similarity between plants and bombs. However, the spatial distribution of the random points of impact of flying-bomb hits on London during World War II provides a remarkable verification of the Poisson distribution.

An area of 144 square kilometres in south London was divided into 576 squares, each 1/4 square kilometre in area. A count was made of the number of flying-bombs which fell in each square. The following table gives the distribution of 537 flying-bombs counted in the whole area.

Table 9.6   Showing the frequency distribution of flying-bomb
hits in 576 units of area in south London

Number of flying-bombs in a square (w)	Observed frequency	Expected frequency
0	229	226·74
1	211	211·39
2	93	98·54
3	35	30·62
4	7	7·14
$w \geqslant 5$	1	1·57
Total	576	576·00

Source: R.D. Clarke (1946), *Journal of the Institute of Actuaries*, Vol. 72, p. 48.

In this example, the distribution of the flying-bombs is considered over both space and time. It is assumed that the flying-bombs hit randomly, and the probability of w of them falling in a square is

$$Q(w) = e^{-\mu}\mu^{w}/w! , \quad \text{for } w \geqslant 0,$$

where μ is the mean number of flying-bombs per square. The observed average number of flying-bombs per square is $537/576 = 0·9323$, so that the expected frequency of squares with w flying-bombs is

$$576 e^{-0·9323}(0·9323)^{w}/w! , \quad \text{for } w \geqslant 0.$$

These expected frequencies conform very well with the observations made.

Example 9.8

One of the most important applications of the Poisson distribution arises when the random occurrence of a phenomenon is considered over time. This is well exemplified by a classic experiment concerning radioactive disintegration.

A radioactive substance emits α-particles at random instants of time. As the decay continues, the rate of emission of particles declines, but for a substance such as radium it takes years before a measurable decrease may be observed. Therefore, over relatively short periods, the conditions under which the emission of particles takes place can be regarded as constant.

The following data pertain to a total of 10,094 particles emitted by a radioactive substance in 2,608 time-intervals of 7·5 seconds each.

Table 9.7   Showing the frequency distribution of α-particles emitted
by a radioactive substance in 2,608 time-intervals

Number of particles emitted in a time-interval (w)	Observed frequency	Expected frequency
0	57	54·40
1	203	210·52
2	383	407·36
3	525	525·50
4	532	508·42
5	408	393·52
6	273	253·82
7	139	140·32
8	45	67·88
9	27	29·19
$w \geqslant 10$	16	17·07
Total	2608	2608·00

Source: Rutherford, Chadwick and Ellis, *Radiations from Radioactive Substances*, Cambridge, 1920, p. 172.

The average number of particles emitted per time-interval is $10094/2608 = 3·870$. Therefore, on the basis of the Poisson model, the expected frequency of the emission of w particles per time-interval is

$$2608\, e^{-3·870}\, (3·870)^{w}/w! , \quad \text{for } w \geqslant 0.$$

The over-all agreement between observation and expectation is good.

As remarked earlier, this example is particularly important because it brings out the idea of random events occurring in a temporal sequence. This concept underlies the *theory of stochastic processes* which covers, among much else, the statistical study of such diverse phenomena as industrial and traffic accidents, queues at service counters, telephone calls at an exchange, and the growth of biological populations.

9.12   Some applications of the Poisson distribution

Since the Poisson distribution depends upon only one parameter, it is relatively simple to draw probabilistic conclusions when it can be reasonably assumed that a random phenomenon has this distributional form with a specified mean value. We illustrate this practical use of the Poisson distribution by a few simple examples.

Example 9.9

A farmer produces seed potatoes for export which are despatched in large consignments to customers overseas. However, the potatoes should be free from infection, and the export regulations require that a consignment should only be allowed to pass if a random sample of 100 potatoes from it is found to be completely free from infection.

It is known from the past experience of the farmer that there is a 0·2 per cent rate of infection in the consignments offered for inspection. If it may be assumed that the size of the consignments is large relative to the sample size so that the effect of sampling from a finite population is negligible, find the probability that a random consignment will be passed as free of infection. Hence show that the probability that of a batch of ten consignments at least three will not be found free of infection is $0 \cdot 2666$.

Also, prove that in order to reduce this probability of the rejection of consignments to $0 \cdot 15$ the farmer needs approximately to halve his present rate of infection.

Since the rate of infection of potatoes is $0 \cdot 2$ per cent, the probability of an infected potato in a consignment is $p = 0 \cdot 002$. Hence the probability of finding r infected potatoes in a random sample of 100 is

$$B(r) = \binom{100}{r}(0 \cdot 002)^r (1 - 0 \cdot 002)^{100 - r}, \quad \text{for } 0 \leqslant r \leqslant 100.$$

Now $100p = 0 \cdot 2$ is small so that the binomial probabilities can be approximated by the Poisson distribution. Therefore the probability that a sample is free from infection is

$$Q(0) = (1 - 0 \cdot 002)^{100} \sim e^{-0 \cdot 2} = 0 \cdot 8187.$$

This is the probability of a consignment being passed as free of infection, and the probability of its being rejected on inspection is

$$1 - Q(0) = 0 \cdot 1813.$$

Hence the probability that of ten consignments exactly w will be rejected is

$$P(w) = \binom{10}{w}(0 \cdot 1813)^w (0 \cdot 8187)^{10 - w}, \quad \text{for } 0 \leqslant w \leqslant 10.$$

Therefore the probability that at least three consignments will be rejected is

$$\sum_{w=3}^{10} P(w) = 1 - \sum_{w=0}^{2} P(w)$$

$$= 1 - \sum_{w=0}^{2} \binom{10}{w}(0 \cdot 1813)^w (0 \cdot 8187)^{10 - w}$$

$$= 1 - (0 \cdot 1353 + 0 \cdot 2996 + 0 \cdot 2985) = 0 \cdot 2666.$$

If a per cent is the new rate of infection, then clearly $0 < a < 0 \cdot 2$. Also, by the previous argument, the probability that of ten at least three consignments will be rejected is

$$1 - \sum_{w=0}^{2} \binom{10}{w}(1 - e^{-a})^w (e^{-a})^{10 - w},$$

so that the equation for determining a is

$$\sum_{w=0}^{2} \binom{10}{w}(1 - e^{-a})^w (e^{-a})^{10 - w} = 0 \cdot 85,$$

or

$$e^{-8a}[36 e^{-2a} - 80 e^{-a} + 45] = 0 \cdot 85.$$

This is the required equation for determining a, and it can be solved by inverse interpolation to any degree of accuracy. However, we can obtain a simple approximation by noting that $z = 1 - e^{-a}$ is small. Then the transformed equation for z is

$$(1 - z)^8 [36(1 - z)^2 - 80(1 - z) + 45] = 0 \cdot 85,$$

or $(1 + 8z + 36z^2)(1 - 8z + 28z^2 - 56z^3 + 70z^4 - 56z^5 + 28z^6 - 8z^7 + z^8) = 0 \cdot 85$.

The coefficients of z and z^2 vanish on the left-hand side and so, ignoring

the powers of z greater than three, we have
$$1 - 120z^3 \sim 0 \cdot 85.$$
Therefore $z^3 = 1/800$,   or $z = 0 \cdot 1077$.

Hence $e^{-a} = 0 \cdot 8923$,   so that $a = 0 \cdot 11$ approximately.

Example 9.10

A garage has five similar cars which it hires out to customers by the day. It is known from the past experience of the garage that the daily demand for cars has a Poisson distribution with mean $2 \cdot 24$. Prove that, on any day, the probability that

(i) at least one car remains unused is $0 \cdot 9231$; and

(ii) the demand is not fully met is $0 \cdot 0268$.

(iii) Show that, on the average, the probability of exactly one car not being used on exactly one out of four days is $0 \cdot 3131$.

(iv) Also, show that the proportion of the expected demand which has to be refused is $1 \cdot 69$ per cent, but that the expected proportion of the available stock of cars which remains unused is $55 \cdot 96$ per cent.

(v) One of the customers has an accident and therefore, over a period of time, the damaged car is withdrawn from service for repairs. Prove that with four cars the probability of not satisfying the total demand on any day is $0 \cdot 0769$, and the expected proportion of unsatisfied demand is $5 \cdot 13$ per cent.

The probability of w cars being demanded on any day is
$$\begin{aligned} Q(w) &= e^{-2 \cdot 24}(2 \cdot 24)^w/w! , \qquad \text{for } 0 \leqslant w < \infty, \\ &= 0 \cdot 10646 \; (2 \cdot 24)^w/w! \end{aligned}$$

Hence, using the Poisson recurrence relation, we have

$$\begin{array}{llll} Q(0) &= 0 \cdot 10646; & Q(1) &= 0 \cdot 23847; \\ Q(2) &= 0 \cdot 26709; & Q(3) &= 0 \cdot 19942; \\ Q(4) &= 0 \cdot 11168; & Q(5) &= 0 \cdot 05003; \\ Q(6) &= 0 \cdot 01868; & P(w \geqslant 7) &= 0 \cdot 00817. \end{array}$$

(i) Therefore the probability that at least one car out of the five available remains unused on any day is
$$P(\text{i}) = P(w \leqslant 4) = Q(0) + Q(1) + Q(2) + Q(3) + Q(4) = 0 \cdot 9231.$$

(ii) The probability that the demand is not fully met on any day is
$$P(\text{ii}) = P(w \geqslant 6) = Q(6) + P(w \geqslant 7) = 0 \cdot 0268.$$

(iii) The probability that the demand is for exactly four cars on any day is $Q(4)$. Hence the total required probability is
$$P(\text{iii}) = \binom{4}{1} Q(4)[1 - Q(4)]^3 = 0 \cdot 3131.$$

(iv) With a stock of five cars, the expected demand met is
$$\sum_{w=0}^{4} wQ(w) + 5P(w \geqslant 5) = 2 \cdot 20203.$$

Therefore the expected demand not met is $2\cdot24 - 2\cdot20203 = 0\cdot03797$, and so the percentage unsatisfied expected demand is $3\cdot797/2\cdot24 = 1\cdot69$.

Again, the available stock is five cars, and so the expected unused stock is $5 - 2\cdot20203 = 2\cdot79797$. Hence the expected percentage of unused stock is $279\cdot797/5 = 55\cdot96$.

(v) With four cars, the probability of not completely satisfying demand is
$$P(w \geqslant 5) = 0\cdot0769.$$
The expected satisfied demand is
$$\sum_{w=0}^{3} wQ(w) + 4P(w \geqslant 4) = 2\cdot12515.$$
Therefore the expected percentage of unsatisfied demand is
$$(2\cdot24 - 2\cdot12515)100/2\cdot24 = 11\cdot485/2\cdot24 = 5\cdot13.$$

Example 9.11

An egg marketing organisation supplies eggs to a retailer in crates, each of which contains 600 eggs. Because of the risks in transit, there is a $0\cdot3$ per cent probability of an egg being damaged in a crate. According to the organisation's contract, the retailer accepts a one per cent breakage in a crate, but a refund is allowed for greater damage. Find the probability that the organisation will have to pay a refund on a crate supplied.

The organisation sends 1,000 crates to the retailer in a year, and it is known that, on the average, a refund on a crate amounts to £1. 10s. Prove that the probability of the expected annual refund payable by the organisation being at least £6 is $0\cdot2640$, and that the expected refund for a year is £3. 18s.

The probability of an egg being damaged in a crate is $0\cdot003$. Therefore the probability of there being w damaged eggs in the crate is
$$P(w) = \binom{600}{w}(0\cdot003)^{w}(1 - 0\cdot003)^{600-w}.$$

Now since the damage allowance is six eggs, therefore the probability of a refund being made on a crate is
$$P(w \geqslant 7) = 1 - P(w \leqslant 6)$$
$$= 1 - \sum_{w=0}^{6} \binom{600}{w}(0\cdot003)^{w}(1 - 0\cdot003)^{600-w}$$
$$\sim 1 - \sum_{w=0}^{6} e^{-1\cdot8}(1\cdot8)^{w}/w!$$

using the Poisson approximation for the binomial distribution,
$$= 1 - 0\cdot1653(2\cdot8 + 1\cdot62 + 0\cdot972 + 0\cdot4374 + 0\cdot15464 + 0\cdot0472392) = 0\cdot0026.$$
Thus the probability of a crate requiring a refund is $0\cdot0026$. Hence the probability of r crates requiring refund is
$$P(r) = \binom{1000}{r}(0\cdot0026)^{r}(1 - 0\cdot0026)^{10-r}.$$

Therefore the probability that four or more crates require refund is

$$P(r \geqslant 4) = 1 - P(r \leqslant 3)$$

$$= 1 - \sum_{r=0}^{3} \binom{1000}{r} (0 \cdot 0026)^r (1 - 0 \cdot 0026)^{1000-r}$$

$$\sim 1 - \sum_{r=0}^{3} e^{-2 \cdot 6} (2 \cdot 6)^r / r!$$

$$= 1 - 0 \cdot 07427 \,(1 + 2 \cdot 6 + 3 \cdot 38 + 2 \cdot 9293) = 0 \cdot 2640.$$

Thus the probability of at least four crates requiring refund is $0 \cdot 2640$. But since, on the average, one refund amounts to £1. 10s., hence the probability of the expected refund being at least £6 is $0 \cdot 2640$.

Clearly, the expected number of crates requiring refund is $2 \cdot 6$, and on the average the payment on a crate is £1. 10s. Therefore the expected amount of refund in the year is £3. 18s.

Example 9.12

In the distribution of the number of police prosecutions per motorist in a large city over a given period of time, the frequency of motorists having one, two, or more prosecutions is available, but the number of motorists who did not have a prosecution cannot be enumerated owing to the motorist population of the city fluctuating during that period. This gives rise to a discrete distribution in which the zero group is unobserved.

An observed distribution of this kind can usually be well represented by a *truncated* Poisson distribution with mean μ, that is, a Poisson distribution in which the probability of the zero class is ignored. Show that for such a modified Poisson distribution the variance is never greater than the mean, and that in fact the variance is equal to the mean multiplied by ρ, the probability of a prosecuted motorist having at least two convictions.

The probability of a motorist having w prosecutions is proportional to

$$Q(w) = e^{-\mu} \mu^w / w! \,, \quad \text{for } 1 \leqslant w < \infty.$$

Now $\quad \sum_{w=1}^{\infty} Q(w) = e^{-\mu} \sum_{w=1}^{\infty} \mu^w / w! = e^{-\mu}(e^\mu - 1) = 1 - e^{-\mu}.$

Hence the probability of a prosecuted motorist having w prosecutions is

$$e^{-\mu} \mu^w / (1 - e^{-\mu}) w! \,, \quad \text{for } 1 \leqslant w < \infty.$$

For this probability distribution the rth factorial moment of w about the origin is

$$E[w^{(r)}] = e^{-\mu}(1 - e^{-\mu})^{-1} \sum_{w=1}^{\infty} w^{(r)} \mu^w / w!$$

$$= \mu^r e^{-\mu}(1 - e^{-\mu})^{-1} \sum_{w=r}^{\infty} \mu^{w-r} / (w - r)!$$

$$= \mu^r e^{-\mu}(1 - e^{-\mu})^{-1} \sum_{j=0}^{\infty} \mu^j / j! = \mu^r / (1 - e^{-\mu}).$$

Therefore, in particular,

$$E(w) = \mu / (1 - e^{-\mu}) \quad \text{and} \quad E[w^{(2)}] = \mu^2 / (1 - e^{-\mu}),$$

whence

$$\text{var}(w) = \frac{\mu(\mu+1)}{1 - e^{-\mu}} - \frac{\mu^2}{(1 - e^{-\mu})^2} = \frac{\mu[1 - (\mu+1)e^{-\mu}]}{(1 - e^{-\mu})^2} .$$

Thus

$$E(w) - \text{var}(w) = \frac{\mu}{1 - e^{-\mu}} - \frac{\mu(1 - e^{-\mu} - \mu e^{-\mu})}{(1 - e^{-\mu})^2} = \frac{\mu^2 e^{-\mu}}{(1 - e^{-\mu})^2} > 0.$$

Hence the variance of w is never greater than the mean. Finally,

$$\rho = \text{var}(w)/E(w) = [1 - (\mu+1)e^{-\mu}]/(1 - e^{-\mu}) = 1 - [\mu e^{-\mu}/(1 - e^{-\mu})],$$

which is the probability of $w \geqslant 2$ for the truncated distribution. Hence the result.

9.13 The negative binomial distribution

The equality of the mean and variance is an important characteristic of the Poisson distribution, whereas for the binomial distribution the mean is always greater than the variance. Occasionally, however, observable phenomena give rise to empirical discrete distributions which show a variance larger than the mean. Some of the commonest examples of such behaviour are the frequency distributions of plant density obtained by quadrat sampling when the clustering of plants makes the simple Poisson model inapplicable. It has been shown by different investigators that in such cases the negative binomial distribution provides an excellent model because this distribution has a variance larger than the mean. Bacterial clustering also leads to the negative binomial distribution. This important probability distribution is sometimes also referred to as the Pascal distribution after the French mathematician Blaise Pascal (1623 – 1662), but there seems to be no historical justification for this association. There are many ways of deriving the negative binomial distribution from empirical considerations. We present here an approach which is simple and also reveals an interesting aspect of the association of this distribution with the binomial distribution.

We have seen that the binomial distribution arises when we consider the number of white balls in a random sample of n balls drawn from an infinite population of white and black balls in proportions p and q respectively, where $p + q = 1$. Then the probability of the sample containing w white balls is

$$B(w) = \binom{n}{w} p^w q^{n-w}, \quad \text{for } 0 \leqslant w \leqslant n.$$

Now suppose the roles of w and n are interchanged in the sense that we fix in advance the number of white balls required in the sample. We then sample from the population, one ball at a time, till exactly w white balls are selected, and let n be the total number of balls sampled. This sampling procedure implies that n is a random variable whose value in any particular case ensures that there are exactly w (fixed) white balls in the sample. It thus follows that the probability of n is

$$\begin{aligned}
I(n) &= \text{Probability of } w-1 \text{ white balls in the first } n-1 \text{ balls sampled} \\
&\quad \times \text{Probability that the } n\text{th ball drawn is white} \\
&= \binom{n-1}{w-1} p^{w-1} q^{n-w} \times p = \binom{n-1}{n-w} p^w q^{n-w}.
\end{aligned}$$

Clearly n can now take all the integral values $\geqslant w$, and we have

$$\sum_{n=w}^{\infty} I(n) = \sum_{n=w}^{\infty} \binom{n-1}{n-w} p^w q^{n-w}$$

$$= p^w \sum_{k=0}^{\infty} \binom{w+k-1}{k} q^k, \quad \text{where } n-w = k,$$

$$= p^w (1-q)^{-w} \equiv 1.$$

Hence the probabilities $I(n)$ define a proper probability distribution of an integral random variable n for $w \leqslant n < \infty$. Since w is fixed, we define another discrete random variable $x = n - w$ with the probability distribution

$$I(x) = \binom{x+w-1}{x} p^w q^x, \quad \text{for } 0 \leqslant x < \infty.$$

This distribution is the negative binomial distribution. The name is due to the fact that the point-probabilities are obtained by the expansion of $p^w(1-q)^{-w}$. It is important to note that unlike the Poisson distribution, the negative binomial has two parameters w and p, and this gives it greater flexibility.

Cumulants

The probability-generating function of x is

$$G(\theta) \equiv E(\theta^x) = \sum_{x=0}^{\infty} \binom{x+w-1}{x} p^w (q\theta)^x = [p/(1-q\theta)]^w.$$

Hence the moment-generating function of x about the origin is

$$M_0(t) = [p/(1-qe^t)]^w = [p/(p+q-qe^t)]^w = [1 - \lambda(e^t - 1)]^{-w},$$

where $\lambda \equiv q/p$, whence the cumulant-generating function of x is

$$\kappa(t) \equiv \log M_0(t) = -w \log[1 - \lambda(e^t - 1)]$$

$$= w\left[\lambda(e^t - 1) + \frac{1}{2}\lambda^2(e^t-1)^2 + \frac{1}{3}\lambda^3(e^t-1)^3 + \frac{1}{4}\lambda^4(e^t-1)^4 + \cdots\right]$$

$$= w\left[\lambda\left(t + \frac{t^2}{2!} + \frac{t^3}{3!} + \frac{t^4}{4!} + \cdots\right) + \frac{1}{2}\lambda^2 t^2\left(1 + \frac{t}{2!} + \frac{t^2}{3!} + \cdots\right)^2 + \right.$$

$$\left. + \frac{1}{3}\lambda^3 t^3\left(1 + \frac{t}{2!} + \cdots\right)^3 + \frac{1}{4}\lambda^4 t^4 + \cdots\right]$$

$$= w\left[\lambda t + (\lambda + \lambda^2)\frac{t^2}{2!} + (\lambda + 3\lambda^2 + 2\lambda^3)\frac{t^3}{3!} + (\lambda + 7\lambda^2 + 12\lambda^3 + 6\lambda^4)\frac{t^4}{4!} + \cdots\right]$$

$$= w\left[\frac{q}{p}t + \frac{q}{p^2}\frac{t^2}{2!} + \frac{q(1+q)}{p^3}\frac{t^3}{3!} + \frac{q(1+4q+q^2)}{p^4}\frac{t^4}{4!} + \cdots\right].$$

Therefore the first four cumulants of the distribution of x are

$$\kappa_1 = wq/p; \quad \kappa_2 = wq/p^2; \quad \kappa_3 = wq(1+q)/p^3; \quad \kappa_4 = wq(1+4q+q^2)/p^4.$$

Hence the coefficients of skewness and kurtosis of the distribution are

$$\gamma_1 = (1+q)/(wq)^{\frac{1}{2}}; \quad \gamma_2 = (1+4q+q^2)/wq.$$

The distribution of x is always positively skew and leptokurtic. The mean of the distribution is wq/p and this is clearly less than the variance wq/p^2. As remarked earlier, this is an important feature of the distribution.

Example 9.13

The first two columns of the table below give the frequency distribution of accidents in five weeks to 647 women working on high explosive shells.

Table 9.8   Showing the frequency distribution of the number of accidents in five weeks to 647 women working on high explosive shells

Number of accidents per woman (x)	Observed frequency	Expected Poisson frequency	Expected negative binomial frequency
0	447	406·32	443·00
1	132	188·99	138·57
2	42	43·96	44·37
3	21	6·82	14·32
4	3	0·79	4·64
$x \geqslant 5$	2	0·12	2·10
Total	647	647·00	647·00

Source: Major Greenwood and G. Udny Yule (1920), *Journal of the Royal Statistical Society*, Vol. 83, p. 255.

The mean and variance of the observed distribution are found to be 0·4652 and 0·6919 respectively. As the variance is considerably larger than the mean, it is likely that the Poisson distribution will not provide a good model for this frequency distribution. However, we can verify this simply if we assume that the expected frequency of women having x accidents in the five weeks is

$$647 \, Q(x) \;=\; 647 \, e^{-0\cdot4652} (0\cdot4652)^{x}/x! \, , \quad \text{for } x \geqslant 0.$$

As can be seen from the third column of Table 9.8, these expected frequencies do not correspond well with the observed frequencies.

If we next consider a negative binomial distribution, then by equating the mean and variance of the model and the observed values, we obtain the equations

$$wq/p = 0\cdot4652 \quad \text{and} \quad wq/p^2 = 0\cdot6919.$$

The solution of these equations gives

$$p = 0\cdot6724 \quad \text{and} \quad w = 0\cdot9548.$$

Hence, on the basis of this model, the expected frequency of women having x accidents during the period of five weeks is

$$647 \, I(x) \;=\; 647 \binom{x - 0\cdot0452}{x} (0\cdot6724)^{0\cdot9548} (0\cdot3276)^{x}, \quad \text{for } x \geqslant 0.$$

These values are given in the last column of Table 9.8, and they are clearly in excellent agreement with the data. A possible reason for the failure of the Poisson distribution is that the accident proneness varies with the women. It can be shown (but we do not consider this here) that a negative binomial distribution arises if the variation in the accident proneness of the women is taken into account.

9.14   The Poisson distribution as limit of the negative binomial distribution

As we have seen, there is a characteristic difference between the negative binomial and Poisson distributions in that the mean and variance are equal for the latter distribution, whilst for the former the variance is necessarily larger than the mean. Despite this, there is a relation between the two distributions; and, indeed, a simple limiting process shows that the Poisson distribution may also be regarded as a limit of the negative binomial.

In earlier notation the mean of the negative binomial distribution is wq/p. Now suppose that $w \to \infty$ and $q \to 0$ such that $wq/p \to \mu$, a finite quantity. We may thus set $q/p = \mu/w$, so that the probability-generating function of the distribution of x may be written as

$$G(\theta) = [p/(1-q\theta)]^w = [p/\{p - q(\theta-1)\}]^w = [1 - \mu(\theta-1)/w]^{-w}.$$

Hence

$$\lim_{w \to \infty} G(\theta) = e^{\mu(\theta-1)},$$

which is the probability-generating function of a Poisson distribution with mean μ. This means that if the parameter w of the negative binomial distribution is large compared with its mean, then the simpler Poisson distribution should give an adequate approximation. In other words, if w is large compared with $\mu = wq/p$, then

$$I(x) = \binom{x+w-1}{x} p^w q^x \sim e^{-\mu}\mu^x/x! , \quad \text{for } x \geqslant 0.$$

This is a useful result and, taken in conjunction with our initial derivation of the Poisson distribution, it shows how, under suitable conditions, the Poisson distribution can be regarded as a limiting form of the Bernoulli and negative binomial distributions.

From the theoretical point of view, the technique used in deducing these Poisson approximations is even more important than the results themselves, and it is useful to recapitulate the essential principle involved. We have regarded the probability- or moment-generating function of a random variable as a mathematically concise method of representing a probability distribution. Besides, we have used a basic theorem that if two random variables have the same probability- or moment-generating function then they have the same distribution. The limiting distributional form is arrived at by considering the limit of the generating function under suitable restrictions on the parameters of the probability distribution whose limiting form is to be deduced. We thus derived the Poisson distribution as a limit of the binomial distribution under the conditions that $np \to \mu$, a finite number, when $p \to 0$ and $n \to \infty$. In the same way, the limiting form of the negative binomial distribution is deduced when $wq/p \to \mu$, a finite number, as $w \to \infty$ and $q \to 0$. We now use this method to show that, under certain conditions, both the binomial and Poisson distributions approach normality. These results are of considerable practical value, and they also exemplify the central position of the normal distribution in statistical theory.

9.15   The normal distribution as limit of the binomial distribution

The normal distribution was first derived by Abraham de Moivre (1667 —

1754) as the limit to which the binomial distribution tends when the binomial parameter p is fixed but n, the sample size, tends to infinity. De Moivre considered this limiting behaviour in the particular case of $p = \frac{1}{2}$, and his derivation is given in his famous work *The Doctrine of Chances* (1718). Pierre Simon Laplace (1749 – 1827), in his *Théorie analytique des probabilités* (1812), extended de Moivre's result by proving that the approach to normality was also valid for $p \neq \frac{1}{2}$. This general result is therefore known as the *de Moivre-Laplace theorem*, and it is of fundamental importance in probability theory. There are many interesting mathematical points associated with this theorem, but we shall not consider them. Our concern here is to show in a simple way the theoretical relation between the normal and binomial distributions when in the latter distribution p is fixed and $n \to \infty$, so that $np \to \infty$ also.

We recall that the point-probabilities defining the binomial distribution are

$$B(w) = \binom{n}{w} p^w q^{n-w}, \quad \text{for } 0 \leqslant w \leqslant n \text{ and } p+q = 1,$$

whence
$$E(w) = np; \quad \text{var}(w) = npq.$$

To prove the theorem, we now consider a new random variable z defined by the equation
$$z = (w - np)/(npq)^{\frac{1}{2}}.$$

Clearly, $E(z) = 0$ and $\text{var}(z) = 1$. We now show further that in the limit as $n \to \infty$ and p finite, z is also normally distributed. In essence, this is the de Moivre-Laplace theorem.

The moment-generating function of z is

$$
\begin{aligned}
M(t) &\equiv E(e^{tz}) \\
&= E[\exp(w-np)t/\sigma], \quad \text{where for convenience we have set } \sigma \equiv (npq)^{\frac{1}{2}}, \\
&= e^{-npt/\sigma} E(e^{wt/\sigma}) \\
&= e^{-npt/\sigma} \sum_{w=0}^{n} \binom{n}{w} p^w q^{n-w} e^{wt/\sigma} \\
&= e^{-npt/\sigma} \sum_{w=0}^{n} \binom{n}{w} (pe^{t/\sigma})^w q^{n-w} \\
&= e^{-npt/\sigma} (q + pe^{t/\sigma})^n \\
&= (qe^{-pt/\sigma} + pe^{qt/\sigma})^n \\
&= \left[q \sum_{r=0}^{\infty} (-pt/\sigma)^r/r! + p \sum_{r=0}^{\infty} (qt/\sigma)^r/r! \right]^n \\
&= \left[q\left\{ 1 - pt/\sigma + \frac{1}{2!}(pt/\sigma)^2 - \frac{1}{3!}(pt/\sigma)^3 + \frac{1}{4!}(pt/\sigma)^4 - \cdots \right\} + \right. \\
&\qquad \left. + p\left\{ 1 + qt/\sigma + \frac{1}{2!}(qt/\sigma)^2 + \frac{1}{3!}(qt/\sigma)^3 + \frac{1}{4!}(qt/\sigma)^4 + \cdots \right\} \right]^n \\
&= \left[1 + \frac{t^2}{2!}\frac{1}{n} + \frac{t^3}{3!}\frac{q-p}{n^{3/2}(pq)^{\frac{1}{2}}} + \frac{t^4}{4!}\frac{q^2 - qp + p^2}{n^2 pq} + \cdots \right]^n.
\end{aligned}
$$

Therefore

$$\log M(t) = n \log\left[1 + \frac{t^2}{2!}\frac{1}{n} + \frac{t^3}{3!}\frac{q-p}{n^{3/2}(pq)^{\frac{1}{2}}} + \frac{t^4}{4!}\frac{1-3pq}{n^2 pq} + \dots\right]$$

$$= \frac{t^2}{2!} + \frac{t^3}{3!}\frac{q-p}{(npq)^{\frac{1}{2}}} + \frac{t^4}{4!}\frac{1-6pq}{npq} + O(n^{-3/2}).$$

Hence $\qquad \lim_{n\to\infty}\log M(t) = \frac{1}{2}t^2,$ or $\quad \lim_{n\to\infty}M(t) = \exp\frac{1}{2}t^2.$

But $\exp\frac{1}{2}t^2$ is the moment-generating function of a unit normal variable, whence we infer that in the limit z has a unit normal distribution.

We further observe that the expansion for $\log M(t)$ may alternatively be written as

$$\log M(t) = \frac{1}{2}t^2 + \frac{t^3}{3!}\gamma_1(w) + \frac{t^4}{4!}\gamma_2(w) + O(n^{-3/2}),$$

where $\gamma_1(w)$ and $\gamma_2(w)$ are the coefficients of skewness and kurtosis of the probability distribution of w. If $p = q = \frac{1}{2}$, making the distribution of w symmetrical, then

$$\log M(t) = \frac{1}{2}t^2 - \frac{t^4}{4!}\frac{2}{n} + O(n^{-2}),$$

since it is easily seen that the coefficients of $t^{2r+1}/(2r+1)!$ ($r \geqslant 0$) all vanish for $p = q$. It is clear that in this case $\log M(t)$ approaches the limit $\frac{1}{2}t^2$ faster than when $p \neq q$.

The main theorem has some important practical consequences. Suppose n is moderately large and p finite. Then for any two integers j and k such that $0 < j \leqslant k < n$,

$$P(j \leqslant w \leqslant k) = \sum_{w=j}^{k}\binom{n}{w}p^w q^{n-w}$$

$\qquad$ = the area of the histogram of the binomial distribution from the lower class-limit $j - 0\cdot5$ to the upper class-limit $k + 0\cdot5$

$\qquad \sim$ the corresponding area under the normal curve with mean np and variance npq.

Accordingly, we have the useful approximation that

$$P(j \leqslant w \leqslant k) \sim \frac{1}{\sqrt{2\pi npq}}\int_{j-0\cdot5}^{k+0\cdot5}\exp[-\tfrac{1}{2}(x-np)^2/npq]\,dx$$

$$= \frac{1}{\sqrt{2\pi}}\int_{(j-np-0\cdot5)/(npq)^{\frac{1}{2}}}^{(k-np+0\cdot5)/(npq)^{\frac{1}{2}}}\exp-\tfrac{1}{2}z^2\,dz,$$

or $\quad P(j \leqslant w \leqslant k) \sim \Phi[(k-np+0\cdot5)/(npq)^{\frac{1}{2}}] - \Phi[(j-np-0\cdot5)/(npq)^{\frac{1}{2}}],$

in standard notation for the distribution function of a unit normal variable. Also, in particular,

$$B(w) = \binom{n}{w}p^w q^{n-w} \sim \Phi[(w-np+0\cdot5)/(npq)^{\frac{1}{2}}] - \Phi[(w-np-0\cdot5)/(npq)^{\frac{1}{2}}].$$

Furthermore, if np is an integer r, then for $w = r$

$$B(r) = \binom{n}{r}p^r q^{n-r} \sim \Phi[0\cdot5/(npq)^{\frac{1}{2}}] - \Phi[-0\cdot5/(npq)^{\frac{1}{2}}] = 2\Phi[0\cdot5/(npq)^{\frac{1}{2}}] - 1.$$

These approximations facilitate quick approximate calculation of the binomial point-probabilities and their sums when n is large and p finite. It is important to note that although the limiting distribution is theoretically attained only for n infinite, the approximations give excellent results even for relatively small values of n. This is the reason for the general use of these approximations in statistical analysis.

Example 9.14

The following table gives a numerical comparison of the binomial probabilities and the normal approximation for $n = 100$ and $p = 0.3$. The sample

Table 9.9 Showing a comparison of the binomial probabilities for $n = 100$ and $p = 0.3$ with the normal approximation

Binomial variable w	Exact binomial probability	Normal approximation
$9 \leqslant w \leqslant 11$	0·00001	0·00003
$12 \leqslant w \leqslant 14$	0·00015	0·00033
$15 \leqslant w \leqslant 17$	0·00201	0·00283
$18 \leqslant w \leqslant 20$	0·01430	0·01599
$21 \leqslant w \leqslant 23$	0·05907	0·05895
$24 \leqslant w \leqslant 26$	0·14887	0·14447
$27 \leqslant w \leqslant 29$	0·23794	0·23405
30	0·08672	0·08648
$31 \leqslant w \leqslant 33$	0·23013	0·23405
$34 \leqslant w \leqslant 36$	0·14086	0·14447
$37 \leqslant w \leqslant 39$	0·05889	0·05895
$40 \leqslant w \leqslant 42$	0·01702	0·01599
$43 \leqslant w \leqslant 45$	0·00343	0·00283
$46 \leqslant w \leqslant 48$	0·00049	0·00033
$49 \leqslant w \leqslant 51$	0·00005	0·00003

size is only moderate and there is an appreciable amount of skewness in the binomial distribution. Nevertheless, the general agreement obtained is sufficiently close for most practical purposes.

9.16   The normal distribution as limit of the Poisson distribution

The method used to derive the limiting form of the binomial distribution can also be used to show that under certain conditions the normal distribution is a limiting form of the Poisson distribution. The point-probabilities of the Poisson distribution are

$$Q(w) = e^{-\mu}\mu^{w}/w! , \quad \text{for } w \geqslant 0,$$

and $E(w) = \mathrm{var}(w) = \mu$. As before, we define a standardised random variable z by

$$z = (w - \mu)/\sqrt{\mu},$$

so that $E(z) = 0$ and $\mathrm{var}(z) = 1$. We now prove that as $\mu \to \infty$, the limiting distribution of z is unit normal.

The moment-generating function of z is

$$\begin{aligned} M(t) &\equiv E(e^{tz}) = e^{-t\sqrt{\mu}}E(e^{tw/\sqrt{\mu}}) \\ &= e^{-t\sqrt{\mu}} \sum_{w=0}^{\infty} e^{-\mu}\mu^{w}e^{tw/\sqrt{\mu}}/w! \\ &= e^{-t\sqrt{\mu}}e^{-\mu} \sum_{w=0}^{\infty} (\mu e^{t/\sqrt{\mu}})^{w}/w! \end{aligned}$$

$$= e^{-t\sqrt{\mu}}e^{-\mu}\exp(\mu e^{t/\sqrt{\mu}})$$

$$= e^{-t\sqrt{\mu}}\exp[\mu(e^{t/\sqrt{\mu}} - 1)].$$

Hence $$\log M(t) = -t\sqrt{\mu} + \mu(e^{t/\sqrt{\mu}} - 1)$$

$$= -t\sqrt{\mu} + \sum_{r=1}^{\infty} \frac{t^r}{r!} \mu^{-(r-2)/2}$$

$$= \frac{t^2}{2!} + \frac{t^3}{3!}\frac{1}{\sqrt{\mu}} + \frac{t^4}{4!}\frac{1}{\mu} + O(\mu^{-3/2}).$$

Therefore $$\lim_{\mu\to\infty} \log M(t) = \tfrac{1}{2}t^2, \quad \text{or} \quad \lim_{\mu\to\infty} M(t) = \exp\tfrac{1}{2}t^2,$$

so that as $\mu \to \infty$, z is unit normal.

The practical interest of this result lies in the fact that, although the limiting distribution is strictly true for only μ infinite, the Poisson distribution approaches approximate normality for values of μ which are rather small. To use this approximating distribution, suppose j and k are two integers such that $0 < j \leqslant k < n$. Then

$$P(j \leqslant w \leqslant k) = \sum_{w=j}^{k} e^{-\mu}\mu^w/w!$$

= the area of the histogram of the Poisson distribution from the lower class-limit $j - 0\cdot5$ to the upper class-limit $k + 0\cdot5$

~ the corresponding area under the normal curve with mean μ and variance μ.

Accordingly, we have the approximation that

$$P(j \leqslant w \leqslant k) \quad \sim \frac{1}{\sqrt{2\pi\mu}} \int_{j-0\cdot5}^{k+0\cdot5} \exp[-\tfrac{1}{2}(x-\mu)^2/\mu]\,dx$$

$$= \frac{1}{\sqrt{2\pi}} \int_{(j-\mu-0\cdot5)/\sqrt{\mu}}^{(k-\mu+0\cdot5)/\sqrt{\mu}} \exp-\tfrac{1}{2}z^2\,dz,$$

or $$P(j \leqslant w \leqslant k) \sim \Phi[(k-\mu+0\cdot5)/\sqrt{\mu}] - \Phi[(j-\mu-0\cdot5)/\sqrt{\mu}],$$

in standard notation. Also, in particular,

$$Q(w) = e^{-\mu}\mu^w/w! \sim \Phi[(w-\mu+0\cdot5)/\sqrt{\mu}] - \Phi[(w-\mu-0\cdot5)/\sqrt{\mu}];$$

and if μ is an integer r, then for $w = r$

$$Q(r) = e^{-r}r^r/r! \sim \Phi(0\cdot5/\sqrt{r}) - \Phi(-0\cdot5/\sqrt{r}) = 2\Phi(0\cdot5/\sqrt{r}) - 1.$$

Example 9.15

Table 9.10 opposite gives a numerical comparison of the Poisson probabilities and the normal approximation when $\mu = 8$. This is not a large value, but the general agreement is reasonably good.

9.17   Some applications of the normal approximations to the binomial and Poisson distributions

Example 9.16

It is known from a national housing survey that 15 per cent of the houses in the country have some form of central heating. On the basis of

Table 9.10   Showing a comparison of the Poisson probabilities for $\mu = 8$ with the normal approximation

Poisson variable w	Exact Poisson probability	Normal approximation
$0 \leqslant w \leqslant 1$	0·00302	0·00945
$2 \leqslant w \leqslant 3$	0·03936	0·04503
$4 \leqslant w \leqslant 5$	0·14887	0·13254
$6 \leqslant w \leqslant 7$	0·26175	0·24141
$w = 8$	0·13960	0·14048
$9 \leqslant w \leqslant 10$	0·22336	0·24141
$11 \leqslant w \leqslant 12$	0·12033	0·13254
$13 \leqslant w \leqslant 14$	0·04654	0·04503
$15 \leqslant w \leqslant 16$	0·01354	0·00945
$17 \leqslant w \leqslant 18$	0·00307	0·00323
$19 \leqslant w \leqslant 20$	0·00056	0·00010

this information, find the probability that in a random sample of 500 houses taken from a large city

(i)   exactly 80 houses have central heating;

(ii)   at least 420 houses are without central heating;

(iii)   fewer than 60 houses have central heating;   and

(iv)   not more than 60 houses have central heating.

(i) Let w be the random variable denoting the number of houses in the sample having central heating. Then the probability distribution of w is

$$P(w) \;=\; \binom{500}{w}(0{\cdot}15)^{w}(0{\cdot}85)^{500-w}, \quad \text{for } 0 \leqslant w \leqslant 500.$$

Therefore the probability of exactly 80 houses in the sample having central heating is

$$
\begin{aligned}
P(\text{i}) \;&=\; \binom{500}{80}(0{\cdot}15)^{80}(0{\cdot}85)^{420}\\
&\sim\; \Phi[(80 - 75 + 0{\cdot}5)/(500 \times 0{\cdot}15 \times 0{\cdot}85)^{\frac{1}{2}}] \;-\\
&\qquad\qquad -\; \Phi[(80 - 75 - 0{\cdot}5)/(500 \times 0{\cdot}15 \times 0{\cdot}85)^{\frac{1}{2}}]\\
&=\; \Phi(5{\cdot}5/7{\cdot}984) - \Phi(4{\cdot}5/7{\cdot}984)\\
&=\; \Phi(0{\cdot}689) - \Phi(0{\cdot}564) \;=\; 0{\cdot}7546 - 0{\cdot}7137 \;=\; 0{\cdot}0409.
\end{aligned}
$$

(ii) The probability that at least 420 houses are without central heating is

$$
\begin{aligned}
P(\text{ii}) \;&=\; \text{Probability that not more than 80 houses have central heating}\\
&=\; P(w \leqslant 80)\\
&=\; \sum_{w=0}^{80} \binom{500}{w}(0{\cdot}15)^{w}(0{\cdot}85)^{500-w}\\
&\sim\; \Phi[(80 - 75 + 0{\cdot}5)/7{\cdot}984] \;=\; \Phi(0{\cdot}689) \;=\; 0{\cdot}7546.
\end{aligned}
$$

(iii) The probability that fewer than 60 houses have central heating is

$$
\begin{aligned}
P(\text{iii}) \;&=\; P(w \leqslant 59)\\
&=\; \sum_{w=0}^{59} \binom{500}{w}(0{\cdot}15)^{w}(0{\cdot}85)^{500-w}\\
&=\; \Phi[59 - 75 + 0{\cdot}5)/7{\cdot}984] \;=\; 1 - \Phi(1{\cdot}941) \;=\; 1 - 0{\cdot}9739 \;=\; 0{\cdot}0261.
\end{aligned}
$$

(iv) The probability that not more than 60 houses have central heating is

$$P(iv) = P(w \leqslant 60)$$
$$\sim \Phi[(60 - 75 + 0\cdot5)/7\cdot984] = 1 - \Phi(1\cdot816) = 1 - 0\cdot9653 = 0\cdot0347.$$

Example 9.17

In a certain country, the dental profession recommend the compulsory fluoridation of water to reduce the incidence of dental decay, but the government will not legislate for this unless it has strong assurance that at least 55 per cent of the adult population support such action. Accordingly, to assess public opinion, the national dental association decides to carry out a countrywide survey. If the true proportion of the population in favour of compulsory fluoridation is 60 per cent, how large a sample should be taken in the survey so that the chance of the sample proportion being not more than 55 per cent is $0\cdot1$ per cent?

Suppose that a sample of n adults is taken and let w of these be in favour of fluoridation. Then

$$P(w) = \binom{n}{w} (0\cdot6)^{w}(0\cdot4)^{n-w}.$$

Therefore the equation for determining n is

$$P(w \leqslant 0\cdot55n) = \sum_{w=0}^{0\cdot55n} \binom{n}{w} (0\cdot6)^{w}(0\cdot4)^{n-w} = 0\cdot001.$$

Hence, using the normal approximation, the equation for n may be written as

$$\Phi[(0\cdot55n - 0\cdot6n + 0\cdot5)/(n \times 0\cdot6 \times 0\cdot4)^{\frac{1}{2}}] = 0\cdot001,$$

so that

$$(- 0\cdot05n + 0\cdot5)/(0\cdot24n)^{\frac{1}{2}} = - 3\cdot09,$$

which gives

$$n^2 - 936\cdot62n + 100 = 0.$$

Therefore $n = \frac{1}{2}[936\cdot62 \pm \{(936\cdot62)^2 - 400\}^{\frac{1}{2}}] = \frac{1}{2}(936\cdot62 \pm 936\cdot41)$,

whence the value for n is $936\cdot52$, or 937 to the nearest integer.

Example 9.18

Ball-bearings produced at a factory are supplied to a customer in large batches. In order to check the quality of the product, the customer inspects a fixed number of ball-bearings from each batch received. This inspection process shows that, over a period of time, eight per cent of the samples had at most two defective items each, and five per cent of the samples had at least ten defective items. If it is assumed that the quality of the production remained the same over the inspection period and the batch size is large compared with the sample size, determine the sample size and the proportion of defective items in the batches received by the customer.

Let the sample size be n and the proportion defective in the batches p. If w is the number of defective items in a sample, then on the average

$$P(w \leqslant 2) = \sum_{w=0}^{2} \binom{n}{w} p^{w}(1 - p)^{n-w} = 0\cdot08,$$

and

$$P(w \geqslant 10) = \sum_{w=10}^{n} \binom{n}{w} p^{w}(1 - p)^{n-w} = 0\cdot05.$$

Hence, using the normal approximation,

$$\Phi[(2 - np + 0\cdot5)/\{np(1 - p)\}^{\frac{1}{2}}] = 0\cdot08,$$

and
$$1 - \Phi[(10 - np - 0\cdot5)/\{np(1 - p)\}^{\frac{1}{2}}] = 0\cdot05,$$

or
$$\Phi[(9\cdot5 - np)/\{np(1 - p)\}^{\frac{1}{2}}] = 0\cdot95.$$

Therefore
$$(2\cdot5 - np)/[np(1 - p)]^{\frac{1}{2}} = -1\cdot405, \qquad (1)$$

and
$$(9\cdot5 - np)/[np(1 - p)]^{\frac{1}{2}} = 1\cdot645,$$

so that
$$(2\cdot5 - np)/(9\cdot5 - np) = -1\cdot405/1\cdot645,$$

which gives on solution
$$np = 5\cdot7246.$$

Hence, from (1),

$$(2\cdot5 - 5\cdot7246)^2 = (1\cdot405)^2 \times 5\cdot7246(1 - p),$$

so that
$$1 - p = 10\cdot3980/11\cdot3005 = 0\cdot9201.$$

Therefore $p = 0\cdot0799$ and $n = 71\cdot65 = 72$ to the nearest integer.

Example 9.19

According to post office regulations, the charge for a local call in an area is 6d irrespective of its duration. As a concession to offices and other frequent users of the telephone, the post office agrees to a different system of charging for calls depending upon the hourly intensity of demand on the telephone service. Accordingly, it is agreed that if there are up to ten calls in any hour from the same office telephone, each will be charged at 3d, if there are between eleven and twenty in the hour then at 6d, and if there are more than twenty then at a flat rate of 9d. If the hourly intensity of calls emanating from an office telephone has a Poisson distribution with mean 15, prove that, according to the new system of charging, the reduction to the office in the cost per call is, on the average, 2·24 per cent approximately.

Suppose w is a random variable denoting the number of calls in an hour emanating from the office telephone. Then w has a Poisson distribution with mean 15. Therefore

$$P(w \leqslant 10) = \sum_{w=0}^{10} e^{-15}(15)^w/w!$$

$$\sim \Phi[(10 - 15 + 0\cdot5)/\sqrt{15}], \quad \text{using the normal approximation,}$$

$$= \Phi(-4\cdot5/3\cdot873) = 1 - \Phi(1\cdot162) = 1 - 0\cdot8774 = 0\cdot1226.$$

Similarly,

$$P(11 \leqslant w \leqslant 20) = \sum_{w=11}^{20} e^{-15}(15)^w/w!$$

$$\sim \Phi[(20 - 15 + 0\cdot5)/\sqrt{15}] - \Phi[(11 - 15 - 0\cdot5)/\sqrt{15}]$$

$$= \Phi(1\cdot420) + \Phi(1\cdot162) - 1 = 0\cdot9222 + 0\cdot8774 - 1 = 0\cdot7996.$$

Finally,

$$P(w \geqslant 21) = \sum_{w=21}^{\infty} e^{-15}(15)^w/w!$$

$$\sim 1 - \Phi[(21 - 15 - 0\cdot5)/\sqrt{15}] = 1 - \Phi(1\cdot420) = 0\cdot0778.$$

Hence the expected cost of a call in any hour is

$$3 \times 0\cdot1226 + 6 \times 0\cdot7996 + 9 \times 0\cdot0778 = 5\cdot8656d,$$

whence the percentage decrease in the cost per call is

$$100(6 - 5\cdot8656)/6 = 2\cdot24 \text{ approximately.}$$

9.18   The multinomial distribution

This is a very general discrete probability distribution which arises quite naturally by an extension of the sampling experiment leading to the binomial distribution. We now consider that the infinite population consists of balls of $k (\geqslant 2)$ different colours which exist in proportions $p_1, p_2, ..., p_k$ respectively such that

$$0 \leqslant p_i \leqslant 1 \quad \text{and} \quad \sum_{i=1}^{k} p_i \equiv 1.$$

The experiment consists of drawing a sample of N balls from the population and noting the number of balls of each of the k colours in the sample. Suppose there are n_i balls of the ith colour $(i = 1, 2, ..., k)$ where, of course,

$$0 \leqslant n_i \leqslant N \quad \text{and} \quad \sum_{i=1}^{k} n_i \equiv N.$$

The n_i are k random variables, but they are not all independent since their sum must be necessarily N, the sample size. The condition

$$\sum_{i=1}^{k} n_i \equiv N$$

is said to be a *linear constraint* on the k random variables so that we have, in effect, only $k - 1$ independent random variables. The multinomial distribution arises by considering the joint distribution of the n_i.

Suppose a particular sequence of N independent drawings from the population is given by n_1 balls of the first colour; then n_2 balls of the second colour; and so on till n_k balls of the kth colour. The probability of obtaining this particular *ordering* of the N drawings of the sample is

$$p_1^{n_1} p_2^{n_2} \cdots p_k^{n_k}. \tag{1}$$

But any rearrangement of the order in which the different balls occur also leads to the probability (1) and so obviously satisfies the conditions of the problem. Now the number of such possible arrangements is

$$N! \Big/ \prod_{i=1}^{k} n_i!$$

Hence the required probability of the observed sample is

$$P(n_1, n_2, ..., n_k) = N! \prod_{i=1}^{k} p_i^{n_i} / n_i!,$$

where the n_i satisfy the conditions stated above. These probabilities define the multinomial distribution. It is clear that these probabilities arise from an expansion of the multinomial $(p_1 + p_2 + ... + p_k)^N$, since

$$\sum_{n_i} P(n_1, n_2, ..., n_k) = \sum_{n_i} N! \prod_{i=1}^{k} p_i^{n_i} / n_i! = (p_1 + p_2 + ... + p_k)^N \equiv 1,$$

where the summation extends over all the possible values of the n_i.

Example 9.20

For $k = 2$; the multinomial distribution becomes the binomial distribution where

$$n_1 + n_2 = N; \quad p_1 = p; \quad p_2 = 1 - p_1 = q.$$

Example 9.21
For $k = 6$, the multinomial distribution gives the probability of the number of times each of the six faces of a cubical die turns up in N throws. If the observed frequencies of the six faces are $n_1, n_2, \ldots, n_6$, and their respective probabilities are $p_1, p_2, \ldots, p_6$, then the probability of the sample is

$$P(n_1, n_2, \ldots, n_6) = N! \prod_{i=1}^{6} p_i^{n_i}/n_i!$$

In particular, if the die is unbiased, then each $p_i = 1/6$ and so

$$P(n_1, n_2, \ldots, n_6) = N!/6^N \prod_{i=1}^{6} n_i!$$

Example 9.22
A more practically useful example of the multinomial distribution arises quite often in the statistical study of particulate inheritance. Animals or plant progenies obtained from certain types of crossings may, for example, be of four distinct types. In a given sample, the observed frequencies of the types have a multinomial distribution with probabilities p_i. The values of the p_i are deduced appropriately by the laws of Mendelian inheritance and may depend upon one or more parameters. A typical example of such data is given in Table 9.11. The four classes are of the progeny of *Pharbitis* (Morn-

Table 9.11   Showing the observed frequencies and the theoretical probabilities of the four types of progeny obtained by an experimental crossing of *Pharbitis*

Class	AB	Ab	aB	ab
Observed frequency	187	35	37	31
Probability	$(2 + \theta)/4$	$(1 - \theta)/4$	$(1 - \theta)/4$	$\theta/4$

Source: Y. Imai (1931), *Genetics*, Vol. 16, p. 26.

ing Glory) and the probabilities, involving a parameter θ, are based on a certain Mendelian law of inheritance. Accordingly, the probability of the observed sample is

$$\frac{290!}{187!\ 35!\ 37!\ 31!} [(2 + \theta)/4]^{187} [(1 - \theta)/4]^{72} [\theta/4]^{31}.$$

This probability can be used as the basis of statistical analysis to answer questions about the parameter θ which are of interest to a geneticist. An exposition of these ideas will be presented later, and we have cited Imai's data only to give an illustration of an application of the multinomial distribution in a simple but frequent situation in biometry.

Example 9.23
As a final example of the multinomial distribution, we consider a question based on Example 9.19. Find the probability that of four random calls made at the office, two are charged at 6d, and one each at 3d and 9d respectively.
The probability of a call being charged at 3d, 6d or 9d is respectively 0·1226, 0·7996, 0·0778. Hence, using the multinomial distribution for $k = 3$ and $N = 4$, the required probability is

$$\frac{4!}{2! \ 1! \ 1!} (0 \cdot 7996)^2 \times 0 \cdot 1226 \times 0 \cdot 0778 \ = \ 0 \cdot 0732.$$

The multinomial distribution is a *multivariate* distribution since it involves the joint probabilistic behaviour of more than one random variable. A theoretical investigation of the properties of this distribution will be considered later when we have presented some of the salient ideas of joint distributions of random variables.

9.19   Chebyshev's inequality

This classic inequality is an important theorem in probability theory and, in the main, it is due to the Russian mathematician Pafnutii L'vovich Chebyshev (1821 – 1894).

Suppose x is any random variable, discrete or continuous, but having a finite variance σ^2. If the expectation of x is μ and k any positive number, then Chebyshev's inequality states that

$$P(|x - \mu| \geqslant k\sigma) \leqslant 1/k^2, \quad \text{or} \quad P(\mu - k\sigma < x < \mu + k\sigma) \geqslant 1 - 1/k^2.$$

Expressed in words, the inequality states that the probability of a value of x lying outside the range $\pm k\sigma$ on either side of the mean is not greater than $1/k^2$. It is important to note that we have made no assumption about the probability distribution of x except that it has a finite variance and so, necessarily, a finite mean. The proof of this inequality is simple and, in principle, applies equally to discrete and continuous random variables. We assume here that x is a discrete random variable such that the probability

$$P(x) \ = \ p_x, \quad \text{for } x \geqslant 0.$$

Then, by definition,

$$\text{var}(x) \ = \ \sum_{x=0}^{\infty} (x - \mu)^2 p_x \ = \ \sigma^2.$$

Now each term in the summation is non-negative, and so the omission of some terms cannot increase the sum of the remaining series. Therefore, if we delete all terms (if any) for which

$$|x - \mu| < k\sigma,$$

we obtain

$$\sigma^2 \geqslant \sum_{x}^{*} (x - \mu)^2 p_x,$$

where the asterisk indicates that the summation extends over only those values of x for which $|x - \mu| \geqslant k\sigma$. Hence we again cannot increase this sum by replacing $(x - \mu)^2$ by $k^2 \sigma^2$. Therefore

$$\sigma^2 \geqslant \sum_{x}^{*} (k\sigma)^2 p_x,$$

so that

$$1/k^2 \geqslant \sum_{x}^{*} p_x \ = \ P(|x - \mu| \geqslant k\sigma).$$

Hence $\qquad\qquad P(|x - \mu| \geqslant k\sigma) \leqslant 1/k^2,$

or, equivalently in terms of the complementary events,

$$P(|x - \mu| < k\sigma) \geqslant 1 - 1/k^2.$$

If $0 < k < 1$, then the inequality does not give any useful information about the probabilistic behaviour of the random variable x. For then $1/k^2 >$ 1, and the inequality only asserts the obvious fact that the required probability is less than a number greater than unity. However, for $k > 1$, the inequality provides some useful information. Thus for $k = 2$, we have

$$P(|x-\mu| \geqslant 2\sigma) \leqslant \tfrac{1}{4},$$

and for $k = 3$,

$$P(|x-\mu| \geqslant 3\sigma) \leqslant \tfrac{1}{9},$$

The general nature of the inequality is clear: the greater the deviations of x from its expectation, the smaller is the probability of the occurrence of such deviations. The main interest of the inequality lies in its complete generality, apart from the existence of the variance of x. However, a price has to be paid for such generality. Indeed, the inequality can be considerably sharpened if some further assumptions are made about the nature of the probability distribution of x. We shall not go into such mathematical niceties. A simple example illustrates this point adequately. Suppose that x is normally distributed with mean μ and variance σ^2. Then we know from the tables of the normal probability integral that

$$P(|x-\mu| \geqslant 2\sigma) = 0\cdot04550; \quad P(|x-\mu| \geqslant 3\sigma) = 0\cdot00270.$$

These values are considerably smaller than $\tfrac{1}{4}$ and $\tfrac{1}{9}$ respectively given by the inequality.

For our present purpose, we have introduced this inequality because its application leads directly to another remarkable theorem, due to James Bernoulli.

9.20   Bernoulli's theorem

We have seen that the binomial distribution is defined by the point-probabilities

$$B(w) = \binom{n}{w} p^w q^{n-w}, \quad \text{for } 0 \leqslant w \leqslant n,$$

where p and q $(= 1-p)$ are the proportions of white and black balls in the infinite population sampled. Then for the random variable w denoting the number of white balls in a random sample of n balls,

$$E(w) = np \quad \text{and} \quad \text{var}(w) = npq.$$

Now $p^* = w/n$ is a function of the random variable w and, quite clearly,

$$E(p^*) = p \quad \text{and} \quad \text{var}(p^*) = pq/n.$$

Hence, by Chebyshev's inequality,

$$P[|p^* - p| \geqslant k(pq/n)^{\frac{1}{2}}] \leqslant 1/k^2$$

or, if we set $\delta = k(pq/n)^{\frac{1}{2}}$, then

$$P[|p^* - p| \geqslant \delta] \leqslant pq/n\delta^2.$$

But for $0 \leqslant p \leqslant 1$, $pq \leqslant \tfrac{1}{4}$, so that

$$P[|p^* - p| \geqslant \delta] \leqslant 1/4n\delta^2.$$

Therefore for any positive number ϵ, however small, we can always find n so large that

$$P[|p^* - p| \geqslant \delta] < \epsilon, \text{ for any given } \delta > 0.$$

This is Bernoulli's theorem, and it may also be stated as follows:

$$\lim_{n \to \infty} P[|p^* - p| \geqslant \delta] = 0, \quad \text{for any given } \delta > 0.$$

This means that p^*, the observed relative frequency of white balls in the sample, *converges in probability* or *stochastically* to the unknown value p as n tends to infinity. It is worth recalling that we have already used this idea in our earlier intuitive approach to probability. Bernoulli's theorem now gives a theoretical justification for using p^* as an "estimate" of the parameter p. This is a central idea underlying the principles of *statistical estimation*, and we shall return to its fuller consideration in a subsequent chapter.

CHAPTER 10

INDIRECT METHODS IN PROBABILITY

Mathematical preliminaries

The combined use of the theorems of probability and the distributions of random variables provide a theoretical basis for the solution of a wide variety of statistical problems. However, there are many problems which are not amenable to such a direct approach, and we have to resort to *indirect* methods for their solution. In general, these problems arise when the probabilistic behaviour of a phenomenon is considered over a period of time. As we have mentioned earlier, this is the central idea behind the theory of stochastic processes, which is by now an important branch of theoretical statistics. In this chapter, we shall be concerned with some of the simplest ideas of an interesting but mathematically difficult subject partly because of their intrinsic importance and partly because this is an effective method for deepening the understanding of the concepts of probability which we have studied so far. The scope of the ideas discussed here is based on the use of *difference equations* and *generating functions*. We therefore first consider these mathematical preliminaries and then show how they are used to solve some simple problems in probability.

10.1 Difference equations

Suppose $y = f(n)$ is a function of an integral variable n, and it is known that y takes the values $y_0, y_1, y_2, \ldots$ corresponding to the equispaced values of the argument $n = 0, 1, 2, \ldots$. A difference equation is a relation between the finite differences of $f(n)$ at one or more general values of the argument n. For example, the following are difference equations:

$$\Delta^2 y_{n+2} - 2\Delta y_{n+1} + 4y_n = 0, \tag{1}$$

$$(\Delta y_n)^2 + 3\Delta^2 y_{n+1} - 5 = 0. \tag{2}$$

Since the rth differences of the function values y_n, that is $\Delta^r y_n$, can always be expressed in terms of $y_n, y_{n+1}, y_{n+2}, \ldots, y_{n+r}$, it is possible to express a difference equation in an alternative way in terms of the function values without directly involving the difference operator Δ. This is how the term *difference equation* has arisen. Thus, for example, since

$$\Delta y_{n+1} = y_{n+2} - y_{n+1} \quad \text{and} \quad \Delta^2 y_{n+2} = y_{n+4} - 2y_{n+3} + y_{n+2},$$

we can also write the difference equation (1) as

$$y_{n+4} - 2y_{n+3} + y_{n+2} - 2(y_{n+2} - y_{n+1}) + 4y_n = 0,$$

or
$$y_{n+4} - 2y_{n+3} - y_{n+2} + y_{n+1} + 4y_n = 0. \qquad (3)$$

Generally, therefore, any functional relation of the form

$$F(n, y_n, y_{n+1}, \ldots, y_{n+r}) = 0 \qquad (4)$$

is a difference equation. Such expressions are frequently found also under the more common name of *recurrence relations*. When an expression for y_n has been found to satisfy (4), it is called a *solution* of the difference equation.

10.2   Derivation of difference equations

We have seen in Chapter 4 that if the coefficients of a power series satisfy a recurrence relation of the type (3), then it is possible to sum the series by using a generating function. This argument can be reversed and we can show that a difference equation can be obtained in this way. It is of some interest to illustrate this derivation of a difference equation, though our main concern is to show how to solve some of the difference equations which arise from specific probability problems.

Example 10.1

Suppose the function $e^x/(1-x)$ is expanded as a power series in x so that

$$e^x/(1-x) = \sum_{r=0}^{\infty} u_r x^r,$$

and we assume that the series is convergent for $|x| < 1$.

Then

$$e^x = (1-x) \sum_{r=0}^{\infty} u_r x^r,$$

so that

$$\sum_{r=0}^{\infty} x^r/r! = u_0 + \sum_{r=1}^{\infty} (u_r - u_{r-1}) x^r.$$

Hence, equating the coefficients of x^r on both sides, we have

$$u_0 = 1, \quad \text{and} \quad u_r - u_{r-1} = 1/r!, \quad \text{for } r \geqslant 1.$$

Therefore, on summation,

$$\sum_{r=1}^{n} (u_r - u_{r-1}) = \sum_{r=1}^{n} 1/r!,$$

so that

$$u_n - u_0 = \sum_{r=1}^{n} 1/r!, \quad \text{or} \quad u_n = \sum_{r=0}^{n} 1/r!$$

Therefore

$$\lim_{n \to \infty} u_n = \sum_{r=0}^{\infty} 1/r! = e.$$

We notice that in this case the determination of u_n has involved solving the difference equation

$$u_r - u_{r-1} = 1/r!, \quad \text{for } r \geqslant 1,$$

with the *initial condition* $u_0 = 1$.

Example 10.2

As another example, consider the expansion

$$(4 - 3x^2)/(x^2 - 7x + 6) = \sum_{r=0}^{\infty} u_r x^r,$$

where the series is assumed to be convergent for $|x| < 1$. Therefore, multiplying both sides by $x^2 - 7x + 6$, we have

$$4 - 3x^2 = (x^2 - 7x + 6)\sum_{r=0}^{\infty} u_r x^r$$

$$= 6u_0 + (6u_1 - 7u_0)x + \sum_{r=2}^{\infty}(6u_r - 7u_{r-1} + u_{r-2})x^r.$$

Hence, equating coefficients of x^r on both sides, we obtain

$$6u_0 = 4,$$
$$6u_1 - 7u_0 = 0,$$
$$6u_2 - 7u_1 + u_0 = -3,$$

and $$6u_r - 7u_{r-1} + u_{r-2} = 0, \quad \text{for } r \geqslant 3.$$

Solution of the first three equations gives

$$u_0 = 2/3, \quad u_1 = 7/9, \quad \text{and} \quad u_2 = 8/27.$$

For $r \geqslant 3$, we have

$$6u_r - 7u_{r-1} + u_{r-2} = 0,$$

or $$6(u_r - u_{r-1}) - (u_{r-1} - u_{r-2}) = 0,$$

or $$6\,\Delta u_{r-1} - \Delta u_{r-2} = 0,$$

or $$\Delta(6u_{r-1} - u_{r-2}) = 0.$$

Therefore $$6u_{r-1} - u_{r-2} = c, \text{ a constant.}$$

Hence, taking $r = 3$, we obtain

$$c = 6u_2 - u_1 = 1.$$

Accordingly, for $r \geqslant 3$, we have the difference equation

$$6u_{r-1} - u_{r-2} = 1, \quad \text{or} \quad 6u_r - u_{r-1} = 1, \quad \text{for } r \geqslant 2.$$

We shall see later how a standard method may be used to solve this difference equation. However, we solve it here by a convenient device.

Multiplying through by 6^{r-1}, we have

$$6^r u_r - 6^{r-1} u_{r-1} = 6^{r-1}.$$

If we now set $v_r = 6^r u_r$, then

$$v_r - v_{r-1} = 6^{r-1}, \quad \text{for } r \geqslant 2.$$

Therefore, on addition,

$$\sum_{r=2}^{n}(v_r - v_{r-1}) = \sum_{r=2}^{n} 6^{r-1},$$

or $$v_n - v_1 = 6\sum_{r=2}^{n} 6^{r-2} = 6\sum_{s=0}^{n-2} 6^s = (6^n - 6)/5.$$

Therefore $$v_n = v_1 + (6^n - 6)/5, \quad \text{for } n \geqslant 2.$$

Now $v_1 = 6u_1 = 14/3$, and thus

$$v_n = \frac{1}{5} 6^n + \left[\frac{14}{3} - \frac{6}{5}\right]$$

$$= \frac{1}{5} 6^n + \frac{52}{15}, \qquad \text{for } n \geqslant 2.$$

Therefore
$$u_n = \frac{1}{5} + \frac{52}{15} 6^{-n}, \qquad \text{for } n \geqslant 2,$$

and
$$\lim_{n \to \infty} u_n = \frac{1}{5}.$$

We observe that the formula for u_n also holds for $n = 1$, but not for $n = 0$.

This method may also be used in the expansion of a trigonometric function in terms of an infinite series of sines and cosines of multiple angles, as the next example shows.

Example 10.3

Suppose $(4 + \cos x)/(17 + 8\cos x) = \sum_{r=0}^{\infty} u_r \cos rx$, where we assume that the infinite series is convergent. We consider a cosine series because the left-hand side is an even function of x. Since $2\cos x \cos rx = \cos(r-1)x + \cos(r+1)x$, we have

$$4 + \cos x = (17 + 8\cos x)\sum_{r=0}^{\infty} u_r \cos rx$$

$$= 17\sum_{r=0}^{\infty} u_r\cos rx + 4\sum_{r=0}^{\infty} u_r[\cos(r-1)x + \cos(r+1)x]$$

$$= 17\sum_{r=0}^{\infty} u_r\cos rx + 4u_0\cos x + 4\sum_{r=0}^{\infty} u_{r+1}\cos rx + 4\sum_{r=1}^{\infty} u_{r-1}\cos rx$$

$$= 17u_0 + 4u_1 + (8u_0 + 17u_1 + 4u_2)\cos x +$$
$$+ \sum_{r=2}^{\infty}(4u_{r-1} + 17u_r + 4u_{r+1})\cos rx.$$

Therefore, equating coefficients of $\cos rx$ on both sides, we have

$$17u_0 + 4u_1 = 4, \qquad\qquad (1)$$

$$8u_0 + 17u_1 + 4u_2 = 1, \qquad\qquad (2)$$

and
$$4u_{r-1} + 17u_r + 4u_{r+1} = 0, \quad \text{for } r \geqslant 2. \qquad (3)$$

The standard method for solving such difference equations gives the general solution of (3) as

$$u_r = A(-4)^r + B(-1/4)^r,$$

where A and B are arbitrary constants. Now in order that the infinite series be convergent we must have $A = 0$, and so the required solution of (3) is
$$u_r = B(-1/4)^r.$$

Therefore $u_0 = B, \quad u_1 = (-1/4)B, \quad u_2 = (-1/4)^2 B$.

Substituting the values of u_0 and u_1 in (1) gives $B = 1/4$. For this value of B, (2) is also satisfied. Thus all the relations (1) to (3) are consistent, and we have the expansion

$$(4 + \cos x)/(17 + 8\cos x) = \frac{1}{4}\sum_{r=0}^{\infty}(-1/4)^r \cos rx.$$

As a simple check, we observe that for $x = 0$ the right-hand side series is

$$\frac{1}{4}\sum_{r=0}^{\infty}(-1/4)^r = \frac{1}{5},$$

so that the expansion gives the correct answer.

Difference equations also arise in the solution of differential equations of some kinds. We illustrate the method by the following example.

Example 10.4

Solve the differential equation

$$(24x^2 - 25x + 6)y'' + 2(48x - 25)y' + 48y = 0,$$

given that $y(0) = 1$ and $y'(0) = 3$.

Assume that

$$y = \sum_{r=0}^{\infty}u_r x^r,$$

where Taylor's expansion of y at $x = 0$ gives $u_r = y^{(r)}(0)/r!$ Then

$$y = u_0 + u_1 x + u_2 x^2 + \ldots + u_r x^r + \ldots,$$
$$y' = u_1 + 2u_2 x + 3u_3 x^2 + \ldots + ru_r x^{r-1} + \ldots,$$
$$y'' = 2u_2 + 6u_3 x + \ldots + r(r-1)u_r x^{r-2} + \ldots.$$

Hence the coefficient of x^r in the differential equation is

$$24r(r-1)u_r - 25r(r+1)u_{r+1} + 6(r+1)(r+2)u_{r+2} + 96\,ru_r -$$
$$- 50(r+1)u_{r+1} + 48u_r = 0.$$

This reduces to

$$(r+1)(r+2)(6u_{r+2} - 25u_{r+1} + 24u_r) = 0.$$

Therefore the u_r satisfy the difference equation

$$6u_{r+2} - 25u_{r+1} + 24u_r = 0, \quad \text{for } r \geqslant 0.$$

The general solution of this difference equation is

$$u_r = A(8/3)^r + B(3/2)^r,$$

where A and B are arbitrary constants. To determine them, we note that

$$u_0 = y(0) = 1; \quad u_1 = y'(0) = 3.$$

Hence $A + B = 1; \quad (8/3)A + (3/2)B = 3.$

Solving these simultaneous equations, we obtain $A = 9/7$ and $B = -2/7$. Thus

$$u_r = \frac{9}{7}(8/3)^r - \frac{2}{7}(3/2)^r,$$

and so $y = \frac{9}{7}\sum_{r=0}^{\infty}(8x/3)^r - \frac{2}{7}\sum_{r=0}^{\infty}(3x/2)^r.$

We can sum these series for $|x| < 3/8$ to obtain

$$y = (9/7)(1 - 8x/3)^{-1} - (2/7)(1 - 3x/2)^{-1}$$
$$= (6 - 7x)/(3 - 8x)(2 - 3x).$$

Difference equations may also be obtained from a direct statement of a mathematical problem. We illustrate this by evaluating a determinant which has a certain type of symmetry. More general determinants of this kind whose elements are algebraic expressions have important statistical applications.

Example 10.5

Evaluate the $n \times n$ determinant

$$D_n = \begin{vmatrix} 13 & 6 & 0 & 0 & . & . & . \\ 6 & 13 & 6 & 0 & . & . & . \\ 0 & 6 & 13 & 6 & . & . & . \\ 0 & 0 & 6 & 13 & . & . & . \\ . & . & . & . & . & . & . \\ . & . & . & . & . & . & . \\ . & . & . & . & . & . & . \end{vmatrix} .$$

Expanding D_n by the first row, we find that

$$D_n = 13D_{n-1} - 36D_{n-2},$$

or, changing n to $n+2$, we have the difference equation

$$D_{n+2} - 13D_{n+1} + 36D_n = 0, \quad \text{for } n \geqslant 3.$$

It can be shown that the general solution of this difference equation is

$$D_n = A \, 3^{2n} + B \, 2^{2n},$$

where A and B are arbitrary constants. These are evaluated simply by noting that $D_1 = 13$ and $D_2 = 133$. Thus A and B are determined from the simultaneous equations

$$D_1 = 4A + 9B = 13; \quad D_2 = 16A + 81B = 133.$$

Therefore $A = -4/5$ and $B = 9/5$, so that

$$D_n = (3^{2n+2} - 2^{2n+2})/5.$$

As a somewhat different method of using a difference equation, consider the following statistical problem.

Example 10.6

Suppose x is a continuous random variable having a probability density function proportional to

$$e^{-x} \cos^{\nu} x, \quad \text{for } -\pi/2 \leqslant x \leqslant \pi/2,$$

where $\nu \geqslant 0$ is a parameter. Show that if

$$g(x) = [(\nu+2)\cos 2x - \sin 2x] \cos^2 x,$$

then

$$E[g(x)] = (\nu+1)(\nu+2)(\nu+4)/[1 + (\nu+4)^2].$$

$$\text{Let} \quad I(\nu) = \int_{-\pi/2}^{\pi/2} e^{-x} \cos^{\nu} x \, dx,$$

so that the probability density function of x is

$$e^{-x} \cos^{\nu} x / I(\nu).$$

Now integrating by parts twice gives

$$I(\nu) = \int_{-\pi/2}^{\pi/2} e^{-x} \cos^{\nu} x \, dx$$

$$= \left[-e^{-x}\cos^\nu x\right]_{-\pi/2}^{\pi/2} - \nu \int_{-\pi/2}^{\pi/2} e^{-x}\cos^{\nu-1} x \sin x\, dx$$

$$= -\nu\left[-e^{-x}\cos^{\nu-1} x \sin x\right]_{-\pi/2}^{\pi/2} - \nu \int_{-\pi/2}^{\pi/2} e^{-x}\left[-(\nu-1)\cos^{\nu-2} x \sin^2 x + \cos^\nu x\right] dx$$

$$= \nu \int_{-\pi/2}^{\pi/2} e^{-x}\left[(\nu-1)\cos^{\nu-2} x - \nu\cos^\nu x\right] dx .$$

Hence $\qquad I(\nu) = \nu(\nu-1)I(\nu-2) - \nu^2 I(\nu)$,

or $\qquad I(\nu) = [\nu(\nu-1)/(1+\nu^2)] I(\nu-2)$,

or $\qquad I(\nu+2) = [(\nu+1)(\nu+2)/\{1+(\nu+2)^2\}] I(\nu)$, $\qquad$ (1)

which is the required recurrence relation for $I(\nu)$.

Finally, since

$$g(x) = [(\nu+2)\cos 2x - \sin 2x]\cos^2 x$$

$$= 2(\nu+2)\cos^4 x - (\nu+2)\cos^2 x - 2\sin x \cos^3 x,$$

therefore

$$E[g(x)] = \frac{1}{I(\nu)} \int_{-\pi/2}^{\pi/2} e^{-x}\cos^\nu x\left[2(\nu+2)\cos^4 x - (\nu+2)\cos^2 x - 2\cos^3 x \sin x\right] dx$$

$$= \frac{1}{I(\nu)}\left[2(\nu+2) I(\nu+4) - (\nu+2) I(\nu+2) - 2\int_{-\pi/2}^{\pi/2} e^{-x}\cos^{\nu+3} x \sin x\, dx\right].$$

But

$$\int_{-\pi/2}^{\pi/2} e^{-x}\cos^{\nu+3} x \sin x\, dx = \left[-e^{-x}(\nu+4)^{-1}\cos^{\nu+4} x\right]_{-\pi/2}^{\pi/2} - (\nu+4)^{-1}\int_{-\pi/2}^{\pi/2} e^{-x}\cos^{\nu+4} x\, dx$$

$$= -I(\nu+4)/(\nu+4) .$$

Hence

$$E[g(x)] = [2(\nu+2) I(\nu+4) - (\nu+2) I(\nu+2) + 2(\nu+4)^{-1} I(\nu+4)]/I(\nu)$$

$$= \frac{1}{I(\nu)}\left[\frac{2\{1+(\nu+2)(\nu+4)\}}{\nu+4} I(\nu+4) - (\nu+2) I(\nu+2)\right].$$

Therefore, using (1) by putting $\nu+2$ for ν, we obtain

$$E[g(x)] = \frac{1}{I(\nu)}\left[\frac{2\{1+(\nu+2)(\nu+4)\}}{\nu+4}\frac{(\nu+3)(\nu+4)}{\{1+(\nu+4)^2\}} I(\nu+2) - (\nu+2) I(\nu+2)\right]$$

$$= \frac{I(\nu+2)}{I(\nu)}\left[\frac{2(\nu+3)\{1+(\nu+2)(\nu+4)\} - (\nu+2)\{1+(\nu+4)^2\}}{1+(\nu+4)^2}\right]$$

$$= \frac{(\nu+1)(\nu+2)}{\{1+(\nu+2)^2\}}\left[\frac{(\nu+4) + 2(\nu+2)(\nu+3)(\nu+4) - (\nu+2)(\nu+4)^2}{1+(\nu+4)^2}\right]$$

using (1),

$$= (\nu+1)(\nu+2)(\nu+4)/[1+(\nu+4)^2] .$$

10.3   Formal derivation of a difference equation

In general, suppose $y_n = f(n, c)$, where c is a constant. Then

$$y_{n+1} = f(n+1, c).$$

The elimination of c between these two equations gives a relation between n, y_n, and y_{n+1}, which may be formally denoted as

$$F(n, y_n, y_{n+1}) = 0.$$

This is a difference equation of the *first order*. The order is the interval between the y values with the largest and smallest suffixes, in this case $(n+1) - n = 1$. More generally, when

then
$$\begin{aligned}
y_n &= f(n, c_1, c_2, \ldots, c_k), \text{ the } c_i \text{ being constants}, \\
y_{n+1} &= f(n+1, c_1, c_2, \ldots, c_k), \\
y_{n+2} &= f(n+2, c_1, c_2, \ldots, c_k), \\
&\cdots\cdots \\
y_{n+k} &= f(n+k, c_1, c_2, \ldots, c_k).
\end{aligned}$$

There are $k+1$ equations, and if the k constants $c_1, c_2, \ldots c_k$ are eliminated from them, we can formally obtain an equation of the type

$$F(n, y_n, y_{n+1}, \ldots, y_{n+k}) = 0.$$

This is a difference equation of the *kth order*, since the highest suffix of y is $n+k$ and the lowest n. Thus the elimination of the k constants from the equation $y_n = f(n, c_1, c_2, \ldots, c_k)$ leads to a kth-order difference equation, and we may therefore expect that conversely the general solution of a kth-order difference equation will contain k arbitrary constants. It can be shown that this is indeed true under rather general conditions on the difference equation, and in particular in the case of "linear difference equations with constant coefficients" to which we shall restrict ourselves from Section 10.4 onwards.

Example 10.7

Suppose $y_n = cn + 3c^3$, where c is a constant.

Clearly, $c = y_{n+1} - y_n$.

Accordingly, elimination of c gives

$$y_n = n(y_{n+1} - y_n) + 3(y_{n+1} - y_n)^3,$$

or $3y_{n+1}^3 - 9y_n y_{n+1}^2 + 9y_n^2 y_{n+1} - 3y_n^3 + ny_{n+1} - (n+1)y_n = 0,$

which is the required difference equation.

Example 10.8

Suppose $y_n = c_1 3^n + c_2 5^n$, where c_1 and c_2 are constants.

Then $y_{n+1} = 3c_1 3^n + 5c_2 5^n$ and $y_{n+2} = 9c_1 3^n + 25c_2 5^n$.

Elimination of c_1 and c_2 from the above three equations gives

$$\begin{vmatrix} y_n & 1 & 1 \\ y_{n+1} & 3 & 5 \\ y_{n+2} & 9 & 25 \end{vmatrix} = 0, \quad \text{or} \quad \begin{vmatrix} y_n & 1 & 0 \\ y_{n+1} & 3 & 2 \\ y_{n+2} & 9 & 16 \end{vmatrix} = 0.$$

Expansion of the determinant gives
$$30y_n - 1(16y_{n+1} - 2y_{n+2}) = 0,$$
or
$$15y_n - 8y_{n+1} + y_{n+2} = 0,$$
which is the required difference equation.

Example 10.9
Suppose $y_{n+1} = \alpha y_n^\beta$, where α and β are constants.

Then
$$y_{n+2} = \alpha y_{n+1}^\beta; \qquad y_{n+3} = \alpha y_{n+2}^\beta.$$
Therefore
$$y_{n+1}/y_{n+2} = (y_n/y_{n+1})^\beta; \qquad y_{n+2}/y_{n+3} = (y_{n+1}/y_{n+2})^\beta$$
or, in terms of logarithms,
$$\log(y_{n+1}/y_{n+2}) = \beta \log(y_n/y_{n+1}); \qquad \log(y_{n+2}/y_{n+3}) = \beta \log(y_{n+1}/y_{n+2}).$$
Hence
$$(\log y_{n+1} - \log y_{n+2})^2 = (\log y_n - \log y_{n+1})(\log y_{n+2} - \log y_{n+3}),$$
which is the required difference equation for y_n.

10.4   Linear difference equations with constant coefficients

We have seen that it is a relatively simple matter to derive a difference equation. However, no general method exists for the solution of difference equations. We shall consider here only the simplest types of difference equations which can be solved by elementary methods. These are known as *linear difference equations with constant coefficients*.

When the terms of a difference equation involve y_n, y_{n+1}, ..., y_{n+k} to the first degree, the equation is said to be *linear* and of the kth order. Furthermore, if the coefficients of the y_r are constants then the linear difference equation is also said to have *constant coefficients*. The solution of such equations is quite simple, and these equations have a useful role in probability applications.

Solution of first-order difference equations
Consider the simplest difference equation of this type, namely,
$$y_{n+1} - y_n = 0, \qquad \text{or} \quad \Delta y_n = 0. \tag{1}$$
The alternative *operational form* of the difference equation implies that
$$y_n = A, \quad \text{an arbitrary constant.} \tag{2}$$
It is convenient at this stage to introduce the *shift operator* E defined by the relation
$$E = 1 + \Delta.$$
Then, since $y_{r+1} = (1 + \Delta)y_r$, we can write alternatively
$$y_{r+1} = E y_r,$$
and, more generally, the relation
$$y_{r+s} = (1 + \Delta)^s y_r = E^s y_r.$$
In this notation, the difference equation (1) can also be written as

$$(E - 1)y_n = 0, \quad \text{with the solution (2).}$$

Again, the difference equation

$$y_{n+1} - \lambda y_n = 0, \quad \text{where } \lambda \neq 0 \text{ is a constant,} \tag{3}$$

can be solved simply by rewriting the difference equation as

$$y_{n+1}/\lambda^{n+1} - y_n/\lambda^n = 0, \quad \text{or} \quad \Delta(y_n/\lambda^n) = 0.$$

Therefore $y_n = B\lambda^n$,   where B is an arbitrary constant.   (4)

But (3) can also be written in the operational form

$$(E - \lambda)y_n = 0. \tag{5}$$

We thus infer that (5) has the solution (4).

Solution of second-order difference equations

The preceding approach can be extended to solve second-order linear difference equations with constant coefficients. Any difference equation of this kind may be written as

$$y_{n+2} + 2\lambda y_{n+1} + \mu y_n = 0, \tag{6}$$

where λ and μ are non-zero constants. This can be written in the operational form

$$(E^2 + 2\lambda E + \mu)y_n = 0. \tag{7}$$

The solution of this equation now depends upon the nature of the roots of the quadratic equation

$$z^2 + 2\lambda z + \mu = 0. \tag{8}$$

Hence three cases arise according as the roots of (8) are (i) real and unequal; (ii) complex; and (iii) real and equal. We present the solutions in the three cases separately.

(i) *Roots real and unequal*

If α and β are the roots of (8), then the operational form (7) may be rewritten as

$$(E - \alpha)(E - \beta)y_n = 0. \tag{9}$$

Now if y_n is a function of n which satisfies the subsidiary equation

$$(E - \alpha)y_n = 0, \tag{10}$$

then it also satisfies (9). Similarly, if y_n satisfies

$$(E - \beta)y_n = 0, \tag{11}$$

then it also satisfies (9). It therefore follows that we can derive two independent solutions of (9) by solving the two subsidiary equations (10) and (11). The solutions of (10) and (11) are

$$y_n = A\alpha^n \quad \text{and} \quad y_n = B\beta^n,$$

where A and B are arbitrary constants. Hence

$$y_n = A\alpha^n + B\beta^n$$

satisfies (9), and since this contains two arbitrary constants it is the general solution. The generalisation of this method to difference equations of order higher than two is obvious.

Example 10.10

To solve the difference equation

$$y_{n+2} - 7y_{n+1} + 12y_n = 0,$$

consider the equation in the operational form

$$(E^2 - 7E + 12)y_n = 0, \qquad \text{or} \quad (E - 3)(E - 4)y_n = 0.$$

Therefore the general solution of the difference equation is

$$y_n = A\,3^n + B\,4^n,$$

where A and B are arbitrary constants.

Example 10.11

Again, consider the difference equation

$$y_{n+3} - 3y_{n+2} - 4y_{n+1} + 12y_n = 0.$$

This has the operational form

$$(E^3 - 3E^2 - 4E + 12)y_n = 0, \qquad \text{or} \quad (E - 2)(E + 2)(E - 3)y_n = 0.$$

Hence the general solution of the given difference equation is

$$y_n = A\,2^n + B(-2)^n + C\,3^n,$$

where A, B, and C are arbitrary constants.

(ii) *Roots complex*

The above method can also be used for solving the difference equation (6) when the operational equation (7) does not have real roots, that is, when $\lambda^2 - \mu < 0$. If the roots of the quadratic equation (8) are

$$\alpha = a + ib \qquad \text{and} \qquad \beta = a - ib,$$

where a and b are real, then $a = -\lambda$ and $b = \sqrt{\mu - \lambda^2}$. Therefore the general solution of the difference equation (6) is

$$\begin{aligned}
y_n &= A\alpha^n + B\beta^n \\
&= A(a + ib)^n + B(a - ib)^n \\
&= (a^2 + b^2)^{n/2}[A(\cos n\theta + i\sin n\theta) + B(\cos n\theta - i\sin n\theta)],
\end{aligned}$$

where $\tan\theta \equiv b/a = -(\mu - \lambda^2)^{\frac{1}{2}}/\lambda$.

Thus
$$\begin{aligned}
y_n &= (a^2 + b^2)^{n/2}[(A + B)\cos n\theta + i(A - B)\sin n\theta] \\
&= \mu^{n/2}[Ae^{in\theta} + Be^{in\theta}] \\
&= \mu^{n/2}\tfrac{1}{2}C[e^{i(n\theta+\phi)} + e^{-i(n\theta+\phi)}] \\
&= C\mu^{n/2}\cos(n\theta + \phi),
\end{aligned}$$

where C and ϕ are arbitrary constants.

(iii) *Roots equal*

In this case the difference equation (6) may be written in the more convenient form

$$y_{n+2} - 2\lambda y_{n+1} + \lambda^2 y_n = 0, \tag{12}$$

with the operational form
$$(E - \lambda)^2 y_n = 0. \tag{13}$$
To solve this difference equation, we make the substitution
$$y_n = \lambda^n z_n,$$
where z_n is a new dependent variable. Then the difference equation (12) may be written as
$$\lambda^{n+2} z_{n+2} - 2\lambda^{n+2} z_{n+1} + \lambda^{n+2} z_n = 0,$$
or $z_{n+2} - 2z_{n+1} + z_n = 0,$ or $\Delta^2 z_n = 0.$

Hence $z_n = An + B$, where A and B are arbitrary constants.

Therefore $y_n = \lambda^n (An + B).$

An easy generalisation of this result gives that if a difference equation of the kth order has the operational form
$$(E - \lambda)^k y_n = 0,$$
then its general solution is
$$y_n = \lambda^n \sum_{r=0}^{k-1} A_r n^r,$$
where the A_r are arbitrary constants.

Example 10.12

Consider the solution of the difference equation
$$y_{n+4} - 5y_{n+3} + 4y_{n+2} + 3y_{n+1} + 9y_n = 0.$$
The operational equation is
$$(E^4 - 5E^3 + 4E^2 + 3E + 9)y_n = 0, \quad \text{or} \quad (E - 3)^2(E^2 + E + 1)y_n = 0.$$
Now the roots of the quadratic $z^2 + z + 1 = 0$ are $(-1 \pm i\sqrt{3})/2$.

Hence, using the results of (ii) and (iii), the required solution is
$$y_n = (A_0 + A_1 n)3^n + A_2 \cos(n\pi/3 + \phi),$$
where A_0, A_1, A_2, and ϕ are arbitrary constants.

Initial conditions

We have so far considered methods for the general solution of linear difference equations with constant coefficients, but in statistical applications interest usually centres on solutions which satisfy certain *initial* or *boundary conditions*. The procedure in such cases is to determine the arbitrary constants in the general solution so that the initial conditions are satisfied. This then leads to the required solution of the difference equation. The following examples illustrate this procedure.

Example 10.13

Solve the difference equation
$$6y_{n+2} + y_{n+1} - y_n = 0,$$
given that $y_0 = 0$ and $y_1 = 4$.

The operational form of the difference equation is
$$(6E^2 + E - 1)y_n = 0, \quad \text{or} \quad (3E - 1)(2E + 1)y_n = 0.$$

Therefore the general solution is

$$y_n = A(1/3)^n + B(-1/2)^n,$$

where A and B are arbitrary constants.

Next, using the initial conditions $y_0 = 0$ and $y_1 = 4$, the equations for A and B are

$$A + B = 0; \qquad \frac{1}{3}A - \frac{1}{2}B = 4.$$

Therefore $A = 24/5$ and $B = -24/5$. Hence the required solution satisfying the boundary conditions is

$$y_n = \frac{24}{5}\left[(1/3)^n - (-1/2)^n\right].$$

Example 10.14

The successive terms of a sequence $y_0, y_1, y_2, \ldots$ are known to satisfy the difference equation

$$y_{n+2} + \lambda y_{n+1} + \mu y_n = 0,$$

where λ and μ are unknown constants. Determine the expression for y_n if it is known that

$$y_0 = 0, \quad y_1 = 1, \quad y_2 = 4, \quad \text{and} \quad y_3 = 7.$$

The difference equation is of second order and so it has two arbitrary constants in its general solution. We need two initial conditions to determine these constants and, in addition, two more to determine λ and μ. Substituting $n = 0$ and $n = 1$ successively in the difference equation, we have

$$4 + \lambda = 0 \quad \text{and} \quad 7 + 4\lambda + \mu = 0,$$

whence $\lambda = -4$ and $\mu = 9$. Therefore the difference equation is

$$y_{n+2} - 4y_{n+1} + 9y_n = 0,$$

with the operational form

$$(E^2 - 4E + 9)y_n = 0.$$

Now the roots of the quadratic equation $z^2 - 4z + 9 = 0$ are $2 \pm i\sqrt{5}$.

Hence the general solution of the given difference equation is

$$y_n = A(2 + i\sqrt{5})^n + B(2 - i\sqrt{5})^n,$$

A and B being arbitrary constants.

Now $y_0 = 0$ and $y_1 = 1$, so that the equations for determining A and B are

$$A + B = 0; \qquad A(2 + i\sqrt{5}) + B(2 - i\sqrt{5}) = 1.$$

Therefore

$$A = -i/2\sqrt{5} \quad \text{and} \quad B = i/2\sqrt{5}.$$

Thus

$$y = \frac{i}{2\sqrt{5}}\left[(2 - i\sqrt{5})^n - (2 + i\sqrt{5})^n\right]$$

$$= \frac{i}{2\sqrt{5}}\left[(\cos n\theta - i\sin n\theta) - (\cos n\theta + i\sin n\theta)\right]3^n$$

$$= \frac{3^n}{\sqrt{5}}\sin n\theta, \qquad \text{where } \tan\theta \equiv \frac{1}{2}\sqrt{5}.$$

10.5   Non-homogeneous linear difference equations

So far we have restricted ourselves to the solution of linear difference equations in which the right-hand side is zero. These are known as *homogeneous* linear difference equations. We can readily extend the methods of solution to *non-homogeneous* difference equations in which the right-hand side is a constant. We shall not consider here the more general form of non-homogeneous difference equations in which the right-hand side is a function of the independent variable n.

Suppose the difference equation is

$$y_{n+2} + 2\lambda y_{n+1} + \mu y_n = c,$$

where λ, μ, and c are constants. To reduce this difference equation to a homogeneous form, we make the transformation

$$y_n = z_n + \gamma,$$

where γ is a constant so chosen that

$$z_{n+2} + \gamma + 2\lambda(z_{n+1} + \gamma) + \mu(z_n + \gamma) = c$$

reduces to

$$z_{n+2} + 2\lambda z_{n+1} + \mu z_n = 0.$$

Clearly, therefore, $\gamma = c/(1 + 2\lambda + \mu)$, provided $1 + 2\lambda + \mu \neq 0$. Hence, if

$$z_n = A\alpha^n + B\beta^n$$

is the general solution of the homogeneous difference equation, then the general solution of the original difference equation is

$$y_n = c/(1 + 2\lambda + \mu) + A\alpha^n + B\beta^n,$$

where A and B are arbitrary constants.

When $1 + 2\lambda + \mu = 0$, then an appropriate choice of γ is impossible. In this case the difference equation may be written as

$$y_{n+2} + 2\lambda y_{n+1} - (1 + 2\lambda)y_n = c.$$

To solve this equation, we make the transformation

$$y_n = z_n + \delta n,$$

where δ is a constant so chosen that

$$z_{n+2} + \delta(n+2) + 2\lambda[z_{n+1} + \delta(n+1)] - (1+2\lambda)[z_n + \delta n] = c$$

is reduced to the homogeneous form

$$z_{n+2} + 2\lambda z_{n+1} - (1 + 2\lambda)z_n = 0.$$

Thus $\delta = c/2(1+\lambda)$. Hence, if the general solution of this difference equation is

$$z_n = A\alpha^n + B\beta^n,$$

then

$$y_n = cn/2(1 + \lambda) + A\alpha^n + B\beta^n.$$

This method can obviously be extended to higher order non-homogeneous linear difference equations.

10.6   Solution by the use of generating functions

The method of solving a homogeneous linear difference equation with constant coefficients depends upon the factorisation of the operational equation. However, when the operational equation is of degree greater than two, it is not always possible to factorise it simply into all its linear factors. In such cases, we can use another general and powerful method of solving the difference equation. This method rests upon the use of a generating function and is due to Laplace. We have already introduced this method in a slightly different form to sum a power series whose coefficients satisfy a recurrence relation.

Consider, for example, the solution of the difference equation

$$y_n - y_{n-1} + \lambda\mu^r y_{n-r-1} = 0, \quad \text{for} \quad n \geqslant r+1,$$

where λ and μ are positive parameters. This is a $(r+1)$th-order difference equation, and suppose its initial conditions are

$$y_\nu = 1, \quad \text{for} \quad 0 \leqslant \nu \leqslant r-1, \quad \text{and} \quad y_r = \mu^r.$$

To solve this difference equation by the method of generating functions, we assume that the generating function is

$$G(\theta) = \sum_{\nu=0}^{\infty} y_\nu \theta^\nu,$$

so that y_ν is the coefficient of θ^ν in $G(\theta)$. Now

$$G(\theta) = y_0 + y_1\theta + y_2\theta^2 + \ldots + y_{r-1}\theta^{r-1} + y_r\theta^r + y_{r+1}\theta^{r+1} + \ldots,$$

$$\theta\, G(\theta) = y_0\theta + y_1\theta^2 + \ldots + y_{r-2}\theta^{r-1} + y_{r-1}\theta^r + y_r\theta^{r+1} + \ldots,$$

$$\lambda\mu^r\,\theta^{r+1}G(\theta) = y_0\lambda\mu^r\,\theta^{r+1} + \ldots.$$

Hence, on summation,

$$(1 - \theta + \lambda\mu^r\theta^{r+1})\, G(\theta) = 1 - (1 - \mu^r)\theta^r,$$

the other terms on the right-hand side being evidently zero because of the initial conditions and the difference equation for $n \geqslant r+1$.

Therefore

$$G(\theta) = \frac{1 - (1 - \mu^r)\theta^r}{1 - \theta + \lambda\mu^r\theta^{r+1}}.$$

Thus $G(\theta)$ is a rational function of θ, and when expanded as a power series in θ, then the coefficient of θ^n is y_n.

This method is completely general and can be applied to all kinds of difference equations.

10.7   Simultaneous linear difference equations

The methods developed so far have been concerned with the solution of difference equations involving only one dependent variable. We next consider the solution of linear difference equations when two or more dependent variables are involved. The general principle is to derive a linear equation in one of the dependent variables which is solved first. The solutions for the other dependent variables are then obtained in terms of the first dependent variable. This is best explained by the following example.

Example 10.15

Suppose we have the simultaneous difference equations

$$x_{n+1} + \lambda y_{n+1} + \mu y_n = 1, \tag{1}$$

and

$$x_{n+1} + x_n + y_{n+1} = 0. \tag{2}$$

These involve the two dependent variables x_n and y_n. Also, λ and μ are constants. Then, from (2),

$$y_{n+1} = -(x_{n+1} + x_n),$$

and so

$$y_n = -(x_n + x_{n-1}).$$

Hence, from (1), elimination of y_{n+1} and y_n gives

$$x_{n+1} - \lambda(x_{n+1} + x_n) - \mu(x_n + x_{n-1}) = 1,$$

or

$$(1-\lambda)x_{n+1} - (\lambda+\mu)x_n - \mu x_{n-1} = 1.$$

Therefore, writing n for $n-1$, we have the difference equation in x_n

$$(1-\lambda)x_{n+2} - (\lambda+\mu)x_{n+1} - \mu x_n = 1.$$

This is a second-order non-homogeneous linear difference equation, and if we make the transformation

$$x_n = z_n + \gamma,$$

then we obtain the homogeneous equation

$$(1-\lambda)z_{n+2} - (\lambda+\mu)z_{n+1} - \mu z_n = 0,$$

where $\gamma = 1/(1 - 2\lambda - 2\mu)$. If the general solution of this homogeneous equation is

$$z_n = A\alpha^n + B\beta^n,$$

A and B being arbitrary constants, then

$$x_n = 1/(1 - 2\lambda - 2\mu) + A\alpha^n + B\beta^n. \tag{3}$$

Finally,

$$\begin{aligned}
y_n &= -(x_n + x_{n-1}) \\
&= -2/(1-2\lambda-2\mu) - A\alpha^n - B\beta^n - A\alpha^{n-1} - B\beta^{n-1} \\
&= -2/(1-2\lambda-2\mu) - A(1+\alpha)\alpha^{n-1} - B(1+\beta)\beta^{n-1}. \tag{4}
\end{aligned}$$

Equations (3) and (4) together constitute the general solution of the given simultaneous difference equations.

10.8 Pseudo-non-linear difference equations

The methods presented above provide a variety of techniques for the solution of linear difference equations of the kind that arise in statistical applications. However, not all difference equations of interest are linear. Indeed, non-linear difference equations are of frequent occurrence, and for many of these equations no methods for their solution exist. On the other hand, there are other difference equations which may be termed *pseudo-non-linear* in the sense that though they are non-linear, nevertheless they can be transformed into linear difference equations. The transformation may be effected through a change of either the dependent or the independent variable. The former transformation is usually the simpler one to carry out, and we shall be concerned with only this here. There is no standard method for making such a transformation as much depends upon the character of the difference equation being transformed. We illustrate some of the simpler devices used by the following examples.

Example 10.16

Solve the difference equation

$$y_{n+3}^2 - 3y_{n+2}^2 + 3y_{n+1}^2 - y_n^2 = 0.$$

This is a third-order non-linear difference equation. The transformation $z_n = y_n^2$ reduces the equation to the directly solvable form

$$z_{n+3} - 3z_{n+2} + 3z_{n+1} - z_n = 0.$$

This linear equation has the operational form

$$(E - 1)^3 z_n = 0,$$

whence the general solution of the difference equation in z_n is

$$z_n = A_0 + A_1 n + A_2 n^2,$$

where the A_r are arbitrary constants. Therefore the general solution of the original difference equation is

$$y_n^2 = A_0 + A_1 n + A_2 n^2.$$

Example 10.17

Solve the difference equation

$$y_n = \sqrt{[y_{n-1} + \sqrt{\{y_{n-2} + \sqrt{(y_{n-3} + \ldots)\}}]}}.$$

Here

$$y_n^2 = y_{n-1} + \sqrt{\{y_{n-2} + \sqrt{(y_{n-3} + \ldots)\}}}$$

$$= y_{n-1} + y_{n-1}.$$

Therefore $y_n^2 = 2y_{n-1}$, or $2 \log y_n = \log 2 + \log y_{n-1}$.

If we next make the transformation $\log y_n = z_n$, then

$$2z_n - z_{n-1} = \log 2.$$

This is a non-homogeneous linear equation of the first order. To solve it we make the transformation $w_n + \gamma = z_n$, where $\gamma = \log 2$. Then the difference equation for w_n is

$$2w_n - w_{n-1} = 0.$$

Therefore $w_n = A2^{-n}$, so that $z_n = A2^{-n} + \log 2$.

Hence $\log y_n = A2^{-n} + \log 2$, or $\frac{1}{2} y_n = \exp(A2^{-n})$,

where A is an arbitrary constant. Thus the general solution is

$$y_n = 2 B^{2^{-n}},$$

where B is an arbitrary constant.

In particular, given that $y_1 = 1$, we have $B = \frac{1}{4}$.

Therefore

$$y_n = 2\left(\tfrac{1}{4}\right)^{2^{-n}} = 2(2^{-2})^{2^{-n}} = 2[2^{-1/2^{n-1}}] = 2^{(1-1/2^{n-1})}.$$

Example 10.18

Solve the difference equation

$$y_n y_{n+2}^3 = y_{n+1}^3 y_{n+3},$$

where $\quad y_0 = 1/\sqrt{2\pi}, \quad y_1 = 1/\sqrt{2\pi e}, \quad$ and $\quad y_2 = 1/\sqrt{2\pi e^4}.$

By taking logarithms, the difference equation may be written as

$$\log y_n + 3 \log y_{n+2} = 3 \log y_{n+1} + \log y_{n+3}.$$

Next put $w_n = \log y_n$, which gives the linear difference equation

$$w_{n+3} - 3w_{n+2} + 3w_{n+1} - w_n = 0$$

or, in operational form,

$$(E - 1)^3 w_n = 0.$$

Hence $\qquad\qquad w_n = A_0 + A_1 n + A_2 n^2,$

so that $\qquad\qquad y_n = \exp(A_0 + A_1 n + A_2 n^2)$

$$= B \exp(A_1 n + A_2 n^2),$$

where A_1, A_2, and B are arbitrary constants.

Now $\qquad\qquad y_0 = 1/\sqrt{2\pi}, \quad$ so that $\quad B = 1/\sqrt{2\pi}.$

Hence $\qquad\qquad y_1 = e^{(A_1 + A_2)}/\sqrt{2\pi} = 1/\sqrt{2\pi e},$

and $\qquad\qquad y_2 = e^{(2A_1 + 4A_2)}/\sqrt{2\pi} = 1/\sqrt{2\pi e^4}.$

Therefore $\qquad A_1 + A_2 = -\frac{1}{2} \quad$ and $\quad 2A_1 + 4A_2 = -2,$

which give $\qquad\qquad A_1 = 0 \quad$ and $\quad A_2 = -\frac{1}{2}.$

Hence $\qquad\qquad y_n = e^{-\frac{1}{2}n^2}/\sqrt{2\pi}.$

Example 10.19

Solve the simultaneous difference equations

$$x_{n+1} = x_n^2 + 2x_n y_n + \tfrac{2}{3}y_n^2 + \tfrac{1}{3}x_n z_n,$$

$$y_{n+1} = x_n y_n + \tfrac{4}{3}y_n^2 + \tfrac{2}{3}x_n z_n + y_n z_n,$$

$$z_{n+1} = \tfrac{2}{3}y_n^2 + \tfrac{1}{3}x_n z_n + 2y_n z_n + z_n^2,$$

with the conditions that $x_n + y_n = p$ and $y_n + z_n = q$, for all $n \geqslant 0$, where p and q are constants such that $p + q = 1$.

We observe that there are five equations involving the three unknowns x_n, y_n, and z_n; but as these equations are consistent it is possible to obtain a solution for the unknowns. We have

$$x_{n+1} - p^2 = x_n^2 + 2x_n y_n + \tfrac{2}{3}y_n^2 + \tfrac{1}{3}x_n z_n - (x_n + y_n)^2$$

$$= \tfrac{1}{3}(x_n z_n - y_n^2).$$

Similarly,

$$y_{n+1} - pq = -\tfrac{1}{3}(x_n z_n - y_n^2),$$

and

$$z_{n+1} - q^2 = \tfrac{1}{3}(x_n z_n - y_n^2).$$

Clearly, $x_n z_n - y_n^2$ is a quantity of intrinsic importance, and if we set

$$D_n = x_n z_n - y_n^2, \quad \text{for} \quad n \geqslant 0,$$

then $\qquad\qquad D_{n+1} = x_{n+1} z_{n+1} - y_{n+1}^2$

$$= [p^2 + \tfrac{1}{3}D_n][q^2 + \tfrac{1}{3}D_n] - [pq - \tfrac{1}{3}D_n]^2$$

$$= \tfrac{1}{3}D_n.$$

We thus have the simple linear difference equation

$$D_{n+1} = \tfrac{1}{3}D_n,$$

so that
$$D_n = \left(\tfrac{1}{3}\right)^n D_0.$$

Therefore
$$x_n = p^2 + \left(\tfrac{1}{3}\right)^n D_0; \qquad y_n = pq - \left(\tfrac{1}{3}\right)^n D_0;$$
$$z_n = q^2 + \left(\tfrac{1}{3}\right)^n D_0.$$

Applications

10.9 Deterministic and stochastic difference equations

A difference equation is a functional relation between the values of one or more dependent variables for values of a discrete independent variable, and we have seen how such equations can be solved in a number of cases. In considering the applications of such equations, we need to identify two distinct elements which are associated with the definition of the dependent variable or variables. The first of these elements is simply the physical component of variation, and the second refers to the probabilistic nature of the variation. In statistical applications both the elements are invariably present, and the difference equations occurring in such situations are known as *stochastic* difference equations. On the other hand, when no probabilistic component is present, then the difference equations are said to be *deterministic* in character. It is worth while to consider a few examples of deterministic difference equations before introducing the additional element of probability. Some of the simplest examples of deterministic difference equations occur in problems of interest and capital accumulation, and we now use them for illustrative purposes.

10.10 Difference equations in interest and capital accumulation

(i) *Simple interest*

An elementary example of a deterministic difference equation arises in calculating the amount due after a given period of time when a certain sum of money is deposited at a fixed rate of simple interest. If a sum of money earns simple interest at the rate ρ (that is the percentage yearly rate is 100ρ), the amount accumulated at any interest date is equal to the amount accumulated one year before that date plus the interest earned in that year on the *initial* principal invested.

Accordingly, if S_n denotes the sum accumulated after n years and if S_0 is the initial sum invested, then

$$S_{n+1} = S_n + \rho S_0,$$

or
$$S_{n+1} - S_n = \rho S_0, \quad \text{for } n \geqslant 0.$$

This is a first-order difference equation which is deterministic in character since no uncertainty is implied in its derivation. The dependent variable S_n represents the amount accumulated after n years so that the unit of variation of the independent variable n is a year. To solve the above difference equation, we sum for $0 \leqslant n \leqslant N - 1$ for some given integer N. Thus

$$\sum_{r=0}^{N-1} (S_{n+1} - S_n) = N\rho S_0,$$

so that $\qquad S_N - S_0 = N\rho S_0,$ or $S_N = (1 + N\rho)S_0.$

This is the well-known simple interest formula giving the amount accumulated after N years for a given initial sum S_0 and annual interest rate ρ.

(ii) *Compound interest*

If the sum S_0 is deposited at a fixed compound interest rate ρ, then the interest at the end of each year is calculated on the total amount accumulated at the beginning of the year. The annual interest accumulated thus changes from year to year but in a definite manner. In this case, the difference equation for S_n, the sum accumulated after n years, is

$$\begin{aligned} S_{n+1} &= S_n + \rho S_n \\ &= (1 + \rho)S_n. \end{aligned}$$

This is again a first-order difference equation with the solution

$$S_{n+1} = (1 + \rho)^{n+1} S_0.$$

Hence the amount accumulated after N years is

$$S_N = (1 + \rho)^N S_0, \quad \text{for} \quad N \geqslant 0.$$

(iii) *Amortisation of a debt*

Amortisation is a method of debt repayment by periodic instalments with the residual debt accumulating compound interest at a fixed rate. Suppose the original debt is A and the compound interest rate ρ per annum. Also, assume that the yearly repayment is R, which consists partly of the interest due and partly of the repayment of the amount A borrowed. If P_n is the outstanding amount after n repayments have been made, then

$$\begin{aligned} P_{n+1} &= P_n + \rho P_n - R \\ &= (1 + \rho)P_n - R, \quad \text{for} \quad n \geqslant 0. \end{aligned} \qquad (1)$$

This is a non-homogeneous first-order linear difference equation, and its solution is obtained readily by rewriting it as

$$P_{n+1} - \gamma = (1 + \rho)(P_n - \gamma), \quad \text{where} \quad \gamma = R/\rho.$$

Hence $\qquad P_{n+1} - \gamma = (1 + \rho)^{n+1}(P_0 - \gamma).$

But it $P_0 = A$, and so the amount due after n repayments is

$$\begin{aligned} P_n &= R/\rho + (1 + \rho)^n(A - R/\rho) \\ &= (1 + \rho)^n A - (R/\rho)[(1 + \rho)^n - 1]. \end{aligned}$$

If the debt is to be fully amortised in N equal instalments of R, then $P_N = 0$. Therefore

$$(1 + \rho)^N A = (R/\rho)[(1 + \rho)^N - 1],$$

so that $\qquad R = \dfrac{\rho A}{1 - (1 + \rho)^{-N}}.$

Hire-purchase agreement

In a hire-purchase agreement, the customer also agrees to repay the price of the article bought by N equal instalments, but the hire-purchase company charges interest at some rate on the full amount A for the entire period of repayment. If the hire-purchase company's stated rate of interest is r per annum, then the total repayment made under the agreement is $A + rNA$. Thus, comparing this with the amount payable under amortisation, we have

$$A(1 + rN) = \frac{\rho NA}{1 - (1 + \rho)^{-N}},$$

so that

$$r = \frac{(1 + \rho)^{-N} + \rho N - 1}{N[1 - (1 + \rho)^{-N}]}.$$

If we now assume that ρN is small, we have approximately

$$r = (N + 1)\rho/2N \sim \tfrac{1}{2}\rho, \quad \text{or} \quad \rho = 2r.$$

This means that by charging interest on the original amount A, the hire-purchase company effectively doubles the rate of compound interest which would be chargeable if the debt were correctly amortised.

Unequal repayments

If the repayments are not constant but of increasing amounts R, $2R$, $3R, \ldots$, then

$$P_{n+1} = P_n + \rho P_n - (n + 1)R,$$

or

$$P_{n+1} - (1 + \rho)P_n = -(n + 1)R, \quad \text{for} \quad n \geqslant 0. \tag{2}$$

The right-hand side of this linear difference equation involves the independent variable n. It is possible to solve such an equation by using a generating function, but we shall solve it here by reducing it to a homogeneous form. To do this, we rewrite the difference equation as

$$[P_{n+1} - \{\alpha + \beta(n + 1)\}] - (1 + \rho)[P_n - \{\alpha + \beta n\}] = 0,$$

where α and β are so chosen that

$$(1 + \rho)(\alpha + \beta n) - [\alpha + \beta(n + 1)] \equiv (n + 1)R, \quad \text{for} \quad n \geqslant 0.$$

Hence, equating constants and the coefficients of n on both sides, we have

$$[(1 + \rho) - 1]\beta = R; \quad \alpha[(1 + \rho) - 1] - \beta = R.$$

Therefore

$$\beta = R/\rho \quad \text{and} \quad \alpha = (1 + \rho)R/\rho^2.$$

If we now make the transformation

$$w_n = P_n - (\alpha + \beta n),$$

then we have the simple difference equation

$$w_{n+1} = (1 + \rho)w_n$$
$$= (1 + \rho)^{n+1} w_0, \quad \text{where } w_0 \text{ is an arbitrary constant.}$$

Hence

$$P_{n+1} = w_{n+1} + \alpha + \beta(n + 1)$$
$$= (1 + \rho)^{n+1} w_0 + (1 + \rho)R/\rho^2 + R(n + 1)/\rho,$$

or

$$P_n = (1 + \rho)^n w_0 + (1 + \rho)R/\rho^2 + Rn/\rho,$$

which is the general solution. But $P_0 = A$, the original sum borrowed, so that

$$w_0 = A - (1 + \rho)R/\rho^2.$$

Hence $P_n = (1 + \rho)^n [A - (1+\rho)R/\rho^2] + (1 + \rho)R/\rho^2 + Rn/\rho$

$\quad\quad\quad = A(1 + \rho)^n - (R/\rho^2)[(1 + \rho)\{(1 + \rho)^n - 1\} - \rho n].$

Finally, if the debt is fully amortised after N repayments, then $P_N = 0$. This gives

$$R = \frac{\rho^2(1 + \rho)^N A}{(1 + \rho)[(1 + \rho)^N - 1] - \rho N}.$$

(iv) *Evaluation of an annuity*

An annuity is a common method of making a saving for the future by means of periodic payments, usually at equispaced intervals of time. These payments accumulate compound interest at a fixed rate, and the whole amount is payable after a stipulated number of years. Suppose a constant sum R is paid each year and the rate of interest is ρ. If S_n is the amount accumulated after n payments, then

$$S_{n+1} = S_n(1 + \rho) + R, \quad \text{for } n \geqslant 0.$$

This equation is of the same form as (1) in (iii) with R replaced by $-R$. Therefore its general solution is

$$S_{n+1} = (-R/\rho) + (1 + \rho)^{n+1}(S_0 + R/\rho).$$

But $S_0 = 0$, so that

$$S_{n+1} = (R/\rho)[(1 + \rho)^{n+1} - 1],$$

or $S_n = (R/\rho)[(1 + \rho)^n - 1], \quad \text{for } n \geqslant 0.$

Therefore the accumulated sum after N payments have been made is

$$S_N = (R/\rho)[(1 + \rho)^N - 1],$$

and if this is required to be equal to a stipulated amount A, then

$$R = \frac{\rho A}{(1 + \rho)^N - 1}.$$

Unequal payments

If the annual payments of the annuity are $R, 2R, 3R, \ldots$, then

$$S_{n+1} = (1 + \rho)S_n + (n + 1)R, \quad \text{for } n \geqslant 0 \quad \text{and} \quad S_0 = 0.$$

This equation is of the same form as (2) in (iii) with $-R$ replaced by R. Hence the general solution of S_n is

$$S_n = w_0(1 + \rho)^n - (Rn/\rho) - R(1 + \rho)/\rho^2.$$

But since $S_0 = 0$, we have

$$w_0 = R(1 + \rho)/\rho^2.$$

Therefore $S_n = (R/\rho^2)[(1 + \rho)\{(1 + \rho)^n - 1\} - \rho n];$

and if the annuity is again to provide a sum A after N payments, then

$$R = \frac{A\rho^2}{(1 + \rho)[(1 + \rho)^N - 1] - \rho N}.$$

Stochastic difference equations

10.11   Games of chance

We shall now consider the more pertinent problem of how difference equations arise in the study of chance phenomena. Such is the diversity of the occurrence of stochastic difference equations that it is possible to approach the subject from many angles in relation to the particular requirements of applied sciences in which this type of probability argument is used. However, we shall not follow this interesting but somewhat more diffused approach and shall restrict our presentation to the context of games of chance. As mentioned earlier, the study of such games led initially to the development of the modern theory of probability and, indeed, some of the well-known historical problems provide instructive illustrations of the use of difference equations. This approach does not imply any triviality in the subject-matter presented because the structure of a game of chance can be readily identified with a parallel and physically more meaningful phenomenon in applied science. Thus, for example, the heads and tails of a coin tossing experiment, which could be regarded as the structural basis of a game of chance, can be identified with the two sexes of a biological population, or with the transmission and failure of a signal in communication theory, or with the success and failure of an animal in a learning situation in experimental psychology. The main difference — and this is of some consequence to us — is that games of chance provide the simplest elucidation of the underlying probability argument.

Formally, a game of chance is the resultant of the outcomes of one or more *contingent* events, that is, events whose occurrence is based on the play of chance. In other words, a game of chance is a random phenomenon whose realisations in a temporal sequence are subject to the laws of probability. Accordingly, the definition of a game of chance requires the specification of

   (i) the contingent events and their associated probabilities, which constitute the structure of the game;

  (ii) the rule to define unambiguously the termination of the game; and

 (iii) the rule to identify the final result as a success, a failure or a draw for a player playing the game against an opponent.

As an example of a game of chance, consider an experiment of rolling two unbiased dice. A game may be defined by the rules that two players A and B alternately roll the dice, A starting the sequence of trials. A wins the game if he throws a total of seven points before B throws a total of six points, B winning in the contrary event. The trials continue until either A or B wins the series. A simple problem associated with this game is that of the determination of the initial probability of success for each of the two players.

A game of chance is said to be *finite* if the game is so defined that it necessarily terminates after a finite number of trials. On the other hand, an *unlimited* game is one which *theoretically* involves an unending sequence of trials. It is worth emphasising at this stage that not all games necessarily require the indirect argument of a difference equation for their solution, as some of the simpler games can be solved by a simple enumeration of the outcomes and their associated probabilities. We shall first consider these direct solutions of some simple games and then lead up to problems where the use of difference equations is the natural method of solution.

10.12   Bernoulli trials

Suppose an experiment can result in any one of two mutually exclusive events, which may be conveniently denoted as success (S) and failure (F). If the trials of the experiment are independent and the probabilities $P(S)$ and $P(F)$ are constant for all trials, then the trials are known as *Bernoulli trials*. We have already come across such trials in the derivation of the binomial distribution from a sampling experiment in which n balls are drawn from an infinite population of white and black balls in proportions p and q respectively $(p + q = 1)$. We may regard an experiment as the drawing of a ball from the population so that the events success and failure are identified with, for example, the colour of the ball drawn being white and black respectively. Therefore

$$P(S) = p \quad \text{and} \quad P(F) = q.$$

When the total number of successes is considered in a fixed number n of trials, then we are led to the binomial distribution. On the other hand, if we consider the number of trials required to realise a fixed number of successes, then the distribution of the number of trials is a negative binomial. Our interest now is in neither of these two random variables, but in the *sequence* of successes and failures which arises when the trials are continued indefinitely. Such a sequence has an interest because it leads to a variety of simple games of chance.

In the above specification of Bernoulli trials we have identified the success or failure outcomes of a trial with simple events – the colour of ball drawn being white or black. In the same way, Bernoulli trials also result from a coin tossing experiment in which the two alternatives of each trial are the simple events – a head (H) or a tail (T). However, it is not essential to the definition of Bernoulli trials that the possible outcomes of a trial should be simple events. We could quite legitimately construct Bernoulli trials on the basis of a more elaborate experimental structure in which the alternatives success and failure are identified with compound events. The basic elements in the definition of Bernoulli trials are that, firstly, there are only two possible and mutually exclusive outcomes of any trial and, secondly, the trials are independent with constant probabilities for the possible outcomes. For example, consider the experiment of rolling two unbiased dice. There are 36 simple events defining the possible outcomes of a trial. If we define success (S) to be the compound event that the sum of points obtained at a trial is seven, and the non-occurrence of this event a failure (F), then the trials as defined are Bernoullian in character. This is so because the trials are independent and the probabilities

$$P(S) = P(7) = 1/6 \quad \text{and} \quad P(F) = 1 - P(7) = 5/6$$

are constant for all trials. More generally, any event which has a constant probability and its complementary event can be used to generate a sequence of Bernoulli trials.

10.13 Finite games

Two players A and B contest a series of games, and their respective probabilities of winning each game are p and q $(p+q = 1)$; the games cannot be drawn. The players agree to play a fixed number n of games, and the player who wins a majority of the games wins the series.

This is a simple example of a finite game based on Bernoulli trials, and we wish to determine the initial probability of each of the two players winning the series. As we shall see, the probabilities depend upon whether n is odd or even, and we consider the two cases separately.

Case I: $n = 2m + 1$, m an integer ≥ 0

The probability that A wins exactly r of the $2m + 1$ games played is

$$\binom{2m+1}{r} p^r q^{2m+1-r}.$$

Therefore the total probability of A winning the series is

$$P(A) = \sum_{r=m+1}^{2m+1} \binom{2m+1}{r} p^r q^{2m+1-r}. \tag{1}$$

Similarly, or by interchanging p and q, the total probability of B winning the series is

$$P(B) = \sum_{r=m+1}^{2m+1} \binom{2m+1}{r} q^r p^{2m+1-r}$$

$$= \sum_{r=m+1}^{2m+1} \binom{2m+1}{2m+1-r} q^r p^{2m+1-r}$$

or, putting $2m+1-r = s$,

$$= \sum_{s=0}^{m} \binom{2m+1}{s} p^s q^{2m+1-s}. \tag{2}$$

Clearly,

$$P(A) + P(B) = (p + q)^{2m+1} \equiv 1.$$

It thus follows that if n is an odd integer, the outcome of the whole series of games must definitely be a win for either A or B.

Case II: $n = 2m$, m an integer ≥ 1

We now have

$$P(A) = \sum_{r=m+1}^{2m} \binom{2m}{r} p^r q^{2m-r}, \tag{3}$$

and

$$P(B) = \sum_{r=m+1}^{2m} \binom{2m}{r} q^r p^{2m-r}$$

$$= \sum_{s=0}^{m-1} \binom{2m}{s} p^s q^{2m-s}. \tag{4}$$

The sum of the probabilities (3) and (4) is

$$(p + q)^{2m} - \binom{2m}{m}(pq)^m = 1 - \binom{2m}{m}(pq)^m.$$

This means that the outcome of the series is now not a certain event in the sense that it will not necessarily lead to a win for either A or B. There is a non-trivial probability

$$\binom{2m}{m}(pq)^m$$

for each of A and B to win exactly one half of the number of games played, so that the outcome of the entire series would be regarded as a draw. The interesting point is that although each game is definitely decided as a win for either A or B, there is a possibility that the whole series may end in a draw.

Problem of points

As another example of a finite game, we have the famous "Problem of points", which was originally considered by Fermat, Pascal, and Montmort. In this problem, the probability is required that A will win α (fixed) games *before B* wins β (fixed) games, the other conditions of play for the individual games remaining the same as in the previous problem.

Here the exact number of games to be played is not fixed in advance, although, quite evidently, the series must be determined after $\alpha + \beta - 1$ games at most and not before at least α games have been played. In general, A can win the series in r games ($\alpha \leqslant r \leqslant \alpha + \beta - 1$) if A wins just α out of the first r games, provided that the games won by A include the rth game, for otherwise A would have won in fewer than r games. This means that the probability of A winning in exactly r games

$$= \text{Probability of } A \text{ winning } \alpha - 1 \text{ games out of the first } r - 1 \text{ games} \times \text{Probability of } A \text{ winning the } r\text{th game}$$

$$= \binom{r-1}{\alpha-1} p^{\alpha-1} q^{r-\alpha} \times p = \binom{r-1}{\alpha-1} p^{\alpha} q^{r-\alpha}.$$

Since there are β mutually exclusive ways in which A can win, for $\alpha \leqslant r \leqslant \alpha + \beta - 1$, the total probability for A winning the series is

$$P(A) = \sum_{r=\alpha}^{\alpha+\beta-1} \binom{r-1}{\alpha-1} p^{\alpha} q^{r-\alpha}$$

$$= \sum_{r=\alpha}^{\alpha+\beta-1} \binom{r-1}{r-\alpha} p^{\alpha} q^{r-\alpha}$$

$$= p^{\alpha} \sum_{t=0}^{\beta-1} \binom{\alpha+t-1}{t} q^{t}. \tag{1}$$

Similarly, or by interchanging (α, β) and (p, q), the total probability for B winning the series is

$$P(B) = \sum_{r=\beta}^{\alpha+\beta-1} \binom{r-1}{\beta-1} q^{\beta} p^{r-\beta}$$

$$= q^{\beta} \sum_{t=0}^{\alpha-1} \binom{\beta+t-1}{t} p^{t}. \tag{2}$$

Aliter

Another instructive way of looking at the above problem is as follows. Suppose that $\alpha + \beta - 1$ games are played. Then A wins the series if B wins in all no more than $\beta - 1$ games. There are thus β distinct possibilities and so

$$P(A) = \sum_{r=0}^{\beta-1} \binom{\alpha+\beta-1}{r} p^{\alpha+\beta-1-r} q^{r}. \tag{3}$$

Similarly,

$$P(B) = \sum_{r=0}^{\alpha-1} \binom{\alpha+\beta-1}{r} q^{\alpha+\beta-1-r} p^{r}. \tag{4}$$

Since (by the change of variable $r = \alpha + \beta - 1 - s$) the right-hand side of (4) can be rewritten as

$$\sum_{s=\beta}^{\alpha+\beta-1} \binom{\alpha+\beta-1}{\alpha+\beta-1-s} q^s p^{\alpha+\beta-1-s}, \quad \text{or} \quad \sum_{s=\beta}^{\alpha+\beta-1} \binom{\alpha+\beta-1}{s} p^{\alpha+\beta-1-s} q^s,$$

it is clear that these expressions for $P(A)$ and $P(B)$ represent the first β and the last α terms of the binomial expansion

$$\sum_{t=0}^{\alpha+\beta-1} \binom{\alpha+\beta-1}{t} p^{\alpha+\beta-1-t} q^t$$

of $(p + q)^{\alpha+\beta-1}$, whence $P(A) + P(B) \equiv 1$.

The equivalence of the two expressions for $P(A)$ [and similarly of those for $P(B)$] is an algebraic identity arrived at on purely probabilistic grounds, and it is instructive, but not quite straightforward, to establish this algebraically. It is sufficient to consider (1) and (3). Now, from (1),

$$P(A) = p^\alpha \sum_{t=0}^{\beta-1} \binom{\alpha+t-1}{t} q^t$$

$$= p^\alpha \sum_{t=0}^{\beta-1} \left[\text{coefficient of } x^t \text{ in } (1 + qx)^{\alpha+t-1}\right]$$

$$= p^\alpha \sum_{t=0}^{\beta-1} \left[\text{coefficient of } x^{\beta-1} \text{ in } x^{\beta-1-t}(1 + qx)^{\alpha+t-1}\right]$$

$$= p^\alpha \times \text{coefficient of } x^{\beta-1} \text{ in } x^{\beta-1}(1 + qx)^{\alpha-1} \sum_{t=0}^{\beta-1} x^{-t}(1 + qx)^t$$

$$= p^\alpha \times \text{coefficient of } x^{\beta-1} \text{ in } (1 + qx)^{\alpha-1}[(1 + qx)^\beta - x^\beta]/(1 - px)$$

$$= p^\alpha \times \text{coefficient of } x^{\beta-1} \text{ in } (1 + qx)^{\alpha+\beta-1}(1 - px)^{-1}$$

$$= p^\alpha \times \text{coefficient of } x^{\beta-1} \text{ in } \left[\sum_{r=0}^{\alpha+\beta-1} \binom{\alpha+\beta-1}{r}(qx)^r \times \sum_{s=0}^{\infty} (px)^s\right],$$

for $|px| < 1$,

$$= p^\alpha \sum_{r=0}^{\beta-1} \binom{\alpha+\beta-1}{r} q^r p^{\beta-1-r},$$

or $\qquad P(A) = \sum_{r=0}^{\beta-1} \binom{\alpha+\beta-1}{r} p^{\alpha+\beta-1-r} q^r$, which is the same as (3).

10.14   Unlimited games

The finite games that we have considered were associated with the Bernoulli trials of a single event whose occurrence and non-occurrence were identified with the success of the two players A and B respectively. The rule for terminating the series of trials ensured that the number of trials made was finite. We now extend this argument to games which require *theoretically* an unlimited number of trials. However, we shall distinguish two types of games which, though simple in the sense that they are directly solvable, indicate an important and characteristic difference. The first type of games are those in which the initial state *recurs* as the trials are continued by the players, and the second type consists of games in which there is *no reversion* to the initial state. A simple example of a game in which there is reversion to the initial state is a football match. The two teams revert to their initial position whenever the goals scored by the sides are equal in number. In contrast, the play in a bridge rubber never reverts to the initial state since each deal played adds points to the score already made.

Reversion to initial state

Consider two independent events X_1 and X_2 such that the constant probabilities for their occurrence at a trial are p_1 and p_2 respectively. The actual description of the events is irrelevant to our development, but as a concrete illustration we may describe the events as follows:

X_1 : In one rolling of two unbiased dice the sum of points obtained is seven.

X_2 : At least four heads occur in five consecutive tosses of an unbiased penny.

Then $P(X_1) \; = \; \dfrac{1}{6}$ and $P(X_2) \; = \; \binom{5}{4}\left(\dfrac{1}{2}\right)^5 + \binom{5}{5}\left(\dfrac{1}{2}\right)^5 \; = \; \dfrac{3}{16}.$

It is clear that repeated trials of X_1 and X_2 are Bernoullian in character, and we may use them to define an unlimited game between two players A and B in the following way.

The trials of X_1 and X_2 are made alternately, starting with X_1. A wins the series if X_1 occurs first, whilst B wins if X_2 occurs first. We wish to determine the initial probability for each of A and B to win the series.

Let x be the required probability that A wins the series. To evaluate x, we determine a simple equation involving x as an unknown by considering the three possible outcomes of the first two trials as follows:

(i) X_1 occurs at the first trial and A wins.

(ii) X_1 does not occur at the first trial, and X_2 occurs at *its* first trial. In this case A loses.

(iii) Neither X_1 nor X_2 occur at their respective first trials and — this is important — the game *reverts* to its initial state when a second trial of X_1 is to be made.

Now, using the notation of complementary events, we can write

$$\Omega \; = \; X_1 + \overline{X}_1 X_2 + \overline{X}_1 \overline{X}_2 \,,$$

where Ω denotes the sample space of the outcomes of the first two trials. Hence, if Y denotes the event that A wins the series, we have

$$Y \; = \; YX_1 + Y\overline{X}_1 X_2 + Y\overline{X}_1 \overline{X}_2 \,.$$

Therefore

$$
\begin{aligned}
P(Y) \; &= \; P(YX_1) + P(Y\overline{X}_1 X_2) + P(Y\overline{X}_1\overline{X}_2) \\
&= \; P(Y|X_1)\,P(X_1) + P(Y|\overline{X}_1 X_2)\,P(\overline{X}_1 X_2) + P(Y|\overline{X}_1\overline{X}_2)\,P(\overline{X}_1\overline{X}_2) \\
&= \; P(Y|X_1)\,P(X_1) + P(Y|\overline{X}_1 X_2)\,P(\overline{X}_1)\,P(X_2) + \\
&\qquad + P(Y|\overline{X}_1\overline{X}_2)\,P(\overline{X}_1)\,P(\overline{X}_2),
\end{aligned}
$$

since the independence of X_1 and X_2 ensures that $(\overline{X}_1, X_2)$ and $(\overline{X}_1, \overline{X}_2)$ are also independent. But

$$P(Y|X_1) \; = \; 1, \quad P(Y|\overline{X}_1 X_2) \; = \; 0, \quad \text{and} \quad P(Y|\overline{X}_1\overline{X}_2) \; = \; x.$$

Hence, if we write $P(X_1) = p_1$ and $P(X_2) = p_2$, we have

$$x \; = \; 1.p_1 + 0.q_1 p_2 + x.q_1 q_2, \quad \text{where} \quad q_1 \; = \; 1 - p_1 \quad \text{and} \quad q_2 \; = \; 1 - p_2,$$

so that $$x = p_1/(1 - q_1q_2). \tag{1}$$

Similarly, if y denotes the probability of B winning the series, we have

$$y = 0 \cdot p_1 + 1 \cdot q_1p_2 + y \cdot q_1q_2,$$

so that $$y = q_1p_2/(1 - q_1q_2).$$

Clearly, $x + y \equiv 1$. It is thus practically certain that the series will be terminated ultimately by a win for one of the two players.

Aliter

In the solution of this problem we have used the kind of argument which, in a more general situation, leads to a difference equation. However, the method is, in principle, indirect since the evaluation of x (or y) is made to depend upon the solution of an equation. The fact that the equation is extremely simple to solve does not make the method direct. As a matter of fact, the direct solution of this problem is also quite straightforward. For, the probability that A wins at the sth ($s \geqslant 1$) trial is the probability that neither X_1 nor X_2 occur in the first $s - 1$ times and then X_1 occurs at its sth trial. Since these cases are mutually exclusive, therefore the probability that A wins in r trials is

$$\sum_{s=1}^{r} (q_1q_2)^{s-1} p_1 = p_1[1 - (q_1q_2)^r]/(1 - q_1q_2).$$

Hence, as $r \to \infty$, the limiting value of this probability is $p_1/(1 - q_1q_2)$, which is the value obtained for x before.

Aliter

There is yet another way of solving the above problem which, in principle, means arriving at the equation (1) for x in two stages. Suppose x_{-1} is the probability of A winning the series when X_1 did not occur at the first trial and before the second trial is made. Then, by considering what happens initially when the first trial takes place, we have

$$x = p_1 \cdot 1 + q_1 \cdot x_{-1}.$$

Similarly, after the first trial resulted in the non-occurrence of X_1, the two possible outcomes of the second trial give

$$x_{-1} = p_2 \cdot 0 + q_2 \cdot x.$$

Elimination of x_{-1} from the above two linear equations leads to (1), as before. This technique of introducing subsidiary unknowns simplifies the determination of the equation for the required unknown. In this method, we have implicitly introduced another basic idea of stochastic difference equations. The unknowns x and x_{-1} are probabilities of the same event (A winning the series) but considered at different stages of the series, namely, at the outset and after a first trial which did not lead to a win for A.

A variation in the rules

The fact that in the above problem $x + y \equiv 1$ means that it is practically certain that if the series is sufficiently prolonged, it will end in a win for one of the two players. However, not all unlimited games have this limiting certainty, and we show next that a change in the rules of the present game leads to a situation in which there is a positive probability that the series will not be decided in favour of either of the two players.

Suppose, then, the trials of X_1 and X_2 are to be made *simultaneously* instead of *alternately*, the series being regarded as drawn if both X_1 and X_2 occur for the first time at the same trial. With this modification, the equation for x becomes

$$x = p_1p_2.0 + p_1q_2.1 + q_1p_2.0 + q_1q_2.x,$$

so that $x = p_1q_2/(1 - q_1q_2).$

Similarly, $y = q_1p_2/(1 - q_1q_2).$

Therefore we now have

$$x + y = (p_1q_2 + p_2q_1)/(1 - q_1q_2)$$
$$= 1 - p_1p_2/(1 - q_1q_2) < 1.$$

Hence it is no longer certain that A or B will ultimately win the series. There is, in fact, a positive probability $p_1p_2/(1 - q_1q_2)$ that the series will be drawn. This can also be proved directly. Thus, if z is the probability at the outset for a drawn series, then

$$z = p_1p_2.1 + p_1q_2.0 + q_1p_2.0 + q_1q_2.z,$$

so that $z = p_1p_2/(1 - q_1q_2),$ as before.

No reversion to initial state

Not all games revert to their initial state. As an example, suppose A and B play a series of games with the rule that the winner of the series is the player who wins two consecutive games, it being assumed that no game can be drawn. For such a sequence of independent trials the situation is never the same as it was initially since at every stage of the series either A or B has just won a game. Let p and q be the constant probabilities of A and B respectively winning a game ($p + q = 1$). If x_1 and x_{-1} denote the probabilities of A winning the series when he has just won or lost a game then, initially, the probability of A winning the series is

$$P(A) = p.x_1 + q.x_{-1}. \tag{1}$$

To determine an equation for x_1, consider the state of the series when A has just won a game. Then, since A can either win or lose the next game, we have

$$x_1 = p.1 + q.x_{-1}. \tag{2}$$

In the same way, by considering the state of the series when A has just lost a game, we have

$$x_{-1} = p.x_1 + q.0. \tag{3}$$

The solution of (2) and (3) gives

$$x_1 = p/(1 - pq) \quad \text{and} \quad x_{-1} = p^2/(1 - pq),$$

whence, from (1),

$$P(A) = p^2(1 + q)/(1 - pq). \tag{4}$$

Similarly, the initial probability of B winning the series is

$$P(B) = q^2(1 + p)/(1 - pq). \tag{5}$$

As a check, we have

$$P(A) + P(B) = [p^2 + q^2 + pq(p + q)]/(1 - pq) \equiv 1, \quad \text{as expected.}$$

This means that, if the series is sufficiently prolonged, it is practically certain that either A or B will win the series.

A change in the rules

It is instructive to consider what happens if the players agree that the winner of the series is the player who wins not two but three consecutive games. If x_r and x_{-r} denote the probabilities of A winning the series when he has just won or lost r games ($r = 1, 2$), then the initial probability for A to win the series is again given by (1). However, the equations for determining x_1 and x_{-1} are

$$x_2 = p \cdot 1 + q \cdot x_{-1},$$
$$x_1 = p \cdot x_2 + q \cdot x_{-1},$$
$$x_{-1} = p \cdot x_1 + q \cdot x_{-2},$$

and

$$x_{-2} = p \cdot x_1 + q \cdot 0.$$

The solution of these linear equations by successive elimination is straightforward, and we readily obtain

$$x_1 = \frac{p^2}{1 - pq(2 + pq)} \; ; \quad x_{-1} = \frac{p^3(1 + q)}{1 - pq(2 + pq)} .$$

Therefore, using (1), the initial probability of A now winning the series is

$$P(A) = \frac{p^3 + p^3 q(1 + q)}{1 - pq(2 + pq)} = \frac{p^3(1 + q + q^2)}{1 - pq(2 + pq)} . \tag{6}$$

By symmetry, the initial probability of B winning the series is

$$P(B) = \frac{q^3(1 + p + p^2)}{1 - pq(2 + pq)} . \tag{7}$$

It is again seen that $P(A) + P(B) \equiv 1$, so that it is practically certain that in the long run one of the players must win the series.

Sum of points won

We next consider another kind of variation in the games played by A and B. Suppose the players keep scores of the number of games they have won, and agree that the winner of the series is the player who first reaches a score of two points ahead of his opponent. It is clear that under this system of play the initial state recurs whenever the scores are equal. Moreover, a player can now win two consecutive games without winning the series. Thus, for instance, if the first three games are won by A, B, and B respectively, then A may still go on to win the series. We wish to determine the initial probability of A winning the series.

Suppose y_1 and y_{-1} are the respective probabilities of A winning the series when he is one game ahead and one game behind B. In this notation, we may also write y_0 as the initial probability of A winning the series, that is, when his score is equal to that of B. Then, by considering the three possible stages of the game for A, we have

$$y_0 = p \cdot y_1 + q \cdot y_{-1}, \tag{8}$$
$$y_1 = p \cdot 1 + q \cdot y_0,$$

and

$$y_{-1} = p \cdot y_0 + q \cdot 0.$$

A simple solution of these equations gives

$$y_{-1} = py_0 ; \quad y_1 = p + qy_0 .$$

Hence, using (8), the initial probability of A winning the series is

$$P(A) = p^2/(p^2 + q^2) . \tag{9}$$

By symmetry, the initial probability of B winning the series is

$$P(B) = q^2/(p^2 + q^2) . \tag{10}$$

As a check, we verify that $P(A) + P(B) \equiv 1$, so that it is practically certain that in the long run the series will be decided in favour of one of the two players.

A variation in the rules

The above method is quite general and, as another illustration, we extend it to the case when the player requires three points more than his opponent to win the series. If y_r and y_{-r} denote the probabilities of A winning the series when he is r points ahead or behind B $(r = 1, 2)$, then the initial probability of A winning the series is again given by (8), but with the additional equations

$$y_1 = p \cdot y_2 + q \cdot y_0 ,$$
$$y_2 = p \cdot 1 + q \cdot y_1 ,$$
$$y_{-1} = p \cdot y_0 + q \cdot y_{-2} ,$$

and
$$y_{-2} = p \cdot y_{-1} + q \cdot 0 .$$

Therefore $y_{-1} = py_0/(1 - pq) ; \quad y_1 = (p^2 + qy_0)/(1 - pq) . \tag{11}$

Hence, from (8),

$$y_0 = [p(p^2 + qy_0) + pqy_0]/(1 - pq),$$

so that the initial probability of A winning the series is now

$$P(A) = p^3/(p^3 + q^3) . \tag{12}$$

By symmetry, the initial probability of B winning the series is

$$P(B) = q^3/(p^3 + q^3), \tag{13}$$

and, as before, $P(A) + P(B) \equiv 1.$

Effect of change of rules

We observe that the change of rules from "two consecutive wins" to "two points more than opponent" favours the player who has the greater probability of winning each game. For, with $p \neq 0$, $q \neq 0$, we have from (4) and (9) that

$$\frac{p^2}{p^2 + q^2} > \frac{p^2(1 + q)}{1 - pq}$$

if
$$(1 - pq) > (1 + q)(p^2 + q^2),$$

which reduces to
$$(1 - p)(1 - 2p) < 0,$$

Hence, since $1 - p > 0$, we must have $p > \frac{1}{2}$. This is a result which is not immediately obvious.

Furthermore, a similar result also holds if we consider the change from "three consecutive wins" to "three points more than opponent" rule. Using (6) and (12), we have

$$\frac{p^3}{p^3 + q^3} > \frac{p^3(1 + q + q^2)}{1 - pq(2 + pq)} .$$

This inequality holds if
$$1 - pq(2 + pq) > (p^3 + q^3)(1 + q + q^2),$$
which reduces to
$$(1 - p)^2(1 - 2p) < 0.$$
As before, the inequality is satisfied if $p > \frac{1}{2}$.

One point to win

If we require the probability of A winning the series when he has scored one point more than B under the "three points more than opponent" rule, we have from (11)
$$y_1 = \frac{p^2 + qy_0}{1 - pq}, \quad \text{where} \quad y_0 = \frac{p^3}{p^3 + q^3} = \frac{p^3}{1 - 3pq}.$$
Therefore
$$y_1 = \frac{p^2(1 - 2pq)}{(1 - pq)(1 - 3pq)}. \tag{14}$$

Similarly, the probability of A winning the series when his score is one point less than B's is
$$y_{-1} = \frac{py_0}{1 - pq} = \frac{p^4}{(1 - pq)(1 - 3pq)}. \tag{15}$$

If we interchange p and q in (14) and (15), we have B's corresponding probabilities as
$$\frac{q^2(1 - 2pq)}{(1 - pq)(1 - 3pq)}; \quad \frac{q^4}{(1 - pq)(1 - 3pq)}.$$

Thus the probability of A winning the series when he is one point behind B is $p^4/(1 - pq)(1 - 3pq)$, and the probability of B winning the series when he is one point ahead of A is $q^2(1 - 2pq)/(1 - pq)(1 - 3pq)$. These two probabilities refer to the *same stage* of the series and their sum is
$$\frac{p^4 + q^2(1 - 2pq)}{(1 - pq)(1 - 3pq)} \equiv 1, \quad \text{on simplification.}$$

10.15 General use of difference equations

The solution of the preceding simple problems of unlimited games by the application of the fundamental theorems of probability is possible because in each case we are led *either* to an equation in which the required probability occurs as the only unknown, *or* to a finite number of linear equations involving several unknown probabilities from which a simple equation for the required probability can be obtained easily. The latter equations are, in fact, linear difference equations, but because of their simplicity a solution is possible by direct elimination of the unknowns other than the particular unknown whose value is required. We next consider a more general class of problems where, although a difference equation can be obtained, it does not at once yield the required probability because it is not immediately possible to eliminate the other unknowns involved. The simpler method of solving such difference equations is based on the techniques for their solution already indicated.

In general, there are two different ways in which unknown probabilities can occur in an equation. Firstly, if we confine our attention to some *particular event* and consider the probabilities of its occurrence at *different stages* of the process, then we may denote these probabilities by u_0, u_1, u_2, ..., and the equation will state a relation between u_n, u_{n+1}, u_{n+2}, ... Secondly, if we consider the probabilities of the occurrence of a *number of events* corresponding to different values of an integral variable at the *same stage* in a process, then we may denote these probabilities by v_0, v_1, v_2, ..., and the equation will give a relation between v_m, v_{m+1}, v_{m+2}, ... In either case, the equation is a difference equation and its solution in many cases can be obtained by standard methods.

Dependent trials

Consider a series of trials X_1, X_2, X_3, ..., in which the probability of success at each stage depends upon the result of the *previous* trial only. This means that

$$P(X_r|X_{r-1}) = p_1 \quad \text{and} \quad P(X_r|\overline{X}_{r-1}) = p_2, \quad \text{for } r \geqslant 2,$$

where p_1 and p_2 are constant probabilities. If the probability of success in the first trial is p, we wish to determine the probability of success at the nth $(n \geqslant 2)$ trial.

Suppose u_n is the required probability. Now since at any trial the event may or may not occur, we can define the sample space Ω of the $(n-1)$th trial as

$$\Omega = X_{n-1} + \overline{X}_{n-1}, \quad \text{for } n \geqslant 2.$$

Therefore
$$X_n = X_n X_{n-1} + X_n \overline{X}_{n-1},$$

so that
$$P(X_n) = P(X_n X_{n-1}) + P(X_n \overline{X}_{n-1})$$
$$= P(X_n|X_{n-1})P(X_{n-1}) + P(X_n|\overline{X}_{n-1})P(\overline{X}_{n-1}).$$

Hence, using the definition of u_n, we have

$$u_n = p_1 . u_{n-1} + p_2 . (1 - u_{n-1}),$$

which is the required difference equation for u_n. This is a non-homogeneous first-order difference equation and to obtain its solution we rewrite it as

$$u_n - \gamma = (p_1 - p_2)(u_{n-1} - \gamma),$$

where $\gamma \equiv p_2/(1 - p_1 + p_2)$.

Hence
$$u_n - \gamma = (p_1 - p_2)^{n-1}(u_1 - \gamma)$$

or, since $u_1 = p$,

$$u_n = p_2/(1 - p_1 + p_2) + (p_1 - p_2)^{n-1}[p - p_2/(1 - p_1 + p_2)].$$

We observe that as $|p_1 - p_2| < 1$,

$$\lim_{n \to \infty} u_n = p_2/(1 - p_1 + p_2),$$

which is the limiting probability of success. Thus, if $p = p_2/(1 - p_1 + p_2)$, the probability of success is constant throughout all trials.

10.16 Forward and backward difference equations

In the above problem, the probabilities u_n were all probabilities considered at the *same stage*, namely at the outset, and they referred to *different events* corresponding to different values of n. In some cases, the probabilities all refer to the *same event*, but are considered at *different stages* of the process. It is important to distinguish carefully between these two types because, whereas in the first type a difference equation is obtained by considering *what must have happened* at previous trials in order that the event (whose probability is u_n) shall occur, in the second type a difference equation is derived by considering *what must happen next* after the event whose probability is required has occurred. The first is the *backward* and the second the *forward* difference equation. A further clarification of the difference in the argument used in deriving the two types of equations is obtained by the solution of a classical problem which can be regarded as of either type, depending upon the definition of the unknown probability used.

Problem I

In a sequence of Bernoulli trials, the probability of success at a trial is a constant p and that of a failure q ($p+q = 1$). Two points are scored for each success and one for each failure. Find the probability when the score is n ($\leqslant N$) that it will at some stage be exactly N.

(i) Forward approach

Let u_n be the required probability. Then, quite clearly, u_n refers to the realisation of a specific event (the score is exactly N at some stage), but the probability is defined for variable n, that is, for different stages of the process. Suppose a score of n has been attained, and we consider what could happen next. If X and $\overline{X}$ respectively are the events that the next trial is a success or a failure then, since the trials are Bernoullian in character,

$$P(X) = p \quad \text{and} \quad P(\overline{X}) = q,$$

and we may write the sample space of the outcomes of this trial as

$$\Omega = X + \overline{X}.$$

Next, suppose Y is the event that the total score at some stage is exactly N. Then

$$Y = YX + Y\overline{X},$$

so that
$$P(Y) = P(YX) + P(Y\overline{X})$$
$$= P(Y|X)P(X) + P(Y|\overline{X})P(\overline{X}).$$

But
$$P(Y|X) = u_{n+2} \quad \text{and} \quad P(Y|\overline{X}) = u_{n+1}.$$

Therefore
$$u_n = p \cdot u_{n+2} + q \cdot u_{n+1}.$$

This is a second-order homogeneous difference equation. The roots of the auxiliary equation $1 = px^2 + qx$ are 1 and $-1/p$, and as these are unequal,

$$u_n = A + B(-1/p)^n,$$

where A and B are arbitrary constants to be determined by the boundary conditions $u_N = 1$ and $u_{N-1} = q$. Thus A and B are given by the equations

$$1 = A + B(-1/p)^N \quad \text{and} \quad q = A + B(-1/p)^{N-1},$$

whence
$$A = 1/(1+p) \quad \text{and} \quad B = -(-p)^{N+1}/(1+p).$$

Hence
$$u_n = \frac{1 - (-p)^{N-n+1}}{1+p}.$$

(ii) Backward approach

We first observe that the probability of attaining an exact score of N

points when the score is n must be the same as the probability of attaining a score of $N - n$ when the score is zero, that is, at the outset, because in each case it is the probability of getting an additional score of $N - n$ points.

Let v_m be the probability at the outset of obtaining exactly m points so that the required probability is v_m for $m = N - n$. Clearly, v_m always represents a probability at the outset, but for different values of m it refers to *different* events. To obtain the backward difference equation for v_m, we have to consider the distinct events which could lead to a score of m at a further trial. In fact, the two mutually exclusive and exhaustive possibilities are

(a) by reaching a score of $m - 2$ and then scoring 2;   and
(b) by reaching a score of $m - 1$ and then scoring 1.

If Z_{-2}, Z_{-1}, and Z_0 are the events representing the attainment of scores of $m - 2$, $m - 1$, and m respectively, then

$$Z_0 = Z_{-2}X + Z_{-1}\overline{X}.$$

Therefore
$$P(Z_0) = P(Z_{-2}X) + P(Z_{-1}\overline{X})$$
$$= P(Z_{-2})P(X) + P(Z_{-1})P(\overline{X}),$$

since the trials are independent.

Thus
$$v_m = p \cdot v_{m-2} + q \cdot v_{m-1}.$$

This is again a homogeneous second-order linear difference equation. The roots of the auxiliary equation $x^2 = p + qx$ are 1 and $-p$. Therefore

$$v_m = A + B(-p)^m,$$

where A and B are arbitrary constants. The boundary conditions are $v_0 = 1$ and $v_1 = q$, whence we find that

$$A = 1/(1 + p) \quad \text{and} \quad B = p/(1 + p).$$

Hence
$$v_m = \frac{1 - (-p)^{m+1}}{1 + p},$$

and so, putting $m = N - n$, we finally obtain the required probability as

$$v_{N-n} = \frac{1 - (-p)^{N-n+1}}{1 + p}, \quad \text{as before .}$$

We conclude with the caution that it is important to understand the argument used in deriving the difference equations for u_n and v_m. A little thought will show that

$$u_n \neq p \cdot u_{n-2} + q \cdot u_{n-1},$$

and
$$v_m \neq p \cdot v_{m+2} + q \cdot v_{m+1}.$$

It is easy for a beginner to confuse the "forward" and "backward" approaches in obtaining a difference equation for any particular problem. The proper procedure is to define clearly the unknown probability, which is the dependent variable of a difference equation, and to tackle the problem on its own merits. As the above example illustrates, it is not the problem itself but the definition of the dependent variable which indicates the particular method to be used.

10.17 Further games based on Bernoulli trials

We consider next two well-known and important problems based on a sequence of Bernoulli trials. The results are presented in the context of games of chance between two players, but they can be more appropriately regarded as two general theorems in probability theory.

Problem II

Consider a sequence of Bernoulli trials in which the constant probabilities of success or failure at any trial are p and q ($p+q = 1$) respectively. The trials may be regarded as games between two players A and B with the understanding that A wins a game when a trial is a success whilst B wins in the contrary event. It is agreed that the series of games will be won by A if he wins α games more than B before B wins β games more than A, and by B in the opposite event. Find the initial probability of A winning the series.

Let u_n be the probability of A winning the series considered at the stage when he has already won n games more than B, where $-\beta \leqslant n \leqslant \alpha$. This is a valid definition because the conditions of winning the series are concerned only with the difference between the number of games won by A and B respectively and not with the absolute numbers themselves. To determine a forward difference equation for u_n we proceed as follows.

Suppose X is the event that the next trial is a success and Y the event that A wins the series. Then

$$P(X) = p, \quad P(\overline{X}) = q, \quad P(Y|X) = u_{n+1}, \quad \text{and} \quad P(Y|\overline{X}) = u_{n-1},$$

where $\overline{X}$ denotes the event complementary to X.

Now
$$Y = YX + Y\overline{X},$$
so that
$$P(Y) = P(Y|X)P(X) + P(Y|\overline{X})P(\overline{X}).$$
Therefore
$$u_n = p \cdot u_{n+1} + q \cdot u_{n-1}.$$

This difference equation holds for $-\beta < n < \alpha$, it being understood that u_{n+1} is undefined for $n = \alpha$, and u_{n-1} for $n = -\beta$. The roots of the auxiliary equation $x = px^2 + q$ are 1 and q/p, which are equal only if $p = q = \frac{1}{2}$. Hence

$$u_n = A + B(q/p)^n, \quad \text{if } p \neq q,$$
$$= A' + B'n, \quad \text{if } p = q,$$

where A, B, A', B' are arbitrary constants to be determined by the boundary conditions $u_\alpha = 1$ and $u_{-\beta} = 0$. Therefore, for the case $p \neq q$, the equations for A and B are

$$1 = A + B(q/p)^\alpha \quad \text{and} \quad 0 = A + B(q/p)^{-\beta},$$

whence
$$A = 1 / \left[1 - (q/p)^{\alpha+\beta}\right]; \quad B = -(q/p)^\beta / \left[1 - (q/p)^{\alpha+\beta}\right].$$
Therefore
$$u_n = \frac{1 - (q/p)^{n+\beta}}{1 - (q/p)^{\alpha+\beta}}, \quad \text{if } p \neq q. \tag{1}$$

Similarly, for $p = q$, the equations for A' and B' are

$$1 = A' + B'\alpha \quad \text{and} \quad 0 = A' - B'\beta,$$

so that $\qquad$ $A' = \beta/(\alpha + \beta);\quad B' = 1/(\alpha + \beta).$

Therefore $\qquad$ $u_n = (\beta + n)/(\alpha + \beta),\quad \text{if } p = q.$ $\qquad$ (2)

Initially, and at every subsequent stage when A and B have won an equal number of games, we have $n = 0$, and the required probability of A winning at this stage is

$$u_0 = \frac{1 - (q/p)^{\beta}}{1 - (q/p)^{\alpha+\beta}},\quad \text{if } p \neq q,$$
$$= \beta/(\alpha + \beta),\quad \text{if } p = q. \qquad (3)$$

Checks

    (i) If $p = 1$, $u_0 = 1$, and A evidently wins every game.

    (ii) If $p \to 0$, $u_0 = p^{\alpha}(p^{\beta} - q^{\beta})/(p^{\alpha+\beta} - q^{\alpha+\beta}) \to 0$, and A loses every game.

    (iii) If $p \neq 0$, $q \neq 0$, $\alpha = 0$, and $\beta \neq 0$, then $u_0 = 1$ in both expressions of (3).

    (iv) If $\alpha \neq 0$ and $\beta = 0$, then $u_0 = 0$ in both expressions of (3).

    (v) If $\alpha = \beta$ then

$$u_0 = 1/[1 + (q/p)^{\alpha}],\quad \text{if } p \neq q,$$
$$= \tfrac{1}{2},\quad \text{if } p = q.$$

    (vi) Finally, we can use (1) to determine the probability of B winning the series when he is n games behind A. To do this, we simply interchange (α, β), (p, q), and write $-n$ for n in (1). Therefore the probability of B winning the series when he is n games behind A is

$$\frac{1 - (p/q)^{-n+\alpha}}{1 - (p/q)^{\alpha+\beta}}.$$

But at this stage the probability of A winning the series is u_n as given by (1). The sum of the two probabilities, which represent the chance of A and B to win at a particular stage of the process (when A is n games ahead) is clearly unity. This is another useful check.

Applications

The results obtained can be used to provide solutions to a number of applications in which the sequence of Bernoulli trials can be identified with physically meaningful processes. We consider here two famous applications.

(i) *Gambler's ruin*

This is a classical problem in which two gamblers with limited capital repeatedly make payments to each other according to whether an event does or does not occur. In a general situation of this type the payments may vary, and it is not essential that the trials be necessarily Bernoullian in character. However, our solution is only applicable in the particular case when the trials are Bernoullian and the stake paid by the loser to the winner at a trial is one unit. Thus, in our example, A will be ruined if B wins β more games than A before A wins α more games than B. In this sense, A and B have initial capital of β and α units respectively. The initial probability of A escaping ultimate ruin is given by (3).

As a further consequence of this gambling situation, suppose B has unlimited capital. Then

$$\lim_{a \to \infty} u_0 = 1 - (q/p)^{\beta}, \quad \text{if } p > q,$$
$$= 0, \quad \text{if } p \leqslant q.$$

This means that if $p > \frac{1}{2}$, A has a positive probability of never being ruined even if B is "infinitely" rich. On the other hand, if $p \leqslant \frac{1}{2}$, then it is certain that A will not escape ultimate ruin.

Craps and double-sixes

The principle that p be less than $\frac{1}{2}$ is always ensured in devising games of chance when B is not an individual gambler but a gaming establishment. This is clearly illustrated by the well-known game of craps and the famous problem of the Chevalier de Méré.

In craps, a gambler plays against an establishment. According to the rules the gambler repeatedly rolls two unbiased dice. He wins if the first trial results in the sums of points 7 or 11 or, alternatively, if the first sum is $4, 5, 6, 8, 9$ or 10, and the same sum reappears before the sum 7 reappears. This implies that the gambler loses if the first throw results in the sum 2, 3, or 12, or, alternatively, if the first sum is $4, 5, 6, 8, 9$ or 10, and the sum 7 reappears before the first sum reappears.

If S denotes the score obtained at any one throw, then $2 \leqslant S \leqslant 12$. A simple enumeration of the 36 sample points corresponding to a single throw of the two dice gives

$$P(S = 2) = P(S = 12) = 1/36; \quad P(S = 3) = P(S = 11) = 1/18;$$
$$P(S = 4) = P(S = 10) = 1/12; \quad P(S = 5) = P(S = 9) = 1/9;$$
$$P(S = 6) = P(S = 8) = 5/36; \quad P(S = 7) = 1/6.$$

(a) The probability of winning on the first throw is $1/6 + 1/8 = 2/9$.

(b) If the first throw gives $S = 4$, then the probability of continuing the game is

$$1 - P(S = 4) - P(S = 7) = 3/4.$$

Hence the conditional probability of winning ultimately in this case is

$$\sum_{r=0}^{\infty} (1/12)(3/4)^r = 1/3.$$

This is also the conditional probability of winning if the first throw gives $S = 10$.

(c) Similarly, if the first throw gives $S = 5$, then the probability of continuing the game is

$$1 - P(S = 5) - P(S = 7) = 13/18.$$

Therefore the conditional probability of winning ultimately in this case is

$$\sum_{r=0}^{\infty} (1/9)(13/18)^r = 2/5.$$

This is also the conditional probability of winning ultimately if the first throw gives $S = 9$.

(d) Finally, if the first throw gives $S = 6$, then the probability of continuing the game is

$$1 - P(S = 6) - P(S = 7) = 25/36.$$

Therefore the conditional probability of winning ultimately in this case is

$$\sum_{r=0}^{\infty} (5/36)(25/36)^r = 5/11.$$

This is also the conditional probability of winning if the first throw gives $S = 8$. Hence, using the results of (a), (b), (c), and (d), the total probability of the gambler winning is

$$2/9 + 2[(1/12)(1/3) + (1/9)(2/5) + (5/36)(5/11)] = 244/495 = 0\cdot493,$$

which is just less than $\frac{1}{2}$.

We have already mentioned the problem of the Chevalier de Méré. The probability of obtaining at least one double-six in 24 throws of a pair of unbiased dice is

$$1 - (35/36)^{24} = 0\cdot491,$$

which is again just less than $\frac{1}{2}$. As the two examples illustrate, a safety margin of the order of one per cent is enough to keep the croupiers in business.

(ii) *Random walk in one dimension*

A particle is initially located at the origin and at each move jumps one unit along the x axis, to the right with probability p and to the left with probability q. Equation (3) on p. 268 gives the probability that the particle will reach tne point $(\alpha, 0)$ without having reached the point $(-\beta, 0)$.

Problem III

This problem is also based on a sequence of Bernoulli trials, but interest now centres on *consecutive* successes and failures. As before, we assume that the probabilities of success or failure at any one trial are p and q respectively $(p+q = 1)$. Two players A and B play with the rule that A will win the series of trials if a run of at least α successes occurs before a run of at least β failures, B winning the series in the opposite event. We wish to determine the initial probability of A winning the series.

Let X be the event that a run of at least α successes precedes the first run of at least β failures, so that A wins if X occurs. Suppose u_n is the probability of X occurring considered at the stage when a run of exactly n successes has just occurred $(n < \alpha)$, no run of at least α successes or of β failures having occurred. In the same way, let v_n be the probability of X to occur when a run of exactly n failures has just occurred $(n < \beta)$, no run of at least β failures or α successes having occurred.

If now a run of n successes has just occurred and the series is still undecided, there being no earlier occurrence of any run of at least α successes or β failures, then the next trial could be either a success or a failure. Hence the forward difference equation for u_n is

$$u_n = p \cdot u_{n+1} + q \cdot v_1, \quad \text{if} \quad 0 < n < \alpha, \quad \text{and} \quad u_\alpha = 1. \tag{1}$$

Similarly, considering the stage when a run of n failures has just occurred, we have

$$v_n = q \cdot v_{n+1} + p \cdot u_1, \quad \text{if} \quad 0 < n < \beta, \quad \text{and} \quad v_\beta = 0. \tag{2}$$

We now solve (1) and (2) to obtain u_1 and v_1. Since (1) is a first-order difference equation for u_n, its solution is deduced by rewriting it as

$$u_n - v_1 = p(u_{n+1} - v_1), \quad \text{for } 1 \leqslant n \leqslant a - 1.$$

Hence
$$u_n - v_1 = p^{a-n}(u_a - v_1)$$
$$= p^{a-n}(1 - v_1), \quad \text{since } u_a = 1.$$

If we now put $n = 1$, then

$$u_1 = v_1 + p^{a-1}(1 - v_1). \tag{3}$$

Similarly, from (2), we obtain

$$v_n - u_1 = q^{\beta-n}(v_\beta - u_1)$$
$$= q^{\beta-n}(-u_1), \quad \text{since } v_\beta = 0.$$

Therefore
$$v_n = u_1(1 - q^{\beta-n}),$$
whence
$$v_1 = u_1(1 - q^{\beta-1}). \tag{4}$$

Solving the simultaneous equations (3) and (4), we find

$$u_1 = \frac{p^{a-1}}{p^{a-1} + q^{\beta-1}(1 - p^{a-1})}; \quad v_1 = \frac{p^{a-1}(1 - q^{\beta-1})}{p^{a-1} + q^{\beta-1}(1 - p^{a-1})}.$$

Now it is clear that at the outset the probability of A winning the series is $pu_1 + qv_1$, since the first trial can be either a success or a failure. Therefore the required probability of A winning the series is

$$P(A) \equiv pu_1 + qv_1$$
$$= \frac{p^a + qp^{a-1}(1 - q^{\beta-1})}{p^{a-1} + q^{\beta-1}(1 - p^{a-1})}$$
$$= \frac{p^{a-1}(1 - q^\beta)}{p^{a-1} + q^{\beta-1}(1 - p^{a-1})}. \tag{5}$$

Checks
(i) If $p = 1$, $P(A) = 1$, whilst if $p = 0$, $P(A) = 0$.
(ii) If $a = 1$, $P(A) = 1 - q^\beta$, which correctly indicates that A will win the series unless B wins every one of the first β games.
(iii) If (p,q) and (a, β) are interchanged, then (5) gives the probability that B wins β consecutive games before A wins a run of a games. This probability is

$$P(B) = \frac{q^{\beta-1}(1 - p^a)}{q^{\beta-1} + p^{a-1}(1 - q^{\beta-1})}.$$

Hence, as a check, we have $P(A) + P(B) \equiv 1$.

Corollary

An interesting consequence can be deduced from the general results of Problems II and III.

Suppose A and B play a series of games with constant probabilities p and q respectively of winning each game ($p \neq q$, $p \neq 0$, $q \neq 0$, $p + q = 1$), and consider the following alternative rules for deciding the winner of the series.

(a) the winner is the player who first wins a consecutive games; and

(b) the winner is the player who first wins α games more than his opponent.

Then the corollary states that for $p > q$, A has a greater probability of winning the series under rule (b) than under rule (a). This generalises the results already obtained for $\alpha = 2$ and 3.

Putting $\alpha = \beta$ in (5) of Problem III, we obtain the probability of A winning the series under (a) as

$$\frac{p^{\alpha-1}(1 - q^{\alpha})}{p^{\alpha-1} + q^{\alpha-1}(1 - p^{\alpha-1})} .$$

Again, putting $\alpha = \beta$ in (3) of Problem II, we have the probability of A winning the series under (b) as

$$\frac{p^{\alpha}}{p^{\alpha} + q^{\alpha}} .$$

To establish the corollary, we have now to prove that for $p > q$

$$\frac{p^{\alpha}}{p^{\alpha} + q^{\alpha}} > \frac{p^{\alpha-1}(1 - q^{\alpha})}{p^{\alpha-1} + q^{\alpha-1}(1 - p^{\alpha-1})} ,$$

that is, $p(p^{\alpha-1} + q^{\alpha-1} - p^{\alpha-1}q^{\alpha-1}) > (1 - q^{\alpha})(p^{\alpha} + q^{\alpha})$,

which is equivalent to the simplified inequality

$$p - q > p^{\alpha+1} - q^{\alpha+1} .$$

This inequality is true because

$$p - q = (p - q)(p + q)^{\alpha}$$
$$> (p - q)(p^{\alpha} + p^{\alpha-1}q + \ldots + q^{\alpha}) = p^{\alpha+1} - q^{\alpha+1} ,$$

so that $p - q > p^{\alpha+1} - q^{\alpha+1}$,   whence the corollary.

10.18   Expectation by indirect methods

We have so far considered the evaluation of probabilities by indirect methods, and we next show that it is also possible to use similar techniques to evaluate expectations of different random variables. A particularly important case is that of the expected number of trials required for the occurrence of some event which we may conveniently describe as a "success". It is not necessary for our present argument that the probability of success at each trial should be the same. Thus, for example, the experiment might consist of rolling two unbiased dice till a double-six (success) appears for the first time. It is clear that the probability of the first success at the rth trial is

$$(1/36)(35/36)^{r-1}, \quad \text{for } r \geqslant 1,$$

which, as a function of r, varies from trial to trial.

In general, let X be the event that success is *not* achieved in the first $r-1$ trials and let Y be the event that success is achieved for the first time at the rth trial.

If $P(X) = f(r - 1)$ and $P(Y|X) = p_r$,

then $P(YX) = P(Y|X) P(X) = f(r - 1)p_r$.

This is the probability that exactly r trials will be made for the first success to occur at the rth trial, with the convention that $f(0) = 1$. Hence the expected number of trials for the first success is

$$\sum_{r=1}^{\infty} r f(r-1) p_r \;=\; \sum_{r=0}^{\infty} (r+1) f(r) p_{r+1}, \quad \text{if this sum exists.}$$

This is a direct evaluation of the expected number of trials for a success to be realised. Another simpler way of determining this expectation is to note that the probability that exactly r trials are required for a success is

$$f(r-1) - f(r).$$

Therefore an alternative expression for the expected number of trials is

$$\sum_{r=1}^{\infty} r[f(r-1) - f(r)] \;=\; \sum_{r=0}^{\infty} (r+1) f(r) - \sum_{r=1}^{\infty} r f(r) \;=\; \sum_{r=0}^{\infty} f(r).$$

As an example, we calculate the expected number of trials required for a double-six in rolling two unbiased dice. The probability

$$f(r) \;=\; (35/36)^r,$$

so that $$\sum_{r=0}^{\infty} f(r) \;=\; \sum_{r=0}^{\infty} (35/36)^r \;=\; 36.$$

10.19 Expectation by difference equations

An effective method for the evaluation of expectations is by the use of difference equations. The argument is simple, and it is best illustrated by the following general case in Bernoulli trials.

Consider a sequence of Bernoulli trials with probability of success p and failure q $(p+q = 1)$. Let T be the expected number of trials required for an event X to occur. Also let T_1 and T_{-1} be the expected number of trials required for the occurrence of X *after* the first trial according as the first trial is a success or a failure. Then, since the first trial is certainly made, we have

$$T \;=\; 1 + p.T_1 + q.T_{-1}, \tag{1}$$

which is the required difference equation. This result leads to a simple solution of many problems.

Example 10.20

Let X be the event "the realisation of exactly r successes in the Bernoulli trials defined above". If N_r denotes the required expected number of trials, then

$$T_1 \;=\; N_{r-1} \quad \text{and} \quad T_{-1} \;=\; N_r.$$

Hence (1) gives $N_r \;=\; 1 + p.N_{r-1} + q.N_r,$ for $r > 1$,

or $N_r \;=\; N_{r-1} + 1/p.$

This is a non-homogeneous first-order difference equation, and we readily obtain the solution $$N_r \;=\; N_1 + (r-1)/p.$$

To obtain N_1, we have the equation

$$N_1 \;=\; 1 + q.N_1 + p.0,$$

since after one success we require no further trials, so that $N_1 = 1/p$.

Hence $$N_r \;=\; r/p.$$

It is worth while to compare this derivation with the direct calculation of the expectation. Thus, if x is the random variable denoting the number of trials required for obtaining r (fixed) successes, then

$$P(x = n) = \binom{n-1}{r-1} p^r q^{n-r}, \quad \text{for } n \geqslant r.$$

Hence, by definition,

$$
\begin{aligned}
E(x) &= \sum_{n=r}^{\infty} n \binom{n-1}{r-1} p^r q^{n-r} \\
&= \sum_{n=r}^{\infty} r \binom{n}{r} p^r q^{n-r} \\
&= r p^r \sum_{n=r}^{\infty} \binom{n}{n-r} q^{n-r} \\
&= r p^r \sum_{t=0}^{\infty} \binom{r+t}{t} q^t = r p^r (1-q)^{-(r+1)} = r/p,
\end{aligned}
$$

so that $E(x) = r/p,$ as before.

We end this chapter with two further examples illustrating the use of difference equations in the evaluation of probabilities and expectations.

Example 10.21

(i) In a game of chance a player has probabilities 1/3, 5/12, and 1/4 of scoring 0, 1, and 2 points respectively at each trial, the game terminating on the first realisation of a zero score at a trial. Assuming that the trials are independent, prove that u_n, the probability of the player obtaining a score of n points, satisfies the difference equation

$$12u_n - 5u_{n-1} - 3u_{n-2} = 0, \quad \text{for } n \geqslant 2.$$

Hence obtain an explicit expression for u_n, and use it to show that if the player repeatedly plays this game, then his expected score per game is 11/4.

(ii) Suppose the rules are changed so that the game does not end on the first realisation of a zero score at a trial but continues indefinitely. In this case, show that the probability of the player obtaining a score of exactly n points at some stage of play is

$$v_n = \frac{8}{11} + \frac{3}{11}\left(-\frac{3}{8}\right)^n, \quad \text{for } n \geqslant 0.$$

(i) If X is the random variable representing the score obtained at a trial, then

$$P(X = 0) = 1/3; \qquad P(X = 1) = 5/12; \qquad P(X = 2) = 1/4.$$

Now considering the possible states of the game which could lead to an exact score of n points, we have

$$u_n = \frac{5}{12} \cdot u_{n-1} + \frac{1}{4} \cdot u_{n-2},$$

or $12u_n - 5u_{n-1} - 3u_{n-2} = 0, \quad \text{for } n \geqslant 2.$

This is a homogeneous second-order linear difference equation. The auxiliary equation is

$$12x^2 - 5x - 3 = 0, \quad \text{or} \quad (3x+1)(4x-3) = 0,$$

so that the roots are 3/4 and $-1/3$. Hence the general solution is

$$u_n = A(3/4)^n + B(-1/3)^n,$$

where A and B are arbitrary constants which are determined to satisfy the boundary conditions $u_0 = 1/3$ and $u_1 = (5/12)(1/3) = 5/36$. Therefore the equations for determining A and B are

$$\frac{1}{3} = A + B \quad \text{and} \quad \frac{5}{36} = \frac{3}{4}A - \frac{1}{3}B,$$

whence $\qquad\qquad A = 3/13; \quad B = 4/39.$

Therefore $\qquad u_n = \frac{3}{13}\left(\frac{3}{4}\right)^n + \frac{4}{39}\left(-\frac{1}{3}\right)^n, \quad$ for $n \geqslant 0$

The score S obtained in one play of the game is a random variable such that the probability $\quad P(S = n) = u_n, \quad$ for $n \geqslant 0.$

These equations define a proper probability distribution since

$$\sum_{n=0}^{\infty} u_n \equiv 1.$$

Hence

$$E(S) = \sum_{n=0}^{\infty} nu_n = \frac{3}{13}\sum_{n=0}^{\infty} n(3/4)^n + \frac{4}{39}\sum_{n=0}^{\infty} n(-1/3)^n$$
$$= (3/13)(3/4)(1 - 3/4)^{-2} + (4/39)(-1/3)(1 + 1/3)^{-2}$$
$$= 11/4.$$

(ii) The backward difference equation for v_n is

$$v_n = \frac{1}{3}\cdot v_n + \frac{5}{12}\cdot v_{n-1} + \frac{1}{4}\cdot v_{n-2},$$

or $\qquad\qquad 8v_n - 5v_{n-1} - 3v_{n-2} = 0, \quad$ for $n \geqslant 2.$

The auxiliary equation is $8x^2 - 5x - 3 = 0$, with the roots 1 and $-3/8$. Therefore

$$v_n = A + B(-3/8)^n,$$

where A and B are again arbitrary constants to be determined by the boundary conditions. Clearly; $v_0 = 1$, but

$$v_1 = (5/12)\sum_{r=0}^{\infty}(1/3)^r = (5/12)(3/2) = 5/8.$$

Hence the equations for A and B are

$$1 = A + B \quad \text{and} \quad \frac{5}{8} = A - \frac{3}{8}B.$$

Their solution gives $A = \frac{8}{11}$, $B = -\frac{3}{11}$, so that

$$v_n = \frac{8}{11} + \frac{3}{11}\left(-\frac{3}{8}\right)^n, \quad \text{for } n \geqslant 0.$$

Example 10.22
   (i) In a sequence of n Bernoulli trials in which the probability of success (S) is p and that of failure (F) is q $(p+q = 1)$, show that y_n, the probability that the pattern SF does not occur in the entire sequence, satisfies the difference equation

$$y_n - y_{n-1} + pq\,y_{n-2} = 0, \quad \text{for } n \geqslant 2.$$

Hence obtain explicit expressions for y_n when $p \neq q$ and $p = q$.

(ii) Also, prove that the expected number of trials necessary for the realisation of r consecutive repetitions of the pattern SF is

$$\frac{1 - (pq)^r}{(1 - pq)(pq)^r} \geqslant \frac{4}{3}(4^r - 1).$$

(i) By definition,

$$y_{n-1} - y_n = \text{Probability of the realisation of the pattern at the } n\text{th trial}$$
$$= pq \, y_{n-2},$$

so that

$$y_n - y_{n-1} + pq \, y_{n-2} = 0, \quad \text{for } n \geqslant 2, \quad \text{and} \quad y_0 = y_1 = 1.$$

The auxiliary equation $x^2 - x + pq = 0$ has the roots p, q. Therefore

$$y_n = A p^n + B q^n, \quad \text{for } p \neq q,$$

where A and B are constants determined by the equations

$$1 = A + B; \quad 1 = Ap + Bq.$$

The solution of these equations gives

$$A = p/(p - q); \quad B = -q/(p - q).$$

Hence

$$y_n = (p^{n+1} - q^{n+1})/(p - q), \quad \text{for } p \neq q \quad \text{and} \quad n \geqslant 0.$$

When $p = q = \frac{1}{2}$, then $y_n = (A' + B'n)(\frac{1}{2})^n$,

where A' and B' are arbitrary constants. The boundary conditions $y_0 = y_1 = 1$ now give

$$1 = A' \quad \text{and} \quad 1 = (A' + B')\tfrac{1}{2}, \quad \text{so that} \quad B' = 1.$$

Hence

$$y_n = (n + 1)(\tfrac{1}{2})^n, \quad \text{for } p = q \quad \text{and} \quad n \geqslant 0.$$

(ii) Let x_m be the expected number of *additional* trials necessary when a stage has been reached where the first m letters of the sequence are in agreement with the actual realisation, for $0 \leqslant m \leqslant 2r$. Clearly $x_{2r} = 0$ and, according to this definition of x_m, we have to determine x_0.

Now for any integer $k < r$, we have

$$x_{2k} = 1 + px_{2k+1} + qx_0, \qquad (1)$$

and

$$x_{2k+1} = 1 + px_1 + qx_{2k+2}. \qquad (2)$$

Putting $k = 0$ in (1), we obtain

$$x_0 = 1 + px_1 + qx_0, \quad \text{so that} \quad px_1 = px_0 - 1.$$

Hence, from (2), elimination of x_1 gives

$$x_{2k+1} = px_0 + qx_{2k+2};$$

and, using this in (1), we have the required difference equation

$$x_{2k} = 1 + p^2x_0 + pqx_{2k+2} + qx_0$$
$$= 1 + (1 - pq)x_0 + pqx_{2k+2}.$$

To solve this difference equation, we rewrite it as

$$x_{2k} - a = pq(x_{2k+2} - a), \quad \text{where} \quad a \equiv \frac{1 + (1 - pq)x_0}{1 - pq}.$$

Therefore

$$x_{2k} - a = \frac{1}{pq}(x_{2k-2} - a) = \frac{1}{(pq)^2}(x_{2k-4} - a) = \ldots = \frac{1}{(pq)^k}(x_0 - a).$$

Hence, putting $k = r$, we have

$$x_{2r} - a = \frac{1}{(pq)^r}(x_0 - a);$$

and, since $x_{2r} = 0$,

$$x_0 = [1 - (pq)^r]a = \frac{1 - (pq)^r}{(1 - pq)(pq)^r}.$$

If $\theta \equiv pq$, so that $0 \leqslant \theta \leqslant \frac{1}{4}$, we have

$$x_0 = \frac{1 - \theta^r}{(1 - \theta)\theta^r} = \frac{1}{\theta^r}(\theta^{r-1} + \theta^{r-2} + \ldots + 1)$$

$$= \frac{1}{\theta} + \frac{1}{\theta^2} + \ldots + \frac{1}{\theta^r} \geqslant 4 + 4^2 + \ldots + 4^r$$

$$= \frac{4}{3}(4^r - 1).$$

Therefore

$$x_0 = \frac{1 - \theta^r}{(1 - \theta)\theta^r} \geqslant \frac{4}{3}(4^r - 1).$$

PART III

STATISTICAL INFERENCE

CHAPTER 11

PRINCIPLES AND METHODS OF ESTIMATION

11.1 The problem of estimation

Our preliminary study of the ideas of probability, random variables, and probability distributions provides the basis for considering one of the central ideas of statistical inference, that is, of *estimation*. Formally, estimation is that part of statistical theory which is concerned with the question of how to use a finite number of observations to provide the "best possible" evaluation of the unknown parameters occurring in the mathematical definition of the population from which the observations are obtained. In the strict deterministic sense a statistical population is unknowable; and even in the restricted sense when the mathematical form of the population has been stipulated, there is still the uncertainty about the magnitude of its parameters. Clearly, no finite set of observations from an infinite population can give *all* the information about the parameters, and the problem of estimation arises only because of this inevitable limitation of measurable and so necessarily finite experience. In a sense this is the only problem of statistics, at least in so far as its theory is concerned with the methodology of general inferences from finite sets of observations. Any estimation procedure is subject to uncertainty, and the "best possible" evaluation of population parameters has meaning only on the basis of certain axioms or postulates which have their theoretical justification on grounds of probability or convenience or both. No theory of estimation is "absolutely" right, and a method of estimation has a validity only in relation to the stated principles of the theory from which it is deduced.

Generalities apart, possibly the best way of understanding the problem of estimation is to consider some simple examples. Suppose we have a penny about which nothing is known as to whether it is unbiased or, if biased, the magnitude of its bias. It is obvious that in this situation the only experimental method of obtaining any knowledge about the bias of the penny is to carry out a number of tosses and to observe the outcomes of the trials. However, in making these trials which would provide information about the penny, we have to make some assumptions to ensure the proper application of the laws of probability. Firstly, we need to assure ourselves that the penny has two distinct faces, which are conveniently denoted by head (H) and tail (T). Secondly, we agree that a head or a tail are the only two possible outcomes of a trial. In other words, we exclude the theoretically conceivable possibilities such as, for example, that at a trial the penny may just roll away or stand on its edge. Thirdly, we assume that the trials are performed honestly so that there is no deliberate attempt to manipulate their outcomes. This implies that, as far as possible, the trials are made in exactly the same manner so that chance is allowed to determine their actual outcomes. Under these assumptions, the trials will be Bernoullian in character, the actual realisation being a random sequence of H's and T's. Now in any finite period of time only a finite number of trials can be carried out, but it is conceptually possible to visualise an infinite sequence of trials. We postulate that in this conceptual population of H's and T's the relative frequencies of H and T give the true probabilities of the penny turning up heads or tails. Accordingly, we agree to write $P(H) = p$ and $P(T) = q$, where $p + q = 1$. This population contains all the information about the bias of the penny or, equivalently, about the parameter p. Now if we have actually observed the outcomes of a finite number of trials (say n), then the simple estimation problem is to determine what, if any, evaluation can be made of p on the basis of the observations obtained. In more usual terminology, the problem is to *estimate* p given the sample observations. If in the n trials w heads were observed, then we have used the sample relative frequency of heads $p^* = w/n$ as an "estimate" of p. This is intuitively reasonable, but the point now is whether this procedure has any formal justification. This is where the theory of estimation comes in, and its argument depends upon the fact that, under the stated assumptions, the observed realisation of w heads in n trials is a random variable having a binomial distribution. Indeed, this is the probability model which underlies the estimation problem. However, if the experimental conditions under which the trials were conducted were such that the assumptions leading to Bernoulli trials are substantially inapplicable, then the estimation problem cannot even be specified in terms of the binomial distribution. For example, suppose the trials are so carried out that the probability of obtaining a head changes from trial to trial; then there are, in general, as many unknown probabilities as there are trials in the observed sequence of H's and T's. This is a completely different situation, and the resulting estimation problem is no longer a simple one.

To sum up, our initial estimation problem, given that the n trials are Bernoullian in character, is *either*

(i) What is the "best possible" evaluation of the parameter p on the basis of the sample information alone?

or (ii) What is the formal justification for using the sample relative frequency of heads p^* as the estimate of p?

The answers to both these questions depend substantially on the fact that, under the stated assumptions, the probability of w is

$$B(w) = \binom{n}{w} p^w q^{n-w}, \quad \text{for } 0 \leqslant w \leqslant n.$$

The above argument can be extended to the case when the specified population has more than one unknown parameter. Thus, suppose a factory produces light bulbs, and the manufacturer wishes to know the life-time characteristics of his product for advertisement. Here a primary assumption is that the manufacturing process is *stable*, so that the variations in the life-time of bulbs produced can be used to infer the quality of the future product. The estimation problem can be defined if we further make an assumption about the model for representing the life-time of the bulbs produced. If, for example, it is assumed that the life-time of the bulbs is a normally distributed random variable with mean μ and variance σ^2 (these are the two parameters of the probability distribution), then we require the estimates of μ and σ^2 on the basis of the observed life-times of n randomly selected bulbs. If these life-times are denoted by $x_1, x_2, \ldots, x_n$, then these observations are the sample information from a conceptually infinite number of life-times of bulbs produced and to be produced by the factory *under similar conditions*. In such a situation, we have used the sample average $\bar{x}$ and the sample variance s^2 as "estimates" of μ and σ^2 respectively. We again seek a formal justification for this procedure, and to determine the sense in which these estimates are the "best possible". This depends upon the fact that, under the stated conditions, the joint probability density of the sample observations is

$$\left(\frac{1}{\sigma\sqrt{2\pi}}\right)^n \exp - \sum_{i=1}^{n} (x_i - \mu)^2/2\sigma^2.$$

It follows from the preceding discussion that the problem of estimation may be thought of in four distinct stages. Firstly, there is the mathematical specification of the distributional behaviour of a random variable. This implies the indication of the parameters occurring in the definition of the probability distribution. Secondly, there is the random sample of observations from the population conceptually representing the behaviour of the random variable. Thirdly, estimates of the population parameters are obtained by using the sample information. Finally, having obtained the estimates, comes the determination of the sense in which the estimates represent the "best possible" use of the sample information. As we shall see, this implies a consideration of the sampling distributions of the estimates.

It is sufficient for our immediate purpose to understand that any estimate of a parameter is a function of the sample observations, each of which is a random variable. We therefore consider first some important properties of *functions of random variables*. In this we are, in fact, extending the ideas presented in our earlier consideration of the properties of a function of a single random variable.

11.2 Some properties of functions of random variables

Suppose $x_1, x_2, \ldots, x_n$ are n observations such that

$$E(x_i) = \mu_i \quad \text{and} \quad \text{var}(x_i) = \sigma_i^2, \quad \text{for } i = 1, 2, \ldots, n.$$

This means that each observation is a random variable and has a probability distribution with its assigned mean and variance. Now any function of these random variables is also a random variable. For example,

$$\sum_{i=1}^{n} x_i, \quad \prod_{i=1}^{n} x_i, \quad \sum_{i=1}^{n} (x_i - \mu_i)^2$$

are all random variables. A particular function of any two observations x_i, x_j $(i \neq j)$ is

$$(x_i - \mu_i)(x_j - \mu_j),$$

and given the n observations we could define $\binom{n}{2}$ such functions. These are all random variables. We now assume that the given n observations are such that

$$E(x_i - \mu_i)(x_j - \mu_j) = 0, \quad \text{for all } i \neq j. \tag{1}$$

It is to be noted that this expectation is defined over the *joint distribution* of the two random variables x_i and x_j. Roughly speaking, the condition (1) implies that the value of x_i has no effect on the value of x_j and vice versa. If (1) holds, then x_i and x_j are said to be *uncorrelated random variables* or, briefly, *uncorrelated*. This condition necessarily holds if the x_i are independent observations.

Furthermore, if the observations are independent, then

$$E\left[\prod_{i=1}^{n} x_i\right] = \prod_{i=1}^{n} E(x_i) = \prod_{i=1}^{n} \mu_i.$$

More generally, if $g_1(x_1), g_2(x_2), \dots, g_n(x_n)$ are functions of the independent observations $x_1, x_2, \dots, x_n$ respectively, then

$$E\left[\prod_{i=1}^{n} g_i(x_i)\right] = \prod_{i=1}^{n} E[g_i(x_i)].$$

This means that if the x_i are independent random variables so also are the $g_i(x_i)$ and the product rule for evaluating expectations holds.

Given a set of arbitrary constants a_i $(i = 1, 2, \dots, n)$ not all zero simultaneously, we define a *linear function of the observations* as

$$X = \sum_{i=1}^{n} a_i x_i.$$

Then X is a random variable whose probability distribution depends upon the probability distributions of the x_i. We shall not consider the derivation of the probability distribution of X, but two theorems are of particular importance.

Theorem 1

This theorem refers to the expectation of X. Since X is a random variable, we have

$$E(X) = E\left[\sum_{i=1}^{n} a_i x_i\right]$$

$$= \sum_{i=1}^{n} E(a_i x_i) = \sum_{i=1}^{n} a_i E(x_i),$$

so that

$$E(X) = \sum_{i=1}^{n} a_i \mu_i.$$

A proof of this theorem depends upon the joint probability distribution of the n random variables $x_1, x_2, \dots, x_n$, and we shall not consider it. An intuitive acceptance of the theorem may be obtained from the fact that since the expectation of a random variable is a sum, the expectation of the sum of a number of random variables will be the sum of their expectations. This is implied in the equality

$$E\left[\sum_{i=1}^{n} a_i x_i\right] = \sum_{i=1}^{n} E(a_i x_i).$$

Theorem 2

This theorem pertains to the variance of X. We have from the definition of the variance of a random variable

$$\begin{aligned}
\text{var}(X) &= E[X - E(X)]^2 \\
&= E\left[\sum_{i=1}^{n} a_i x_i - \sum_{i=1}^{n} a_i \mu_i\right]^2 \\
&= E\left[\sum_{i=1}^{n} a_i(x_i - \mu_i)\right]^2 \\
&= E\left[\sum_{i=1}^{n} a_i^2(x_i - \mu_i)^2 + \sum_{i=1}^{n}\sum_{j \neq i} a_i a_j(x_i - \mu_i)(x_j - \mu_j)\right].
\end{aligned}$$

We have thus expressed $\text{var}(X)$ as the expectation of the sum of certain functions of the random variables x_i. Hence, using Theorem 1,

$$\text{var}(X) = \sum_{i=1}^{n} a_i^2 E(x_i - \mu_i)^2 + \sum_{i=1}^{n}\sum_{j \neq i} a_i a_j E(x_i - \mu_i)(x_j - \mu_j)$$

whence, using (1) and recalling that by definition $E(x_i - \mu_i)^2 = \sigma_i^2$, we obtain

$$\text{var}(X) = \sum_{i=1}^{n} a_i^2 \sigma_i^2.$$

Specification of a random sample

Suppose $x_1, x_2, \ldots, x_n$ are independent observations of a random variable x having mean μ and variance σ^2. Then these observations constitute a *random sample of size n*. It now follows that

(i) $E(x_i) = \mu$;

(ii) $\text{var}(x_i) = \sigma^2$; and

(iii) $E(x_i - \mu)(x_j - \mu) = 0$, for $i \neq j = 1, 2, \ldots, n$.

With this specification of a random sample, we can use Theorems 1 and 2 to evaluate the mean and variance of certain important functions of the sample observations.

Expectation of the sample mean

The sample mean is $\bar{x} = \sum_{i=1}^{n} x_i / n$. Hence, putting $a_i = 1/n$ in Theorem 1, we obtain

$$E(\bar{x}) = \mu.$$

Variance of the sample mean

If we next put $a_i = 1/n$ in Theorem 2, then

$$\text{var}(\bar{x}) = \sigma^2/n.$$

Hence, using the definition of the variance of a random variable, we also have

$$\text{var}(\bar{x}) = E(\bar{x} - \mu)^2 = \sigma^2/n.$$

Expectation of the sample variance

We have defined the sample variance as

$$s^2 = \frac{1}{n-1} \sum_{i=1}^{n} (x_i - \bar{x})^2$$

$$= \frac{1}{n-1} \sum_{i=1}^{n} [(x_i - \mu) - (\bar{x} - \mu)]^2$$

$$= \frac{1}{n-1} \sum_{i=1}^{n} [(x_i - \mu)^2 - 2(\bar{x} - \mu)(x_i - \mu) + (\bar{x} - \mu)^2]$$

$$= \frac{1}{n-1} \left[\sum_{i=1}^{n} (x_i - \mu)^2 - 2(\bar{x} - \mu) \sum_{i=1}^{n} (x_i - \mu) + n(\bar{x} - \mu)^2 \right]$$

$$= \frac{1}{n-1} \left[\sum_{i=1}^{n} (x_i - \mu)^2 - n(\bar{x} - \mu)^2 \right].$$

Therefore

$$E(s^2) = \frac{1}{n-1} \left[\sum_{i=1}^{n} E(x_i - \mu)^2 - nE(\bar{x} - \mu)^2 \right], \quad \text{by Theorem 1,}$$

$$= \frac{1}{n-1} \left[\sum_{i=1}^{n} \text{var}(x_i) - n \, \text{var}(\bar{x}) \right]$$

$$= \frac{1}{n-1} [n\sigma^2 - n(\sigma^2/n)] = \sigma^2.$$

Hence $$E(s^2) = \sigma^2.$$

Expectation of the sample standard deviation

Although $E(s^2) = \sigma^2$, it does *not* follow that $E(S) = \sigma$. In fact, since the variance of a random variable is non-negative, we must have $\text{var}(s) \geqslant 0$. But, by definition,

$$\text{var}(s) = E[s - E(s)]^2$$
$$= E(s^2) - E^2(s)$$
$$= \sigma^2 - E^2(s) \geqslant 0.$$

Therefore $$\sigma^2 \geqslant E^2(s),$$
so that $$\sigma \geqslant E(s).$$

Expectation and variance of the mean, variance, and standard deviation of a random sample from a normal population

Suppose next that $x_1, x_2, \ldots, x_n$ are independent observations from a normal population with mean μ and variance σ^2. It then follows from the general results proved above that the sample mean $\bar{x}$ has expectation μ and variance σ^2/n. Moreover, since the observations are from a normal population, a much more powerful result can be deduced. In fact, it is known that in this case the random variable $\bar{x}$ is *also* normally distributed with mean μ and variance σ^2/n. This is a fundamental theorem in statistical theory, but its proof will not be considered here. This result means that the probability distribution of $\bar{x}$ is

$$\sigma^{-1}(n/2\pi)^{\frac{1}{2}} e^{-n(\bar{x}-\mu)^2/2\sigma^2} \, d\bar{x}, \quad \text{for } -\infty < \bar{x} < \infty,$$

so that $z = (\bar{x} - \mu)\sqrt{n}/\sigma$ is a unit normal variable.

Hence $$E(z^{2r}) = \frac{(2r)!}{2^r r!}; \quad E(z^{2r+1}) = 0,$$

for integral values of $r \geqslant 0$.

We now use these results to obtain the variance of the sample variance s^2 and of its positive square root s, the sample standard deviation. Since

$E(s^2) = \sigma^2$, we have, by definition,

$$\text{var}(s^2) = E(s^2 - \sigma^2)^2$$
$$= E(s^4) - \sigma^4. \tag{1}$$

To evaluate $E(s^4)$, we observe first that

$$(n - 1)s^2 = \sum_{i=1}^{n} (x_i - \bar{x})^2 = \sum_{i=1}^{n} (x_i - \mu)^2 - n(\bar{x} - \mu)^2.$$

If we now set $y_i = (x_i - \mu)/\sigma$ and $\bar{y} = (\bar{x} - \mu)/\sigma$, then

$$(n - 1)s^2 = \sigma^2 \left[\sum_{i=1}^{n} y_i^2 - n\bar{y}^2 \right],$$

where the y_i are independent unit normal variables, and $\bar{y}$ is a normal variable with zero mean and variance $1/n$. Next, we have

$$(n - 1)^2 s^4 = \sigma^4 \left[\left\{ \sum_{i=1}^{n} y_i^2 \right\}^2 - 2n\bar{y}^2 \sum_{i=1}^{n} y_i^2 + n^2\bar{y}^4 \right]$$

$$= \sigma^4 \left[\left\{ \sum_{i=1}^{n} y_i^4 + \sum_{i=1}^{n} \sum_{j \neq i} y_i^2 y_j^2 \right\} - \frac{2}{n} \left\{ \sum_{i=1}^{n} y_i \right\}^2 \sum_{i=1}^{n} y_i^2 + n^2\bar{y}^4 \right]$$

$$= \sigma^4 \left[\left\{ \sum_{i=1}^{n} y_i^4 + \sum_{i=1}^{n} \sum_{j \neq i} y_i^2 y_j^2 \right\} - \frac{2}{n} \left\{ \sum_{i=1}^{n} y_i^2 + \sum_{i=1}^{n} \sum_{j \neq i} y_i y_j \right\} \times \right.$$

$$\left. \times \left\{ \sum_{i=1}^{n} y_i^2 \right\} + n^2\bar{y}^4 \right]$$

$$= \sigma^4 \left[\left\{ \sum_{i=1}^{n} y_i^4 + \sum_{i=1}^{n} \sum_{j \neq i} y_i^2 y_j^2 \right\} - \frac{2}{n} \left\{ \sum_{i=1}^{n} y_i^4 + \sum_{i=1}^{n} \sum_{j \neq i} y_i^2 y_j^2 + \right. \right.$$

$$\left. \left. + \sum_{i=1}^{n} \sum_{j \neq i} y_i^3 y_j + \sum_{i=1}^{n} \sum_{j \neq k \neq i} y_i^2 y_j y_k \right\} + n^2\bar{y}^4 \right]. \tag{2}$$

But $E(y_i) = E(\bar{y}) = 0$; $E(y_i^2) = \text{var}(y_i) = 1$; $E(\bar{y}^2) = \text{var}(\bar{y}) = 1/n$;
$E(y_i^3) = 0$; $E(y_i^4) = 3$; $E(\bar{y}^4) = 3/n^2$.

Hence, taking expectations in (2) and remembering that the y_i are independent random variables, we have

$$E[(n - 1)^2 s^4] = \sigma^4 \left[\{3n + n(n - 1)\} - \frac{2}{n} \{3n + n(n - 1) + 0 + 0\} + n^2 \frac{3}{n^2} \right]$$

$$= (n^2 - 1)\sigma^4.$$

Therefore

$$E(s^4) = \frac{(n + 1)\sigma^4}{n - 1},$$

whence, from (1),

$$\text{var}(s^2) = \frac{2\sigma^4}{n - 1}.$$

Finally, we use the results for the mean and variance of s^2 to derive approximations for the mean and variance of s.

If we set $w = s^2$, then $E(w) = \sigma^2$ and $\text{var}(w) = 2\sigma^4/(n - 1)$. Now,

$$s = \sqrt{w} = [(w - \sigma^2) + \sigma^2]^{\frac{1}{2}}$$

$$= \sigma[1 + (w - \sigma^2)/\sigma^2]^{\frac{1}{2}}$$

$$= \sigma \left[1 + \frac{w - \sigma^2}{2\sigma^2} - \frac{(w - \sigma^2)^2}{8\sigma^4} + \cdots \right].$$

Hence, for large n, we have

$$E(s) \sim \sigma\left[1 - \frac{1}{8\sigma^4} \times \frac{2\sigma^4}{n-1}\right] = \sigma\left[1 - \frac{1}{4(n-1)}\right] \sim \sigma\left[1 - \frac{1}{4n}\right].$$

Also, $\mathrm{var}(s) = E(s^2) - E^2(s)$

$$\sim \sigma^2 - \sigma^2\left[1 - \frac{1}{4n}\right]^2 \sim \sigma^2 - \sigma^2\left[1 - \frac{1}{2n}\right] = \frac{\sigma^2}{2n}.$$

The exact results for the mean and variance of s are cumbersome expressions, and the above approximations are good enough for most statistical purposes.

11.3   The methods of estimation

There are several different methods of estimation, and of these, two are based on separate theories of estimation. The justification of the theories rests on mathematics beyond the scope of this book and we shall not consider it, although an intuitive understanding of the reasoning behind the methods will be presented. However, our main concern is in the formal application of the methods to particular cases of estimation. It is worth stressing that, in general, the methods lead to rather different estimates if used in the same situation. This divergence is to be expected since the methods emanate from different theories of estimation. On the other hand, in some of the common but important examples of estimation, all the methods lead to effectively the same results. This agreement, though welcome, is quite fortuitous and should not blur the basic differences which are characteristic of the estimation procedures used.

For our purposes, there are three main methods of estimation, namely,

   (i) the method of maximum likelihood,

  (ii) the method of least squares,

 (iii) the method of moments,

and we now consider them.

11.4   The method of maximum likelihood

This is undoubtedly the most useful method of estimation, and it is based on a theory due to R.A. Fisher. It is a versatile technique and is applicable equally to the estimation of parameters of the probability distributions of discrete and continuous random variables. In this exposition we present the simplest form of the method and its underlying logic.

Suppose $x_1, x_2, \ldots, x_n$ are independent observations of a random variable x having the probability density function (point-probability) $f(x, \theta)$, where θ is a parameter or a function of a parameter. We assume that the range of variation of x does not depend upon θ. This is not an essential limitation for the application of the method of maximum likelihood, but we shall not consider the more general case when the range of variation of x depends upon the parameter to be estimated. Now, since the observations are independent, the joint probability density (probability) of the sample is

$$L = \prod_{i=1}^{n} f(x_i, \theta).$$

Clearly, once the sample observations have been made, both n and the x_i are fixed, so that L involves only one unknown and that is the parameter θ. As a function of θ, L is called the *likelihood*, and it varies over the permissible range of variation of θ. There is a small conceptual difficulty here in taking θ to be a *variable*. It is true that as a population parameter θ has some fixed but unknown value. However, as a parameter with a physical meaning, θ could theoretically lie within some range. For example, if θ is a binomial proportion, then it can have any value in the range $(0, 1)$. In the same way, if θ is the variance of a normal population, then θ must be non-negative. It is in this sense that the likelihood L is considered to be a function of θ defined over the permissible range of variation of the parameter. According to Fisher's theory, the maximum-likelihood estimate of θ is obtained simply by maximising L with respect to variation in θ and for fixed $x_1, x_2, \ldots, x_n$. This implies that the maximum-likelihood estimate is such that it *maximises* the probability of obtaining the observed sample. This is the basic argument of the method of maximum likelihood.

Now the maximisation of L is equivalent to that of $\log L$, and it is usually convenient to maximise $\log L$ since, in general, L will occur as a product of a certain number of terms. We thus have

$$\log L = \sum_{i=1}^{n} \log f(x_i, \theta),$$

$$\frac{d \log L}{d\theta} = \sum_{i=1}^{n} \frac{1}{f(x_i, \theta)} \frac{df(x_i, \theta)}{d\theta}.$$

The maximum-likelihood estimate $\hat{\theta}$ is a root of the equation

$$\sum_{i=1}^{n} \frac{1}{f(x_i, \hat{\theta})} \left[\frac{df(x_i, \theta)}{d\theta} \right]_{\theta = \hat{\theta}} = 0. \tag{1}$$

It is also known from general theory that there always exists a root of this equation which will maximise $\log L$, that is,

$$\left[\frac{d^2 \log L}{d\theta^2} \right]_{\theta = \hat{\theta}} < 0.$$

The formal solution for a maximum-likelihood estimate is extremely simple, but the main practical difficulty in using this estimation procedure is that the resulting equation (1) can prove rather difficult to solve. In fact, in many cases, the only method of solving (1) is by some iterative process such as inverse interpolation. This difficulty becomes a major problem when L involves more than one unknown parameter so that the maximum-likelihood estimates are sought as a solution of a set of simultaneous equations analogous to (1). Generally, if the probability density function (point-probability) of x is $f(x, \theta_1, \theta_2, \ldots, \theta_k)$, where $\theta_1, \theta_2, \ldots, \theta_k$ are independent parameters, then the likelihood of the sample observations $x_1, x_2, \ldots, x_n$ is

$$L = \prod_{i=1}^{n} f(x_i, \theta_1, \theta_2, \ldots, \theta_k),$$

and the maximum-likelihood estimates of the parameters are obtained as a solution of the k simultaneous equations

$$\left[\frac{\partial \log L}{\partial \theta_j}\right]_{\theta_j = \hat{\theta}_j} = 0, \quad \text{for } j = 1, 2, \ldots, k.$$

Variance of the estimate

It is obvious that $\hat{\theta}$, the root of (1), must be some function of the sample observations $x_1, x_2, \ldots, x_n$ and n, the sample size, and so, in principle, $\hat{\theta}$ can be evaluated in any given case. We may, therefore, write the solution of (1) as

$$\hat{\theta} = g(x_1, x_2, \ldots, x_n ; n),$$

where g is some function of the observations and n. However, the sample observations are a particular but chance set of realisations of the random variable x. Indeed, for any fixed sample size n, another sample from the same population would consist of a different set of observations, say $x_1', x_2', \ldots, x_n'$, and so, quite obviously, lead in general to a different value for $\hat{\theta}$. In other words, as the sample observations are random variables, $\hat{\theta}$ is a function of the random variables. Therefore $\hat{\theta}$ itself is a random variable. R.A. Fisher called such a random variable a *statistic*. This is a fundamental idea, and it lies at the heart of the relationship between a population parameter and the statistic based on a random sample which estimates the parameter.

Since $\hat{\theta}$ is a random variable, it must necessarily have a probability distribution, which is, in fact, the sampling distribution of $\hat{\theta}$. This distribution gives the probabilistic behaviour of $\hat{\theta}$ when the estimate is obtained from different random samples of the *same* size n. We know that a random variable does not always have a finite variance, but this is largely true of the statistics with which we shall be concerned. It is a particularly convenient property of the method of maximum likelihood that it gives estimates which have a finite variance. We now show how this variance can be evaluated.

The second derivative $d^2 \log L / d\theta^2$ of the sample likelihood L will, in general, involve the observations $x_1, x_2, \ldots, x_n$, the parameter θ, and the sample size n. Now if, for fixed n and θ, we consider the expectation of this derivative for variation over the random variables $x_1, x_2, \ldots, x_n$, then it is known that

$$\text{var}(\hat{\theta}) = -\frac{1}{E\left[\dfrac{d^2 \log L}{d\theta^2}\right]} . \tag{2}$$

This formula is not exact, though in many simple cases it does, indeed, give the exact variance of $\hat{\theta}$. Besides, (2) invariably gives an approximation for $\text{var}(\hat{\theta})$, which is adequate for all practical purposes, if the sample size n is sufficiently large.

The variance of an estimate is a fundamental property in judging how far an estimate may be regarded as the "best possible". We shall return to this idea when we study the properties of estimates obtained by different methods of estimation.

11.5 Applications of maximum-likelihood estimation

Example 11.1: Estimation in the binomial distribution

Suppose p is the unknown probability of success in a binomial distribution, and our problem is to derive its maximum-likelihood estimate. If w successes were observed in a random sample of n trials, then the probability of the realisation of this result is

$$B(w) = \binom{n}{w} p^w (1-p)^{n-w},$$

where theoretically w could have had any integral value in the range $(0, n)$. The actual outcome w is a chance event, but once w and n have been fixed then, as a function of p,

$$L = \binom{n}{w} p^w (1-p)^{n-w}$$

is the sample likelihood. Clearly, p could have any possible value in the range $(0, 1)$, and we regard that particular value of p as the maximum-likelihood estimate of p which maximises L or, equivalently, $\log L$.

We have

$$\log L = \log\binom{n}{w} + w \log p + (n-w) \log(1-p),$$

so that

$$\frac{d \log L}{dp} = \frac{w}{p} - \frac{n-w}{1-p}.$$

Hence the equation for $\hat{p}$, the maximum-likelihood estimate of p, is

$$\frac{w}{\hat{p}} - \frac{n-w}{1-\hat{p}} = 0,$$

which gives $\hat{p} = w/n$. The observed relative frequency of success is thus seen to be the maximum-likelihood estimate of p. This result is happily in agreement with our intuitive ideas, and we also observe that

$$E(\hat{p}) = \frac{1}{n} E(w) = p.$$

To obtain the variance of $\hat{p}$, we have

$$\frac{d^2 \log L}{dp^2} = -\frac{w}{p^2} - \frac{n-w}{(1-p)^2}.$$

Therefore

$$-E\left[\frac{d^2 \log L}{dp^2}\right] = \frac{E(w)}{p^2} + \frac{E(n-w)}{(1-p)^2}$$

$$= \frac{n}{p} + \frac{n}{1-p} = \frac{n}{pq},$$

where $p + q = 1$. Hence

$$\mathrm{var}(\hat{p}) = \frac{pq}{n}.$$

This result, although obtained by using the formula for the variance of a maximum-likelihood estimate, is exact. This is easily verified since

$$\mathrm{var}(\hat{p}) = \mathrm{var}(w/n) = \mathrm{var}(w)/n^2 = pq/n.$$

Example 11.2: Estimation in the negative exponential distribution

The probability density function of a random variable x having a negative exponential distribution is

$$\frac{1}{\lambda}e^{-x/\lambda}, \quad \text{where } 0 \leqslant x < \infty, \quad \text{and the parameter } \lambda > 0.$$

We also recall that $E(x) = \lambda$ and $\text{var}(x) = \lambda^2$, so that as for the binomial
distribution, both the mean and the variance of x are functions of the same
parameter λ.

To estimate λ, suppose $x_1, x_2, \ldots, x_n$ are n independent observations
from this population. Then the joint probability density of the sample is

$$\lambda^{-n}\exp - \sum_{i=1}^{n} x_i/\lambda,$$

and so the logarithm of the sample likelihood is

$$\log L = -n \log \lambda - \frac{n\bar{x}}{\lambda}, \quad \text{where } \bar{x} \text{ is the sample mean.}$$

Therefore
$$\frac{d \log L}{d\lambda} = -\frac{n}{\lambda} + \frac{n\bar{x}}{\lambda^2}.$$

Hence the equation for $\hat{\lambda}$, the maximum-likelihood estimate of λ, is

$$-\frac{n}{\hat{\lambda}} + \frac{n\bar{x}}{\hat{\lambda}^2} = 0,$$

which gives $\hat{\lambda} = \bar{x}$. Thus the sample mean is the maximum-likelihood estimate of the population mean λ. Evidently, we also have

$$E(\hat{\lambda}) = E(\bar{x}) = \lambda.$$

For the variance of $\hat{\lambda}$, we have

$$\frac{d^2 \log L}{d\lambda^2} = \frac{n}{\lambda^2} - \frac{2n\bar{x}}{\lambda^3},$$

so that
$$-E\left[\frac{d^2 \log L}{d\lambda^2}\right] = -\frac{n}{\lambda^2} + \frac{2n E(\bar{x})}{\lambda^3} = \frac{n}{\lambda^2}.$$

Hence
$$\text{var}(\hat{\lambda}) = \frac{\lambda^2}{n}.$$

We observe that in this case also the formula for the variance of a maximum-
likelihood estimate gives the exact result. This is readily verified, since

$$\text{var}(\hat{\lambda}) = \text{var}(\bar{x}) = \text{var}(x)/n = \lambda^2/n.$$

Example 11.3: Estimation in the normal distribution

The probability density function of a normally distributed random variable x is

$$\frac{1}{\sigma\sqrt{2\pi}} e^{-(x-\mu)^2/2\sigma^2}, \quad \text{for } -\infty < x < \infty,$$

where μ and σ^2 are the two independent parameters such that

$$E(x) = \mu \quad \text{and} \quad \text{var}(x) = \sigma^2.$$

We are concerned here with the joint estimation of two parameters by the
method of maximum likelihood. This problem is of great importance.

If $x_1, x_2, \ldots, x_n$ are independent observations from the specified normal population, then their joint probability density is

$$(\sigma\sqrt{2\pi})^{-n} \exp - \sum_{i=1}^{n} (x_i - \mu)^2/2\sigma^2 ,$$

and the logarithm of the sample likelihood is

$$\log L = -\frac{1}{2}n \log(2\pi) - n \log \sigma - \frac{1}{2\sigma^2} \sum_{i=1}^{n} (x_i - \mu)^2.$$

We have now to maximise $\log L$ with respect to both μ and σ (or σ^2). For this purpose, successive differentiation with respect to μ and σ gives

$$\frac{\partial \log L}{\partial \mu} = \frac{1}{\sigma^2} \sum_{i=1}^{n} (x_i - \mu), \tag{1}$$

and

$$\frac{\partial \log L}{\partial \sigma} = -\frac{n}{\sigma} + \frac{1}{\sigma^3} \sum_{i=1}^{n} (x_i - \mu)^2. \tag{2}$$

For $\sigma \neq 0$, the two equations for the estimates $\hat{\mu}$ and $\hat{\sigma}^2$ are obtained from (1) and (2) as

$$\sum_{i=1}^{n} (x_i - \hat{\mu}) = 0 ; \qquad -n + \frac{1}{\hat{\sigma}^2} \sum_{i=1}^{n} (x_i - \hat{\mu})^2 = 0.$$

These immediately give

$$\hat{\mu} = \bar{x} ; \qquad \hat{\sigma}^2 = \frac{1}{n} \sum_{i=1}^{n} (x_i - \bar{x})^2.$$

We thus see that though the maximum-likelihood estimate of μ is the sample mean $\bar{x}$, the maximum-likelihood estimate of σ^2 is not the sample variance

$$s^2 = \sum_{i=1}^{n} (x_i - \bar{x})^2/(n - 1).$$

In fact,

$$\hat{\sigma}^2 = \frac{(n - 1)s^2}{n} .$$

This is a fundamental result.

Since $\hat{\mu}$ and $\hat{\sigma}^2$ have been determined by a joint estimation procedure, the earlier formula for the variance of a maximum-likelihood estimate does not apply. There is a more general formula which gives the variances in this case, but we shall not go into this. It follows immediately from the results proved earlier that

(i) $E(\hat{\mu}) = E(\bar{x}) = \mu$, $\quad$ var$(\hat{\mu}) = var(\bar{x}) = \sigma^2/n$; $\quad$ and

(ii) $E(\hat{\sigma}^2) = E[(n - 1)s^2/n] = \dfrac{(n - 1)\sigma^2}{n}$,

var$(\hat{\sigma}^2) = var[(n - 1)s^2/n] = \dfrac{(n - 1)^2}{n^2} \times \dfrac{2\sigma^4}{n - 1} \sim \dfrac{2\sigma^4}{n}$, $\quad$ for large n.

11.6   The method of least squares

This important method of estimation is based on the theory of *linear estimation*, whose basic principles are due to Karl Friedrich Gauss (1777 – 1855) and Andrei Andreevich Markov (1856 – 1922). We present the method in its simplest form, though greater generality can be obtained by relaxing some of our conditions.

Suppose $x_1, x_2, \ldots, x_n$ are independent observations such that

$$E(x_\nu) = \sum_{i=1}^{k} a_{i\nu}\theta_i, \quad \text{for } \nu = 1, 2, \ldots, n, \quad \text{and } k < n;$$

and $\text{var}(x_\nu) = \sigma^2$,   where

(i) the $a_{i\nu}$ are known constants;

(ii) the θ_i are unknown but independent parameters;   and

(iii) σ^2 is an unknown parameter which is independent of the θ_i.

It is to be observed that each $E(x_\nu)$ is expressed as a linear function of the k parameters θ_i; and by the independence of the θ_i we mean that no linear relation of the type

$$\sum_{i=1}^{k} c_i \theta_i = 0,$$

where the c_i are constants which are not all simultaneously zero, can be found amongst the θ_i. Furthermore, the independence of σ^2 means that it does not depend upon the θ_i, but this condition is not necessary for a formal application of the method of least squares. What is essential to the method is that the x_ν should all have equal variance. The further restriction on the variance of the observations is necessary to justify certain methods of statistical analysis which emanate from the theory of linear estimation. We shall return to this in a subsequent chapter.

Given the observations, the method of least squares is based on the principle that estimates of the θ_i can be obtained by minimising the sum of the squares of the deviations of the observations from their expectations, that is, by minimising

$$\Omega = \sum_{\nu=1}^{n} [x_\nu - E(x_\nu)]^2 = \sum_{\nu=1}^{n} \left[x_\nu - \sum_{i=1}^{k} a_{i\nu}\theta_i \right]^2$$

with respect to variations in the θ_i. Thus the estimates θ_i^* of the θ_i are obtained as a solution of the equations

$$\left[\frac{\partial \Omega}{\partial \theta_r} \right]_{\theta_r = \theta_r^*} = 0,$$

or $\sum_{\nu=1}^{n} a_{r\nu}\left[x_\nu - \sum_{i=1}^{k} a_{i\nu}\theta_i^* \right] = 0, \quad \text{for } r = 1, 2, \ldots, k.$

There are k simultaneous linear equations, and it is known that a solution of these equations must be of the form

$$\theta_i^* = \sum_{\nu=1}^{n} b_{i\nu} x_\nu, \quad \text{for } i = 1, 2, \ldots, k,$$

where the $b_{i\nu}$ are constants depending upon the $a_{i\nu}$ and n but do not involve the observations x_ν. In other words, the estimates θ_i^* are linear functions of the observations. This explains the term "linear estimation".

An important property of all least-squares estimates is that

$$E(\theta_i^*) = \sum_{\nu=1}^{n} b_{i\nu} E(x_\nu) \equiv \theta_i.$$

Also, since the x_ν are independent observations,

$$\begin{aligned}
\text{var}(\theta_i^*) &= \left[\text{var} \sum_{\nu=1}^{n} b_{i\nu} x_\nu \right] \\
&= \sum_{\nu=1}^{n} b_{i\nu}^2 \, \text{var}(x_\nu) \\
&= \sigma^2 \sum_{\nu=1}^{n} b_{i\nu}^2, \quad \text{for } i = 1, 2, \ldots, k.
\end{aligned}$$

If the variance σ^2 is an independent parameter, then the theory of linear estimation also provides that an appropriate estimate of σ^2 is

$$s_e^2 = M/(n - k),$$

where M is obtained from Ω by substituting θ_i^* for θ_i. Thus

$$M = \sum_{\nu=1}^{n} \left[x_\nu - \sum_{i=1}^{k} a_{i\nu} \theta_i^* \right]^2,$$

and the denominator $n - k$ is the difference between the number of observations and the number of parameters θ_i. It is known that in all cases

$$E(s_e^2) = \sigma^2.$$

This apparently involved method of estimation is, in fact, rather simple to apply as the following examples illustrate.

11.7 Applications of the method of least squares

Example 11.4: Estimation in the binomial distribution

In earlier notation, suppose w successes were obtained in a sample of n observations so that

$$E(w) = np \quad \text{and} \quad \text{var}(w) = npq,$$

where p is the unknown parameter denoting the probability of success at a trial and $p + q = 1$. Here w is the single random variable corresponding to the x_ν, and both its mean and variance depend upon the one parameter p. To obtain the least-squares estimate of p, we have

$$\Omega = (w - np)^2 \quad \text{and} \quad \frac{d\Omega}{dp} = -2n(w - np).$$

Therefore the equation for p^*, the least-squares estimate of p, is

$$w - np^* = 0, \quad \text{or} \quad p^* = w/n.$$

We thus observe that the least-squares estimate of p is exactly the same as the maximum-likelihood estimate. As already pointed out, this is an exception rather than the rule.

Example 11.5: Estimation in the negative exponential distribution

If $x_1, x_2, \ldots, x_n$ are independent observations from a negative exponential distribution with mean λ, then

$$E(x_i) = \lambda \quad \text{and} \quad \text{var}(x_i) = \lambda^2, \quad \text{for } i = 1, 2, \ldots, n.$$

Then for the least-squares estimate of λ, we have

$$\Omega = \sum_{i=1}^{n} (x_i - \lambda)^2 \quad \text{and} \quad \frac{d\Omega}{d\lambda} = -2 \sum_{i=1}^{n} (x_i - \lambda).$$

Therefore the least-squares estimate of λ is obtained from the equation

$$\sum_{i=1}^{n} (x_i - \lambda^*) = 0, \quad \text{so that} \quad \lambda^* = \bar{x}, \quad \text{the sample mean.}$$

In this case also, the least-squares estimate is the same as the maximum-likelihood estimate.

Example 11.6 : Estimation in the normal distribution

If $x_1, x_2, \ldots, x_n$ are independent observations from a normal population with mean μ and variance σ^2, then

$$E(x_i) = \mu \quad \text{and} \quad \text{var}(x_i) = \sigma^2, \quad \text{for } i = 1, 2, \ldots, n.$$

Thus for the least-squares estimate of μ, we have

$$\Omega = \sum_{i=1}^{n} (x_i - \mu)^2 \quad \text{and} \quad \frac{d\Omega}{d\mu} = -2 \sum_{i=1}^{n} (x_i - \mu).$$

Therefore the equation for the least-squares estimate of μ is

$$\sum_{i=1}^{n} (x_i - \mu^*) = 0, \quad \text{so that} \quad \mu^* = \bar{x}, \quad \text{the sample mean.}$$

We thus see that the least-squares estimate of μ is the same as that obtained by the method of maximum likelihood. However, the least-squares estimate of σ^2 is

$$s_e^2 = \sum_{i=1}^{n} (x_i - \bar{x})^2/(n - 1).$$

This is, in fact, the sample variance as already defined, and it differs from the maximum-likelihood estimate $\hat{\sigma}^2$.

Example 11.7 : General linear estimation

As a somewhat more complicated example of estimation by least squares, suppose $x_1, x_2, \ldots, x_n$ are independent observations such that

$$E(x_\nu) = \theta_1 + \nu\theta_2 \quad \text{and} \quad \text{var}(x_\nu) = \sigma^2, \quad \text{for } \nu = 1, 2, \ldots, n,$$

where θ_1, θ_2 are independent parameters, and σ^2 does not depend upon θ_1 and θ_2. It is to be noted that here we are not making any further assumptions about the distributional behaviour of the x_ν.

For determining the least-squares estimates of θ_1 and θ_2, we have

$$\Omega = \sum_{\nu=1}^{n} (x_\nu - \theta_1 - \nu\theta_2)^2,$$

so that

$$\frac{\partial\Omega}{\partial\theta_1} = -2 \sum_{\nu=1}^{n} (x_\nu - \theta_1 - \nu\theta_2);$$

$$\frac{\partial\Omega}{\partial\theta_2} = -2 \sum_{\nu=1}^{n} \nu(x_\nu - \theta_1 - \nu\theta_2).$$

If we now set $\bar{x} = \sum_{\nu=1}^{n} x_\nu/n$ and $\bar{x}_w = 2 \sum_{\nu=1}^{n} \nu x_\nu/n(n + 1)$, then the equations for the least-squares estimates θ_1^* and θ_2^* can be written as

$$\bar{x} = \theta_1^* + \tfrac{1}{2}(n + 1)\theta_2^*; \qquad \bar{x}_w = \theta_1^* + \tfrac{1}{3}(2n + 1)\theta_2^*.$$

A straightforward solution of these simultaneous equations gives

$$\theta_1^* = \frac{2(2n + 1)\bar{x} - 3(n + 1)\bar{x}_w}{n - 1}; \qquad \theta_2^* = \frac{6(\bar{x}_w - \bar{x})}{n - 1}.$$

Also, the least-squares estimates of σ^2 is

$$s_e^2 = \sum_{\nu=1}^{n} (x_\nu - \theta_1^* - \nu\theta_2^*)^2 /(n - 2).$$

To obtain the expectations of θ_1^* and θ_2^*, we first observe that

$$\bar{x} = \frac{1}{n} \sum_{\nu=1}^{n} x_\nu,$$

so that

$$E(\bar{x}) = \frac{1}{n} \sum_{\nu=1}^{n} E(x_\nu) = \frac{1}{n} \sum_{\nu=1}^{n} (\theta_1 + \nu\theta_2) = \theta_1 + \tfrac{1}{2}(n + 1)\theta_2.$$

Thus

$$E(\bar{x}) = \theta_1 + \tfrac{1}{2}(n + 1)\theta_2. \tag{1}$$

In the same way, since

$$\bar{x}_w = \frac{2}{n(n + 1)} \sum_{\nu=1}^{n} \nu x_\nu,$$

we have

$$E(\bar{x}_w) = \frac{2}{n(n + 1)} \sum_{\nu=1}^{n} \nu E(x_\nu)$$

$$= \frac{2}{n(n + 1)} \sum_{\nu=1}^{n} \nu(\theta_1 + \nu\theta_2)$$

$$= \frac{2}{n(n + 1)} \left[\theta_1 \frac{n(n + 1)}{2} + \theta_2 \frac{n(n + 1)(2n + 1)}{6} \right],$$

or

$$E(\bar{x}_w) = \theta_1 + \tfrac{1}{3}(2n + 1)\theta_2. \tag{2}$$

We now use (1) and (2) to obtain the expectations of θ_1^* and θ_2^*.

Since $\qquad \theta_1^* = [2(2n + 1)\bar{x} - 3(n + 1)\bar{x}_w]/(n - 1),$

we have $\quad E(\theta_1^*) = [2(2n + 1)E(\bar{x}) - 3(n + 1)E(\bar{x}_w)]/(n - 1) = \theta_1.$

Again, since $\qquad \theta_2^* = 6(\bar{x}_w - \bar{x})/(n - 1),$

we have $\qquad E(\theta_2^*) = 6[E(\bar{x}_w) - E(\bar{x})]/(n - 1) = \theta_2.$

To obtain the variances of θ_1^* and θ_2^*, it is convenient first to express these estimates *explicitly* as linear functions of the x_ν. Thus

$$\theta_1^* = \frac{1}{n - 1} \left[2(2n + 1)\frac{1}{n} \sum_{\nu=1}^{n} x_\nu - 3(n + 1) \frac{2}{n(n + 1)} \sum_{\nu=1}^{n} \nu x_\nu \right]$$

$$= \frac{2}{n(n - 1)} \sum_{\nu=1}^{n} (2n + 1 - 3\nu) x_\nu,$$

which is the required explicit expression for θ_1^* in terms of the x_ν. Hence, since the x_ν are independent observations, we have

$$\mathrm{var}(\theta_1^*) = \frac{4}{n^2(n - 1)^2} \sum_{\nu=1}^{n} (2n + 1 - 3\nu)^2 \, \mathrm{var}(x_\nu)$$

$$= \frac{4\sigma^2}{n^2(n - 1)^2} \sum_{\nu=1}^{n} [(2n + 1)^2 - 6(2n + 1)\nu + 9\nu^2]$$

$$= \frac{4\sigma^2}{n^2(n - 1)^2} \left[n(2n + 1)^2 - 6(2n + 1) \frac{n(n + 1)}{2} + 9\frac{n(n + 1)(2n + 1)}{6} \right]$$

$$= \frac{2(2n + 1)\sigma^2}{n(n - 1)}.$$

Similarly, since $\qquad \theta_2^* = \dfrac{6(\bar{x}_w - \bar{x})}{n - 1} = \dfrac{6}{n(n - 1)} \sum_{\nu=1}^{n} \left[\dfrac{2\nu}{n+1} - 1 \right] x_\nu,$

we have
$$
\begin{aligned}
\operatorname{var}(\theta_2^*) &= \frac{36\sigma^2}{n^2(n-1)^2} \sum_{\nu=1}^{n} \left[\frac{2\nu}{n+1} - 1 \right]^2 \\
&= \frac{36\sigma^2}{n^2(n-1)^2} \sum_{\nu=1}^{n} \left[\frac{4\nu^2}{(n+1)^2} - \frac{4\nu}{n+1} + 1 \right] \\
&= \frac{36\sigma^2}{n^2(n-1)^2} \left[\frac{4}{(n+1)^2} \frac{n(n+1)(2n+1)}{6} - \frac{4}{n+1} \frac{n(n+1)}{2} + n \right] \\
&= \frac{12\sigma^2}{n(n^2-1)} .
\end{aligned}
$$

It is a little complicated to verify that s_e^2, as defined in this case, also has expectation σ^2. We shall consider a method for proving this subsequently.

11.8   The method of moments

This method of estimation is largely due to Karl Pearson (1857 — 1936). It suffers from the limitation that it is not based on any theory of estimation, and the estimates obtained by its use have, in general, only simplicity and intuitive reasonableness to commend them. However, in many of the simpler situations, the method gives estimates which are the same as those obtained by least squares or the maximum-likelihood approach. In the past, the method has been much criticised for its undoubted limitations; but, more recently, the extreme simplicity of moment estimates in certain situations where the maximum-likelihood procedure leads to intractable equations has placed the method in a somewhat more favourable light.

Suppose k ($\geqslant 1$) parameters $\theta_1, \theta_2, \ldots, \theta_k$ occur in the probability distribution of a random variable x. Then, to estimate them by the method of moments, we first evaluate the mean of the distribution of x and the next $k-1$ central moments, or moments about the origin of x. In general, these k moments will be functions of the parameters $\theta_1, \theta_2, \ldots, \theta_k$. We may thus write

$$
\begin{aligned}
E(x) &= \mu_1'(\theta_1, \theta_2, \ldots, \theta_k), \\
\text{and} \qquad E(x^r) &= \mu_r'(\theta_1, \theta_2, \ldots, \theta_k), \\
\text{or} \qquad E[(x - \mu_1')^r] &= \mu_r(\theta_1, \theta_2, \ldots, \theta_k), \quad \text{for } 2 \leqslant r \leqslant k.
\end{aligned}
$$

Next, given the independent observations $x_1, x_2, \ldots, x_n$ from the population, the corresponding moments for the sample are calculated. Finally, by equating the population and sample moments, we deduce k *estimating equations* for the parameters which, on solution, provide the moment estimates $\tilde{\theta}_1, \tilde{\theta}_2, \ldots, \tilde{\theta}_k$.

We have already used this method to estimate the mean and variance of a normal distribution, and we recall that the estimates so obtained were the same as the least-squares ones. It is not difficult to verify that this method again leads to the standard estimates for the parameters of the binomial and the negative exponential distributions. We next illustrate the use of the method of moments by an example in which the moment estimates are not immediately obvious.

Example 11.8 : Estimation in the lognormal distribution

A continuous random variable x has the *lognormal distribution* when its probability distribution is of the form

$$\frac{1}{\sigma\sqrt{2\pi}}x^{-1}\exp\left[-(\log x-\mu)^2/2\sigma^2\right]dx, \quad \text{for } 0 \leqslant x < \infty,$$

where $\sigma > 0$ and μ are parameters. The name "lognormal" is due to the fact that the random variable $y = \log x$ is normally distributed with mean μ and variance σ^2.

To apply the method of moments, we need first to evaluate the mean and variance of x which, as we shall see, are functions of μ and σ. For any integer r,

$$E(x^r) = \frac{1}{\sigma\sqrt{2\pi}}\int_0^\infty x^{r-1}\exp[-(\log x-\mu)^2/2\sigma^2]\,dx.$$

To evaluate the integral, we make the transformation

$$z = (\log x-\mu)/\sigma, \quad \text{so that} \quad dx = \sigma x\,dz, \quad \text{and} \quad -\infty < z < \infty.$$

Hence

$$\begin{aligned}
E(x^r) &= \frac{1}{\sqrt{2\pi}}\int_{-\infty}^\infty \exp[r(z\sigma+\mu)-\tfrac{1}{2}z^2]\,dz \\
&= \frac{1}{\sqrt{2\pi}}\int_{-\infty}^\infty \exp[-\tfrac{1}{2}(z-r\sigma)^2+r\mu+\tfrac{1}{2}r^2\sigma^2]\,dz \\
&= e^{r\mu+\frac{1}{2}r^2\sigma^2}\frac{1}{\sqrt{2\pi}}\int_{-\infty}^\infty e^{-\frac{1}{2}w^2}\,dw, \quad \text{where } w = z-r\sigma, \\
&= e^{r\mu+\frac{1}{2}r^2\sigma^2},
\end{aligned}$$

by using the value of the unit normal integral. Therefore, in particular,

$$E(x) = e^{\mu+\frac{1}{2}\sigma^2},$$

and
$$\text{var}(x) = e^{2\mu+2\sigma^2}-e^{2\mu+\sigma^2} = e^{2\mu+\sigma^2}(e^{\sigma^2}-1).$$

If $\bar{x}$ and s^2 are the sample mean and variance calculated from n independent observations from the lognormal population, then the estimating equations for the moment estimates of μ and σ^2 are

$$\bar{x} = e^{\tilde{\mu}+\frac{1}{2}\tilde{\sigma}^2}, \tag{1}$$

and
$$s^2 = e^{2\tilde{\mu}+\tilde{\sigma}^2}(e^{\tilde{\sigma}^2}-1). \tag{2}$$

Elimination of $\tilde{\mu}$ from (1) and (2) gives

$$s^2/\bar{x}^2 = e^{\tilde{\sigma}^2}-1,$$

so that
$$e^{\tilde{\sigma}^2} = (s^2+\bar{x}^2)/\bar{x}^2, \tag{3}$$

whence
$$\tilde{\sigma}^2 = \log[(s^2+\bar{x}^2)/\bar{x}^2].$$

Again, from (1)
$$e^{\tilde{\mu}} = \bar{x}\,e^{-\frac{1}{2}\sigma^2},$$

whence, using (3),
$$e^{\tilde{\mu}} = \bar{x}\left[\frac{\bar{x}^2}{s^2+\bar{x}^2}\right]^{\frac{1}{2}} = \bar{x}^2/(s^2+\bar{x}^2)^{\frac{1}{2}}.$$

Therefore
$$\tilde{\mu} = \log[\bar{x}^2/(s^2+\bar{x}^2)^{\frac{1}{2}}].$$

11.9   Properties of estimates

We have seen that an estimate of a population parameter is invariably a function of the sample observations. Accordingly, an estimate is a random

variable having a probability distribution. It is the behaviour of this probability distribution which provides a basis for assessing the merits (or de-
merits) of an estimate obtained by any method of estimation; and the behaviour of the probability distribution of an estimate may, in essentials, be
considered in terms of its mean and variance. This is the method which we
shall adopt. However, we note the fact that in this we are confining ourselves to estimates which necessarily have finite variance.

11.10   Properties of maximum-likelihood estimates

(i) Unbiasedness and consistency

If a statistic T, based on n independent observations, is an estimate
of a parameter θ such that $E(T) = \theta$, then T is said to be an *unbiased*
estimate of θ. On the other hand, if $E(T) = \theta(1 + c)$, where $c \neq 0$ may or
may not involve θ, then T is said to be a *biased* estimate of T and, relative to 1, the bias in T is c.

In general, there are two ways of looking at this property of unbiasedness of an estimate. Firstly, this means that if the statistic T were to be
calculated not from a finite number of observations but from all the conceptually infinite number constituting the population sampled, then T would be
identical with θ. This implies that when T is calculated from a finite
sample, the difference between T and θ (whatever its unknown magnitude)
is due only to the *finiteness* of the sample and not to the *method* of evaluating
T. Secondly, the fact that $E(T) = \theta$ means that, whatever the unknown magnitude of θ, the probability distribution of T is necessarily centred about
θ, irrespective of the value of n. In this we have made a statement about
the location of not one probability distribution of T but of an infinite sequence of probability distributions of T, each distribution being specified
by a given value of n, the number of observations used for calculating T.
This quite naturally leads to the basic idea of what is meant by the sampling distribution of a statistic T.

Suppose n is fixed and T is calculated from n independent observations
$x_1, x_2, \ldots, x_n$. We may indicate this by writing T as $T(x_1, x_2, \ldots, x_n)$. Next,
suppose another sample is taken which consists of the n independent
observations, say $x_1', x_2', \ldots, x_n'$. In general, the two values of T will differ
from each other. This process of evaluating T from different samples, each
of n independent observations, can, at least conceptually, be continued indefinitely since n is finite and the population sampled contains an infinite
number of observations. In this way, we obtain an infinite sequence of
values of T, say $T_1, T_2, T_3, \ldots$, each value being derived from a sample of n independent observations. This infinite sequence, in its totality, constitutes
a *derived population*, which is the random sampling distribution of the statistic T based on samples of size n. The fact that $E(T) = \theta$ signifies that
the mean of this derived population is θ. Furthermore, what has been
obtained is *one* probability distribution of values of T based on samples
of size n and, indeed, n is a parameter of this probability distribution. We
can therefore legitimately think of sampling distributions of T for each possible value of n. All these probability distributions have the same mean θ,

but their other characteristics such as, for example, the variance, are, in general, different. This then is the second way of viewing the concept of unbiasedness of a statistic estimating a parameter. As an example, we have shown that the sample mean is invariably an unbiased estimate of the population mean.

The method of maximum likelihood does not always lead to unbiased estimates. Thus, for example, the maximum-likelihood estimate of the variance of a normal population is not unbiased. It is known that the method of maximum likelihood almost always provides what Fisher termed *consistent* estimates. Formally, we define that if a statistic T *converges in probability* to the value θ as the sample size n approaches infinity, then T is said to be a consistent estimate of θ. The notion of "convergence in probability" is subtle and it is linked with the probability distribution of the statistic T. In general, there is a rather close parallelism between unbiased and consistent estimates of the commoner parameters occurring in statistical theory such as, for example, the mean of a normal population and the parameter p denoting the probability of success in a binomial distribution. Nevertheless, it is to be emphasised that consistency is a very different concept from unbiasedness, and it is also derived from a different theory of estimation. (Unbiasedness is derived from the theory of least squares.) As such, a consistent estimate may or may not be unbiased. Conversely, an unbiased estimate may or may not be consistent. Despite this, there exist estimates (such as the sample mean) which are both unbiased and consistent.

To explain further the notion of consistency, we consider a specific example of

    (i) a consistent estimate which is also unbiased;   and

    (ii) a consistent estimate which is not unbiased.

As an example of (i), consider the sample mean $\bar{x}$ as an estimate of the mean μ of a normal population with variance σ^2. Clearly $\bar{x}$ is an unbiased estimate of μ since $E(\bar{x}) = \mu$. We have also seen that $\bar{x}$ is the maximum-likelihood estimate of μ and as such is consistent. We next consider what this consistency implies in relation to the sampling distribution of $\bar{x}$. We recall that $\mathrm{var}(\bar{x}) = \sigma^2/n$, where n is the sample size. If we consider different values of n, say $n_1, n_2, n_3, \ldots$, then correspondingly we get a sequence of sampling distributions such that each distribution of the sequence has mean μ, but the variances of the distributions are $\sigma^2/n_1, \sigma^2/n_2, \sigma^2/n_3, \ldots$ respectively. As it happens, we also know that each of these distributions is normal, so that we may denote the sequence of distributions as $N(\mu, \sigma^2/n_1)$, $N(\mu, \sigma^2/n_2)$, $N(\mu, \sigma^2/n_3)$, $\ldots$. However, the normality of these distributions is not relevant to our present argument. The essential point is that as n increases, the variances of these distributions decrease and, in the limit as $n \to \infty$, $\sigma^2/n \to 0$. Accordingly, as n increases, we have a sequence of distributions which are all centred at μ, but which have steadily decreasing variances. The limiting form of these distribution is a *line distribution* which passes throught the point μ with probability one. This signifies that with variance zero, the limiting distribution of the sample mean $\bar{x}$ is

$$P(\bar{x} = \mu) = 1 \quad \text{and} \quad P(\bar{x} \neq \mu) = 0.$$

This is the main idea underlying the concept of convergence in probability, which ensures the consistency of an estimate. Consistency therefore implies that in the limit as $n \to \infty$, the estimate must become identical with the parameter estimated.

As an example of (ii), consider the estimation of σ^2, the variance of a normal distribution. We have seen that the maximum-likelihood estimate of σ^2 is not unbiased since $E(\hat{\sigma}^2) = (n - 1)\sigma^2/n \neq \sigma^2$. Also, we recall that $\text{var}(\hat{\sigma}^2) = 2(n - 1)\sigma^4/n^2$. For fixed n, the random sampling distribution of $\hat{\sigma}^2$ is not normal, but it is known to depend upon the two parameters n and σ^2, and we formally denote this distribution as $G(n, \sigma^2)$. Hence, following our previous argument, by varying n we obtain another sequence of sampling distributions

$$G(n_1, \sigma^2), \ G(n_2, \sigma^2), \ G(n_3, \sigma^2), \ \ldots,$$

with means

$$(n_1 - 1)\sigma^2/n_1, \ (n_2 - 1)\sigma^2/n_2, \ (n_3 - 1)\sigma^2/n_3, \ \ldots,$$

and variances

$$2(n_1 - 1)\sigma^4/n_1^2, \ 2(n_2 - 1)\sigma^4/n_2^2, \ 2(n_3 - 1)\sigma^4/n_3^2, \ \ldots \ .$$

For increasing n the means of these distributions steadily tend to σ^2, and their variances decrease to zero in the limit as $n \to \infty$. Thus, in this case, a *two-fold* limiting process is taking place. Firstly, the distributions $G(n, \sigma^2)$ converge to a line distribution and, secondly, the *varying* means of the distributions tend to σ^2. Accordingly, in the limit as $n \to \infty$, we again have a line distribution passing through the point σ^2 with probability one. Thus consistency ensures unbiasedness in the limit.

(ii) Efficiency

In general, there are many consistent estimates of a parameter based on the same number of observations. For example, consider the sample mean $\bar{x}$ and the *weighted mean*

$$\bar{x}_w = 2 \sum_{\nu=1}^{n} \nu x_\nu / n(n + 1)$$

as estimates of the mean μ of a normal population with variance σ^2. Both $\bar{x}$ and $\bar{x}_w$ are based on the same sample of n observations $x_1, x_2, \ldots, x_n$. Now

$$E(\bar{x}_w) = \frac{2}{n(n + 1)} \sum_{\nu=1}^{n} \nu E(x_\nu) = \frac{2}{n(n + 1)} \sum_{\nu=1}^{n} \nu\mu = \mu.$$

Also,

$$\text{var}(\bar{x}_w) = \frac{4}{n^2(n + 1)^2} \sum_{\nu=1}^{n} \nu^2 \text{var}(x_\nu) = \frac{4\sigma^2}{n^2(n + 1)^2} \frac{n(n + 1)(2n + 1)}{6}$$

$$= \frac{\sigma^2}{n}\left[\frac{4n + 2}{3n + 3}\right].$$

It is also known that $\bar{x}_w$ is normally distributed with mean μ and variance $(4n + 2)\sigma^2/3n(n + 1)$, but this is not relevant to our present argument. We note that $\bar{x}_w$ is another consistent estimate of μ, since $\text{var}(\bar{x}_w) \to 0$ as $n \to \infty$, and $E(\bar{x}_w) = \mu$. Therefore the natural question is that of a choice between $\bar{x}$ and $\bar{x}_w$ as consistent estimates of the parameter μ. For this purpose Fisher introduced a second criterion, that of the *efficiency* of an estimate.

Formally, of all consistent estimates of a parameter based on the same number of observations, the estimate having minimum variance is known as an *efficient* estimate. As an example, since $\text{var}(\bar{x}_w) > \text{var}(\bar{x})$, for all $n > 1$, the "better" (and not necessarily an efficient) estimate of μ is $\bar{x}$ as compared with $\bar{x}_w$. However, it is also known that $\bar{x}$ has, among the class of all consistent estimates of μ, the least variance. In view of this, $\bar{x}$ is an efficient estimate of μ and, accordingly, $\bar{x}_w$ is an *inefficient* estimate of μ.

The *efficiency* of an inefficient estimate is defined as the ratio of the variances of the efficient and inefficient estimates based on the same number of observations. Thus the efficiency of $\bar{x}_w$ is

$$\text{eff}(\bar{x}_w) = \frac{\sigma^2}{n} \bigg/ \frac{\sigma^2}{n}\left[\frac{4n + 2}{3n + 3}\right] = \frac{3n + 3}{4n + 2} \sim \frac{3}{4}, \quad \text{for large } n.$$

It is to be noted that the efficiency of an inefficient estimate is always less than unity. The percentage loss in efficiency by using an inefficient estimate is

$$100 \times (1 - \text{efficiency of the inefficient estimate}).$$

Thus there is a 25 per cent loss in efficiency in using $\bar{x}_w$ rather than $\bar{x}$ as an estimate of μ. A physical interpretation of such a loss can be seen as follows. Every sample contains a certain amount of information about the parameter estimated, and the loss in efficiency by the use of an inefficient estimate is the loss of information contained in the sample. In other words, by using $\bar{x}_w$ as an estimate of μ, we have in effect thrown away a quarter of the sample observations.

It is also possible to have more than one efficient estimate for a parameter, and it follows that the efficiency of any efficient estimate must be unity. Fisher proposed another criterion (that of *sufficiency*) to choose between efficient estimates but, as this is largely of theoretical interest, we shall not consider it here.

It is a useful property of the method of maximum likelihood that it almost always gives efficient estimates. In practically all the cases of estimation with which we shall be concerned, it can be assumed that the maximum-likelihood estimates are both consistent and efficient. The exceptions to this rule fall well outside the scope of this work.

11.11   Properties of least-squares estimates

We have already, in passing, referred to some of the properties of least-squares estimates. However, it is now appropriate to consider fully their main properties. In general, the method of least squares is used when the independent observations $x_1, x_2, \ldots, x_n$ are such that

$$E(x_\nu) = \sum_{i=1}^{k} a_{i\nu} \theta_i \quad (k < n)$$

and $$\text{var}(x_\nu) = \sigma^2, \qquad \text{for } \nu = 1, 2, \ldots, n,$$

where the symbols have their earlier meaning. An important point in least squares as an estimating procedure is that no assumption need be made about the distribution of the observations apart from the specifications concerning their means and variances.

The estimates θ_i^* are linear functions of the observations x_ν, and they are always unbiased so that

$$E(\theta_i^*) = \theta_i, \quad \text{for } i = 1, 2, \ldots, k.$$

Indeed, unbiasedness is the primary property of all least-squares estimates. Furthermore, amongst the class of all linear unbiased estimates of a parameter, the least-squares estimate has minimum variance. Thus, for example, $\text{var}(\theta_i^*)$ is less than the variance of any other linear function of the n observations which is also an unbiased estimate of θ_i. In this sense, the least-squares estimates are "best". A convenient mnemonic for these estimates is BLUE (*best linear unbiased estimates*). It therefore follows that amongst the class of linear unbiased estimates of a parameter, the least-squares estimate is efficient.

As remarked earlier, it is not essential for the application of the method of least squares that σ^2 should be independent of the θ_i. The method rests on the equality of the variances of the observations. Thus, for example, it is perfectly valid to use the method to estimate a binomial parameter or the mean of a negative exponential distribution, though in both cases the variances of the observations are functions of the parameter occurring in their respective expectations. However, in the most important applications of the method, which lead to the *analysis of variance* (to be considered in Chapter 15), σ^2 is an independent parameter. In such cases, the method also provides an unbiased estimate of σ^2 defined as

$$s_e^2 = \sum_{\nu=1}^{n} \left[x_\nu - \sum_{i=1}^{k} a_{i\nu}\, \theta_i^* \right]^2 \bigg/ (n - k).$$

This estimate is, of course, a quadratic function of the x_ν since the θ_i^* are linear in the x_ν. A minimal property of this estimate is deduced from the fact that

$$M = \sum_{\nu=1}^{n} \left[x_\nu - \sum_{i=1}^{k} a_{i\nu}\, \theta_i^* \right]^2$$

is the absolute minimum of

$$\Omega = \sum_{\nu=1}^{n} \left[x_\nu - \sum_{i=1}^{k} a_{i\nu}\, \theta_i \right]^2$$

for variations with respect to the θ_i.

11.12 Properties of moment estimates

In some of the important probability distributions such as the normal and the binomial, the moment estimates of the parameters are found to be identical with the least-squares and (or) the maximum-likelihood estimates. When the moment estimates are characteristically different, as in the lognormal distribution, then the major recommendation of the estimates is their ease of calculation. Such estimates are usually consistent but not unbiased. However, their most serious drawback is that they are invariably inefficient. Nevertheless, when the loss in efficiency is not great, the inefficiency could,

on occasion and in a practical sense, be compensated by the ease of calculation of the estimates.

11.13   Standard error of an estimate

The variance of a random variable is a statistical measure of the fluc-
tuations in the observable values of the random variable. This also applies
a fortiori to the variance of an estimate of a population parameter, and the
smaller the variance the more precise is the estimate. In assessing the
relative precisions of estimates, we are implicitly concerned with estimates
obtained by using a given number of observations. This is an important
point because with estimates whose variance is proportional to the reciprocal of the sample size, the variance could be made as small as we please
by choosing the sample size sufficiently large. Consequently, the concept
of efficiency is defined in relative terms, and the efficiency can be determined once the variances of the estimates are evaluated. However, the
efficiency of an estimate is a theoretical concept which measures relatively
the dispersion of two sampling distributions. This enables a theoretical
discrimination to be made between different consistent or unbiased estimates
of a parameter.

Now, once the choice of an estimate of a parameter has been made, then
we are led to a consideration of the distributional behaviour of the estimate
in random sampling; and in this connection the variance of the estimate
again appears as an important characteristic. Unfortunately, this variance
generally involves either the parameter estimated and (or) some other parameters occurring in the specification of the population sampled. For example,
suppose the binomial parameter p is to be estimated. Then the maximum-
likelihood estimate of p is $\hat{p}$, the relative frequency of successes in the
observed sample. Also, $\text{var}(\hat{p}) = pq/n$, where $p+q = 1$ and n is the number
of trials made. Thus $\text{var}(\hat{p})$ is unknown as it is a function of the parameter
p, although we know that, irrespective of the value of p, $0 \leqslant \text{var}(\hat{p}) \leqslant 1/4n$.
However, an exact determination of $\text{var}(\hat{p})$ is not possible. Similarly, consider the sample mean $\bar{x}$ based on n independent observations from a normal population with mean μ and variance σ^2. This is the maximum-likelihood estimate of μ, and $\text{var}(\bar{x}) = \sigma^2/n$, which is again unknown because of
the parameter σ^2. In this case we can only state the inequality $0 < \sigma^2/n < \infty$,
which is of no practical use whatsoever.

These two elementary examples illustrate a fundamental difficulty in
studying the sampling distribution of an estimate θ^* of a population parameter θ. It is obvious that, like θ, the variance of the estimate θ^* must
also be estimated. The usual method is to obtain, if readily possible, an
unbiased estimate of $\text{var}(\theta^*)$. As we shall see, this is not always possible,
and in the absence of an unbiased estimate of $\text{var}(\theta^*)$, it is usual to accept
a consistent estimate of $\text{var}(\theta^*)$.

The positive square root of the estimate of $\text{var}(\theta^*)$ is called the *standard error of the estimate*, and we write this as

$$\text{s.e.}(\theta^*) \;=\; [\text{estimate } \text{var}(\theta^*)]^{\frac{1}{2}}.$$

11.14   Applications

We consider next a few examples of determining the standard errors of estimates when it is simple to derive unbiased estimates of the variances of the estimates of the parameters.

Example 11.9: Binomial distribution

If w successes are observed in n Bernoulli trials, then the maximum-likelihood estimate of the success parameter p is

$$\hat{p} = w/n, \quad \text{and} \quad \text{var}(\hat{p}) = pq/n, \quad \text{where } p + q = 1.$$

To obtain an unbiased estimate of $\text{var}(\hat{p})$, we use the factorial moments

$$E(w) = np; \quad E[w(w - 1)] = n(n - 1)p^2.$$

It then follows that

$$\frac{w}{n} - \frac{w(w - 1)}{n(n - 1)} = \frac{w(n - w)}{n(n - 1)}$$

is an unbiased estimate of $p - p^2 = pq$. Therefore an unbiased estimate of $\text{var}(\hat{p})$ is

$$w(n - w)/n^2(n - 1) = \hat{p}(1 - \hat{p})/(n - 1).$$

Hence $\text{s.e.}(\hat{p}) = [\hat{p}\hat{q}/(n - 1)]^{\frac{1}{2}}, \quad \text{where } \hat{p} + \hat{q} = 1.$

As an alternative procedure, since $E(\hat{p}) = p$, we may consider $\hat{p}\hat{q}/n$ as a possible estimate of pq/n. But we have

$$E[\hat{p}(1 - \hat{p})/n] = \frac{n - 1}{n} E[\hat{p}(1 - \hat{p})/(n - 1)] = \frac{pq}{n}(1 - 1/n).$$

Therefore $\hat{p}\hat{q}/n$ is not an unbiased estimate of pq/n, though for large n the bias is negligible.

Example 11.10 : Negative exponential distribution

If $x_1, x_2, \ldots, x_n$ are independent observations of a random variable x having a negative exponential distribution with mean λ, then it is known that the sample mean $\bar{x}$ is the maximum-likelihood estimate of λ and $\text{var}(\bar{x}) = \lambda^2/n$. Now the sample variance

$$s^2 = \sum_{i=1}^{n} (x_i - \bar{x})^2/(n - 1)$$

is an unbiased estimate of the population variance λ^2. Hence an unbiased estimate of $\text{var}(\bar{x})$ is s^2/n. Therefore

$$\text{s.e.}(\bar{x}) = s/\sqrt{n}.$$

As another possible estimate of $\text{var}(\bar{x})$, consider $\bar{x}^2/n$. Then

$$E(\bar{x}^2/n) = \frac{1}{n} E(\bar{x}^2) = \frac{1}{n}[\text{var}(\bar{x}) + E^2(\bar{x})] = \frac{1}{n}(\lambda^2/n + \lambda^2) = \frac{\lambda^2}{n}(1 + 1/n).$$

Hence $\bar{x}^2/n$ is not an unbiased estimate of $\text{var}(\bar{x})$, though the bias is negligible for large n.

Example 11.11: Normal distribution

Suppose $x_1, x_2, \ldots, x_n$ are independent observations of a normally distributed random variable x with mean μ and variance σ^2. Then the sample mean $\bar{x}$ and the sample variance s^2 are the least-squares estimates of μ and σ^2 respectively. Also,

$$\text{var}(\bar{x}) = \sigma^2/n \quad \text{and} \quad \text{var}(s^2) = 2\sigma^4/(n-1).$$

Clearly, since $E(s^2) = \sigma^2$, an unbiased estimate of $\text{var}(\bar{x})$ is s^2/n.
Therefore
$$\text{s.e.}(\bar{x}) = s/\sqrt{n}.$$

Again,
$$\frac{2\sigma^4}{n-1} = \text{var}(s^2) = E(s^4) - E^2(s^2) = E(s^4) - \sigma^4,$$

so that
$$E(s^4) = \frac{(n+1)\sigma^4}{n-1}.$$

Hence an unbiased estimate of σ^4 is $(n-1)s^4/(n+1)$. Therefore
$$\frac{2}{n-1} \times \frac{(n-1)s^4}{n+1} = \frac{2s^4}{n+1}$$

is an unbiased estimate of $\text{var}(s^2)$, whence
$$\text{s.e.}(s^2) = s^2[2/(n+1)]^{\frac{1}{2}}.$$

As another possible estimate of $\text{var}(s^2)$, consider $2s^4/(n-1)$. But
$$E[2s^4/(n-1)] = \frac{2}{n-1} E(s^4) = \frac{2}{n-1} \times \frac{(n+1)\sigma^4}{n-1}$$
$$= \frac{2\sigma^4}{n-1}\left[1 + \frac{2}{n-1}\right].$$

Thus $2s^4/(n-1)$ is not an unbiased estimate of $\text{var}(s^2)$, though the bias is negligible for large n.

An important general conclusion is suggested by the above three examples. If $g(\theta)$ is the variance of θ^*, an unbiased estimate of θ based on n observations, then, in general, $g(\theta^*)$ is not an unbiased estimate of $g(\theta)$, but the bias is $O(n^{-1})$ relative to unity. Accordingly, for large n, the bias in $g(\theta^*)$ is negligible, and such estimates are generally used to evaluate the standard errors of estimates of population parameters obtained from large samples. In fact, it can be shown that, in general, the estimate $g(\theta^*)$ of $g(\theta)$ is consistent.

We conclude this chapter with a few more examples of estimation and practical applications.

Example 11.12: Estimation in the hypergeometric distribution

We recall that the hypergeometric distribution arises by sampling from a finite collection containing W white and $N-W$ black balls. The random variable of interest is w, the number of white balls observed in a sample of n balls drawn without replacement from the given collection. Our problem now is to estimate W, the number of white balls in the collection or, equivalently, $p = W/N$, the proportion of white balls. It is assumed that N is known.

(i) The simplest method of estimating p is by moments. We have proved that the rth factorial moment of w is
$$E[w^{(r)}] = n^{(r)} W^{(r)} / N^{(r)},$$

so that, in particular,
$$E(w) = np; \quad \text{var}(w) = npq(N-n)/(N-1), \quad \text{where } p+q = 1.$$

Hence the moment estimate of p is $\hat{p} = w/n$, that is, the relative frequency of white balls observed in the sample. This estimate is evidently unbiased, and it can be shown, though this is not quite straightforward, that $\hat{p}$ is also the maximum-likelihood estimate of p.

Clearly,
$$\text{var}(\hat{p}) \;=\; \text{var}(w/n) \;=\; pq(N-n)/n(N-1).$$

To obtain the standard error of $\hat{p}$, we first determine an unbiased estimate of $\text{var}(\hat{p})$. Expressed in terms of W, N, and n, we have

$$\text{var}(\hat{p}) \;=\; \frac{W(N-W)(N-n)}{nN^2(N-1)} \;=\; \frac{(N-n)[(N-1)W - W^{(2)}]}{nN^2(N-1)}.$$

Since it follows immediately from the first and second factorial moments of w that the unbiased moment estimates of W and $W^{(2)}$ are wN/n and $w^{(2)}N^{(2)}/n^{(2)}$, therefore an unbiased estimate of $\text{var}(\hat{p})$ is

$$\frac{N-n}{nN^2(N-1)}\left[\frac{wN^{(2)}}{n} - \frac{w^{(2)}N^{(2)}}{n^{(2)}}\right]$$

$$=\; \frac{(N-n)\,w(n-w)}{Nn^2(n-1)} \;=\; \frac{w(1-w/n)(1-n/N)}{n(n-1)} \;=\; \frac{\hat{p}\hat{q}(1-n/N)}{n-1},$$

where $\hat{p} + \hat{q} = 1$. Furthermore, since n is the number of balls in the sample and N the total number in the collection, the quantity $f = n/N$ is the proportion of the balls in the collection included in the sample. This proportion is called the *sampling fraction* and, in terms of f, we may write alternatively
$$\text{var}(\hat{p}) \;=\; Npq(1-f)/n(N-1),$$

and the unbiased estimate of this variance as
$$\hat{p}\hat{q}(1-f)/(n-1).$$

Therefore $\text{s.e.}(\hat{p}) \;=\; [\hat{p}\hat{q}(1-f)/(n-1)]^{\frac{1}{2}}.$

As a check, we note that when $N \to \infty$, $f \to 0$ for finite n, and we obtain the corresponding results for the binomial distribution.

(ii) If we wish to estimate not p but W then, since $W = Np$, we infer immediately that an unbiased estimate of W is
$$\hat{W} \;=\; N\hat{p} \;=\; w/f.$$

Also, $\text{var}(\hat{W}) \;=\; N^2\,\text{var}(\hat{p}) \;=\; N^2(N-n)\,pq/n(N-1),$

and an unbiased estimate of $\text{var}(\hat{W})$ is
$$N^2\hat{p}\hat{q}(1-f)/(n-1) \;=\; w(n-w)(1-f)/(n-1)f^2.$$

Therefore $\text{s.e.}(\hat{W}) \;=\; [w(n-w)(1-f)/(n-1)f^2]^{\frac{1}{2}}.$

The white and black balls of the collection represent a conceptual dichotomy of a finite population, and they can be identified with a wide variety of physically meaningful variables in practical situations. Our next example is a simple illustration of this from sociometry.

Example 11.13: A housing survey
A town has 50,000 houses, and in a survey of housing conditions interest centres on determining the proportion of houses in the town which have bathrooms satisfying certain specified sanitary conditions. In a random

sample of 1,000 houses inspected, a total of 379 were found to have satisfactory bathrooms. Given these data, the hypergeometric distribution can be used to estimate the proportion, or the total number, of houses in the town with bathrooms of approvable standard.

Each ball of the collection is now identified with one of the $N = 50,000$ houses in the town, the white and black colours corresponding to houses with bathrooms of approvable and non-approvable standards respectively. The sample size is $n = 1,000$, and so if p is the true proportion of houses in the town with satisfactory bathrooms, then the sample estimate of p is

$$\hat{p} = \frac{379}{1000} = 0\cdot379.$$

The sampling fraction $f = 0\cdot02$, and so the unbiased estimate of $\text{var}(\hat{p})$ is

$$\frac{0\cdot379\,(1 - 0\cdot379)(1 - 0\cdot02)}{999} = 0\cdot0002309.$$

Therefore $\qquad\qquad \text{s.e.}(\hat{p}) = 0\cdot0152.$

The sample estimate of the number of houses with satisfactory bathrooms is

$$\hat{W} = N\hat{p} = 18,950,$$

and the standard error of the estimate $\hat{W}$ is

$$\text{s.e.}(\hat{W}) = N \times \text{s.e.}(\hat{p}) = 760.$$

The standard errors of the estimates $\hat{p}$ and $\hat{W}$ are numerical measures of the uncertainty in the estimates, and it may well be argued that a better procedure would have been to inspect all the 50,000 houses in the town and so obtain the exact number of houses with bathrooms of approvable standards. In answer to this apparently reasonable contention, three points are worth making. Firstly, we have obtained the estimates by a two per cent sample and thereby reduced the cost of inspection considerably. Secondly, complete enumeration of the population does not necessarily ensure an exact result because of the inherent human errors in inspection. Indeed, this is a basic point, and by sampling we have allowed for errors of this kind and also obtained a numerical measure of the resulting uncertainty associated with the estimates. On the other hand, there is no satisfactory method for assessing the magnitude of the errors in a complete enumeration of the population. Thirdly, we shall see from further statistical analysis that there is an almost 95 per cent chance that the true proportion of houses with satisfactory bathrooms lies between 0·33 and 0·44. This is a substantially useful result based on a mere two per cent count of the population.

Example 11.14: Incidence of smoking

In a certain country, the Ministry of Health conducted a survey to determine the incidence of smoking amongst the adult population, an adult for the purposes of the survey being defined as a person of either sex and aged 16 or more years. The age-distribution of the country's population given by a previous census indicates that the adult population of the survey is 28·76 millions approximately. For purposes of statistical analysis, this number may be regarded as effectively infinite. In a sample of 5,000 adults taken for the survey, it was found that 2,965 adults were smokers. Given these data, we wish to estimate the true proportion of smokers in the country and to determine the standard error of the estimate.

If p is the true proportion of smokers in the adult population, then the binomial distribution is applicable. Therefore the maximum-likelihood estimate of p is

$$\hat{p} = \frac{2965}{5000} = 0\cdot5930,$$

and the unbiased estimate of var$(\hat{p})$ is

$$\frac{0\cdot5930\,(1 - 0\cdot5930)}{4999} = 0\cdot00004828.$$

Hence   s.e.$(\hat{p}) = 0\cdot006948.$

This estimation procedure is simple, but it slurs over a fundamental difficulty. In fact, the estimation is valid only if it may be reasonably assumed that there is a constant probability for an adult in the population to be a smoker. This is a rigid assumption, which is exceedingly unlikely to be true in any population, since the propensity to smoke varies with age and also presumably with sex. Accordingly, the adult population cannot, at least *prima facie*, be regarded as homogeneous with respect to the incidence of smoking. Indeed, any well-designed large-scale survey would allow for statistical verification of the homogeneity of the population; and if this assumption is untenable, then a different estimation procedure should be used to estimate the proportion of smokers in the population. We shall not go into the details of doing this, and we have mentioned the difficulty to illustrate a general point: the need to justify the model which provides the basis for estimation is an essential element in a practical situation.

Example 11.15: Estimation in the multinomial distribution

The multinomial distribution occurs frequently as an appropriate model for the analysis of data obtained from plant breeding experiments. Typically, a certain number of plants are raised and it is found that they can be classified into a number of qualitatively distinct classes. The plants are the progeny of some crossing and, depending upon the nature of this crossing, a genetical hypothesis suggests the expected proportions of the plants in the distinct classes.

As an example, suppose a plant breeding experiment gives four kinds of progeny with the observed frequencies n_1, n_2, n_3, n_4 respectively, where $\sum_{i=1}^{4} n_i \equiv N$. On a genetical hypothesis, the expected frequencies in the four classes AB, Ab, aB, ab are $N(2 + \theta)/4$, $N(1 - \theta)/4$, $N(1 - \theta)/4$, $N\theta/4$ respectively, where θ is a parameter with some unknown value in the range $0 < \theta < 1$. The total frequency N is fixed by the sample size, and the observed frequencies n_i are random variables subject to the restriction that their sum is N. If the data have arisen from a biological situation which is, except for errors of sampling, in agreement with the hypothesis giving the expected frequencies, then the probabilities of a random observation falling in the classes AB, Ab, aB, ab are $(2+\theta)/4$, $(1-\theta)/4$, $(1-\theta)/4$, $\theta/4$ respectively.

Hence the joint probability of the observed distribution of frequencies is obtained from the multinomial distribution as

$$\frac{N!}{n_1! \, n_2! \, n_3! \, n_4!} \, [(2 + \theta)/4]^{n_1} [(1 - \theta)/4]^{n_2 + n_3} [\theta/4]^{n_4}.$$

Our problem is to derive an estimate of θ and its standard error. This is an obvious example for maximum-likelihood estimation.

Ignoring constants which vanish on differentiation, the logarithm of the sample likelihood may be written as

$$\log L = n_1 \log(2 + \theta) + (n_2 + n_3) \log(1 - \theta) + n_4 \log \theta.$$

Therefore

$$\frac{d \log L}{d\theta} = \frac{n_1}{2 + \theta} - \frac{n_2 + n_3}{1 - \theta} + \frac{n_4}{\theta},$$

so that the equation for the maximum-likelihood estimate $\hat{\theta}$ is

$$\frac{n_1}{2 + \hat{\theta}} - \frac{n_2 + n_3}{1 - \hat{\theta}} + \frac{n_4}{\hat{\theta}} = 0,$$

which is readily seen to reduce to the quadratic equation

$$2n_4 + (3n_1 + n_4 - 2N)\hat{\theta} - N\hat{\theta}^2 = 0.$$

Since θ is a positive parameter, the appropriate root of this equation gives

$$\hat{\theta} = \frac{3n_1 + n_4 - 2N + [(3n_1 + n_4 - 2N)^2 + 8Nn_4]^{\frac{1}{2}}}{2N}.$$

It can also be shown that $\hat{\theta}$ is not an unbiased estimate of θ as

$$E(\hat{\theta}) = \theta + O(N^{-2}).$$

We can, therefore, regard $\hat{\theta}$ to be unbiased to $O(N^{-1})$, which is adequate for all practical purposes.

Next, to obtain the large-sample variance of $\hat{\theta}$, we have

$$\frac{d^2 \log L}{d\theta^2} = -\frac{n_1}{(2 + \theta)^2} - \frac{n_2 + n_3}{(1 - \theta)^2} - \frac{n_4}{\theta^2},$$

so that

$$-E\left[\frac{d^2 \log L}{d\theta^2}\right] = \frac{E(n_1)}{(2 + \theta)^2} + \frac{E(n_2) + E(n_3)}{(1 - \theta)^2} + \frac{E(n_4)}{\theta^2}$$

$$= \frac{N}{4}\left[\frac{1}{2 + \theta} + \frac{2}{1 - \theta} + \frac{1}{\theta}\right] = \frac{N(1 + 2\theta)}{2\theta(1 - \theta)(2 + \theta)}.$$

Hence

$$\text{var}(\theta) = 2\theta(1 - \theta)(2 + \theta)/N(1 + 2\theta) = h(\theta), \quad \text{say.}$$

It is not easy nor, indeed, is it worthwhile to obtain an unbiased estimate of $\text{var}(\hat{\theta})$. The simplest estimate of $\text{var}(\hat{\theta})$ is

$$h(\hat{\theta}) = 2\hat{\theta}(1 - \hat{\theta})(2 + \hat{\theta})/N(1 + 2\hat{\theta}),$$

whence

$$\text{s.e.}(\hat{\theta}) = [2\hat{\theta}(1 - \hat{\theta})(2 + \hat{\theta})/N(1 + 2\hat{\theta})]^{\frac{1}{2}}.$$

The estimate $h(\hat{\theta})$ of $\text{var}(\hat{\theta})$ is not unbiased, but we next show that the bias is of $O(N^{-1})$. To prove this, we recall that if $E(x) = \mu$ and $g(x)$ is any differentiable function of x, then

$$E[g(x)] \sim g(\mu) + \frac{1}{2}\frac{d^2 g(\mu)}{d\mu^2}\text{var}(x).$$

Now, we may consider that $E(\hat{\theta}) = \theta$ and $\text{var}(\hat{\theta}) = h(\theta)$. Hence, using $h(\hat{\theta})$ corresponding to $g(x)$, we have

$$E[h(\hat{\theta})] \sim h(\theta) + \frac{1}{2}\frac{d^2 h(\theta)}{d\theta^2}\,\text{var}(\hat{\theta}) = h(\theta)\left[1 + \frac{1}{2}\frac{d^2 h(\theta)}{d\theta^2}\right].$$

But straightforward differentiation gives

$$\frac{d^2 h(\theta)}{d\theta^2} = -\frac{4(5 + 3\theta + 6\theta^2 + 4\theta^3)}{N(1 + 2\theta)^3} = -\frac{2}{N}\left[1 + \frac{9}{(1 + 2\theta)^3}\right],$$

so that

$$E[h(\hat{\theta})] = h(\theta)\left[1 - \frac{1}{N}\left\{1 + \frac{9}{(1 + 2\theta)^3}\right\}\right].$$

Thus, relative to unity, the absolute value of the approximate bias in $h(\hat{\theta})$ as an estimate of $\text{var}(\hat{\theta})$ is

$$\frac{1}{N}\left[1 + \frac{9}{(1 + 2\theta)^3}\right].$$

Now $0 < \theta < 1$, and so the absolute value of the bias lies between $4/3N$ and $10/N$, which is negligible for moderately large N.

Numerical application

We use these theoretical results on Imai's data pertaining to a breeding experiment with *Pharbitis*, already quoted in Chapter 9. Since $N = 290$, $n_1 = 187$, $n_2 = 35$, $n_3 = 37$, $n_4 = 31$, the equation for $\hat{\theta}$ is the quadratic

$$62 + 12\hat{\theta} - 290\hat{\theta}^2 = 0,$$

which has the positive root

$$\hat{\theta} = \frac{12 + \sqrt{72064}}{580} = 0\cdot4835.$$

The estimate of $\text{var}(\hat{\theta})$ is

$$h(\hat{\theta}) = \frac{0\cdot9670 \times 0\cdot5165 \times 2\cdot4835}{570\cdot43} = 0\cdot002174,$$

so that

$$\text{s.e.}(\hat{\theta}) = 0\cdot04663.$$

Example 11.16: Estimation in the Poisson distribution

The Poisson distribution with parameter μ is defined by the point-probabilities

$$Q(w) = e^{-\mu}\mu^w/w!, \quad \text{for integral values of } w \geqslant 0.$$

Suppose a random sample of N observations is obtained from this distribution, and it is found that exactly n_r of them correspond to the class $w = r$, where

$$\sum_{r=0}^{\infty} n_r = N.$$

Since the sum of this infinite series is N, a finite number, it is obvious that many of the n_r will be zero. This means that if a finite number of observations (N) are randomly distributed over an infinite number of classes, then only a finite number of classes could possibly have non-zero frequencies. Nevertheless, if the observed sample is from the Poisson distribution, it necessarily follows that the class-frequencies n_r are random variables

such that
$$E(n_r) = Ne^{-\mu}\mu^r/r!, \quad \text{for integral values of } r \geqslant 0.$$

This is an important point. The Poisson variable is w, and the point-probabilities of the distribution give the probabilities for a random observation to fall in any one of the possible infinite number of classes corresponding to integral values of w. However, the randomness of the sample arises out of the chance distribution of the N observations amongst the theoretically infinite number of classes of the Poisson distribution. This is analogous to the sample distribution of observations amongst the k classes of a multinomial distribution. Hence, by an extended form of the multinomial distribution with $k \to \infty$, the probability of the distribution of the N observations of the Poisson distribution is

$$\frac{N!}{\prod\limits_{r=0}^{\infty} n_r!} \prod_{r=0}^{\infty}(e^{-\mu}\mu^r/r!)^{n_r} = \frac{N!}{\prod\limits_{r=0}^{\infty} n_r!} \frac{e^{-N\mu}\mu^{\sum\limits_{r=0}^{\infty} rn_r}}{\prod\limits_{r=0}^{\infty}(r!)^{n_r}}.$$

Therefore, ignoring constants which vanish on differentiation with respect to μ, the logarithm of the sample likelihood is

$$\log L = -N\mu + \left[\sum_{r=0}^{\infty} rn_r\right]\log \mu,$$

$$\frac{d\log L}{d\mu} = -N + \frac{1}{\mu}\sum_{r=0}^{\infty} rn_r.$$

If we now set the sample mean as

$$\bar{w} = \frac{1}{N}\sum_{r=0}^{\infty} rn_r,$$

then the equation for the maximum-likelihood estimate of μ is

$$-N + N\bar{w}/\hat{\mu} = 0, \quad \text{so that } \hat{\mu} = \bar{w}.$$

Thus, as expected, the sample mean $\bar{w}$ is the maximum-likelihood estimate of the population mean μ. Therefore

$$E(\hat{\mu}) = E(\bar{w}) = \mu.$$

Also, since $\text{var}(w) = \mu$, we have

$$\text{var}(\hat{\mu}) = \text{var}(\bar{w}) = \mu/N.$$

To conclude, it is instructive to show directly that $E(\bar{w}) = \mu$. By definition

$$\begin{aligned}
E(\bar{w}) &= \frac{1}{N}\sum_{r=0}^{\infty} E(rn_r) \\
&= \frac{1}{N}\sum_{r=0}^{\infty} rE(n_r) \\
&= \frac{1}{N}\sum_{r=0}^{\infty} rNe^{-\mu}\mu^r/r! \\
&= e^{-\mu}\sum_{r=1}^{\infty}\mu^r/(r-1)! = \mu e^{-\mu}\sum_{t=0}^{\infty}\mu^t/t! = \mu.
\end{aligned}$$

11.15 Pooled estimation by maximum likelihood

So far we have used the method of maximum likelihood to obtain estimates of parameters on the basis of a single random sample containing a finite number of observations. Since the multiplicative rule of probabilities holds for independent observations, it also *a fortiori* holds for independent samples. This extension provides the basis for extending the argument of maximum-likelihood estimation to observations from independent samples. Our final example illustrates the method when samples obtained from two different experiments are used to estimate a parameter.

Example 11.17: Plant breeding

Two varieties of plants were used separately in two similar but independent breeding experiments, and progenies of types A and B were obtained in each case but with different relative frequencies. In the first experiment the observed numbers in the A and B classes were n_1 and $N_1 - n_1$, and in the second experiment n_2 and $N_2 - n_2$ respectively. On a biological hypothesis, the expected probabilities of the A and B progenies in the two experiments were

$$(4 - \theta)/8, \quad (4 + \theta)/8; \quad \text{and} \quad (12 + \theta)(4 - \theta)/64, \quad (4 + \theta)^2/64$$

respectively, θ being an unknown parameter. We use these data to derive the equation for $\hat{\theta}$, the maximum-likelihood estimate of θ, and the large-sample variance of the estimate.

Each of the two samples is from a binomial population, and since the samples are based on independent experiments, the joint probability of the samples is

$$\binom{N_1}{n_1}[(4-\theta)/8]^{n_1}[(4+\theta)/8]^{N_1-n_1} \times \binom{N_2}{n_2}[(12+\theta)(4-\theta)/64]^{n_2}[(4+\theta)^2/64]^{N_2-n_2}.$$

Ignoring constants which vanish on differentiation with respect to θ, the logarithm of the joint likelihood is

$$\begin{aligned}
\log L &= n_1\log(4-\theta) + (N_1 - n_1)\log(4+\theta) + n_2\log[(12+\theta)(4-\theta)] + \\
&\quad + 2(N_2 - n_2)\log(4+\theta) \\
&= (n_1 + n_2)\log(4-\theta) + (N_1 + 2N_2 - n_1 - 2n_2)\log(4+\theta) + \\
&\quad + n_2\log(12+\theta).
\end{aligned}$$

Therefore

$$\frac{d\log L}{d\theta} = -\frac{n_1 + n_2}{4 - \theta} + \frac{N_1 + 2N_2 - n_1 - 2n_2}{4 + \theta} + \frac{n_2}{12 + \theta},$$

so that the required equation for $\hat{\theta}$ is the quadratic

$$-\frac{n_1 + n_2}{4 - \hat{\theta}} + \frac{N_1 + 2N_2 - n_1 - 2n_2}{4 + \hat{\theta}} + \frac{n_2}{12 + \hat{\theta}} = 0.$$

For the variance of $\hat{\theta}$, we have

$$\frac{d^2\log L}{d\theta^2} = -\frac{n_1}{(4-\theta)^2} - \frac{N_1 - n_1}{(4+\theta)^2} - \frac{n_2}{(12+\theta)^2} - \frac{n_2}{(4-\theta)^2} - \frac{2(N_2 - n_2)}{(4+\theta)^2},$$

so that

$$-E\left[\frac{d^2\log L}{d\theta^2}\right] = \frac{N_1}{8(4-\theta)} + \frac{N_1}{8(4+\theta)} + \frac{N_2(4-\theta)}{64(12+\theta)} + \frac{N_2(12+\theta)}{64(4-\theta)} + \frac{2N_2}{64}$$

$$= \frac{N_1}{16 - \theta^2} + \frac{4N_2}{(12 + \theta)(4 - \theta)}$$

$$= \frac{8N_1 + (4 + \theta)(N_1 + 4N_2)}{(16 - \theta^2)(12 + \theta)}$$

Hence

$$\text{var}(\hat{\theta}) = \frac{(16 - \theta^2)(12 + \theta)}{8N_1 + (4 + \theta)(N_1 + 4N_2)} .$$

CHAPTER 12

TESTS OF SIGNIFICANCE IN BINOMIAL SAMPLING

12.1 The concept of a test of significance

Random observations are the raw data of statistics, and the inferential part of the theory of statistics is concerned with the development of an objective and logically consistent methodology for drawing probabilistic conclusions about the conceptual population from which the observations are assumed to be drawn. The probabilistic nature of statistical inference is the fundamental characteristic which makes it completely distinct from the usual deductive processes of mathematical reasoning. This difference is best brought out by examining the principles which lead to *tests of significance* in statistical analysis. In general, any function of the observations is likely to differ from its expected value for one of two reasons. Firstly, the difference may be due to sampling error arising from the inevitable use of a finite number of observations; and, secondly, it may be attributable to the fact that the premises on which the expectation is based are invalid. The purpose of a test of significance is to effect a distinction between these two sorts of deviations. As an elementary example, suppose a penny is tossed 100 times and a total of 65 heads are observed. Ideally, if the penny were, in fact, unbiased we would expect 50 heads in 100 trials. However, it is obvious that such a result can hardly ever be realised because of random variations. Hence the question arises whether the observed relative frequency of $0 \cdot 65$ heads in 100 trials differs from the expectation of $0 \cdot 50$ because of sampling or because, in fact, the penny is not unbiased. In essence, the function of a test of significance is as simple as that indicated in this example. However, the mathematics of deriving a test of significance can become exceedingly difficult, in keeping with the complexity of the random phenomenon which provides the observations. In this chapter, we shall be concerned largely with tests of significance which arise from Bernoulli trials and related probability distributions because, apart from their intrinsic importance, these provide the simplest, and intuitively most acceptable, introduction to the reasoning underlying tests of significance.

Broadly speaking, so far as inference is concerned, statistical operations fall into the two categories of

(i) estimation of parameters, and

(ii) tests of significance.

The two operations are fundamentally different in character though, as will be seen, the success of a test of significance depends upon efficient estimation of a parameter or a number of parameters. Accordingly, estimation is the primary operation; and, in the light of our knowledge of the principles and methods of estimation, it is appropriate to clarify first the distinction between the two main statistical operations. Here, again, the simple example of tossing a penny should suffice. Suppose a penny is tossed n times and w heads are observed. If the probability of obtaining a head with a single toss of the penny is p, then we have seen that the maximum-likelihood estimate of p is $\hat{p} = w/n$. Furthermore, we know that $\hat{p}$ is an unbiased estimate of p, but $\hat{p}$ is not necessarily the "true" value of p. Indeed, $\hat{p}$ is a random variable and its value in any particular case depends upon the chance number of heads observed in any set of n independent trials with the penny. However, insomuch as initially we knew nothing about p, apart from the obvious fact that $0 < p < 1$, the experimental information has been used to provide an estimate, that is an "uncertain" value, of p. Besides, by taking n sufficiently large, we could make var$(\hat{p})$, that is the statistical measure of the uncertainty of $\hat{p}$, as small as we please. On the other hand, suppose it is believed that p is equal to some known value p_0, say, and we wish to test whether or not the result actually obtained by limited sampling, that is the value $\hat{p}$, is consistent with the hypothesis that $p = p_0$. This hypothesis of equality is known as a *null hypothesis*, and it is usually written as $H(p = p_0)$. A test of significance is a mathematical procedure for deciding, on grounds of probability, whether the null hypothesis is reasonable or not in the light of the experimental evidence obtained. Clearly, there is an inherent uncertainty in any conclusion about the null hypothesis which is arrived at by using a test of significance. The following relatively extreme example illustrates the nature of this uncertainty.

Suppose it is believed that the penny is unbiased, so that the null hypothesis is $H(p = 0.5)$. Next, suppose three independent experiments are performed with the following results.

Experiment I

The penny is tossed 100 times and it shows a head every time. If the penny is unbiased, then the probability of the observed event is $1/2^{100}$, which is negligibly small.

Experiment II

The penny is tossed ten times and a head turns up every time. If the penny is unbiased, then the probability of the observed event is $1/2^{10} = 0.001$.

Experiment III

The penny is tossed thrice and three heads are obtained. If the penny is unbiased, then the probability of the observed event is $1/2^3 = 0.125$.

All the three events described in the above experiments have positive probabilities of occurrence if the penny is, in fact, unbiased, but the magnitudes of these probabilities give different degrees of assurance for accepting the null hypothesis. The result of Experiment I is so extraordinary that no rational person would regard it as providing any basis for the continued acceptance of the hypothesis that the penny is unbiased. Nevertheless, the essential point is that such a result is not an impossibility with an unbiased penny. In the same way, the event of Experiment II is also reasonably exceptional, and one would not regard it as providing any safe assurance of the truth of the null hypothesis. However, the result of Experiment III is in a different category and although it has a chance of only 0·125, this probability is not usually judged small enough to throw substantial doubt on the validity of the null hypothesis. This is intuitively acceptable because three heads in three trials is not an exceptional result in the light of common experience with unbiased coins.

The general conclusion from the three experiments is that the magnitude of the probability of the observed event, if the null hypothesis is true, provides an objective basis for formulating an opinion about the truth of the null hypothesis. However, a test of significance is based on the result of a single, suitably defined experiment, and it is for us to set up *in advance* a criterion, which in essence is arbitrary, to accept or reject the null hypothesis in the light of the actual outcome of the experiment. Nevertheless, it is clear that whatever the criterion we decide upon, we cannot entirely exclude the possibility that we might well be wrong in rejecting the null hypothesis, since no result, however small its probability, is a complete impossibility when the null hypothesis is, in fact, true. Similarly, even if the probability of the observed event is large, this does not provide certain evidence that the null hypothesis is necessarily true. It therefore follows that whatever the decision we make about the null hypothesis, it must be an *uncertain* decision in the sense that there will be necessarily a probability, however small, of the decision being, in fact, incorrect. Consequently, quite naturally, the choice of the criterion should be such that it gives a "reasonably small" probability of rejecting the null hypothesis when, in fact, it is true, and a "high" probability of rejecting it when, in fact, it is false. With this general understanding, it is worth while to formulate a test of significance for assessing the agreement between the result expected on the basis of the null hypothesis and the actual observations obtained from an experiment.

Example 12.1: A test of significance

Suppose a penny is tossed ten times. If p is the probability for a head at a trial, then the probability of obtaining w heads is

$$\binom{10}{w} p^w (1 - p)^{10-w}.$$

(i) To test the null hypothesis $H(p = 0·5)$, we make the following criterion: we agree to reject the null hypothesis if the actual outcome of the experiment is $w = 8$, but otherwise to accept it.

Accordingly, the probability of rejecting the null hypothesis is

$$p(w = 8) = \binom{10}{8} p^8 (1-p)^2 = 45 p^8 (1-p)^2.$$

This probability is a function of p, the true but unknown probability of obtaining a head at a trial. Table 12.1 below gives the probabilities of rejecting the null hypothesis, on the basis of the criterion, for different values of p.

Table 12.1   Showing the values of $P(w = 8)$ for different values of p

p	0·30	0·40	0·50	0·60	0·70	0·80	0·90	0·99
$P(w = 8)$	0·0014	0·0106	0·0439	0·1209	0·2335	0·3020	0·1937	0·0042

It is clear from this table that if the null hypothesis $H(p = 0·5)$ is true, then $P(w = 8) = 0·0439$, which is small. This means that, on the basis of the criterion, the risk of rejecting the null hypothesis as false when, in fact, it is true is 0·0439. This is a satisfactory situation. However, if $p = 0·30$ so that the null hypothesis $H(p = 0·5)$ is, in fact, false, then, on the basis of the same criterion, the probability of rejecting the null hypothesis $H(p = 0·5)$ is only 0·0014. Similarly, if $p = 0·99$ so that the null hypothesis $H(p = 0·5)$ is false, then again the probability of rejecting it is 0·0042. Thus, when $|p - 0·5|$ is large so that the null hypothesis $H(p = 0·5)$ is, in fact, false, then the criterion is unlikely to reject it. This is clearly undesirable, and we therefore conclude that the above criterion is an unsatisfactory one.

(ii) As another possibility, suppose we adopt the following criterion to test the null hypothesis $H(p = 0·5)$: we agree to reject the null hypothesis if the actual outcome of the experiment is $w = 8, 9$ or 10, but otherwise to accept it.

Accordingly, the probability of rejecting the null hypothesis is

$$P(w \geqslant 8) = 45p^8(1 - p)^2 + 10p^9(1 - p) + p^{10}.$$

Table 12.2 gives these probabilities of rejecting the null hypothesis $H(p = 0·5)$ for different values of p.

Table 12.2   Showing the values of $P(w \geqslant 8)$ for different values of p

p	0·30	0·40	0·50	0·60	0·70	0·80	0·90	0·99
$P(w \geqslant 8)$	0·0016	0·0123	0·0547	0·1673	0·3828	0·6778	0·9298	0·9999

In this case, the probability of rejecting the null hypothesis $H(p = 0·5)$ when it is, in fact, true is 0·0547, which is again reasonably small. Furthermore, for values of $p > 0·5$, the use of this criterion gives a probability for rejecting the null hypothesis which steadily increases to unity as the difference $p - 0·5$ approaches 0·5. In other words, the larger the deviation of the true value of p from its hypothetical value 0·5, the greater is the probability that the null hypothesis will be rejected. On the other hand, for values of $p < 0·5$, the criterion gives steadily decreasing probabilities for the rejection of the null hypothesis which tend to zero as $0·5 - p$ approaches 0·5. Hence this criterion is a satisfactory procedure if the possible alternatives to the null hypothesis $H(p = 0·5)$ are $p > 0·5$; but the criterion is unsatisfactory if the alternatives to the null hypothesis are $p < 0·5$. The idea of possible alternatives to a null hypothesis under test is basic to an evaluation of the "goodness" of a criterion. Indeed, we are assessing here the credibility of a specific null hypothesis with an implicit knowledge of the possible alternatives if the null hypothesis were found to be unacceptable. This criterion would not be satisfactory if we had no knowledge of the alternatives to the null hypothesis.

This criterion is an example of a *one-sided* or a *one-tailed* test of significance based on the binomial distribution. It is called one-sided or one-tailed because the probabilities $P(w \geqslant 8)$ of rejecting the null hypothesis are obtained from a single tail of the binomial distribution.

(iii) Finally, suppose we adopt yet another criterion to test the null hypothesis $H(p = 0 \cdot 5)$: we agree to reject the null hypothesis if the actual outcome of the experiment is one of the values $w = 0, 1, 2, 8, 9$ or 10, but to accept it otherwise.

This criterion is based on the principle that if the null hypothesis is true, then getting w heads is much the same as getting w tails, that is $n - w$ heads. For any value of p, the probability of rejecting the null hypothesis is now

$$P(w \leqslant 2) + P(w \geqslant 8) = \left[(1-p)^{10} + 10p(1-p)^9 + 45p^2(1-p)^8 \right] + \left[45p^8(1-p)^2 + 10p^9(1-p) + p^{10} \right].$$

Table 12.3 gives the probabilities of rejecting the null hypothesis $H(p = 0 \cdot 5)$ for different values of p.

Table 12.3   Showing the values of $P(w \leqslant 2) + P(w \geqslant 8)$ for different values of p

p	0·01	0·10	0·20	0·30	0·40	0·50	0·60	0·70	0·80	0·90	0·99
$P(w \leqslant 2)$ $+P(w \geqslant 8)$	0·9999	0·9298	0·6779	0·3844	0·1796	0·1094	0·1796	0·3844	0·6779	0·9298	0·9999

We observe from the above table that the probability of rejecting the null hypothesis $H(p = 0 \cdot 5)$ when it is, in fact, true is $0 \cdot 1094$, which is moderately small. However, the probability of rejecting the null hypothesis increases steadily as $|p - 0 \cdot 5|$ increases. Accordingly, this is a satisfactory criterion whatever the alternatives to the null hypothesis may be. Since this criterion uses both tails of the binomial distribution, it is known as a *two-sided* or a *two-tailed* test of significance. Such a test procedure is used when we have no knowledge about the possible alternatives to the null hypothesis under test.

12.2   The level of significance

The major problem in testing any null hypothesis is the mathematical difficulty of formulating a test of significance which has the two basic qualities, namely, a low probability of rejecting the null hypothesis when it is, in fact, true, and a high probability of rejecting it when it is, in fact, false. But once the test is provided, then we have only to choose an acceptable degree of risk for rejecting the null hypothesis when it is, in fact, true, and to specify whether the test is to be one-sided or two-sided. Thus, if we feel that in the two-sided test considered above the probability of rejecting the null hypothesis when it is, in fact, true (0·1094) was too large, we could, for example, choose another probability by specifying rejection of the null hypothesis only if the observed result is either $w \leqslant 1$ or $w \geqslant 9$. For such a test, the probability of rejecting the null hypothesis when it is, in fact, true is

$$2\left[(1/2)^{10} + 10(1/2)^{10}\right] = 11/512 = 0·0215.$$

In standard statistical terminology, the probability of rejecting the null hypothesis when it is, in fact, true is known as the *level of significance*, and this usually takes a value of 0·05 or 0·01. These are generally termed the five per cent and one per cent levels of significance. The use of these levels of significance is completely conventional, and it is perfectly legitimate to use a test of significance with any preassigned level of significance.

A test of significance is based on a function of the observations, whose sampling distribution is known when the null hypothesis under test is, in fact, true. Given this sampling distribution and the preassigned probability of rejecting the null hypothesis when it is, in fact, true, the entire domain of the test function is divided into two parts — the *region of acceptance* and the one-sided or two-sided *region of rejection*. The decision about the null hypothesis is then made according to whether the observed value of the test function falls in the region of rejection (when the null hypothesis is rejected), or in the region of acceptance (when the null hypothesis is accepted). In Example 12.1, the test function was simply w, the number of heads observed in the n trials with the penny. It is important to understand the implication of the usage — acceptance and rejection of a null hypothesis on the basis of a test of significance. Clearly, a test of significance never proves or disproves a null hypothesis because, whatever the decision made, there is always a possibility of it being wrong. Accordingly, at a certain level of significance, a test of significance makes a null hypothesis more or less likely, depending upon whether the observed result is considered to be consistent with the expectation based on the null hypothesis or not.

When, at a certain level of significance, the null hypothesis is rejected, the observed result is said to be *significant*, and the acceptance of the null hypothesis is equivalently described by saying that the observed result is *not significant*.

12.3 Binomial sampling in industrial inspection of quality

The simple coin-tossing experiment considered above is admittedly artificial, and it does not indicate how the concept of a test of significance can be used in a meaningful practical situation. We now consider a more complicated problem of statistical inference based on binomial sampling, which is typical of a variety of tests of significance used in industrial quality control procedures. The idea of using statistical methods for assessing the quality of a manufactured product was first introduced by W.A. Shewhart at the Bell Telephone Company of America in the 'twenties. By now, quality control procedures are well established in industry throughout the world, and they constitute an increasingly fruitful field for the use of statistical ideas and principles. Such control is applied both in connection with the finished product and at the successive stages of the manufacturing process. We consider here the inspection of the finished product, which provides an interesting example of a test of significance.

For simplicity, our interest is in production processes in which small items such as screws, rivets, washers, ball-bearings, light bulbs, electronic tubes, etc. are manufactured in large numbers by automatic machines. The quality of such items is assessed in terms of one or more characteristics. For example, the quality of a washer may be measured by its thickness and its internal and external diameters, and that of a screw by its length, thickness, and head diameter. The production process is organised to produce items whose characteristics are expected to lie within acceptable dimensions known as *specification limits*. In general, an item produced is assessed in terms of each specified characteristic, and if it satisfies the requirements on all counts, then it is regarded as a *non-defective*, and in the contrary event it is classified as a *defective*. For our present purpose, we ignore the complication of the types of defects in an item inspected, and consider a simple dichotomy of defectives and non-defectives produced. However, this dichotomy is not necessarily based on measurements alone, as it could equally well arise from a qualitative test. For example, cups and saucers produced by a ceramics factory may be classified as damaged and undamaged pieces. Similarly, glass stoppers which fit bottles are non-defectives, and those not fitting the bottles because they are either too big or too small are classed as defectives. In any case, whether the dichotomy is based on a metrical characteristic or, more simply, on a qualitative basis, the machines are visualised as randomly producing items of both kinds — defectives and non-defectives.

Generally, the items produced are grouped into batches whose size may vary considerably from say a dozen to many thousands. For such a product, the problem of inspection could arise in at least two ways. Firstly, the producer may wish to ensure that the outgoing product satisfies certain requirements, that is, there are not more than a certain proportion of defectives in the batches sent to customers. Secondly, a customer may naturally wish to guard himself against the risk of accepting batches which contain more than a certain proportion of defectives. Both these inspection problems are essentially similar, and we shall consider the one from the point of view of the customer. Now, ideally, the customer would wish not to receive a single defective, and he could, at least theoretically, consider 100 per cent inspection of the product received. However, there are several reasons why this is not the best method for guarding against the risk of accepting defectives. Firstly, total inspection would almost certainly be far too costly to be really economical when compared with the saving due to the detection of defectives. Secondly, there are cases in which inspection can only be done by the destruction of the product. For example, the life-time of electronic tubes or light bulbs can be determined only in this way. This is also true of the tensile strength of steel cables and the extensibility of yarn. Thirdly, even

with 100 per cent inspection, defectives are still known to get through because there is always the human error in inspection.

In the absence of total inspection, some sampling method must be used
whereby the customer can make a decision to accept, reject or completely
inspect a batch on the basis of inspection of relatively few items from it.
For obvious reasons of economy, the customer would like to reject or completely inspect "bad" batches (that is, batches containing more than an
acceptable minimum of defectives), and accept only the "good" ones (that
is, batches containing not more than an acceptable maximum of defectives).
However, because the customer bases his decision on the results of a
sample (that is, a fraction of a batch), he will occasionally accept a bad
batch and reject or completely inspect a good one. We observe that this is
similar to the uncertainty in making a decision about a null hypothesis on
the basis of a test of significance. Our problem is to devise a *sampling
inspection scheme* for the customer which will attempt to minimise the
chance of acceptance of bad batches and of rejection or complete inspection
of good batches. One of the fundamental criteria by which such an inspection scheme is judged is its *operating characteristic* or *O.C. curve*. This
curve is the graph of the proportion of defectives in a batch (p) against the
probability that batches with a given proportion of defectives will be
accepted without further inspection. In general, the graph representing an
O.C. curve has the form indicated in the figure below.

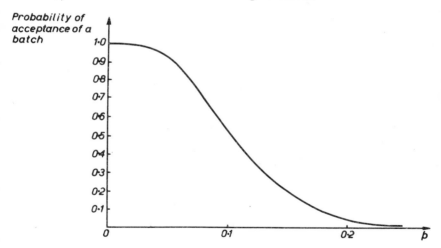

Fig. 12.1   Operating characteristic curve of a single inspection scheme

This shows that the greater the proportion of defectives in a batch (p),
the smaller is the probability of its acceptance. Clearly, the customer ex-
pects to get a small proportion of defectives in each batch but, at the same
time, he wants to guard himself against the risk of accepting a batch having
more than a certain proportion of defectives. Accordingly, suppose p_0 is
the highest proportion of defectives in a batch regarded as being definitely
good, and p_1 the lowest proportion of defectives in a batch considered to be

definitely bad. Evidently, we must have $p_0 < p_1$. Then a "reasonable" sampling scheme will satisfy the following two conditions:

    (i) For a batch with proportion of defectives p_0, the probability of acceptance is $1 - \alpha_0$, where α_0 is small.

    (ii) For a batch with proportion of defectives p_1, the probability of acceptance is α_1, which is small.

Thus the O.C. curve of the sampling scheme will pass through the points $(p_0, 1 - \alpha_0)$ and (p_1, α_1). The four constants p_0, p_1, α_0, and α_1 define in physical terms the risks that the customer is prepared to take in forming his inspection procedure. Single sampling schemes are the simplest inspection schemes based on these requirements. The formal procedure of such a scheme is to inspect a random sample of n items from the batch and then to accept the batch if c or less defectives are found. The batch is rejected or completely inspected if the sample contains more than c defectives. The theoretical problem is to determine n and c in terms of p_0, p_1, α_0, and α_1.

    Suppose a batch contains N items and let p be the proportion of defectives in it. Therefore there are Np defectives and $Nq\,(p+q = 1)$ non-defectives in the batch. Then the probability of accepting the batch

    = the probability of finding c or less defectives in the batch

$$= \sum_{w=0}^{c} \binom{Np}{w} \binom{Nq}{n-w} \Big/ \binom{N}{n},$$

by the hypergeometric distribution, where the white and black balls correspond to defectives and non-defectives in the batch. Hence, by the stated specifications, the equations for determining n and c are

$$\left. \begin{aligned} 1 - \alpha_0 &= \sum_{w=0}^{c} \binom{Np_0}{w} \binom{Nq_0}{n-w} \Big/ \binom{N}{n} \quad (p_0 + q_0 = 1), \\[2mm] \alpha_1 &= \sum_{w=0}^{c} \binom{Np_1}{w} \binom{Nq_1}{n-w} \Big/ \binom{N}{n} \quad (p_1 + q_1 = 1). \end{aligned} \right] \tag{1}$$

and

If N is sufficiently large so that the effect of sampling from a finite population may be ignored approximately, then the hypergeometric probabilities can be replaced by the corresponding binomial probabilities, and the equations for n and c take the simpler form

$$\left. \begin{aligned} 1 - \alpha_0 &= \sum_{w=0}^{c} \binom{n}{w} p_0^w q_0^{n-w}, \\[2mm] \alpha_1 &= \sum_{w=0}^{c} \binom{n}{w} p_1^w q_1^{n-w}. \end{aligned} \right] \tag{2}$$

and

    Before we solve these equations for n and c given p_0, p_1, α_0, and α_1, it is instructive to attempt a solution of the converse problem. Suppose, then, that $n = 40$ and $c = 3$. We show that with these values of n and c, it is possible to obtain reasonable values for p_0, p_1, α_0, and α_1 which satisfy the requirements of a single inspection scheme. The probability for the acceptance of a batch with proportion of defectives p is now

$$P(w \leqslant 3) = \sum_{w=0}^{3} \binom{40}{w} p^{w}(1-p)^{40-w}.$$

Table 12.4 below gives the values of $P(w \leqslant 3)$ corresponding to values of p in the range $0 \cdot 01 \leqslant p(0 \cdot 01) \leqslant 0 \cdot 20$.

Table 12.4   Showing the values of $P(w \leqslant 3)$ for different values of p

p	$P(w \leqslant 3)$	p	$P(w \leqslant 3)$
0·01	0·9993	0·11	0·3452
0·02	0·9918	0·12	0·2768
0·03	0·9686	0·13	0·2184
0·04	0·9252	0·14	0·1698
0·05	0·8619	0·15	0·1302
0·06	0·7827	0·16	0·0984
0·07	0·6937	0·17	0·0735
0·08	0·6007	0·18	0·0542
0·09	0·5092	0·19	0·0395
0·10	0·4231	0·20	0·0285

We observe from this table that

$$\text{if } p_0 = 0 \cdot 03, \text{ then } \alpha_0 = 0 \cdot 0314;$$

and $\qquad$ if $p_1 = 0 \cdot 18$, then $\alpha_1 = 0 \cdot 0542$.

Hence, if these were the specifications of the risks of the customer, then $n = 40$ and $c = 3$ are the required values for the inspection scheme satisfying the requirements. This scheme is based on a one-sided test which uses the lower tail of the binomial distribution. As p increases, the probability in this tail decreases steadily and vice versa. The upper tail of the distribution is not used because an improvement in quality beyond a certain point is not of interest to the customer to make his decision about the batch.

The above solution is not what we required, since our initial problem was to determine n and c from the simultaneous equations (2) for given values of p_0, p_1, α_0, and α_1. This is a more difficult problem, and a solution of (2) can only be obtained by trial and error. To make the problem specific, suppose the probability of rejecting the batch, when the proportion of defectives in it is $0 \cdot 02$, is $0 \cdot 05$. Also, let the probability of accepting the batch be $0 \cdot 01$ when its proportion of defectives is $0 \cdot 05$. This means that we are now given

$$\begin{array}{ll} p_0 = 0 \cdot 02, & \alpha_0 = 0 \cdot 05; \\ p_1 = 0 \cdot 05, & \alpha_1 = 0 \cdot 01; \end{array} \Bigg\} \qquad (3)$$

and we have to determine n and c from (2).

To obtain approximate values of n and c, we can use the normal approximation for the binomial distribution to rewrite (2) as

$$1 - \alpha_0 \sim \Phi[(c - np_0 + 0 \cdot 5)/\{np_0(1 - p_0)\}^{\frac{1}{2}}],$$

and $\qquad \alpha_1 \sim \Phi[(c - np_1 + 0 \cdot 5)/\{np_1(1 - p_1)\}^{\frac{1}{2}}].$

If we next use the numerical values of p_0, p_1, α_0, and α_1 specified in (3), then we obtain from the table of the normal integral

$$1 \cdot 645 \; = \; \frac{c - 0 \cdot 02n + 0 \cdot 5}{\sqrt{0 \cdot 0196n}}, \tag{4}$$

and $$- 2 \cdot 236 \; = \; \frac{c - 0 \cdot 05n + 0 \cdot 5}{\sqrt{0 \cdot 0475n}}. \tag{5}$$

Elimination of c from (4) and (5) gives

$$1 \cdot 645\sqrt{0 \cdot 0196n} + 2 \cdot 236\sqrt{0 \cdot 0475n} \; = \; 0 \cdot 03n,$$

so that $\quad 0 \cdot 03\sqrt{n} \; = \; 0 \cdot 7374, \quad$ or $\quad \sqrt{n} \; = \; 24 \cdot 58.$

Therefore $\qquad\qquad\qquad n \; = \; 604 \cdot 18.$

Hence (4) gives

$$c \; = \; 1 \cdot 645 \times 0 \cdot 14\sqrt{n} + 0 \cdot 02n - 0 \cdot 5 \; = \; 17 \cdot 24.$$

Thus we may take $n = 605$ and $c = 17$ or 18. Admittedly, these are approximations but they should be useful for most practical purposes. If greater accuracy is required, then we could use these approximate values of n and c in (2) to derive closer approximations; but we do not consider the details of doing this here.

Batch size finite

The normal approximation for the binomial distribution generally holds reasonably well in such cases, and the more questionable approximation is that of assuming the batch size to be infinite so that the hypergeometric probabilities in (1) may be replaced by the corresponding binomial ones to deduce (2). We now consider the effect of a finite batch size on the values of n and c of the inspection scheme. To do this, we recall that one of the important consequences of sampling from a finite population is the reduction of the binomial variance. In fact, for a batch of size N,

$$E(w) \; = \; np \quad \text{and} \quad \text{var}(w) \; = \; npq(N - n)/(N - 1),$$

where w is the number of defectives in a sample of size n taken from the batch. It is known, but we do not prove it here, that in this case also

$$\frac{w - np}{\text{s.d.}(w)} \; \sim \; N(0, 1).$$

Hence, by an argument analogous to the one used for the binomial distribution, we may write

$$P(w \leqslant c) \; \sim \; \Phi[(c - np + 0 \cdot 5)/\{npq(N - n)/(N - 1)\}^{\frac{1}{2}}].$$

Accordingly, using (1), the approximate equations for n and c are

$$1 - \alpha_0 \; \sim \; \Phi[(c - np_0 + 0 \cdot 5)/\{np_0 q_0 (N - n)/(N - 1)\}^{\frac{1}{2}}],$$

and $\qquad\qquad \alpha_1 \; \sim \; \Phi[(c - np_1 + 0 \cdot 5)/\{np_1 q_1 (N - n)/(N - 1)\}^{\frac{1}{2}}].$

As an example, suppose $N = 2,000$. Then, using the values in (3), we have the modified forms of equations (4) and (5) as

$$1 \cdot 645[0 \cdot 0196n(N - n)/(N - 1)]^{\frac{1}{2}} \; = \; c - 0 \cdot 02n + 0 \cdot 5, \tag{6}$$

and $\quad - 2 \cdot 326[0 \cdot 0475n(N - n)/(N - 1)]^{\frac{1}{2}} \; = \; c - 0 \cdot 05n + 0 \cdot 5. \tag{7}$

Elimination of c from (6) and (7) gives

$$0 \cdot 7374[(N - n)/(N - 1)]^{\frac{1}{2}} \; = \; 0 \cdot 03\sqrt{n}, \quad \text{or} \quad 5438(N - n)/(N - 1) \; = \; 9n.$$

Therefore $n = \dfrac{5438\,N}{9(N-1) + 5438} = 464 \cdot 03$, for $N = 2000$.

Hence, from (6),

$$c = 8 \cdot 7806 + 0 \cdot 2303\,[n(N-n)/(N-1)]^{\frac{1}{2}}$$
$$= 8 \cdot 7806 + 0 \cdot 2303\,\sqrt{35 \cdot 6514} = 10 \cdot 16.$$

Thus for a batch size of $N = 2{,}000$, we may take $n = 465$ and $c = 10$ or 11. We note the considerable change in these values as compared with the corresponding ones obtained for a batch of infinite size. As derived, the difference is attributable to the fact that the hypergeometric variance is smaller than that of the comparable binomial distribution.

Example 12.2: An exact test of significance in binomial sampling

We next consider an exact test of significance in binomial sampling when the sample size is small so that we cannot legitimately use the normal approximation. The problem may be formulated by means of an example as follows.

Ten lengths of yarn were randomly selected from a large quantity, and each length was cut into two equal parts, which were then tested under a given load, their percentage extension being measured. One half of each length was tested immediately and the other half after six washings. Of the ten lengths tested in this way, six gave a higher percentage extension after washing. On the basis of these data, is it reasonable to infer that washing increases the extensibility of the yarn?

This is a simple example of a test of significance, but we shall consider it in detail to analyse the assumptions leading to the binomial model, which provides the formal basis for the statistical analysis to answer the question posed. We visualise an infinite population of yarn lengths similar to the ones used in the experiment. The ten lengths tested are assumed to be a random selection from this infinite population of yarn lengths. Next, we imagine that each of these lengths in the population is divided into two equal parts, and the extension of one half measured without washing and that of the other half after the six experimental washings. Suppose the percentage extensions of any such pair before and after washing are x and y respectively. Then the difference $y - x$ could be positive, negative or, at least theoretically, zero. The totality of lengths constituting the conceptual population provides a corresponding infinite population of differences of the type $y - x$. We assume that in this population

$$P(y - x > 0) = p \quad \text{and} \quad P(y - x \leqslant 0) = 1 - p,$$

where p is a parameter. Thus the binomial population is determined in terms of positive and negative (or zero) differences in percentage extension, a positive difference indicating an increase in percentage extension due to washing, and a negative or zero difference implying a decrease or no change in extension due to washing.

From this n.odel population, we have obtained a sample of ten differences of which six were found to be positive. Hence the probability of the observed result is $\dbinom{10}{6} p^6 (1-p)^4$.

The question posed has meaning in relation to this conceptual population, and the sample data are to be used to provide a probabilistic answer about the conceptual population or, more simply, about the effect of washing on the percentage extension of the yarn. The null hypothesis is now formulated and states that washing does not change the percentage extensibility of the yarn, so that positive and negative differences of the type $y - x$ are equally likely. Hence the null hypothesis is simply $H(p = 0.5)$. Finally, before we can carry out a test of significance, we have to specify whether the test is to be one-sided or two-sided. When it is clear from the way the question is posed that the experimenter has extra-statistical information which suggests that if the null hypothesis is not true then p must be greater than half, then we should use a one-sided test of significance.

To carry out this test, we determine the total probability of obtaining the observed result or any result more extreme than the one we have obtained, on the assumption that the null hypothesis $H(p = 0.5)$ is, in fact, true. This total probability is

$$\sum_{w=6}^{10} \binom{10}{w}\left(\frac{1}{2}\right)^{10} = \frac{386}{1024} = 0.3770.$$

Thus, if the null hypothesis $H(p = 0.5)$ is true, then the probability that the observed result or a more extreme one could happen by chance is 0.3770. Evidently, such a high probability can hardly suggest any incredibility in the null hypothesis under test. More formally, if we were working on the five per cent level of significance, our inference would be that the observed result is not significant and the null hypothesis is accepted.

As another example, suppose the experiment provided measurements from twenty lengths of yarn of which sixteen gave positive differences. In this case, if the null hypothesis $H(p = 0.5)$ is under test, then the probability in the upper tail of the distribution is

$$\sum_{w=16}^{20} \binom{20}{w}\left(\frac{1}{2}\right)^{20} = \frac{6196}{1048576} = 0.0059.$$

The probability is now less than one per cent, and we have little doubt that an exceptional result has been observed if the null hypothesis is, in fact, true. We therefore conclude that the observed result is significant at the one per cent level of significance, and the null hypothesis can be safely rejected. In other words, the experimental evidence now supports the belief that washing increases the extensibility of the yarn. We shall see later how a numerical measure can be attached to this increase.

12.4   The normal approximation in binomial sampling

In the first example considered above only ten yarn lengths were examined experimentally for the change in extensibility due to washing. This was a small sample, and so a valid application of a test of significance necessitated the calculation of the exact probability on the assumption that the null hypothesis under test was true. This procedure must always be followed for small samples. However, in many useful applications the samples are reasonably large, and then a more convenient procedure is to use the normal approximation for the binomial distribution. The term large samples is rather difficult to define precisely but, as a practical and convenient rule, we may regard 50 to be the smallest sample size for which the normal approximation can be reasonably confidently applied without any serious risk of error. Nevertheless, in using such approximate procedures, it is useful to remember that the significance of an observed result should be interpreted rather flexibly. For example, if we are working on a five per cent level of significance, then it would be unwise to discriminate sharply between observed probabilities of, say, 0·046 and 0·062. Of course, small and large probabilities can safely be interpreted as giving adequate evidence for the rejection and acceptance respectively of a null hypothesis; but in the neighbourhood of the level of significance caution is desirable, as in such a situation it may well be the best course to obtain more observations by further experimentation.

We consider next another example where the sample size is large enough for satisfactory use of the normal approximation in binomial sampling.

Example 12.3: Overcrowding in a town

A housing survey conducted in a large town showed that, of a total of 400 families sampled, 28 were resident in overcrowded accommodation. It is believed that less than one-eighth of the families in the town are living in conditions of overcrowding assessed by certain health standards. Do the sample data provide reasonable evidence for this belief?

Here, again, the first problem is to specify clearly the model which provides the basis for a test of significance. To begin with, we assume that the number of families in the town is sufficiently large for us to ignore the effect of sampling from a finite population. Accordingly, we postulate an "infinite" population of families of which an unknown proportion p are overcrowded. The null hypothesis is $H(p = 0·125)$ and its alternatives are $p < 0·125$. Hence we have to use a one-sided test of significance.

Now, if $H(p = 0·125)$ is true, then the expected number of overcrowded families is

$$400 \times 0·125 = 50,$$

and the number actually observed is 28. This is less than the expectation on the null hypothesis, and our purpose is to find out whether or not the discrepancy between observation and expectation is large enough to throw substantial doubt on the validity of the null hypothesis. Also, since the alternatives to the null hypothesis are $p < 0·125$, our interest is in the lower tail of the distribution. Accordingly, if the null hypothesis is true, then the probability of the observed result and of any result more extreme than it, is

$$P(w \leqslant 28) = \sum_{w=0}^{28} \binom{400}{w} \left(\frac{1}{8}\right)^{w} \left(\frac{7}{8}\right)^{400-w},$$

where w is the random variable denoting the number of overcrowded families in the sample. It is clear that a direct evaluation of this probability would be rather tedious, and the normal approximation can be applied conveniently because 400 families were included in the sample and $E(w) = 50$. Hence

$$P(w \leqslant 28) \sim \Phi[(28 - 50 + 0\cdot5)/(400 \times 0\cdot125 \times 0\cdot875)^{\frac{1}{2}}]$$
$$= 1 - \Phi(3\cdot25) = 1 - 0\cdot9994 = 0\cdot0006.$$

Therefore $P(w \leqslant 28) = 0\cdot0006$, and so there is no doubt that the null hypothesis is untenable. In other words, the data provide evidence for the belief that less than one-eighth of the families in the town are overcrowded.

Aliter

Another convenient way of carrying out the above test of significance on the basis of the normal approximation is as follows.

The maximum-likelihood estimate of p is

$$\hat{p} = \frac{28}{400} = 0\cdot07.$$

Now, if the null hypothesis $H(p = 0\cdot125)$ is true, then $E(\hat{p}) = 0\cdot125$ and

$$\mathrm{var}(\hat{p}) = \frac{0\cdot125 \times 0\cdot875}{400} = 0\cdot00027344.$$

Therefore $\quad\quad\quad\quad \mathrm{s.d.}(\hat{p}) = 0\cdot01654.$

Hence, if the null hypothesis under test is true, then

$$z = \frac{\hat{p} - E(\hat{p})}{\mathrm{s.d.}(\hat{p})} \sim N(0, 1).$$

But with our data, $\quad z = \dfrac{0\cdot07 - 0\cdot125}{0\cdot01654} = -3\cdot32,$

and $\quad P(z \leqslant -3\cdot32) = 0\cdot0004$, from the table of the normal integral.

The probability obtained by this method is a shade less than that deduced by the first method because we have neglected the effect of the continuity correction $0\cdot5$. However, this is not of much consequence, and both tests give overwhelming significance. The real interest in using z is that we need not calculate the total probability in order to carry out the test of the null hypothesis on a certain level of significance. For example, suppose we are using a five per cent significance level. Now, if x is a unit normal variable, then it is known from the table of the normal integral that

$$P(x \geqslant 1\cdot645) = 0\cdot05.$$

In other words, for a one-sided test of significance, the five per cent point of the unit normal distribution is $1\cdot645$. Hence, to test the significance of an observed value of z, it is enough to evaluate $|z|$. If this is greater than $1\cdot645$, then the null hypothesis is rejected at the five per cent level of significance. The null hypothesis is accepted if $|z| < 1\cdot645$. In the above example, $|z| = 3\cdot32$ and so the null hypothesis is rejected conclusively.

An exactly similar method may be used for any other significance level and also for two-sided tests of significance.

Use of the upper tail of the distribution

It is worthwhile reconsidering the use of the lower tail of the distribution for testing the null hypothesis $H(p = 0\cdot125)$. As already mentioned, the significance was assessed by evaluating $P(w \leqslant 28)$ because the alternatives to the null hypothesis under test were $p < 0\cdot125$.

On the other hand, suppose we wished to test whether or not the data are consistent with the hypothesis that more than one-eighth of the families in the town are living in overcrowded accommodation. We again have the null hypothesis $H(p = 0 \cdot 125)$ but with the alternatives $p > 0 \cdot 125$. This means that the null hypothesis is now tested by considering

$$P(w \geqslant 28) = \sum_{w=28}^{400} \binom{400}{w} \left(\frac{1}{8}\right)^{w} \left(\frac{7}{8}\right)^{400-w}$$

$$= 1 - \sum_{w=0}^{27} \binom{400}{w} \left(\frac{1}{8}\right)^{w} \left(\frac{7}{8}\right)^{400-w}$$

$$\sim 1 - \Phi\left[\frac{27 - 50 + 0 \cdot 5}{(400 \times 0 \cdot 125 \times 0 \cdot 875)^{\frac{1}{2}}}\right] = 1 - \Phi(3 \cdot 40) = 0 \cdot 9997.$$

Therefore $P(w \geqslant 28) = 0 \cdot 9997$, and so the null hypothesis is certainly tenable.

It is important to understand clearly the implications of the results of the two one-sided tests of significance. The observed proportion of over-crowded families is $\hat{p} = 0 \cdot 07$ and, in the light of this information, the two one-sided tests have assessed the significance of the null hypothesis $H(p = 0 \cdot 125)$ but with the alternatives $p < 0 \cdot 125$ and $p > 0 \cdot 125$ respectively. The first test of significance means that if the alternatives are $p < 0 \cdot 125$, then the observed value $\hat{p} = 0 \cdot 07$ does not provide any evidence for accepting the null hypothesis $H(p = 0 \cdot 125)$. In contrast, the second test of significance indicates that if the alternatives are $p > 0 \cdot 125$, then the estimate $\hat{p} = 0 \cdot 07$ is consistent with the null hypothesis $H(p = 0 \cdot 125)$. The divergence between these two one-sided tests exemplifies how the credibility of a null hypothesis is affected by prior information about the alternatives.

As another instructive illustration, suppose the observed number of overcrowded families in the sample of 400 is 62 and not 28. We again wish to test the null hypothesis $H(p = 0 \cdot 125)$ with the alternatives $p > 0 \cdot 125$. Hence, for a one-sided test of significance, the total probability to be considered is now

$$P(w \geqslant 62) = \sum_{w=62}^{400} \binom{400}{w} \left(\frac{1}{8}\right)^{w} \left(\frac{7}{8}\right)^{400-w}$$

$$= 1 - \sum_{w=0}^{61} \binom{400}{w} \left(\frac{1}{8}\right)^{w} \left(\frac{7}{8}\right)^{400-w}$$

$$\sim 1 - \Phi\left[\frac{61 - 50 + 0 \cdot 5}{(400 \times 0 \cdot 125 \times 0 \cdot 875)^{\frac{1}{2}}}\right] = 1 - \Phi(1 \cdot 738) = 0 \cdot 0411.$$

The observed result is significant at the five per cent level of significance. However, the probability is so close to $0 \cdot 05$ that a definite decision to reject the null hypothesis $H(p = 0 \cdot 125)$ is hardly justifiable except on a rigid interpretation and use of the chosen level of significance. This means that if the alternatives to the null hypothesis $H(p = 0 \cdot 125)$ are $p > 0 \cdot 125$, then the observed result of 62 overcrowded families in a sample of 400 does not provide very strong evidence about the null hypothesis. This is the kind of situation where the most sensible procedure might well be to obtain more observations to reach a firmer conclusion.

Suppose, next, that there is no knowledge about the alternatives to the null hypothesis $H(p = 0 \cdot 125)$, so that if it is not true then $p \neq 0 \cdot 125$. Accordingly, we have to use a two-sided test of significance. The probability of the observed result $w = 62$ and of the more extreme deviations on both sides of the expectation 50 is

$$P(|w - 50| \geq 12) = P(w \geq 62) + P(w \leq 38)$$
$$\sim 2[1 - \Phi(1 \cdot 738)],$$

because of the symmetry of the normal distribution,

$$= 0 \cdot 0822.$$

Accordingly, we now have reasonably conclusive evidence that the observed result is not significant at the five per cent level. The uncertainty about the alternatives to the null hypothesis has decreased the significance of the observed result. This is a characteristic difference between one-sided and two-sided tests of significance.

Finite number of families in the town

Our assumption that there are an infinite number of families in the town is, in principle, unrealistic, though we now show that this makes no material difference to the conclusion of the test of significance based on the sample observation of 28 overcrowded families amongst a total of 400. As an example, suppose the total number of families in the town is $N = 2,000$. Then the hypergeometric distribution is applicable. Therefore, if the null hypothesis $H(p = 0 \cdot 125)$ is true, we have

$$\text{var}(w) = \frac{np(1 - p)(N - n)}{N - 1}$$

$$= \frac{400 \times 0 \cdot 125 \times 0 \cdot 875 \times 1600}{1999} = 35 \cdot 0175,$$

so that $\text{s.d.}(w) = 5 \cdot 9176.$

Hence, to test the null hypothesis $H(p = 0 \cdot 125)$ when the alternatives are $p < 0 \cdot 125$, we now have

$$P(w \leq 28) = \sum_{w=0}^{28} \binom{250}{w} \binom{1750}{400 - w} \bigg/ \binom{2000}{400}, \text{ since } Np = 250,$$

$$\sim \Phi\left[\frac{28 - 50 + 0 \cdot 5}{5 \cdot 9176}\right] = 1 - \Phi(3 \cdot 63) = 0 \cdot 0001.$$

Thus the finiteness of the population of families in the town has only resulted in enhancing the significance of the observed result. This was to be expected since the hypergeometric variance is less than the binomial variance.

12.5   Comparison of two proportions

We have so far considered a single binomial population, and our concern has been to test some hypothesis about its parameter p on the basis of a random sample drawn from the population. This approach can be very simply and usefully extended to test hypotheses about the parameters of two binomial populations. We assume that the samples obtained from the populations are large and that the population parameters are finite, so that the normal approximation is applicable. It is now best to work in the context of a concrete problem.

Example 12.4:   University entrance of boys and girls

Of a random sample of 1,142 boys who had applied for admission to a university, 896 obtained places on passing their A-level examinations. In the same year, of a random sample of 538 girls who had sought university admission, a total of 364 got places after obtaining their A-levels. Is there any evidence that, on the average, boys have a better chance than girls to gain university admission?

Our first step is to specify clearly the populations of boys and girls whose chances of obtaining university places are the subject of the question posed. Firstly, we are confining the enquiry to a particular year. Of all the students that year who took their A-levels, some did not qualify, and some others who qualified chose not to try for university entrance. The remaining students were the ones who successfully completed their A-levels and tried to enter a university. These students are now regarded as grouped according to sex, and we assume that there were in all N_1 boys and N_2 girls. These constitute our two populations and, quite evidently, these must be of finite size. However, we make a simplifying assumption that N_1 and N_2 are sufficiently large for the effect of sampling from a finite population to be ignored.

Suppose the true proportions of boys and girls who got admission to a university are p_1 and p_2 respectively. These are the parameters of the two binomial populations. In the first population, the "successes" are the boys who obtained university places, and the rest are the "failures". Similarly, in the second population, the "successes" and "failures" are the girls who got and did not get university places respectively.

The null hypothesis to be tested is $H(p_1 - p_2 = 0)$, that is, boys and girls have an equal chance of entering a university. It is important to note that we are not specifying what this equal chance is, and the null hypothesis pertains only to the *equality* of the chances. Furthermore, the way the question has been posed suggests that the investigator has some reason to believe that if the null hypothesis is not tenable, then $p_1 - p_2 > 0$, that is, boys are more favourably placed than girls for university entrance. The interest is thus in a one-sided test of significance.

Given the sample information, we readily conclude that the maximum-likelihood estimates of p_1 and p_2 are

$$\hat{p}_1 = \frac{896}{1142} = 0{\cdot}7846; \qquad \hat{p}_2 = \frac{364}{538} = 0{\cdot}6766.$$

Also, $\text{var}(\hat{p}_1) = p_1(1-p_1)/1142;$ $\qquad$ $\text{var}(\hat{p}_2) = p_2(1-p_2)/538.$

However, these variances are unknown, and their unbiased estimates are obtained as follows:

estimate $\text{var}(\hat{p}_1) = \hat{p}_1(1-\hat{p}_1)/1141 = 0{\cdot}0001481,$

and   estimate $\text{var}(\hat{p}_2) = \hat{p}_2(1-\hat{p}_2)/537 = 0{\cdot}0004075.$

Now $\hat{p}_1$ and $\hat{p}_2$ are independent random variables and, as they are obtained from reasonably large samples, we may also regard them as approximately normally distributed with their respective means and variances. Clearly, their difference is necessarily a random variable such that

$$E(\hat{p}_1 - \hat{p}_2) \;=\; p_1 - p_2 \quad \text{and} \quad var(\hat{p}_1 - \hat{p}_2) \;=\; var(\hat{p}_1) + var(\hat{p}_2).$$

Therefore

$$\text{estimate } var(\hat{p}_1 - \hat{p}_2) \;=\; \text{estimate } var(\hat{p}_1) + \text{estimate } var(\hat{p}_2)$$
$$= \; 0 \cdot 0005556.$$

Hence

$$\text{s.e.}(\hat{p}_1 - \hat{p}_2) \;=\; 0 \cdot 02357.$$

There is a general theorem from which we conclude that the random variable $\hat{p}_1 - \hat{p}_2$ has an approximate distribution such that

$$z \;=\; \frac{\hat{p}_1 - \hat{p}_2 - (p_1 - p_2)}{\text{s.e.}(\hat{p}_1 - \hat{p}_2)} \;\sim\; N(0, 1).$$

This important result is, in fact, a large-sample approximation for the sampling distribution of the random variable $\hat{p}_1 - \hat{p}_2$ and provides the basis for the test of significance of the null hypothesis $H(p_1 - p_2 = 0)$. If this hypothesis is true, then $E(\hat{p}_1 - \hat{p}_2) = 0$ and so

$$z \;=\; \frac{\hat{p}_1 - \hat{p}_2}{\text{s.e.}(\hat{p}_1 - \hat{p}_2)} \;\sim\; N(0, 1).$$

But with our data

$$z \;=\; \frac{0 \cdot 7846 - 0 \cdot 6766}{0 \cdot 02357} \;=\; \frac{0 \cdot 1080}{0 \cdot 02357} \;=\; 4 \cdot 58.$$

The observed difference between the proportions is highly significant, since for a one-sided test $P(z \geqslant 4 \cdot 58) = 1 - \Phi(4 \cdot 58)$ is negligibly small. Therefore the null hypothesis is confidently rejected, and we infer that boys have a better chance than girls for obtaining university entrance.

An extension of the null hypothesis

We can easily extend the above analysis to test an even more interesting null hypothesis. Suppose it is known that boys have a better chance than girls to enter a university, and we wish to test whether or not the data are in agreement with the hypothesis that the difference between the chances of boys and girls to gain university admission is greater than eight per cent. Accordingly, the null hypothesis is $H(p_1 - p_2 = 0 \cdot 08)$ with the alternatives $p_1 - p_2 > 0 \cdot 08$. If this null hypothesis is true, then $E(\hat{p}_1 - \hat{p}_2) = 0 \cdot 08$. Hence

$$z \;=\; \frac{\hat{p}_1 - \hat{p}_2 - 0 \cdot 08}{\text{s.e.}(\hat{p}_1 - \hat{p}_2)} \;\sim\; N(0, 1),$$

so that the observed value of z is $0 \cdot 0280/0 \cdot 02357 = 1 \cdot 187$. Therefore, for a one-sided test of significance, we have

$$P(z \geqslant 1 \cdot 187) \;=\; 1 - \Phi(1 \cdot 187) \;=\; 1 - 0 \cdot 8824 \;=\; 0 \cdot 1176.$$

This is a reasonably high probability, and we conclude that the null hypothesis is in agreement with the data. The samples provide good evidence that boys may have an eight per cent better chance than girls to enter a university. However, it is to be noted that this is not the only null hypothesis which is in conformity with the data. We shall see later how we can obtain a measure of all the hypotheses which are in conformity with the data at any preassigned level of significance.

An alternative procedure for testing the null hypothesis $H(p_1 - p_2 = 0)$

This alternative procedure differs from the first method only in so far as we use the null hypothesis itself in deriving the standard error of $\hat{p}_1 - \hat{p}_2$. We have seen that

$$\text{var}(\hat{p}_1 - \hat{p}_2) = \text{var}(\hat{p}_1) + \text{var}(\hat{p}_2) = \frac{p_1(1-p_1)}{1142} + \frac{p_2(1-p_2)}{538}.$$

Now if the null hypothesis under test is true and we write $p_1 = p_2 = p$, say, then

$$\text{var}(\hat{p}_1 - \hat{p}_2) = \frac{1680\,p(1-p)}{1142 \times 538}.$$

To estimate this variance, we have to find an unbiased estimate of $p(1-p)$. To obtain this latter estimate, we argue as follows. If there is no difference in the true proportions of the two populations, then we may legitimately combine them. Accordingly, our pooled sample data state that in a sample of 1,680 boys and girls from this combined population, a total of 1,260 boys and girls obtained university admission. Therefore the maximum-likelihood estimate of p is

$$\hat{p} = \frac{1260}{1680} = 0\cdot 75.$$

Hence an unbiased estimate of $p(1-p)$ is

$$\frac{1680\,\hat{p}\,(1-\hat{p})}{1679} = \frac{315}{1679} = 0\cdot 1876.$$

Using this unbiased estimate of $p(1-p)$, we have

$$\text{estimate var}(\hat{p}_1 - \hat{p}_2) = \frac{1680 \times 0\cdot 1876}{1142 \times 538} = \frac{315\cdot 1680}{614396} = 0\cdot 0005130.$$

Therefore s.e.$(\hat{p}_1 - \hat{p}_2) = 0\cdot 02265$, which is only slightly different from the first value. The test for the null hypothesis now follows as previously, and we have

$$z = \frac{0\cdot 1080}{0\cdot 02265} = 4\cdot 77.$$

The null hypothesis is, as before, confidently rejected.

The second method of calculating s.e.$(\hat{p}_1 - \hat{p}_2)$ is better for testing the significance of the null hypothesis $H(p_1 - p_2 = 0)$, whereas the first method is appropriate for evaluating a "confidence interval" (to be considered later) for the difference $p_1 - p_2 \neq 0$.

CHAPTER 13

PRINCIPLES AND METHODS OF SAMPLING

13.1   The meaning of sampling

Estimation and tests of significance are the twin pillars of statistical inference and, as we have seen, the application of these ideas necessarily involves the availability of samples from the populations about which inferences are to be drawn. In an intuitive sense, the samples are supposed to be small sections of the populations from which they are drawn, and in using the samples it is implicitly assumed that they are microcosmic representations of their respective populations. Such an obvious use of sampling is reasonable and also in conformity with common experience. Thus, this principle is involved in the testing for the quality of wine, cheese, tea, and a host of other foodstuffs. Furthermore, and on a less personal level, it is usual to regard a handful of grain from a sack as representative of the bulk, and to assess the qualities of a length of cloth by a piece cut off a roll of the material. In such cases, little attention is given to the actual process of sampling because it is assumed that the product sampled in this way is substantially homogeneous in quality so that any small portion from it reflects the characteristics of the whole in a sufficient degree to provide the basis for rational judgement and discrimination. However, when we deal with characteristics which are known to vary appreciably in the product in a random manner, then our intuitive notions of sampling are not sufficient, and we have to devise carefully methods of selection which will ensure the representative character of the samples so obtained. Generally speaking, it is this aspect of sampling which lies at the root of the concept of random samples of observations, which are the raw data of statistics; and in this chapter we shall study some of the basic ideas of the statistical procedures of sampling.

Sampling is a big branch of statistics, and its methodology is concerned with the practical details and the theoretical consequences of the general and more specialised techniques for obtaining samples from different types of populations. The problem of obtaining observations from any suitably defined population is a practical one and so, necessarily, the methodology of sampling must be concerned with practical techniques. However, these techniques can only be regarded as appropriate if their theoretical consequences ensure the representative character of the samples drawn by using them. Briefly, this implies that the techniques used must be such as to permit a valid application of the laws of probability to the observations. Indeed, this is the fundamental justification of any process of sampling.

13.2 The concept of the population sampled

Formally, and in its simplest sense, the process of sampling may be defined as a procedure for selecting a number of individuals (constituting a sample) from a population in such a way that any particular property of the sample will correspond as closely as possible to the true value of the property in the population. At the outset, this definition implies the existence of a specific population and that the sampler knows its limits. For example, this is the case when one speaks of a population of adults in a country, or of milch cattle or lambs on a farm at any instant of time. Such populations exist in a definite physical sense. However, not all populations are determinate in this way. Thus, one may properly speak of a population of screws or light bulbs produced by a factory under similar conditions. In this case, the totality of past, present, and future production constitutes a population which is only conceptually existent. Nevertheless, such a population is delimited because it consists of items produced under similar conditions.

An individual such as an adult, a lamb or a screw has, in general, many characteristics which may be expressed quantitatively or, more simply, differentiated qualitatively. For example, the height of adults, the birth-weight of lambs, and the length of screws are all characteristics which are expressible in quantitative terms. On the other hand, attributes such as the sex or marital status of adults, and the breed of lambs are qualitative characteristics. These usually provide measurements in the form of frequencies in qualitatively distinct classes. Now, a statistical population is not concerned with the individuals *per se* whether these be human beings, lambs or screws, but with some one of their many characteristics which can be either measured or differentiated qualitatively. Accordingly, a statistical population is a conceptual totality of measurements — the heights of adults in a country, the birth-weights of lambs on a farm, the head diameters of screws in a batch, etc. We can extend this concept of measurements to qualitative attributes if we agree to regard the frequencies in the distinct classes as measurements. Evidently, there is a one to one correspondence between the individuals constituting a physical population and the measurements which they provide; and we may, for example, speak rather inexactly about a population of screws when correctly we imply a population of the measurements of lengths of screws assessed to a specified degree of accuracy. This is the simplest visualisation of a statistical population which provides the material for the application of sampling techniques.

In the above specification of a statistical population, we have assumed that the physical population consists of distinct individuals such as adults, lambs or screws, that is, the units are discrete entities. However, there also exist physical populations which do not possess this discreteness. For example, a physical population may be a bulk supply of a chemical solution or a mass of various sized lumps of coal in a wagon; and the measurements of interest may be the percentage purity of the chemical solution or the ash content of coal. It is not immediately obvious how such material may, if at all, be conceived of in terms of discrete elements for determining the measurements of chemical purity or ash content. Nevertheless, it is clear that our simpler picture of discrete units as the physical constituents of a population is not adequate. Accordingly, we have to think more generally in terms of the nature of the material sampled. This material may be discrete like light bulbs or lambs, but it can also be continuous like milk or a chemical solution, or something in between like a sack of wheat or a wagon of coal; and any sampling technique should take account of these differences. Furthermore, the measurements of the characteristic of interest can also exhibit discrete or continuous variation. Hence we have a four-fold classification as follows:

(i) The material sampled is continuous and so also is the variation of the measurable characteristic. This is the case in determining the percentage purity of a chemical solution or the fat content of milk.

(ii) The material sampled is continuous, but the variation of the measurable characteristic is discrete. As an example, we may consider the material to be power supplied by a generating station, and the characteristic of interest may be the number of breakdowns in supply during a week.

(iii) The material sampled is discrete, but the variation of the measurable characteristic is continuous. This is the case in the measurement of lengths of screws or the life-times of light bulbs.

(iv) The material sampled is discrete and so also is the variation of the measurable characteristic. Simple examples of this are the number of children in families and the number of grains in ears of wheat.

A complete specification of the method of sampling will depend upon the nature of the material sampled and on the kind of measurements made on it. Occasionally practical difficulties can arise, and these may have to be sorted out on an *ad hoc* basis. For example, there is no difficulty in defining the units to be sampled if we are dealing with discrete entities such as screws, light bulbs, and adults; but in the sampling of farms in a region to determine the cost of milk production, it may sometimes be difficult to define physically what constitutes a farm. Similarly, when the measurable characteristic has continuous variation (height of adults or life-time of light bulbs), a decision is required regarding the accuracy to which the measurements are to be made.

13.3   Size of the population sampled

The physical size of a population is a very important factor affecting the method of sampling. It is usually impossible or practically not worth while, to enumerate completely a very large population, so that sampling is essential. This is frequently the case in quality control inspection procedures in the manufacture of small mass-produced items such as screws, rivets, washers, ball-bearings. etc. In contrast, it is possible and, at times, even essential to enumerate completely a finite population whatever the cost involved. This applies to the inspection of large units such as aeroplanes, cars, tractors, etc.

Apart from this direct consideration of the size of the population, an even more important factor is the spatial spread of large populations. This arises particularly in areal sampling of farms for crop yield estimation, of forests for the estimation of timber content, and in socio-economic and attitudinal surveys of human populations which are naturally clustered in groups such as houses, towns, and districts. We shall not consider here the detailed methods for sampling such spatially dispersed and heterogeneous populations. We have mentioned these problems only to highlight the importance of the size and physical dispersion of a population in devising a suitable method of sampling. However, from a theoretical point of view, the analysis of samples obtained from very large or infinite populations is simpler than that of samples from finite populations. As a classic example, we have seen that the hypergeometric distribution arises by sampling from a finite population, and it leads to the much simpler binomial distribution if the size of the population may be regarded as effectively infinite.

13.4   Reasons for sampling

We now consider some of the important reasons which show the desirability and occasionally the inevitability of sampling as the only means for obtaining information about some characteristic of the population.

(i) The size of the population is an obvious first consideration. When the population is conceptually infinite in size sampling is essential. This is true of the bulk of small items such as screws, rivets, ball-bearings, etc., which are mass-produced by automatic industrial processes. Moreover, even if the population is of finite size, its spatial spread may be so wide that the cost of complete enumeration could become prohibitively large. This applies to all large-scale areal surveys in agriculture for the estimation of crop yields or acreage under crops, and even more so to forestry surveys for determining the yield of timber. Human populations are also widely dispersed, and so sampling is usually unavoidable in attitudinal surveys, in consumer preference research, and in socio-economic surveys to determine the changes in populations over time.

(ii) When the estimation of a characteristic in a population requires destructive testing, sampling must be resorted to whatever the size of the population. This is true of many quality-testing procedures in industry such as the measurement of the life-time of light bulbs, the tensile strength of steel cables, the breaking strength of briquettes, etc. The consumer market guidance provided by the magazine *Which?* is also necessarily based on the sampling of products.

(iii) In certain situations, a complete enumeration of the population is possible, but some aspects of the analysis of the data are only valuable if the results are obtained quickly. In such cases, sampling is carried out after the enumeration of the population. This usually happens in detailed demographic analysis based on samples of a complete quinquennial census of a country's population. Results giving changes in the population are valuable for planning and administrative decisions if they are obtained quickly, even if the answers are not "exact". This also applies to consumer preference surveys for products which are subject to the mercurial dictates of fashion. The limitation of time, and not necessarily cost, is the essential reason for sampling.

(iv) Time also enters in a different way to justify sampling for testing products which are to be used in the future. This is true of drugs meant for human use and of all sorts of agricultural chemicals and fertilizers. Chemically reinforced diets for animal feeding also come within this category. The value of results in sample tests of such products lies wholly in their prognostic value, since no physician or farmer is interested in the actual results obtained by the production research laboratories. This information is valuable in the belief that comparable results will be obtained under similar conditions in the future.

(v) Sampling also assists in an important way the propagation of a new idea amongst a conservative or suspicious community. This is generally true of almost all agricultural innovations which have to face initially the hostility of ignorant farmers. For example, suppose a new chemical fertilizer has been produced which gave decidedly improved potato harvests when tried out on the manufacturer's experimental stations. In introducing this fertilizer to farmers, it must be shown to increase the yield of their fields, and thus to be in conformity with their needs. Such trials on farmers' fields can only be carried out on a sampling basis to gather evidence which will eventually break down the resistance to a new idea. This is a particularly important use of sampling in under-developed countries.

(vi) The physical unavailability of experimental material for trying out a new drug also necessitates sampling. This is usually true of medicines meant for human use because of the risks to patients by the use of products whose effects on persons are essentially unknown despite earlier experimentation with animals. The recent thalidomide tragedy and the suspicions of thrombosis surrounding the use of the "birth pill" may, in part, be due to inadequate or insufficiently well-planned experimentation.

13.5 Bias and precision of sampling

Whatever the nature of the material sampled and the size of the population, the basic principle of a "good" sampling procedure is that the population be well-mixed prior to the drawing of the sample. In general, this ensures that the sample has the desired correspondence with the population in the sense that the sample will be a microcosmic representation of the population and so reflect its main characteristics. Intuitively speaking, this condition for a "good" sampling procedure means that the sampler should not exercise a deliberate choice in the selection of the sample. For example, it would be wrong to pick a few very large and some very small apples from a tree if we wanted a sample to give a fair representation of the size of the fruit on the tree. In the same way, a quantity of milk from the top of a churn will not be representative of the fat content of the milk as a whole, since we know that cream tends to come to the top of the milk. Similarly, sampling a heap of coal by taking a shovelful from the sides will not give a fair sample of the bulk of coal, since large lumps of coal are likely to roll on to the sides of the heap. These three sampling procedures are, in fact, examples of *biased selection*, since the samples have been selected by taking portions of the bulk which are not representative of the whole. And one of the central ideas in sampling is to avoid any deliberate choice which could lead to such bias.

In general, any sampling method may give rise to two sorts of error, namely, *bias* and *sampling error*. The procedure of selection is said to be biased if, on repetition, the measurements made on the selected material provide an answer which is either always larger or always smaller than the true value. For example, the measurement of the fat content of milk will always be too large if the sample of milk is taken from the top of the churn without thoroughly mixing up its contents. In the same way, the measurement will always be too small if the top is drained away and an assessment of the fat content is made from a sample of the milk left at the bottom of the churn. These are obvious examples of biased samples and they are intuitively objectionable, but we shall see that similar biases enter selection procedures which apparently seem quite proper. With a biased procedure the error of measurement made on the average is a *systematic error*. Thus, if repeated assessments of the fat content of milk are made by taking samples from the top of churns, then the average measurement of fat content will be greater than its true value. As another example, if there is an undetected flaw in a chemical balance, then the measurements made by using it will have a systematic error.

Unfortunately, it is not a simple matter to rectify a bias in a selection procedure, partly because the sampler is usually unaware of its existence and partly because the bias, if any, is masked by the sampling error which is inevitable, however perfect the sampling procedure. To illustrate this point, it is convenient to continue with the milk example. Suppose, then, that repeated samples of the milk are taken and the fat content assessed to a certain degree of numerical accuracy. If n assessments $x_1, x_2, ..., x_n$ are made in this way, then it is obvious that, in general, these will be unequal. However, it is not definite whether the differences between the x_i are only due to chance, or that an improper mixture of the milk or an erratic method of assessing the fat content has given results which may be occasionally too large or too small. If there is no bias in the measurements, then each x_i can be regarded as a random measurement of μ, the true fat content of the milk. In other words, the x_i are independent observations from a conceptual population such that

$$E(x_i) = \mu \quad \text{and} \quad \text{var}(x_i) = \sigma^2, \quad \text{say, for } i = 1, 2, ..., n.$$

The population variance σ^2 is a statistical measure of the sampling error inherent in the selection procedure. With this specification of the sample, we know that the least-squares estimate of μ is $\overline{x}$, the sample average, and that

$$E(\overline{x}) = \mu \quad \text{and} \quad \text{var}(\overline{x}) = \sigma^2/n.$$

Now, the smaller the variance of $\overline{x}$, the less is the uncertainty in $\overline{x}$ as an estimate of μ. Indeed, for any fixed σ^2, we could reduce $\text{var}(\overline{x})$ to as small a quantity as we please by making n sufficiently large. Thus in addition to absence of bias, a further requirement of a "good" sampling procedure is usually that it should minimise $\text{var}(\overline{x})$ for a given value of n. Of course, if we took an inefficient estimate of μ, then for the same set of observations we would have an estimate with a variance larger than that of $\overline{x}$.

We therefore conclude that the effect of sampling error in the estimation of a parameter depends upon the size of the sample (n), the sample design, the intrinsic variability of the material sampled (measured by σ^2), and the way in which the results are calculated (estimation of μ by an efficient statistic $\overline{x}$). It is to be noted that an estimation procedure could also introduce a bias in the estimate of a parameter, and this bias is distinct from that of the sampling procedure. For example, we could estimate μ by a biased statistic but calculated from a set of measurements determined by an unbiased selection procedure.

The *precision* of a sampling method is directly related to the control of sampling error. In other words, the precision of alternative sampling methods for obtaining observations to estimate a parameter μ (say) depends upon the intrinsic variability of the observations made by the methods. For example, suppose the observations $x_1, x_2, ..., x_n$ defined above are obtained by sampling method A; and let $y_1, y_2, ..., y_n$ be another set of random observations obtained by method B such that

$$E(y_i) = \mu \quad \text{and} \quad \text{var}(y_i) = \sigma_0^2, \quad \text{for} \quad i = 1, 2, ..., n.$$

Then the corresponding sample means $\overline{x}$ and $\overline{y}$ are unbiased estimates of μ. For $\text{var}(\overline{x}) < \text{var}(\overline{y})$, that is for $\sigma < \sigma_0$, method A is more precise than method B. Also, as compared with $\overline{y}$, the estimate $\overline{x}$ is the more precise one. The essential point is that the difference in the precision of $\overline{x}$ and $\overline{y}$ as estimates of μ is due to the difference in the intrinsic variability of the observations on which the estimates are based. However, this condition alone is not sufficient to conclude that method A is the "better" of the two unless the costs of sampling per observation are the same for the two methods. Indeed, if these costs are different, method B might be the more economical, though less precise, method to use for the estimation of μ. As a simple illustration, suppose it costs twice as much to obtain an observation by method A as by method B. Then, for the same total cost, we can obtain n observations by method A and $2n$ observations by method B. Hence the variances of the corresponding sample means $\overline{x}_A$ and $\overline{y}_B$, say, are σ^2/n and $\sigma_0^2/2n$ respectively. If now, for example, $\sigma_0^2 = \frac{3}{2}\sigma^2$, then it follows that

$$\text{var}(\overline{x}_A) = \frac{4}{3}\text{var}(\overline{y}_B).$$

Thus, for the same total cost, method B gives an estimate of μ with a variance smaller than that of the estimate obtained by method A, despite the fact that the intrinsic variability of the observations obtained by method B

is greater than that of those obtained by method A. It also follows that, if only $3n/2$ observations are obtained by using method B, then the average based on them would have the same variance as $\bar{x}_A$. Consequently method B is the more economical sampling method for the estimation of μ. Considerations of this kind are necessary for a proper choice of a sampling method.

Whatever the detailed specifications of sampling, the *randomness* of the sample is the essential principle which ensures the absence of bias in the selection procedure, and the independence of the observations so that the multiplicative rule of probability is applicable. So far we have used this idea of randomness in a more or less intuitive way despite our earlier specification of a random sample in terms of the independence of the observations. In ordinary usage, the word randomness is a synonym for haphazardness in the sense of an absence of form and order; and in a way this is true of random observations. Nonetheless, as we shall see, randomness for us is a definite statistical concept, and a set of random observations is obtained by a clearly defined procedure. Apparently this a paradoxical statement, but the essential point is that it is not easy to ensure that observations obtained anyhow do not exhibit some underlying order. Like the method in Hamlet's madness, randomness also has to be ensured by careful design!

Before we consider the method of random selection, it is worth while to illustrate how sampling bias may be introduced in selection procedures which apparently seem to be quite "random". We quote some famous cases.

13.6   Some examples of sampling bias

(i) The commonest source of bias in sampling is due to the personal choice of the observer. The following data pertain to an experiment conducted at Rothamsted. About 1,200 flints of varying sizes were spread on a table and twelve observers were instructed to choose three samples of twenty stones each which would represent according to their visual judgement as nearly as possible the size distribution of the collection. The table below gives the mean weights (in oz) of each sample of twenty stones chosen by the twelve observers.

Table 13.1   Showing the mean weights (in oz) of each sample of 20 stones chosen by 12 observers

Observer / Sample	1	2	3	4	5	6	7	8	9	10	11	12
1	1·9	2·4	2·4	1·9	2·2	2·8	2·4	1·6	2·2	2·6	2·4	2·4
2	1·8	3·0	2·4	2·0	2·7	2·6	2·6	2·0	2·2	2·2	2·4	3·0
3	1·7	2·4	2·1	2·0	3·1	2·8	2·5	2·0	2·2	3·1	1·8	2·4
Mean	1·8	2·6	2·3	2·0	2·7	2·7	2·5	1·9	2·2	2·6	2·2	2·6

Mean of all samples: 2·34 oz.   True mean: 1·91 oz.
Source: F.Yates (1936), *Transactions of the Manchester Statistical Society*, p.41.

As pointed out by Yates, the data reveal a consistent tendency, which is common to most observers, to choose flints which were, on the average, larger than those of the whole collection. Of the twelve observers ten selected samples whose average weight was greater than the true mean weight of 1·91 oz and the mean weight of the 36 samples was 2·34 oz, well above the true mean. This tendency is shown consistently by the samples. Thus, of the 30 samples chosen by ten observers, all except two had average weights greater than 1·91 oz, while all three samples of the first observer were less in average weight than the true mean of the collection.

In this example, there was a deliberate attempt on the part of the observers to choose samples "representative" of the whole collection, and they generally tended to over-estimate the size of the collection. If the choice had been truly random we should have had approximately half the sample means greater than the true mean of 1·91 oz and half less than it. This tendency to over-estimate is typical of visual selection, and it reveals one difficulty in obtaining a section "representative" of the whole. The above results of Yates are also in agreement with those obtained by P.V. Sukhatme in Indian studies of "eye estimation" of crop yield forecasts by officials for assessment of revenue payable by farmers on the basis of prospective harvest returns. However, in this case there might have been a genuine propensity in revenue assessment officers to over-estimate the yield to enhance the tax returns.

(ii) Another serious and unfortunately common source of bias in sampling is due to a failure to adhere strictly to the chosen sample. The following example shows this clearly. A sample of households was taken in Syracuse, U.S.A. in 1930 and 1931 for a morbidity study. Table 13.2 gives the distribution of households according to size in the sample and in the census tracts. Households of one were not included in the survey.

Table 13.2   Showing the number and percentage of various sized households in the sample and the census tracts of Syracuse, U.S.A.

Number in household	Original sample		Census tracts	
	Number	Percentage	Number	Percentage
2	254	19·4	1762	26·8
3	338	25·9	1745	26·5
4	307	23·5	1438	21·9
5	201	15·4	853	13·0
6	106	8·1	388	5·9
7	46	3·5	208	3·2
8	25	1·9	96	1·5
9 and over	29	2·2	86	1·3
Total	1306	99·9	6576	100·1

Source: C.N.Kiser (1934), *Journal of the American Statistical Association*, Vol.29, p.250.

It is seen from the above table that the sample contains a greater proportion of large households than are found in the population. Thus, there is a serious 7·4 per cent under-representation of households of two, whereas there is a consistent over-representation of households having four or more persons in the sample. It is therefore clear that the sample does not correctly represent the household size distribution in the population of Syracuse. One reason for this bias is that small households consist of childless couples who go out to work and so are less likely to be contacted when the investigator visits their residences. On the other hand, large households usually have children and so there is greater possibility of some one being contacted by the investigator. Instead of revisiting households where the first visit produced no contact, the investigator proceeded to sample the next household and so unwittingly introduced a serious bias which made a morbidity study on the data suspect. The point is that incidence of disease is, at least to some extent, related to the living conditions in the home and hence, on the whole, large households are expected to be at a disadvantage. Accordingly, the actual sample has a bias in favour of a higher incidence of disease.

Biases of a similar kind occur quite frequently in questionnaire returns because of the failure of some of the persons included in the sample to respond. Accordingly, the returns represent the views of those who had some reason to respond, and there is therefore an under-representation in the views of those who did not have any motivation for response. It is a difficult but essential requirement in questionnaire surveys that, as far as possible, efforts should be made to get a full or nearly full response from all persons originally included in the sample. An unsatisfactory solution of this problem vitiates considerably the results of many surveys based on questionnaires. Newspapers are also frequent offenders in using results of partial response from their readers as if this were correctly representative of the entire population. The readers of any newspaper are a select group and do not represent a true cross-section of the whole population. This is the first element of the bias. Next, the non-response from a section of the readers adds the second element of the bias which is of the kind common to all questionnaire surveys.

(iii) Another somewhat less obvious way in which a sampling procedure.may be biased occurs when the selection of the sample is based on some characteristic associated with the properties of the individual which are of interest. A bias of this kind is usually found in *ad hoc* procedures for sampling human populations, and we illustrate this by an historic example.

Public opinion polls which forecast general election results are now usually based on sound sampling principles when they are presented by professional forecasting organisations. However, in the 1948 U.S. presidential election a forecast based on telephone calls indicated a Republican victory, whereas the Democratic candidate Truman won the election. The bias in the sampling procedure was presumably due to the fact that owners of private telephones were not a representative section of the voting population. Generally speaking, poor people, who tend to vote Democratic, do not have telephones and it was their under-representation in the sample which showed a Republican majority in the forecast.

Assessment of public opinion on the basis of letters to the editors of newspapers is also liable to bias. People who read newspapers and hold sufficiently strong views to express them in writing are not likely to be representative of the population. For example, the ubiquitous floating voter, who is supposed to decide the outcome of elections, is hardly the person to commit his indefinite views to writing for publication.

Many further interesting examples of sampling bias can be found in the hilariously written but eminently sensible book —

How to Lie with Statistics by Darrell Huff

(Gollancz, London: 1954).

Despite its title, this book is not a perjurer's vade-mecum, and it is worth reading to understand the diversity of ways in which interested organisations abuse statistical methods to mislead the public. And in this context it is appropriate to quote what a farsighted writer like H.G. Wells said: "Statistical thinking will one day be as necessary for efficient citizenship as the ability to read and write."

13.7   Methods of sampling

The simplest methods of sampling pertain to populations consisting of discrete elements, the measurements made on them being either of a continuous or of a discrete random variable; and we shall consider these methods first. It is immaterial for our present purpose whether the size of the population is finite or infinite. From the statistical point of view, there are two main methods of drawing a sample, namely,

(i) The method of random sampling; and

(ii) The method of stratified sampling.

13.8   The method of random sampling

The fundamental principle of this method of sampling is that the sample is so chosen as to ensure that every individual element in the population has an *equal chance* of appearing in the sample. The essential idea is that of "equal chance", and it is important to understand what this condition means in relation to the physical context of the sampling procedure. This is illustrated by the following two examples.

Example 13.1: Sampling in a binomial population

Suppose an experiment consists of tossing an unbiased penny four times and observing the number of heads obtained. The characteristic of interest is a discrete random variable w which can take any one of the five possible values $0, 1, 2, 3, 4$. Now, since the penny is unbiased, it follows that w has a binomial distribution with the probabilities

$$P(w=0) = P(w=4) = 1/16; \quad P(w=1) = P(w=3) = 1/4; \quad P(w=2) = 3/8.$$

If this experiment is performed a very large number (N) of times, then the expected frequencies of the five possible outcomes $w = 0, 1, 2, 3, 4$ are $N/16$, $N/4$, $3N/8$, $N/4$, $N/16$ respectively. Of course, in any finite number of experiments these expectations will not be realised, but since N is very large we assume for simplicity that these expectations are attained approximately.

Next, imagine that each of these observed results is written on identical chips and all the N chips are placed in a bag. Then random sampling implies that in a finite selection of n chips from the bag, each chip has an equal chance of being included in the sample irrespective of the particular value of w which is noted on it. In other words, there is no subjective element in the choice of the chips. It is important to note that the equal chance does not refer to the five possible value of w for, in fact, this chance is determined by the probabilities stated above, which are obtained from the binomial distribution.

Furthermore, if r_0, r_1, r_2, r_3, r_4 are the frequencies of the chips in the sample of n corresponding to the five possible values of w, then the r_i are random variables with the joint probability distribution determined from the multinomial distribution as

$$\frac{n!}{\prod\limits_{i=0}^{4} r_i!} (1/16)^{r_0+r_4}(1/4)^{r_1+r_3}(3/8)^{r_2}, \quad \text{for } 0 \leqslant r_i \leqslant n; \quad \sum_{i=0}^{4} r_i \equiv n.$$

Obviously the r_i are not independent random variables since their sum is a constant n; but the random selection of the chips ensures that

$$E(r_i) = n P(w = i), \quad \text{for } 0 \leqslant w \leqslant 4.$$

Thus the random sample of n chips has given rise to a sampling distribution of five dependent random variables. Of course, this joint distribution is not the only one that could be deduced from the values of w found noted on the n chips randomly selected from the bag. In any case, the process of random sampling refers only to the selection of the chips; and, in view of the effectively infinite magnitude of N, this selection is based on n independent choices of the chips such that for each choice all the chips in the bag have an equal chance of being included in the sample.

Example 13.2: Sampling in a normal population

As a second example of random selection, suppose x is a normal variable with mean μ and variance σ^2. There are now an infinite number of possible values of x, and the probability that an observation of x has a value which lies in the interval $(x, x+dx)$ is

$$\frac{1}{\sigma\sqrt{2\pi}} e^{-(x-\mu)^2/2\sigma^2}dx.$$

Hence, of N random observations of x, the expected number in the interval $(x, x+dx)$ is

$$\frac{N}{\sigma\sqrt{2\pi}} e^{-(x-\mu)^2/2\sigma^2}dx.$$

Random sampling from such a population does not imply that, in the selection of an observation, all possible values of x are equally likely but that all elements of the physical population are equally likely irrespective of the magnitude of the observations made on them. This means that all values of x have a probability of being selected according to the specified normal distribution.

Extending this argument to a sample of n observations, we note that

under random sampling all elements of the physical population are selected
on an equal-chance basis. Now, each element provides an observation, and
so all the sample observations are selected in such a way as to ensure that
each has an equal chance of being included in the sample. Besides, there
is an additional restriction which stipulates that all the sample observations
are statistically independent. This implies that the joint probability of the
sample observations is the product of their individual probabilities, which
is in agreement with the multiplicative rule for the probabilities of independent events. This is the case of *simple random sampling* or, briefly, *simple
sampling*. It is generally assumed that in practice random samples are, in
fact, obtained by simple sampling, though this may not be strictly true. However, this is not to be confused with other methods of sampling in which non-
randomness is deliberately introduced in a planned way.

As an example of simple sampling, if $x_1, x_2, ..., x_n$ are random observations from the above normal population, then the joint probability that the
x_i lie in the intervals $(x_i, x_i + dx_i)$ is

$$\frac{1}{(\sigma\sqrt{2\pi})^n} \exp - \sum_{i=1}^{n} (x_i - \mu)^2 / 2\sigma^2 \times \prod_{i=1}^{n} dx_i.$$

Thus the random sample of n observations has given a joint probability distribution of n identically distributed and independent random variables.
This is the fundamental implication of simple sampling from a population of
a continuous random variable. In this sense, random selection of observations is equivalent to the repetition of an experiment under similar conditions: both ensure the independence of the measurements and, therefore, the
application of the multiplicative rule of probabilities.

13.9   Practical techniques of random sampling

From the practical point of view, almost all populations are of finite
size; and in sampling from such a population we seek a method for selecting from it a given number of individual elements which would provide the
appropriate measurements, or a frequency distribution into a certain number
of qualitatively distinct classes. As an example of a metrical characteristic, we may consider the selection of screws from a batch, the measurements of interest being the head diameters of the screws sampled. As an
example of a frequency distribution, we may wish to determine the proportions of single, married, and widowed women in a country. Accordingly, the
sampling problem is to obtain a given number of women from the population,
each of whom is then classified into one of the three qualitative categories
according to marital status. We thus observe that in both examples the
sampling problem is the same — the selection of n (given) elements from a
physical population which is finite for sampling purposes. For such populations, there are two commonly used techniques for random sampling, namely,

   (i) Lottery or ticket sampling; and

   (ii) Random sampling numbers.

(i) *Lottery or ticket sampling*

This technique of random sampling is simple and also intuitively accept-able in its avoidance of any obvious element of subjective bias in selection. Briefly, the technique consists of constructing a model of the population and sampling from the model. Each element of the physical population is identi-fied by some distinctive characteristic or label which is noted on a card. The pack of cards corresponding to the population is well-shuffled and then a number of cards equal to the size of the sample is drawn from the pack. The drawings from the pack may be done systematically (such as drawing every fourth or every tenth card in the pack) provided the pack is initially well-shuffled. Finally, the elements of the physical population which cor-respond to the cards sampled provide the required measurements of the char-acteristic of interest. This is the method of sampling adopted in lotteries, but unless care is exercised in shuffling the pack, the technique is liable to give biased results. Thus, it is known that the usual methods of shuffling a pack of cards in bridge are hardly ever adequate to give a genuinely random distribution of hands for the players.

Despite the risk of bias, this technique of sampling is satisfactory if a small sample is required from a simply enumerated population such as that of children in a school or of houses in a district. When relatively large samples are to be drawn, it is best to use *random sampling numbers* (briefly called random numbers).

(ii) *Random sampling numbers*

The use of random numbers is the standard and safest way of obtaining a random sample from a population. A model of the physical population is first constructed by attaching a number to each element of the population, the numbering being most simply from one onwards. The set of numbers so obtained is the model population and is, in principle, similar to the card population used in lottery sampling. However, the problem of drawing a random sample now reduces to finding a series of random numbers con-veniently obtained from any one of the several standard published tables of random numbers. Many textbooks of statistics now include short tables of random numbers, but it is not always certain that such numbers are genuine-ly random. It is therefore appropriate to use some standard set of tables which have been tested for their randomness. There are three such sets of tables:

(a) Tippett's numbers comprise 41,600 digits taken from census reports combined into fours to make 10,400 four-figure numbers. These numbers are published as Tract for Computers No.15 (Cambridge University Press).

(b) Kendall and Babington Smith's numbers comprise 100,000 digits grouped in twos and fours and in one hundred separate thousands. These numbers were obtained from a machine specially constructed for the purpose, and are published as Tract for Computers No.24 (Cambridge University Press).

(c) Fisher and Yates' numbers comprise 15,000 digits arranged in twos. These numbers were obtained from the fifteenth to nineteenth digits in

A.J. Thompson's tables of logarithms and were subsequently adjusted for an excess of sixes. The numbers are published in Fisher and Yates' *Statistical Tables for Biological, Agricultural and Medical Research* (Oliver and Boyd, Edinburgh).

13.10 The nature of random numbers

Table 13.3   Showing a page from a standard table of random numbers

	1 − 4	5 − 8	9 − 12	13 − 16	17 − 20	21 − 24	25 − 28	29 − 32	33 − 36	37 − 40
1	6886	9122	1975	2241	3962	8931	9711	0242	2535	6351
2	6733	0729	0225	3085	9626	5309	4133	8981	3982	9492
3	5339	7924	6468	9064	5354	2184	0376	0898	5527	8079
4	2722	0835	8191	2864	7433	4786	1400	7168	1602	5902
5	2764	8800	8006	9034	8640	1482	7086	4007	9605	9612
6	6603	2967	0228	0291	6379	5890	9796	5542	6771	0384
7	5032	7568	7073	8815	0679	0620	6268	5126	0477	4728
8	6892	3064	3714	0742	8961	2231	1779	0260	2902	4431
9	1178	7362	3365	5418	6376	1921	7968	1269	7790	0890
10	9468	9328	2013	3949	6810	3064	8586	0697	0653	1811
11	4659	9718	7795	0633	1966	7813	9802	1821	3025	6973
12	4404	9360	9845	1554	8084	2864	1783	5033	2775	3470
13	8910	0714	6269	9811	0884	0611	5962	4319	0348	8949
14	4465	1823	2793	4807	5146	7309	3931	3912	3870	7400
15	4938	3685	8725	1356	8635	4344	2644	6734	0665	2094
16	6501	3589	8696	2123	5752	3693	3501	2218	7426	8438
17	4146	6621	9164	9540	6799	9818	7511	9999	2048	3211
18	3538	6764	8619	0587	4503	3483	4489	8400	2147	2972
19	9701	6455	0889	4371	2674	3272	1814	0668	1921	7922
20	7675	9933	5113	5052	9909	5659	0846	4035	6255	2155
21	2409	8094	8955	2812	1941	4447	4636	4848	0381	8457
22	1611	6555	7885	5885	2410	2551	4610	0424	2266	9153
23	2078	5355	7066	7473	9635	9753	7167	0713	6799	3073
24	5315	9528	3077	4529	9394	2347	1493	3718	9238	5850
25	5805	2261	8074	8398	8424	9095	4210	5869	2435	3581

Thirty-ninth Thousand

Table 13.3 reproduces page 24 from Kendall and Babington Smith's collection of random numbers (arranged here, for convenience of printing, without the subdivision into pairs), and it shows the general characteristic form of such numbers. It is clear that the elements of these numbers are the ten digits $0, 1, 2, ..., 9$, and it seems that these digits occur in a haphazard way. This is not correct intrinsically and, indeed, any apparently haphazardly written set of numbers would hardly ever be truly random. An important criterion of a proper set of random numbers is obtained as follows.

If, starting from any digit on a table of random numbers, we count a given number N of digits in any order, column- or row-wise, then amongst the numbers so selected the digits $0, 1, 2, ..., 9$ should occur with approximately equal frequency. Furthermore, if the observed frequencies of the ten digits are $n_0, n_1, n_2, ..., n_9 \left[\sum_{i=0}^{9} n_i = N \right]$, then the differences between the n_i and the expectation $N/10$ should be attributable to sampling error. It is this uniform distribution of the ten digits which is difficult to ensure throughout any extended table of random numbers. But, on the other hand, it is this property which provides the theoretical justification for the belief that a sample chosen by using a set of random numbers does satisfy the "equal chance" condition for each element of the population sampled. Haphazardness in the conventional sense has nothing to do with random numbers.

Table 13.3 (continued)

	1 − 4	5 − 8	9 − 12	13 − 16	17 − 20	21 − 24	25 − 28	29 − 32	33 − 36	37 − 40
				Fortieth Thousand						
1	3142	3815	5136	7233	7917	0627	3449	8490	4250	9170
2	8146	5234	4131	1641	2750	4564	0095	7993	9137	1743
3	6631	6524	5665	5265	0564	6750	2426	7899	7735	1531
4	0174	3061	1074	9615	3759	2934	2720	5158	3523	1218
5	3113	9443	1321	2008	1325	1975	8030	3440	6024	7575
6	1778	2627	8551	9283	7203	0139	9059	3127	5684	1968
7	9209	8681	6889	6666	2282	9810	1941	8298	5566	2990
8	1073	3365	6022	9386	2391	6958	6514	3228	3886	5532
9	2367	0601	0372	8431	8769	6787	6959	9904	7901	2046
10	7020	6899	9262	1449	2066	5508	9908	5005	8068	9873
11	1881	3874	2325	2380	3195	3290	6870	3451	5643	7813
12	7304	0978	1895	9502	4786	5720	1773	8816	0568	4512
13	1637	7909	4340	8726	2844	2812	2229	5414	8408	8408
14	8815	9363	0698	4638	6269	8930	2461	1218	1587	9404
15	7263	7707	6541	1968	8588	3612	2856	2292	4418	2864
16	4265	7777	2619	4614	2324	4833	5416	8514	3868	1878
17	6891	7683	2513	3363	4255	0011	4744	5879	1412	7830
18	2528	0488	7987	3363	9406	1546	9181	9606	4536	5943
19	7657	7106	5125	4298	9643	4433	2746	4899	7344	3330
20	1147	7149	7753	6151	2797	7908	1041	0620	7503	9597
21	3469	1705	6681	8636	2394	2597	9233	4062	6772	2413
22	5276	1716	0729	6643	1213	8525	4445	0205	1945	3631
23	1481	8490	6758	7014	3585	4863	5941	7846	7339	0496
24	5099	0303	9699	1280	3030	8799	9812	3010	9965	2739
25	3953	8850	1935	0166	8272	8351	1051	4206	2822	8113

Reproduced from *Tables of Random Sampling Numbers* (Tracts for Computers, No.24) by M.G. Kendall and B. Babington Smith (Cambridge University Press) with the kind permission of Professor M.S. Bartlett, University College London.

Before we discuss some examples of the use of random numbers, it is worth while considering how far a set of random numbers from Table 13.3 satisfy the criterion of uniform distribution of the ten digits. The frequency distribution of the 600 random digits taken from the sixth to the twentieth row of the thirty-ninth thousand set of random numbers of Table 13.3 is given in Table 13.4.

The expected frequency in each of the ten classes is 60 on the hypothesis of uniform distribution of the digits. Accordingly, the differences between the observed and expected frequencies are generally small. We shall see later that an appropriate test of significance indicates that differences as large as the ones observed or larger could occur by chance in approximately 86 per cent of the cases. Hence the agreement of the

Table 13.4   Showing the frequency distribution of the 600 random digits taken
from the sixth to the twentieth row of Table 13.3

Digit	Tally marks	Observed frequency
0	卌 卌 卌 卌 卌 卌 卌 卌 卌 卌 卌 ////	59
1	卌 卌 卌 卌 卌 卌 卌 卌 卌 卌 卌 卌 //	62
2	卌 卌 卌 卌 卌 卌 卌 卌 卌 卌 卌 /	56
3	卌 卌 卌 卌 卌 卌 卌 卌 卌 卌 卌 卌 /	61
4	卌 卌 卌 卌 卌 卌 卌 卌 卌 卌 卌 卌	60
5	卌 卌 卌 卌 卌 卌 卌 卌 卌 卌	50
6	卌 卌 卌 卌 卌 卌 卌 卌 卌 卌 卌 卌 卌 ////	69
7	卌 卌 卌 卌 卌 卌 卌 卌 卌 卌 卌	55
8	卌 卌 卌 卌 卌 卌 卌 卌 卌 卌 卌 卌 /	61
9	卌 卌 卌 卌 卌 卌 卌 卌 卌 卌 卌 卌 卌 //	67
Total		600

observed distribution with the hypothesis of uniform distribution of the
digits is excellent.

This "frequency test" for randomness of a given set of numbers is just
one of several methods for testing randomness, but like all tests of statistical hypotheses this does not prove or disprove the randomness of the
numbers tested. It would take us too far from our immediate purposes to discuss the other criteria which are used to test the randomness of numbers.
We have mentioned the frequency test as a simple example of the internal
evidence of the numbers themselves which supports their randomness; and
it is important to remember that the randomness of any given set of numbers
cannot be decided on *a priori* grounds. The generation of a satisfactory extensive table of random numbers is an elaborate and difficult task. Accordingly, it is essential for the sampler to use proper random number tables for
sampling.

13.11   Use of random numbers in sampling

By some ingenuity, random numbers can be used for sampling in a wide
variety of problems. We illustrate this use by two examples in which random
numbers are employed in rather different ways.

Example 13.3: Geometrical probability

A needle of length greater than 2ρ is placed at random on a circle of
radius ρ so that all values between 0 and ρ for its perpendicular distance
from the centre of the circle are equally likely, and the length l of the
chord made by the needle is measured. If the perpendicular distance of the
chord from the centre of the circle is ρx, where $0 < x < 1$, we have

$$\frac{l}{2\rho} = \sqrt{1 - x^2}.$$

For each observed value of l, the fraction x is a random observation of a
uniformly distributed random variable in the interval $(0, 1)$. We use the 50
two-digit random numbers given in Table 13.5 to determine 50 random
values of the ratio $l/2\rho$. This is done simply by regarding each of the

Table 13.5   Showing fifty two-figure random numbers

04	84	29	77	30	02	87	68	37	23
39	01	12	46	79	10	05	79	77	18
05	08	59	24	97	19	90	22	55	01
44	43	48	83	26	09	17	80	27	41
14	92	50	44	06	44	11	53	49	73

given two-digit random numbers as a decimal fraction, and thus as a random value of x. For example, the first two-digit random number in Table 13.5 is 04. This is interpreted as giving an observed value of $x = 0·04$. Hence, corresponding to this value of x. we have

$$\frac{l}{2\rho} = [1 - (0·04)^2]^{\frac{1}{2}} = \sqrt{0·9984} = 0·9992.$$

The difference $1-(0·04)^2$ is evaluated on a desk calculator by subtracting from $1·0000$ (in the products register) the square of $0·04$. The square root of $0·9984$ is obtained from *Barlow's Tables*. In this way, the 50 random observations of $l/2\rho$ are deduced from the random numbers given in Table 13.5. The results are presented in Table 13.6.

Table 13.6   Showing fifty random values of the ratio $l/2\rho$

x	$l/2\rho$	x	$l/2\rho$	x	$l/2\rho$	x	$l/2\rho$	x	$l/2\rho$
0·04	0·9992	0·84	0·5426	0·29	0·9570	0·77	0·6380	0·30	0·9539
0·39	0·9208	0·01	0·9999	0·12	0·9928	0·46	0·8879	0·79	0·6131
0·05	0·9987	0·08	0·9968	0·59	0·8074	0·24	0·9708	0·97	0·2431
0·44	0·8980	0·43	0·9028	0·48	0·8773	0·83	0·5579	0·26	0·9656
0·14	0·9902	0·92	0·3919	0·50	0·8660	0·44	0·8980	0·06	0·9982
0·02	0·9998	0·87	0·4931	0·68	0·7332	0·37	0·9290	0·23	0·9732
0·10	0·9950	0·05	0·9987	0·79	0·6131	0·77	0·6380	0·18	0·9837
0·19	0·9818	0·90	0·4359	0·22	0·9755	0·55	0·8352	0·01	0·9999
0·09	0·9959	0·17	0·9854	0·80	0·6000	0·27	0·9629	0·41	0·9121
0·44	0·8980	0·11	0·9939	0·53	0·8480	0·49	0·8717	0·73	0·6834

Example 13.4: Field sampling

Table 13.7, overleaf, shows the distribution of Colorado beetles in a heavily infested potato field. The experiment was originally designed with 45 rows each of 50 feet length. The ultimate sampling unit was a two-foot length of row, so that the total number of such units was 45 (rows) × 25 (columns), or 1,125 in all. However, in the course of the experiment, plants in some of the units near the boundary and two interior patches were destroyed through carelessness, and so the beetle count was finally conducted in an irregularly shaped area shown in Table 13.7 and consisting of 964 units only.

We now show how the first 4,026 random numbers from Tract for Computers No. 24 (pages 2−4) may be used to draw a random sample (with replacement) of size 100 from the given sampling field. It is convenient to number the rows from 01 to 45, and the columns from 01 to 25. The random numbers are taken in sets of four digits according to the rows of the tables. The first two digits are identified with the number of a row of the

Table 13.7   Showing the distribution of Colorado beetles
in an infested potato field

Row \ Col	01	02	03	04	05	06	07	08	09	10	11	12	13	14	15	16	17	18	19	20	21	22	23	24	25
01	2	8	6	8	0	13	5	6	0	2	2	4	5	9	13	5	6	2	19	9	9	6	4	5	10
02	3	2	4	6	4	7	6	2	5	8	4	4	4	9	0	5	9	7	10	7	4	9	0	12	0
03	9	7	6	6	5	5	5	3	2	2	12	7	7	2	7	4	6	3	2	13	6	9	0	7	3
04	1	8	5	9	7	6	8	6	6	6	5	5	7	3	7	4	4	3	6	7	5	5	0	4	7
05	10	9	4	9	8	19	8	5	4	5	6	0	5	9	5	3	8	0	19	2	5	2	2	19	0
06	5	3	18	0	5	5	7	5	2	17	3	8	9	2	9	8	9	9	3	7	11	2	9	15	5
07	4	4	13	3	0	7	6	9	2	6	7	3	8	6	5	6	0	5	8	5	7	5	3	6	0
08	5	8	4	2	2	7	5	9	7	5	4	2	4	7	6	17	2	2	5	2	3	9	2	0	4
09	8	9	9	6	7	6	9	2	6	0	0	0	0		7	19	7	9	0	2	0	8	4	5	9
10	2	5	5	0	4	8	2	3	0	3	7	3	3		4	9	7	2	8		4	9	2	6	7
11	8	8	4	6	8	9	9	4	5	9	9	3			17	18			4	7	2	2	5	13	
12	7	20	9	0	6	9	4	6	9	0		5			4	3		8	9	4	8	0	3		
13	10	8	0	7	2	4	2	7	3			0	4		3			6	3	5	4	3	9		
14	9	5	2	7	3	0	0	16	2				2			3	7	6	9	9	3				
15	0	4	4	6	8	9	8	2	0						3	15	6	6	0	2	9				
16	8	4	5	9	6	6	9	5	9	8		7		5		8	8	6	3	15	9	6	4		
17	13	4	0	9	2	16	9	2	3	4	3	2	6	7		3	8	3	6	0	6	4	2	4	2
18	7	7	0	5	7	6	0	8	2	3	3	2	0	18	2	7	12	0	0	7	5	9	19	3	5
19	19	2	15	0	4	9	2	7	4	0	0	11	14	7	16	9	8	9	3	2	9	2	3	3	6
20	5	3	3	9	3	4	5	9	6	4	12	8	0	3	3	9	6	3	0	0	2	6	4	9	7
21	3	2	5	8	11	0	8	0	2	0	6	0	2	7	2	0	3	6	0	7	13	8	20	6	5
22	7	13	2	5	0	0	0	4	2	8	6	6	7	9	9	7	9	7	0	3	2	7	0	5	0
23	8	6	8	5	5	12	6	4	5	6	6	0	7	8	8		8	10	9	9	8	5	6	2	6
24	0	6	9	0	8	0	7	2	6	5	6	8	8		8		7	4	2	6	8	8	0	0	
25		1	2	18	5	16	16	2	2	4	9	7			3	0	3	4	2	6	3	6			
26		2	0	16	0	9	5	4	8	7	3			8	9	3	0	13	6	6	5				
27	11		3	19	8	8	0	3	3	12	0		5	6		9	8	0	6	4	3	9	6	9	
28	9		3	8	5	9	9	4	2		7			6	3	9	9	4	18	3	16	5			
29		13	3	0	0	12	6	3		0			0	0	3	9	9	10	6	0	3				
30		4	0	0	0	6	9	4			2	5	5	6	7	7	4	0							
31	11	9	13	6	0	8	9	2			9		9	0	3	8	9	9	4	5					
32	14	12	12	8	2	5	4	2			5		0	5	4	6	3	3	6	5					
33	4	2	6	8	8	8	2	17	3		4		6	0	5	5	11	9	5	0	3	6			
34	4	7	9	4	4	6	4	5	4	8	0	11	2	0	5	9	6	6	0	9	3	3	6	8	
35	0	0	5	0	5	4	2	0	8	4	5	5	6	7	11	7	2	4	7	0	2	19	9	6	7
36	6	6	15	5	8	9	18	3	8	8	5	5	4	9	7	9	5	5	6	0	4	8	6	9	7
37	8	9	9	4	5	18	2	3	9	8	3	6	3	2	9	5	9	3	6	8	7	6	9	7	2
38	2	5	8	4	0	4	6	6	4	2	3	5	2	2	5	14	2	5	0	7	6	5	2	8	4
39		0	0	9	2	2	5	2	3	7	2	19	10	9	3	3	12	3	4	3	6	9	9	5	
40		9	20	5	3	2	7	0	7	0	6	2	14	7	2	2	0	9	4	5	3	5	2	0	
41	6	6	6	9	0	5	15	2	2	6	5	6	13	4	3	7	2	7	8	7	7	2	2	8	
42	5	6	9	6	7	4	10	16		6		0	7	3	0	6	5	7	6	4	9				
43	0	18	4	3	7	20		0	4	2	19	7	4	3	4	7	6	2							
44	6	0	2	5	4		9	7	2	4	11	8	6	0											
45	8		0	3	3	4	15	16																	

sampling field, and the other two digits with a column number. It is thus clear
that we are concerned with four-digit sets in which the first two digits
lie between 01 and 45, and the second two digits between 01 and 25. As an
example, the first ten four-digit random numbers obtained from the tables

are

2315, 7548, 5901, 8372, 5993, 7624, 9708, 8695, 2303, 6744.

Thus the first four-digit number 2315 corresponds to the sampling unit in the 23rd row and the 15th column. Hence the first observation in the enumerated sample is 8. The next relevant four-digit number is 2303, which gives the sample observation 8 from the sampling unit in the 23rd row and the 3rd column. This process is continued by using the four-digit random numbers systematically until 100 sampling units are enumerated. However, if a four-digit random number (for example 2617) corresponds to a destroyed sampling unit in the field, then this number is ignored in the same way as all four-digit random numbers which do not correspond to any sampling unit (for example 7548). Finally, if a four-digit random number (for example 0925) is found more than once in the first 100 relevant four-digit random numbers, then the corresponding sample observation (9 in the case of 0925) is entered once in the sample for each occurrence. This means that sampling is by replacement, so that the effect of the finiteness of the sampling field is eliminated, since for each of the 100 sample units selected a random choice is made from the whole set of 964 units.

It is clear that the essential point in this sampling procedure is to determine systematically 100 four-digit random numbers which correspond to undestroyed units in the field.

Table 13.8   Showing the first 110 four-figure random numbers corresponding to possible units of the sampling field of Table 13.7

2315	2303	4310	3508	4410	1603	1003	2211	0517
2617	0513	1822	0709	2523	2607	0307	0924	0825
2122	1112	3712	1718	4422	3505	3317	1905	3623
1222	2522	3811	4103	*1614*	2802	1212	3804	0610
4324	3205	2319	3510	1501	1301	3225	3615	4124
3010	4105	4105	*4312*	0604	2304	2018	2222	2013
1023	2302	*4413*	2423	2611	4125	1314	1403	3724
1205	*1612*	2322	4106	2520	4020	0317	0416	2325
2304	0819	4115	0925	3913	0504	3514	0423	0720
3701	0925	3302	0801	2724	4307	0722	4320	1303
0109	0501	*3010*	0217	2110	0121	1624	4018	4023
1906	2117	0603	1223	*4306*	0612	*3115*	2919	*1511*
2510	2203							

Table 13.8 gives the first 110 four-digit random numbers corresponding to the units of the sampling field. These were obtained from the first 4,026 random digits of the tables. Of these 110 four-digit random numbers, ten (underlined) correspond to destroyed sampling units, and these have to be ignored. The remaining 100 four-digit random numbers give the sampling units included in the sample. The observations corresponding to these selected units are shown underlined in Table 13.7, and for units sampled twice, the observations are doubly underlined. The 100 sample observations are finally presented in Table 13.9.

Table **13.9**   **Showing the 100 sample observations obtained from Table 13.7**

0	9	9	5	6	4	0	10	9	5	8	18	0	17	8	2	5	5	4	5
5	5	9	9	5	2	3	6	5	4	8	10	0	4	2	0	5	6	3	9
4	9	0	3	0	3	8	2	6	7	6	8	5	5	8	9	5	6	8	2
3	2	6	0	8	6	9	3	2	5	2	0	5	0	4	7	7	6	8	6
7	4	3	19	0	4	5	6	9	9	0	4	2	8	18	7	4	6	2	11

13.12   The method of stratified sampling

In *stratified sampling*, the sample is deliberately drawn in a systematic way, so that each portion of the sample represents a corresponding section of the population. This method applies particularly when it is known or suspected that systematic variations exist between the various sections of the population. The initial division of a heterogeneous population into a number of relatively homogeneous sections known as *strata* is a standard technique for reducing the sampling error, and the method of stratified sampling is extensively used in field and socio-economic surveys.

We shall not consider here the details of stratified sampling, but restrict ourselves to two simple applications to illustrate the way in which this method differs from random sampling. Suppose we wish to obtain a sample of N persons in the country who are sixty or more years of age. The variable of interest is age of the persons. The simplest method is to take a random sample of size N from the available recorded population of persons in the specified age-group. However, we know that the age distribution of the elderly depends considerably on sex, since, on the average, women tend to live longer than men. Accordingly, bearing in mind this characteristic sex difference, we divide the population of persons aged sixty or more years into two groups or strata according to their sex. We then choose random samples of $N/2$ men and $N/2$ women from the two strata separately. The pooled sample of N persons so obtained is no longer wholly random since we have deliberately selected $N/2$ men and $N/2$ women. This is a *stratified sample* with equal numbers of persons taken from the two strata. If there is considerable difference between the age distributions of the sixty-and-over men and women, then this stratified sample would give a better control of sampling error than that obtained by an *unrestricted* or *unstratified* sample of N persons from the combined population of men and women in the sixty and over age-group.

In general, a population may be divided initially into more than two physically distinct strata, and then random sampling is restricted to each stratum separately. For example, in a sample survey to determine the cost of milk production, farms may be grouped into several strata according to their number of heads of milch cattle if, as is presumably the case, the cost of milk production depends upon the number of milch cattle on a farm.

The methods of analysing observations of stratified samples are elaborate and do not concern us here. It is enough for us to understand the principle of stratification as an important device for the reduction of sampling error in large heterogeneous or spatially dispersed populations.

13.13   Sampling from a continuous population

The use of random numbers is the best technique known at the present time for drawing samples from a population of discrete elements, and these numbers may also be used to draw samples from a continuous population specified mathematically. But, as we have mentioned, cases occur in which random numbers cannot be used. The commonest examples arise in sampling from populations of continuous materials such as chemical solutions, milk, flour, etc. In such cases we are usually obliged to fall back on procedures which have intuitive reasonableness in ensuring randomness of selection. The general principle is to devise a representative method of selection from a well-mixed population. For example, to obtain a random sample of milk from a churn, we might stir the contents thoroughly and scoop up a sample haphazardly. The physical continuity of the material should ensure a representative character of the sample taken. Occasionally, when the population is of a manageable size, we can proceed systematically by dividing it into a number of sections and selecting some of them by the standard technique of random sampling. As an example, this approach might possibly be of use in testing the chemical constitution of soil samples for some property of interest. However, no standard technique is known for sampling a population consisting of a bulk of continuous material. Experience is the best guide for developing a suitable method in any particular situation.

CHAPTER 14

LARGE-SAMPLE TESTS OF SIGNIFICANCE
IN CONTINUOUS POPULATIONS

14.1 Continuous variation and the normal distribution

We introduced in Chapter 12 the concept of a test of significance in the context of binomial sampling, and it is our purpose now to extend the application of tests of significance to continuous populations. There is nothing new in this so far as the logical principles underlying a test of significance are concerned. The novelty lies in the kind of null hypotheses which can be tested and, more particularly, in the test functions and their sampling distributions which provide the mathematical justification for the tests. At the outset, there is a characteristic difference between a binomial population and in the specification of a conceptual population of a continuous random variable. As we have seen, a binomial population can be easily visualised in terms of the possible outcomes of an experiment (tossing a penny or rolling a die) and, accordingly, a sample from such a population may be regarded as a finite set of outcomes of the experiments actually performed. In contrast, a population of a continuous random variable is usually specified in a formal mathematical manner, and the sample measurements are postulated as independent observations from the population, obtained by a practical application of the method of random sampling. Of course, in conformity with our definition, we also regard the process of obtaining an observation from the population as the performance of an experiment. Nevertheless, there is no obvious reason why the population of a particular continuous random variable should have some postulated form. This parallelism between the theoretical model and the distributional behaviour of an observable random phenomenon is established on statistical principles which we shall consider in Chapter 17.

From a purely mathematical point of view, we can specify a population of a continuous random variable by any function which satisfies the requirements for a probability density function, and the formal apparatus of tests of significance can be built up on the basis of random samples drawn from the population. However, this approach has little statistical value, since our interest lies in the mathematical formulation of populations which do represent adequately the behaviour of physically meaningful random variables. It is known from empirical considerations that the normal distribution provides a close representation of the behaviour of many continuously varying random phenomena, and so we shall be concerned largely with tests of significance based on the assumption of normality of the populations sampled. Despite this, it is important to stress that many random phenomena cannot be represented at all adequately by a normal distribution. Indeed, non-normality is a real and important aspect of the distributional behaviour of random variables arising in diverse practical situations, but we shall not be much concerned with tests of significance of null hypotheses based on sampling from non-normal continuous populations.

14.2   Tests of significance for the mean of a normal population

Suppose that the behaviour of a continuous random variable x can be represented by a normal distribution with mean μ and variance σ^2. The random variable x may denote the height of adults in a country, the birth-weight of lambs of a particular breed, the length of screws produced by a factory, etc. The physical nature of the random variable x is not of immediate consequence except in so far as it suggests on empirical grounds that the normal distribution is a reasonable representation of its chance behaviour. Our primary interest is in tests of significance of null hypotheses about the parameter μ. As we shall see, these tests depend upon the magnitude of σ, and we may classify them as tests for null hypotheses about μ when

(i) σ is known; and   (ii) σ is unknown.

The class of null hypotheses about μ are the simplest to test when σ is known, and we shall consider them first. To do this, we assume that $x_1, x_2, ..., x_n$ are random observations from the specified normal population, and the null hypotheses about μ have to be tested on the basis of this sample information.

It is known from earlier results that any linear function of the observations

$$X = \sum_{i=1}^{n} c_i x_i,$$

where the c_i are constants not all simultaneously equal to zero, is a random variable such that

$$E(X) = \mu \sum_{i=1}^{n} c_i; \quad \mathrm{var}(X) = \sigma^2 \sum_{i=1}^{n} c_i^2.$$

Furthermore, since the x_i are observations from a normal population, it is also known that X is a normally distributed random variable with the stated mean and variance. This remarkable result is a cornerstone of normal distribution theory, but we shall not prove it here since it requires mathematics beyond the level assumed for this book. However, we shall use this result frequently in the sequel. As an important corollary, if we set $c_i = 1/n$ for all i, then $X = \bar{x}$, the sample average. Thus, as stated in Chapter 11, $\bar{x}$ is normally distributed with mean μ and variance σ^2/n. In other words, the sampling distribution of the mean of n independent observations from a normal population is normal with the same mean as that of the population sampled, but with a variance which is n^{-1} times the variance of the population sampled. Accordingly, the standardised random variable

$$z = (\bar{x} - \mu)\sqrt{n}/\sigma$$

is a unit normal variable. For σ known, the probability distribution of z provides the theoretical basis for a test of significance for any null hypothesis about μ. As already explained. we could use either a two-sided or a one-sided test of significance, depending upon the alternatives to the null hypothesis under test.

(a) *Two-sided test of significance*

Suppose the null hypothesis is $H(\mu = \mu_0)$, where μ_0 is some pre-assigned quantity, and the alternative hypotheses are $\mu \neq \mu_0$. It is clear that in this case we have to use a two-sided test of significance. Now if $H(\mu = \mu_0)$ is true, then

$$z = (\bar{x} - \mu_0)\sqrt{n}/\sigma$$

is a unit normal variable. Hence the null hypothesis is rejected at the $100\eta \, (0 < \eta < 1)$ per cent level of significance if

$$|z| \geqslant \lambda,$$

where λ is the percentage point of the unit normal distribution determined by the equation

$$\frac{1}{\sqrt{2\pi}} \int_{\lambda}^{\infty} e^{-\frac{1}{2}u^2} \, du = \frac{1}{2}\eta,$$

or $P(|u| \geqslant \lambda) = \eta$ because of the symmetry of the unit normal distribution of u about the origin. In other words, the null hypothesis is rejected at the 100η per cent level of significance if

$$either \quad \bar{x} \geqslant \mu_0 + \lambda\sigma/\sqrt{n} \quad or \quad \bar{x} \leqslant \mu_0 - \lambda\sigma/\sqrt{n}.$$

Another way of looking at these two inequalities is that the null hypothesis is acceptable at the 100η per cent level of significance if the sample mean $\bar{x}$ does not deviate from the hypothetical value μ_0 by more than λ times the standard deviation $\sigma/\sqrt{n}$ on either side of μ_0.

In particular, if $\eta = 0{\cdot}05$, then the normal integral table gives $\lambda = 1{\cdot}96$. Similarly, for $\eta = 0{\cdot}01$, $\lambda = 2{\cdot}58$. These values of λ are the commonly known five and one per cent points of the unit normal distribution, and they are often roughly approximated as 2 and 3 respectively.

(b) *One-sided tests of significance*

The above argument can be adapted easily to a one-sided test of significance when the direction of the possible deviations from the null hypothesis $H(\mu = \mu_0)$ is known. Thus, if the alternatives to the null hypothesis are $\mu < \mu_0$, then we use the lower tail of the normal distribution as the region of rejection. Accordingly, the null hypothesis is rejected at the 100η per cent level of significance if

$$z \leqslant -\lambda', \quad or \quad \bar{x} \leqslant \mu_0 - \lambda'\sigma/\sqrt{n},$$

where

$$\frac{1}{\sqrt{2\pi}} \int_{\lambda'}^{\infty} e^{-\frac{1}{2}u^2} \, du = \eta.$$

Similarly, if the alternatives to the null hypothesis are $\mu > \mu_0$, then we use the upper tail of the normal distribution as the region of rejection. Hence the null hypothesis is rejected at the 100η per cent level of significance if

$$z \geqslant \lambda', \quad \text{or} \quad \bar{x} \geqslant \mu_0 + \lambda'\sigma/\sqrt{n}.$$

In particular, we have from the normal integral table that for $\eta = 0{\cdot}05$, $\lambda' = 1{\cdot}645$, and for $\eta = 0{\cdot}01$, $\lambda' = 2{\cdot}326$.

Population variance unknown

The tests of significance (a) and (b) provide valid methods for testing null hypotheses about the population mean μ only if the population variance σ^2 is known. Unfortunately, in any practical situation, this is hardly ever true and so, in practice, it is not possible to evaluate z, as defined above, to test the null hypothesis $H(\mu = \mu_0)$. However, when σ^2 is unknown, we can obtain its unbiased estimate as

$$s^2 = \sum_{i=1}^{n} (x_i - \bar{x})^2/(n - 1).$$

It now seems natural to replace σ by the sample quantity s to define another random variable

$$t = \frac{\bar{x} - \mu_0}{s/\sqrt{n}}$$

for testing the null hypothesis $H(\mu = \mu_0)$. However, there is a fundamental difference between z and t, since σ is an unknown population parameter whereas s is a function of the sample observations and so a random variable. It is therefore clear that the distribution of t cannot be normal when the null hypothesis $H(\mu = \mu_0)$ is true. Nevertheless, there is a remarkable result that if the sample size n is reasonably large, then t is approximately normally distributed and, as $n \to \infty$, the distribution of t becomes exactly normal. For finite n, the normal approximation for the distribution of t holds very well for $n \geqslant 60$. Hence, if we are dealing with large samples, then we may use the tests of significance (a) and (b) with σ replaced by s. These are the generally known large-sample tests of significance for null hypotheses about the mean of a normal population, and they are used quite frequently in statistical analysis.

Approximations of this kind are of great importance in statistical theory, and it is well to understand that when n is moderately large, the normal approximation is really good. For example, if $n = 61$ and the null hypothesis $H(\mu = \mu_0)$ is true, then

$$P(|z| \geqslant 1{\cdot}96) = 0{\cdot}05, \quad \text{whereas} \quad P(|t| \geqslant 2{\cdot}00) = 0{\cdot}05.$$

The difference between the two percentage points is small. In contrast, the difference becomes appreciable when n is small, and we shall consider later how a test of significance for the null hypothesis $H(\mu = \mu_0)$ can be carried out when n is small and σ unknown. Our present purpose is to emphasise that, if $n \geqslant 60$, we may confidently use the normal distribution for testing the null hypothesis $H(\mu = \mu_0)$ whether σ is known or not. A numerical illustration follows.

Example 14.1: Comparison of examination performance

In an annual public examination, 2,772 pupils were given an English language test, and the marks obtained by them had an average of 99·8 and a standard deviation of 17·3. It is also known that the average mark obtained by pupils in similar examinations during the past twenty years was 101. Are this year's marks significantly

(i) different from the past average performance?

(ii) lower than the average obtained in previous years?

It is known that the distribution of marks of unselected groups of pupils is usually closely normal, and so we have reasonable justification to use tests of significance based on the normality of the population sampled. Accordingly, we postulate that there exists a conceptual population of marks of pupils which may be represented by a normal distribution with mean μ and variance σ^2. The observed marks of the 2,772 pupils in this year's examination are regarded as a random sample from this population and, in the notation used above, we have

$$\bar{x} = 99·8, \quad s = 17·3, \quad \text{and} \quad n = 2,772.$$

Here the sample size is really large, and so we can conveniently ignore the fact that σ is unknown.

The past performance of pupils in similar examinations has given an average mark of 101, and the point of the questions asked is simply to test how far present achievement is consistent with past performance. We formulate this as the null hypothesis $H(\mu = 101)$. For question (i), the alternatives to this null hypothesis are $\mu \neq 101$, and we have to use a two-sided test of significance. We therefore have

$$t = \frac{99·8 - 101}{17·3/\sqrt{2772}} = -3·65,$$

which should be approximately a unit normal variable if the null hypothesis under test is true. But

$$P(|t| \geqslant 3·65) = 2[1 - \Phi(3·65)] = 0·0003.$$

Thus the observed difference is highly significant, and so we can confidently reject the null hypothesis $H(\mu = 101)$. This means that it is highly improbable that the average of 99·8 could have arisen by chance from a population whose mean was, in fact, 101. In other words, this year's average mark is almost certainly not comparable with past average achievement.

Another equivalent method of carrying out the above test of significance for the null hypothesis $H(\mu = 101)$ is as follows. If we are working at the five per cent level of significance, then the null hypothesis under test is acceptable only if

$$\left| \frac{\bar{x} - 101}{17·3/\sqrt{2772}} \right| \leqslant 1·96,$$

that is, if either $\quad \bar{x} \leqslant 101 + \dfrac{1·96 \times 17·3}{\sqrt{2772}} = 101·64,$

or $$\bar{x} \geqslant 101 - \frac{1\cdot96 \times 17\cdot3}{\sqrt{2772}} = 100\cdot36.$$

But $\bar{x} = 99\cdot8$, and so the null hypothesis $H(\mu = 101)$ is again rejected. The inequality $100\cdot36 \leqslant \bar{x} \leqslant 101\cdot64$ can be interpreted to mean that if the null hypothesis $H(\mu = 101)$ is, in fact, true then

$$P(100\cdot36 \leqslant \bar{x} \leqslant 101\cdot64) = 0\cdot95.$$

Thus at the five per cent level of significance the observed value $\bar{x} = 99\cdot8$ lies in the region of rejection.

The wording of question (ii) implies that the alternatives to the null hypothesis $H(\mu = 101)$ are $\mu < 101$. We therefore use a one-sided test of significance, and the region of rejection is the lower tail of the unit normal distribution such that, at the five per cent level of significance,

$$P(t \leqslant -1\cdot645) = 0\cdot05.$$

Hence the null hypothesis is rejected if

$$\frac{\bar{x} - 101}{17\cdot3/\sqrt{2772}} \leqslant -1\cdot645,$$

that is, if $$\bar{x} \leqslant 101 - \frac{1\cdot645 \times 17\cdot3}{\sqrt{2772}} = 100\cdot46.$$

But $\bar{x} = 99\cdot8$, and so the null hypothesis $H(\mu = 101)$ is again rejected at the five per cent level of significance. We thus conclude that it is probable that the observed sample mean was obtained by sampling from a population whose mean was less than 101. Equivalently, if the null hypothesis $H(\mu = 101)$ is true, then the observed value of t is $-3\cdot65$ and we obtain

$$P(t \leqslant -3\cdot65) = \Phi(-3\cdot65) = 0\cdot00013$$

from the normal integral table. Therefore the null hypothesis $H(\mu = 101)$ is decisively rejected.

14.3   Class of acceptable null hypotheses

We have so far considered two-sided and one-sided tests of significance for a single null hypothesis $H(\mu = 101)$ which had some special interest in terms of the data. However, it is also possible to think more generally in terms of a class of null hypotheses about the parameter μ, which may all be found acceptable on the basis of the sample information at a given level of significance. To illustrate this approach, suppose we are working at the five per cent level of significance. Now, whatever may be the true value of μ, it is clear that for large n

$$t = \frac{\bar{x} - \mu}{s/\sqrt{n}}$$

is approximately a unit normal variable. Hence, for a two-sided test of significance,

$$P(|t| \geqslant 1\cdot96) = 0\cdot05,$$

so that $$P(|t| \leqslant 1\cdot96) = 0\cdot95.$$

This latter inequality may also be written as

$$P(\bar{x} - 1\cdot96\,s/\sqrt{n} \leqslant \mu \leqslant \bar{x} + 1\cdot96\,s/\sqrt{n}) = 0\cdot95,$$

whence, on substituting the numerical values of $\bar{x}, s$, and n of our example, we have the inequality

$$P(99\cdot16 \leqslant \mu \leqslant 100\cdot44) = 0\cdot95.$$

This formal "probability" statement requires careful interpretation. We note that μ is an unknown but fixed population parameter and not a random variable. Hence the above inequality is not an ordinary probability statement about μ, since μ has no probability distribution. One interpretation of the inequality is that if we consider any null hypothesis $H(\mu = \mu_0)$ such that the postulated value μ_0 of μ lies in the interval $(99\cdot16, 100\cdot44)$ then, on the basis of the given sample information, this null hypothesis will be found acceptable on the five per cent level of significance. Accordingly, all possible values of μ in the interval $(99\cdot16, 100\cdot44)$ jointly provide a class of null hypotheses about μ which are acceptable at the five per cent level of significance. Evidently, the value $\mu = 101$ lies outside the above interval and so the null hypothesis $H(\mu = 101)$ is rejected at the five per cent level of significance. This is another way of carrying out the test of significance of a null hypothesis, and it could be used for any specified level of significance.

There is another deeper interpretation of the "probability" statement about μ, but we shall consider this more appropriately at a later stage.

14.4   Non-normal populations

It is now worth while to consider some general results concerning tests of significance for null hypotheses about the population mean μ when the sample observations $x_1, x_2, ..., x_n$ are drawn from a non-normal population with mean μ and variance σ^2. An exact derivation of a test of significance for the null hypothesis $H(\mu = \mu_0)$ depends upon the mathematical form of the population sampled. There is, however, a large-sample result which is of great generality and of fundamental practical importance.

If the non-normal population sampled has a finite variance σ^2 and n is large, then

$$z = \frac{\bar{x} - \mu}{\sigma/\sqrt{n}}$$

is approximately a unit normal variable. This is one form of a remarkable result in statistical theory, which is known generally as the *Central Limit Theorem*; and it is this theorem which, indeed, gives to the normal distribution its unique position in statistics. An extension of this result also shows that if σ is replaced by s then, again, for large samples

$$t = \frac{\bar{x} - \mu}{s/\sqrt{n}}$$

is approximately a unit normal variable. Hence the tests of significance used in the case of sampling from a normal population can also be used approximately when the population sampled is non-normal but with a finite variance.

It is not easy to indicate the minimum value of n for which these approximations are reasonably appropriate because their validity depends partly upon the magnitude of the skewness and kurtosis of the non-normal population sampled. Accordingly, in a practical sense, these approximations are of little interest unless we have some definite idea of the skewness and kurtosis of the non-normal population sampled. Fortunately, experience suggests that the magnitude of the skewness and kurtosis observed in many non-normal populations is small enough to permit a use of the normal approximations for large samples. Nevertheless, serious non-normality of the sampled population could vitiate their use, and this point should never be forgotten in testing any null hypothesis about μ. In such cases, more elaborate methods are required, but these lie outside the scope of this book. So far as our present purpose is concerned, we conclude that the normal approximations may be used for large samples, provided the results of the tests of significance are interpreted flexibly without a rigid adherence to the level of significance adopted. Indeed, this is the justification of the tests of significance based on the normal approximation to the binomial distribution .

14.5 Comparison of the means of two normal populations

A useful property of the normal distribution provides the theoretical basis for extending the preceding tests of significance to test hypotheses about the means of two normal populations. Suppose x_1 and x_2 are two independent normal variables with means θ_1 and θ_2, and variances σ_1^2 and σ_2^2. If a and b are two constants not both zero simultaneously, then the linear function

$$Y = ax_1 + bx_2$$

defines another random variable such that

$$E(Y) = a\theta_1 + b\theta_2; \quad \text{var}(Y) = a^2\sigma_1^2 + b^2\sigma_2^2.$$

Hence the function

$$v = \frac{Y - E(Y)}{\sqrt{\text{var}(Y)}}$$

is a standardised random variable with zero mean and unit variance. These results follow from our earlier theorems. Besides, it is also known that v is a unit normal variable. This is another very important result in normal distribution theory, and we now use it to develop large-sample tests of significance for null hypotheses about the population means θ_1 and θ_2.

Suppose $x_{11}, x_{12}, \ldots, x_{1n_1}$ are independent observations from a normal population with mean θ_1 and variance σ_1^2. Then the sample average $\bar{x}_1$ is an unbiased estimate of θ_1, and the sample variance

$$s_1^2 = \sum_{i=1}^{n_1}(x_{1i} - \bar{x}_1)^2/(n_1 - 1)$$

is an unbiased estimate of σ_1^2. Similarly, let $x_{21}, x_{22}, \ldots, x_{2n_2}$ be independent observations from another normal population with mean θ_2 and variance σ_2^2. The sample average $\bar{x}_2$ and the sample variance

$$s_2^2 \;=\; \sum_{i=1}^{n_2}(x_{2i} \;-\; \bar{x}_2)^2/(n_2 \;-\; 1)$$

are the unbiased estimates of θ_2 and σ_2^2 respectively.

We now use these sample quantities to derive a test of significance for a general null hypothesis about θ_1 and θ_2, namely, $H(\alpha\theta_1 + \beta\theta_2 = \delta)$, where α, β, and δ are known constants. Now, it is clear that the linear function

$$Z \;=\; \alpha\bar{x}_1 \;+\; \beta\bar{x}_2$$

is an unbiased estimate of $\alpha\theta_1 + \beta\theta_2$ since

$$E(Z) \;=\; \alpha E(\bar{x}_1) \;+\; \beta E(\bar{x}_2) \;=\; \alpha\theta_1 \;+\; \beta\theta_2.$$

Also,

$$\mathrm{var}(Z) \;=\; \alpha^2\,\mathrm{var}(\bar{x}_1) \;+\; \beta^2\,\mathrm{var}(\bar{x}_2) \;=\; \frac{\alpha^2\sigma_1^2}{n_1} \;+\; \frac{\beta^2\sigma_2^2}{n_2}.$$

Hence, since $\bar{x}_1$ and $\bar{x}_2$ are independent normally distributed random variables, we deduce from the distribution of v that

$$z \;=\; \frac{\alpha\bar{x}_1 \;+\; \beta\bar{x}_2 \;-\; (\alpha\theta_1 + \beta\theta_2)}{\left[\dfrac{\alpha^2\sigma_1^2}{n_1} \;+\; \dfrac{\beta^2\sigma_2^2}{n_2}\right]^{\frac{1}{2}}}$$

is a unit normal variable. Accordingly, if the null hypothesis $H(\alpha\theta_1 + \beta\theta_2 = \delta)$ is true, then

$$z \;=\; \frac{\alpha\bar{x}_1 \;+\; \beta\bar{x}_2 \;-\; \delta}{\left[\dfrac{\alpha^2\sigma_1^2}{n_1} \;+\; \dfrac{\beta^2\sigma_2^2}{n_2}\right]^{\frac{1}{2}}}$$

is a unit normal variable. If σ_1 and σ_2 are assumed known, then z can be evaluated and the null hypothesis $H(\alpha\theta_1 + \beta\theta_2 = \delta)$ tested. Thus for a two-sided test of significance, the null hypothesis under test is rejected at the 100η $(0 < \eta < 1)$ per cent level of significance if

$$|z| \geqslant \lambda,$$

where $1 - \Phi(\lambda) = \frac{1}{2}\eta$ in the standard notation for the distribution function of a unit normal variable. Similarly, for a one-sided test of significance, the null hypothesis $H(\alpha\theta_1 + \beta\theta_2 = \delta)$ is rejected at the 100η per cent level of significance if

$$z \leqslant -\lambda', \quad \text{or} \quad z \geqslant \lambda'$$

depending upon whether the alternative hypotheses are $\alpha\theta_1 + \beta\theta_2 < \delta$ or $\alpha\theta_1 + \beta\theta_2 > \delta$, and where $1 - \Phi(\lambda') = \eta$.

If, as is usually the case, the variances σ_1^2 and σ_2^2 are unknown, then we may replace them by s_1^2 and s_2^2 to obtain the modified random variable

$$w \;=\; \frac{\alpha\bar{x}_1 \;+\; \beta\bar{x}_2 \;-\; \delta}{\left[\dfrac{\alpha^2 s_1^2}{n_1} \;+\; \dfrac{\beta^2 s_2^2}{n_2}\right]^{\frac{1}{2}}}. \tag{1}$$

However, if the null hypothesis $H(\alpha\theta_1 + \beta\theta_2 = \delta)$ is true, then w is not exactly normally distributed. Nevertheless, if the sample sizes n_1 and n_2

are large, then the distribution of w may be approximated by a unit normal distribution.

The null hypothesis $H(\alpha\theta_1 + \beta\theta_2 = \delta)$ is known as a *general linear hypothesis* about the population means θ_1 and θ_2; and its meaning is that the population means lie on a specified straight line $\alpha\theta_1 + \beta\theta_2 = \delta$ in the (θ_1, θ_2) plane. For the present, the importance of such a general hypothesis lies in the fact that it leads directly to two particular null hypotheses of importance in statistical analysis. Firstly, if we put $\alpha = 1$ and $\beta = -1$, then the null hypothesis becomes $H(\theta_1 - \theta_2 = \delta)$, which signifies that the means of the two populations differ by a specified amount δ. For σ_1 and σ_2 unknown, the test function for this null hypothesis is

$$w = \frac{\bar{x}_1 - \bar{x}_2 - \delta}{\left[\dfrac{s_1^2}{n_1} + \dfrac{s_2^2}{n_2}\right]^{\frac{1}{2}}}; \tag{2}$$

and for large samples w is approximately a unit normal variable if the null hypothesis under test is, in fact, true. Secondly, if we put $\alpha = 1, \beta = -1$, and $\delta = 0$, then the general linear hypothesis becomes $H(\theta_1 - \theta_2 = 0)$, which signifies that the means of the two normal populations are the same. Furthermore, if the population variances are unknown, then the large-sample test of significance is based on the approximation that

$$w = \frac{\bar{x}_1 - \bar{x}_2}{\left[\dfrac{s_1^2}{n_1} + \dfrac{s_2^2}{n_2}\right]^{\frac{1}{2}}} \tag{3}$$

is a unit normal variable.

If the populations sampled are not normal, then the application of the above large-sample tests of significance is subject to the same sort of reservations as indicated in the case of testing null hypotheses about the mean of one population. Accordingly, these tests may also be used with a flexible interpretation of the level of significance adopted. Apart from this, there is one structural consideration which is of interest, and it pertains to the nature of the random variable w as defined in (1). We observe that the numerator of w is a normally distributed random variable which has zero expectation if the null hypothesis under test is true. The denominator of w is the standard error of the numerator; and the ratio, as defined in (1), is of a type which occurs very frequently in statistical theory. In many cases, the exact sampling distributions of these ratios are known to have a convenient form, which gives rise to the classical *exact tests of significance for linear hypotheses*. These tests of significance are of great importance in statistical analysis, and we shall consider them systematically in Chapter 15. However, so far as the particular test functions (1) to (3) are concerned, their exact sampling distributions do not have a convenient form. Nevertheless, the Central Limit Theorem ensures approximate normality for large samples, and we restrict the use of these tests accordingly.

Example 14.2: Comparison of anthropometric measurements

In an army survey, a random sample of 74,459 Grade I men had an average chest girth of 35·8 inches and a standard deviation of 1·94 inches. In another random sample of 2,146 Grade IV men, the corresponding values were 34·8 and 2·01 inches. Do these data suggest that the average chest girth measurements of Grade I and Grade IV men are significantly different?

Physical measurements such as height, weight, chest girth, etc. of human beings are known to be closely normally distributed, and so we may reasonably use this distribution as a basis for statistical analysis of the data. Accordingly, we postulate that the conceptual population of the chest girth measurements of Grade I men is normal with mean θ_1 and variance σ_1^2. Similarly, we assume that the population of the chest girth measurements of Grade IV men is also normal with mean θ_2 and variance σ_2^2. The two samples provide information about these conceptual populations, and the question posed may be interpreted as the null hypothesis $H(\theta_1 - \theta_2 = 0)$ whose significance is to be tested on the basis of the sample information.

In the notation used for (3) above, the sample data provide

$$n_1 = 74,459, \quad \bar{x}_1 = 35\text{·}8, \quad s_1 = 1\text{·}94;$$

and

$$n_2 = 2,146, \quad \bar{x}_2 = 34\text{·}8, \quad s_2 = 2\text{·}01.$$

Hence

$$\text{s.e.}(\bar{x}_1 - \bar{x}_2) = \left[\frac{(1\text{·}94)^2}{74459} + \frac{(2\text{·}01)^2}{2146} \right]^{\frac{1}{2}}$$

$$= \sqrt{0\text{·}000051 + 0\text{·}001883} = 0\text{·}04398.$$

Accordingly, to test the null hypothesis $H(\theta_1 - \theta_2 = 0)$ with the alternatives $\theta_1 - \theta_2 \neq 0$, we use the two-sided test of significance with the test function

$$w = \frac{35\text{·}8 - 34\text{·}8}{0\text{·}04398} = 22\text{·}74.$$

Since $P(|w| \geqslant 22\text{·}74) = 2[1 - \Phi(22\text{·}74)] \sim 0$, the rejection of the null hypothesis under test is beyond all doubt. We thus conclude that the observed sample difference almost certainly suggests that the mean chest girths of Grade I and Grade IV men are different. If we now set $\theta_1 - \theta_2 = \delta \neq 0$, then the sample data may also be used to find the class of null hypotheses about δ which are acceptable at the five per cent (say) level of significance. This class is simply given by

$$\bar{x}_1 - \bar{x}_2 - 1\text{·}96 \times \text{s.e.}(\bar{x}_1 - \bar{x}_2) \leqslant \delta \leqslant \bar{x}_1 - \bar{x}_2 + 1\text{·}96 \times \text{s.e.}(\bar{x}_1 - \bar{x}_2),$$

or, substituting numerical values, we have

$$0\text{·}9138 \leqslant \delta \leqslant 1\text{·}0862.$$

14.6   Approximate tests of significance based on maximum-likelihood estimates

There is another important class of large-sample tests of significance which are based on estimates obtained by the method of maximum likelihood. Suppose that, under the previously stated assumptions for the application of the maximum-likelihood estimation procedure, $L(\theta)$ is the sample likelihood for the estimation of a parameter θ. Then we recall that the estimate $\hat{\theta}$ is obtained as a root of the equation

$$\left[\frac{d\log L}{d\theta}\right]_{\theta=\hat{\theta}} = 0,$$

and that the large-sample variance of the estimate is

$$\text{var}(\hat{\theta}) = -1\Big/E\left[\frac{d^2\log L}{d\theta^2}\right].$$

It is also known that in many cases the random variable

$$z = \frac{\hat{\theta} - \theta}{\text{s.e.}(\hat{\theta})}$$

may be regarded approximately as a unit normal variable if the sample size is large. Hence the random variable z may be used to test null hypotheses about the parameter θ. In the more common examples, this approach leads to the large-sample tests of significance we have already considered from a more intuitive standpoint. Thus, for example, the large-sample test of significance for any null hypothesis about a binomial parameter may be readily deduced by this method. As a more interesting illustration, we consider the estimation of the parameter θ of the multinomial distribution given in Example 11.15 of Chapter 11. The numerical calculations based on Imai's data give

$$\hat{\theta} = 0.4835 \quad \text{and} \quad \text{s.e.}(\hat{\theta}) = 0.04663,$$

by a direct application of the maximum-likelihood procedure. We now use these results to carry out an approximate test of significance for the null hypothesis $H(\theta = 0.5)$. If this null hypothesis is true, then the test function

$$z = \frac{\hat{\theta} - 0.5}{\text{s.e.}(\hat{\theta})} = -0.354.$$

Accordingly, if the alternatives to the null hypothesis under test are $\theta \neq 0.5$, then the probability for the rejection of the null hypothesis by a two-sided test is

$$P(|z| \geqslant 0.354) = 2[1 - \Phi(0.354)] = 0.7233.$$

Therefore the null hypothesis $H(\theta = 0.5)$ may be safely accepted as in agreement with the sample information about θ.

In this example, the possible values of θ lie in the interval $(0, 1)$ and the test of significance gives a reasonable answer. However, in using this procedure, it is inadvisable to test the significance of null hypotheses which refer to values of θ in close proximity to either extreme of its possible range, since in such cases the normal approximation to the distribution of $\hat{\theta}$ tends to break down. This is an important point to remember.

14.7 Errors of a test of significance

In the preceding sections, we have considered a number of approximate tests of significance for different null hypotheses about population parameters; and before proceeding further with the tests of significance of more complex null hypotheses, it is worth while to recapitulate the substance of the general argument used. At the outset, it is important to note that though the approximations are generally valid for large samples only, the principles on which the tests of significance are based remain unaffected by the mathematical approximation of the sampling distributions of the test functions used. Consequently, for present discussion, we may ignore the approximative nature of these tests of significance.

Given a random sample of observations, a test of significance involves the specification of

(i) the mathematical form of the population sampled;

(ii) the null hypothesis to be tested and its alternatives;

(iii) the appropriate test function and its sampling distribution; and

(iv) the level of significance used to reject the null hypothesis.

As an example, suppose $\bar{x}$ and s^2 are the sample mean and variance based on n independent observations from a normal population with mean μ and variance σ^2. Then, to test the null hypothesis $H(\mu = \mu_0)$, we have used the test function

$$t = \frac{\bar{x} - \mu_0}{s/\sqrt{n}},$$

which has the unit normal distribution if the null hypothesis under test is, in fact, true. Hence, for a two-sided test of significance, the region of rejection of the null hypothesis $H(\mu = \mu_0)$ at the five per cent (say) level of significance is defined by the equation

$$P(|t| \geqslant 1\cdot96) = 0\cdot05.$$

Any observed value of $|t| \geqslant 1\cdot96$ leads to the rejection of the null hypothesis $H(\mu = \mu_0)$, whereas this null hypothesis is accepted if $|t| < 1\cdot96$. This implies that if the observed sample is really drawn from a normal population with mean μ_0, then the probability that t lies outside the interval $(-1\cdot96, 1\cdot96)$ is $0\cdot05$. Accordingly, the probability of getting a value such as this by chance is $0\cdot05$. In other words, the probability of rejecting the null hypothesis $H(\mu = \mu_0)$ when it is, in fact, true is $0\cdot05$, which is small. This is a satisfactory situation. On the other hand, if the null hypothesis $H(\mu = \mu_0)$ is, in fact, false so that $\mu \neq \mu_0$, then the probability that t lies outside the interval $(-1\cdot96, 1\cdot96)$ is greater than $0\cdot05$ and increases as $|\mu - \mu_0|$ increases. Hence the probability that the null hypothesis $H(\mu = \mu_0)$ is rejected when it is, in fact, false is greater than $0\cdot05$, and may be much larger. This is again a satisfactory situation.

In testing a null hypothesis by the stated principles, it is possible to commit two kinds of errors. Firstly, it is possible to reject the null hypothesis as false when, in fact, it is true: this is known as the *Type I error*. Secondly, it is possible to accept the null hypothesis as true when, in fact, it is false: this is known as the *Type II error*. Ideally, we would wish to eliminate completely these two errors in making a decision about any particular null hypothesis but, as we have already pointed out, it is impossible to do this when the decision-making procedure is dependent upon considerations of probability. In this sense, all statistical inference is *uncertain*. As a matter of fact, it is known that these two errors are complementary, so that by decreasing one we tend to increase the other and vice versa. In view of this limitation, it is usual to restrict the probability of making the Type I error to a small value, and then to devise a test procedure which minimises the risk of making the Type II error. Indeed, the level of significance is the probability of making the Type I error and so, in principle, this has to be assigned by the experimenter in the light of his assessment of the consequences of rejecting the null hypothesis when, in fact, it is true. The quantity (1 − the probability of making the Type II error) is known as the *power of the test*. Thus, in general, the problem of devising an optimum test of significance for a null hypothesis is that of determining a test procedure which, for a given level of significance, has the maximum power.

This approach to the testing of hypotheses is due to J. Neyman and E.S. Pearson, and so also is the terminology of the Type I and Type II errors. It is not our purpose here to enter into any further exposition of this theory because it is based on rather advanced mathematics. It is sufficient to remark that, in general, the tests of significance we have considered and the ones to be presented in the next chapters do satisfy the optimal requirements of power. Apart from this, the notion of the power of a test of significance is a concept of fundamental importance in comparing alternative tests for a null hypothesis, though the determination of power of any test is a matter of considerable complexity.

We conclude with an example which shows how the probability of rejecting a null hypothesis alters when the null hypothesis is changed.

Example 14.3 : Average height of men

A random sample of 1,164 men gave an average height of 68·64 inches and a sample variance of 7·3861 square inches. Assuming that height is a normally distributed random variable with mean μ and variance σ^2, we wish to test the null hypotheses that the mean height in the population sampled is (i) 68·5 inches; and (ii) 65·0 inches.

Here the sample average $\bar{x} = 68·64$ and the sample variance $s^2 = 7·3861$ are obtained from a sample of $n = 1,164$ observations.

Therefore
$$\frac{s}{\sqrt{n}} = \sqrt{0·006345} = 0·07966.$$

Hence, to test the null hypothesis $H(\mu = 68·5)$ with the alternatives $\mu \neq 68·5$, we have
$$t = \frac{68·64 - 68·5}{0·07966} = 1·76;$$

and since this observed value of t is less than 1·96, we infer that the null hypothesis under test is acceptable at the five per cent level of significance. Thus the sample data are consistent with the hypothesis that the mean height of men in the population is 68·5 inches. However, if the null hypothesis is $H(\mu = 65·0)$, then we have
$$t = \frac{68·64 - 65·0}{0·07966} = 45·69.$$

The null hypothesis is now rejected decisively, and it is almost certain that a sample average of 68·64 could not have arisen by chance from a population with a mean of 65·0.

The two tests of significance reveal the characteristic manner in which a good test procedure discriminates between acceptable and unacceptable null hypotheses. The general behaviour is clear: the larger the deviation between the sample estimate of the parameter and its hypothetical value, the greater is the probability of the rejection of the null hypothesis. This is in agreement with the behaviour of tests of significance in binomial sampling.

14.8   Confidence intervals

A test of significance is a mathematical procedure for deciding on grounds of probability whether or not a null hypothesis concerning a parameter (or parameters) can be regarded as consistent with the sample observations.   The crucial point is that the null hypothesis is specific and, for example, in the simplest case of a single parameter, it postulates a particular value for the parameter. On the other hand, suppose that in Example 14.3 above we have no reason for specifying any null hypothesis. There is now no question of using the sample data in the usual way to perform a test of significance. Nevertheless, under the assumption of the normality of the distribution of the heights of men, we still have the large-sample approximation that

$$t = \frac{\bar{x} - \mu}{s/\sqrt{n}}$$

is a unit normal variable, whatever the true value of μ may be. Hence we infer that

$$P(|t| \leqslant 1\cdot96) = 0\cdot95,$$

so that

$$P(\bar{x} - 1\cdot96\,s/\sqrt{n} \leqslant \mu \leqslant \bar{x} + 1\cdot96\,s/\sqrt{n}) = 0\cdot95. \tag{1}$$

As explained earlier, this inequality gives the class of null hypotheses about μ which are acceptable at the five per cent level on the basis of the sample data.

Another important way of interpreting the "probability" statement (1) is as follows. For convenience, we first set $\bar{x} - 1\cdot96\,s/\sqrt{n} \equiv a_1$ and $\bar{x} + 1\cdot96\,s/\sqrt{n} \equiv b_1$, so that (1) may be written more compactly as

$$P(a_1 \leqslant \mu \leqslant b_1) = 0\cdot95. \tag{2}$$

We observe that both a_1 and b_1 are functions of the sample observations and so are necessarily random variables. If we take another sample of size n from the same population, then we can similarly derive two other numbers a_2 and b_2 such that

$$P(a_2 \leqslant \mu \leqslant b_2) = 0\cdot95.$$

By continuing this process, we can determine a sequence of pairs of values $(a_1, b_1), (a_2, b_2), (a_3, b_3), \ldots$, each pair being calculated from a random sample of n observations and satisfying one of the "probability" statements

$$P(a_i \leqslant \mu \leqslant b_i) = 0\cdot95, \quad \text{for } i = 1, 2, 3, \ldots. \tag{3}$$

The length of each of these intervals $b_i - a_i$ is evidently a random variable, and the "probability" statements (3) mean that, whatever be the true value of μ, in 95 per cent of the cases its unknown value will be included in the intervals (a_i, b_i). Accordingly, the probability refers to the proportion of random intervals which would include the parameter μ. The pair (a_1, b_1) of numbers determined on the basis of the observed sample gives what is termed a 95 per cent *confidence interval* for the parameter μ. In other words, there is a probability of $0·95$ that the value of μ will be in the interval (a_1, b_1). Of course, it is possible that the true value of μ may lie outside this interval, but this risk is small. We have thus obtained an *interval estimate* of μ, and this is different in nature from the usual *point estimate* $\bar{x}$ obtained by any one of the methods of estimation discussed earlier.

Example 14.3 (concluded)

To illustrate the calculation of confidence intervals, we use the data of Example 14.3. The point estimate of μ is $\bar{x} = 68·64$. The 95 per cent confidence interval for μ is

$$68·64 \pm 1·96 \times 0·07966 = 68·64 \pm 0·16, \text{ or } (68·48, 68·80).$$

Similarly, the 99 per cent confidence interval for μ is

$$68·64 \pm 2·58 \times 0·07966 = 68·64 \pm 0·21, \text{ or } (68·43, 68·85).$$

We observe that the 99 per cent confidence interval for μ is wider than the 95 per cent confidence interval, and this is a characteristic feature of interval estimation. In general, the greater the confidence associated with an interval estimate of a parameter, the wider must be the interval, and vice versa. This is intuitively reasonable, since to increase the measure of confidence implies a decrease in the risk of the interval not including the true value of the parameter.

This concept of an interval estimate of a parameter is another facet of great importance in the theory of statistical inference due to Neyman and Pearson. The practical interest of such an estimate lies in the fact that it sets a limit to the uncertainty of a parameter value on a specified level of probability. In contrast, a point estimate of a parameter gives a single "best" value for the parameter on the basis of the sample information.

We have illustrated the concept of a confidence interval by using the large-sample approximation for the distribution of

$$t = \frac{\bar{x} - \mu}{s/\sqrt{n}}$$

to evaluate the 95 and 99 per cent confidence intervals for the mean μ. The argument is quite general and applies equally to any other level of confidence. Furthermore, this method can also be used to obtain confidence intervals for other parameters when the normal approximation for the sampling distribution of the appropriate test function holds. In general, if T is an unbiased (or consistent) estimate of a parameter θ (or a function of several parameters) such that the standardised random variable

$$u = \frac{T - \theta}{\text{s.e.}(T)}$$

is approximately normally distributed with zero mean and unit variance, then the $100(1 - \eta)$ per cent confidence interval for θ is

$$T \pm c_\eta \times \text{s.e.}(T) \quad (0 < \eta < 1),$$

where c_η is the 100η per cent point of the unit normal distribution. This is a large-sample technique for approximate confidence intervals and it could, for example, be used for a binomial proportion, and the difference between two binomial proportions or that between the means of the two normal populations. We shall consider in the next chapter how exact confidence intervals based on small samples may be evaluated in several cases of practical importance.

Finally, it is to be noted that the $100(1 - \eta)$ per cent confidence interval for θ can also be used to test a null hypothesis about θ at the 100η per cent level of significance. Thus if a hypothetical value θ_0 of θ lies outside the confidence interval

$$T \pm c_\eta \times \text{s.e.}(T),$$

then the null hypothesis $H(\theta = \theta_0)$ is rejected at the 100η per cent level of significance, whilst it is accepted in the opposite event.

CHAPTER 15

EXACT TESTS OF SIGNIFICANCE AND THE ANALYSIS OF VARIANCE

15.1   Statistical analysis of small samples

The approximate normality of the sampling distributions of many test functions for large samples simplifies the statistical analysis of data considerably, but this approximate technique is not applicable to many problems which give rise to small samples. One of the primary reasons for the use of small samples is the cost of experimentation. Besides, in many cases, it is impossible to obtain more than a certain number of individuals for experimentation. This occurs quite frequently in biological work with twins or with animals belonging to the same litter. In fact, the realisation that small samples are inevitable in many experimental conditions led to the search for suitable methods of statistical analysis which gave rise to the modern corpus of statistical methods known as *exact tests of significance* appropriate for samples of any size and so necessarily applicable to small samples. The start of this search for the solution of problems of statistical analysis based on a comparatively few observations was made by William Sealy Gosset (1876 – 1937) in an historic paper published in 1908 under the now celebrated pseudonym of "Student". The immense potentialities of "Student's" result were quickly observed by Ronald Aylmer Fisher (1890 – 1962) who made many fundamental contributions in the years following the publication of "Student's" 1908 paper. Since then developments in the field have been made extensively by mathematical statisticians in different parts of the world, but much of this work lies well beyond the scope of this introduction. Our concern here is only with some of the basic tests of significance developed, in the main, by "Student" and R.A. Fisher. Furthermore, the detailed mathematical justification of the methods is also not considered here, though we indicate the broad outline of the argument as far as it is possible with the extent of mathematics assumed for the study of this book.

The methods considered here are exact in so far as the populations sampled are assumed to be normal. This is the chief restriction theoretically, though it is now known that these methods are approximately applicable even when there is considerable non-normality in the populations sampled. Accordingly, the practical use of these methods is widespread. Nevertheless, serious non-normality of the populations sampled is a real phenomenon and requires more sophisticated methods. We do not consider them here, and we mention this difficulty in order to stress the fundamental fact that every statistical method has its intrinsic limitations which should not be ignored in a practical situation.

The methods considered here are tests of significance pertaining to means, variances, and certain other statistics known as *regression coefficients*. It is a fortunate coincidence that a unified approach is possible. This is based on the sampling distributions of three random variables which are closely interrelated. Accordingly, we first explain (not prove) the structural forms of the three random variables and the interrelationships of their sampling distributions.

15.2   Sampling distributions of χ^2, F, and t

(i) *Helmert's χ^2 distribution*

Suppose $x_1, x_2, \ldots, x_n$ are independent observations from a normal population with mean μ and variance σ^2. Then the probability distribution of any x_i is

$$\frac{1}{\sigma\sqrt{2\pi}}\, e^{-(x_i-\mu)^2/2\sigma^2} dx_i, \quad \text{where} \quad -\infty < x_i < \infty \quad \text{and} \quad 1 \leqslant i \leqslant n.$$

Next, if we make the standardising transformation $z_i = (x_i - \mu)/\sigma$, then each z_i is an independent unit normal variable with the probability distribution

$$\frac{1}{\sqrt{2\pi}}\, e^{-\frac{1}{2}z_i^2} dz_i, \qquad \text{for} \quad -\infty < z_i < \infty.$$

Finally, we define a new random variable

$$\chi^2 \;=\; \sum_{i=1}^{n} z_i^2,$$

that is, χ^2 is the sum of the squares of n independent unit normal variables. Then it is known that χ^2 has the probability distribution

$$\frac{1}{2^{n/2}\Gamma(n/2)}\, e^{-\frac{1}{2}\chi^2}(\chi^2)^{n/2-1}\, d\chi^2, \quad \text{for} \quad 0 \leqslant \chi^2 < \infty .$$

This is the classical χ^2 distribution with n *degrees of freedom*, usually abbreviated as n d.f. The quantity n is an integral parameter of the distribution, and it is related to the fact that χ^2 is a sum of the squares of n independent unit normal variables. It is also easily seen that the random variable $\frac{1}{2}\chi^2$ has the gamma distribution as defined in Chapter 7 with $\lambda = 1$ and $\alpha = n/2$. The χ^2 distribution was discovered by Friedrich Robert Helmert (1843 − 1917) in 1876 but forgotten until Karl Pearson rediscovered it in 1900. This is the first of the three sampling distributions which we require for later work.

There are two particular forms of the χ^2 distribution which are of considerable importance. Firstly, for $n = 1$, we have $\chi^2 = z_1^2$, so that the square of a unit normal variable is distributed as χ^2 with 1 d.f. Secondly, we recall that the sample variance

$$s^2 \;=\; \sum_{i=1}^{n}(x_i - \bar{x})^2/(n-1)$$

is an unbiased estimate of the population variance σ^2. It is also known that the random variable $(n-1)s^2/\sigma^2$ has the χ^2 distribution with $n-1$ d.f. The number of degrees of freedom is in conformity with the fact that the quantity $(n - 1)s^2$ is a sum of squares which involves only $n - 1$ of the deviations $x_i - \bar{x}$ since $\sum_{i=1}^{n} (x_i - \bar{x}) \equiv 0$.

(ii) *Snedecor's F distribution*

If χ_1^2 and χ_2^2 are independent χ^2 variables with ν_1 and ν_2 d.f. respectively, then the random variable

$$F = \nu_2 \chi_1^2 / \nu_1 \chi_2^2$$

has the probability distribution

$$\frac{(\nu_1/\nu_2)^{\nu_1/2} \, F^{\nu_1/2-1} dF}{B(\nu_1/2, \nu_2/2)(1 + \nu_1 F/\nu_2)^{(\nu_1+\nu_2)/2}}, \quad \text{for } 0 \leqslant F < \infty.$$

This is the F distribution with ν_1, ν_2, d.f. and was first obtained by G.W. Snedecor, who named the ratio F in honour of R.A. Fisher. The quantities ν_1 and ν_2 are the two integral parameters of the distribution. It is easily seen that the distribution of $\nu_1 F/\nu_2$ is the beta distribution of the second kind, as defined in Chapter 7, with $\alpha = \nu_1/2$ and $\beta = \nu_2/2$. It also follows from the definition of F that the distribution of $1/F$ is another F distribution with ν_2, ν_1 d.f. It is important for later use to note this symmetry of the F distribution.

Fisher himself had earlier obtained the distribution of the random variable

$$z = \tfrac{1}{2} \log_e F.$$

(Fisher's z should not be confused with our unit normal variable z introduced in the previous chapter.)

Besides, and this is important, Fisher also showed the relation of his z distribution with an extremely versatile method of statistical analysis known as the *analysis of variance*. As we shall see, it is this method which gives unity to all the tests of significance discussed in this chapter. However, it is conventional to use the simpler F instead of Fisher's z in the testing of significance in problems associated with the analysis of variance.

An important particular case of the F distribution arises when we consider the ratio of the sample variances of two independent normal populations. Suppose s_i^2 is the unbiased estimate of the population variance σ_i^2 based on n_i independent observations, for $i = 1, 2$. Then $(n_i - 1)s_i^2/\sigma_i^2$ has the χ^2 distribution with $n_i - 1$ d.f. Also, since the two samples are from independent normal populations, the two χ^2's are also independent, and we conclude that

$$F = \sigma_2^2 s_1^2 / \sigma_1^2 s_2^2$$

has the F distribution with $n_1 - 1, n_2 - 1$ d.f. More particularly, if the two populations have the same variance, then

$$F = s_1^2 / s_2^2,$$

where s_1^2 and s_2^2 are now interpreted as independent unbiased estimates of the same parameter denoting the variance of the two populations. Hence F may be defined as the ratio of two independent estimates of the same variance. This structural form of F is very important, and it has also led to F being referred to as a "variance ratio".

(iii) *"Student's" t distribution*

Suppose u is a unit normal variable and χ^2 is an independent random variable having the χ^2 distribution with ν d.f. Then the ratio

$$t = \frac{u}{\chi/\sqrt{\nu}},$$

where χ is the positive square root of χ^2, is a random variable with the probability distribution

$$\frac{dt}{\sqrt{\nu}\,B(\frac{1}{2},\,\nu/2)\,(1+t^2/\nu)^{(\nu+1)/2}}, \quad \text{for } -\infty < t < \infty.$$

This is the classical "Student's" t distribution with ν d.f. expressed in the form suggested by R.A. Fisher. In his 1908 paper "Student" considered the distribution of $t/\sqrt{\nu}$, but Fisher's form is now used exclusively. The degrees of freedom of χ^2 are also the degrees of freedom of t, and their number is the only parameter of the t distribution. This distribution is symmetrical about the origin and it tends to normality as $\nu \to \infty$. It is also seen that if in the definition of the general F distribution we put $\nu_1 = 1$, then the distribution of F with 1, ν_2 d.f. is the same as that of t^2 with ν_2 d.f. Hence

$$t = +\sqrt{F}$$

whenever the numerator of F has 1 d.f. This is a result of great importance, and we shall use it frequently.

Finally, an important property of the normal distribution leads to a particular case of the t distribution, which has great practical use. If $x_1, x_2, \ldots, x_n$ are independent observations from a normal population with mean μ and variance σ^2, then we know that

(a) the sample mean $\bar{x}$ is normally distributed with mean μ and variance σ^2/n; and

(b) the sample variance s^2 is a random variable such that $(n-1)s^2/\sigma^2$ is distributed as χ^2 with $n-1$ d.f.

We observe that $\bar{x}$ and s^2 are calculated from the same sample observations, but it is a remarkable property of the normal distribution that the probability distributions of $\bar{x}$ and s^2 are independent. Hence

$$t = \frac{\bar{x}-\mu}{\sigma/\sqrt{n}} \bigg/ \frac{s}{\sigma} = \frac{\bar{x}-\mu}{s/\sqrt{n}} \tag{1}$$

has "Student's" t distribution with $n-1$ d.f. This is the historic achievement of "Student". In other words, he showed how the unit normal distribution of the random variable

$$z = \frac{\bar{x}-\mu}{\sigma/\sqrt{n}} \tag{2}$$

is altered *exactly* when σ, a parameter, is replaced by s, a random variable. This technique of replacing an unknown parameter by its estimate is of very general application in statistical theory, and it is called *studentisation* in honour of "Student". We thus say that t, as defined in (1), is a studentised form of z, as defined in (2).

We recall that in the preceding chapter we used the fact that for large n, t, as defined in (1), is approximately a unit normal variable. This is in agreement with the result that the t distribution tends to normality for large degrees of freedom. We also note that

$$t^2 = n(\bar{x} - \mu)^2/s^2$$

has the F distribution with 1, $n - 1$ d.f. We may interpret t^2 as a ratio of $n(\bar{x} - \mu)^2/\sigma^2$, which is a χ^2 with 1 d.f., and an independent random variable s^2/σ^2, which is distributed as $\chi^2/(n - 1)$ with $n - 1$ d.f. This is in agreement with our earlier definition of F.

From the point of view of the applications considered in this chapter, the structural forms of the random variables χ^2, F, and t are of primary importance, and we shall see how a diversity of practical problems are solved by identifying the test functions with one of these three random variables. The probability distributions of these three random variables have been extensively tabulated for practical use, and their percentage points are available in much the same way as those of the unit normal distribution. But there is one important difference: the distributions depend upon their parameters representing the degrees of freedom. Accordingly, the percentage points are tabulated for a wide range of values of the degrees of freedom, and in looking up the tables, care should be exercised to ensure that the tabular values correspond correctly to the appropriate degrees of freedom. A little experience is all that is needed to use the tables correctly, and this will be provided by numerical examples in the following sections of this chapter.

15.3 The analysis of variance

We have shown in Chapter 11 how the method of least squares is used to estimate parameters in a variety of populations. The method, regarded simply as an estimation procedure, is part of a general theory for testing linear hypotheses and, as originally indicated by Fisher, the applications of this theory can be presented in the compact and instructive form of the analysis of variance. A strict mathematical justification of this procedure depends upon the assumption that the sample observations are from one or more normal populations which have the same but unknown variance. We start with the simplest case of testing null hypotheses about the mean of a single normal population and gradually progress to problems of greater complexity.

15.4 Tests of significance for the mean of a normal population

Suppose $x_1, x_2, \ldots, x_n$ are independent observations from a normal population with mean μ and variance σ^2, both these parameters being unknown. We also assume that the sample size n is not necessarily large and may be as small as two. Our first problem is to test the null hypothesis $H(\mu = \mu_0)$, where μ_0 is some preassigned quantity. Now, if $\bar{x}$ and s^2 are the least-

squares estimates of μ and σ^2 respectively, then it is clear from our preceding study of the t distribution that an exact test for the above null hypothesis is obtained by using the fact that

$$t = \frac{\bar{x} - \mu_0}{s/\sqrt{n}}$$

has "Student's" distribution with $n-1$ d.f., if the null hypothesis is true.
Since the t distribution is symmetrical about the origin, the arguments for
one-sided and two-sided tests are analogous to those used with the approximating normal distribution for large n. This method is correct, but we shall
evolve an equivalent test of significance from a somewhat different standpoint because the latter method can be easily generalised to test other null
hypotheses of interest. We indicate the argument from first principles.

Since $E(x_i) = \mu$ and $\text{var}(x_i) = \sigma^2$, for $i = 1, 2, \ldots, n$, therefore the least-
squares estimate of μ is obtained by minimising

$$\Omega = \sum_{i=1}^{n}(x_i - \mu)^2 = \sum_{i=1}^{n}(x_i - \bar{x})^2 + n(\bar{x} - \mu)^2$$

with respect to variation in μ. Differentiating Ω with respect to μ gives

$$\frac{d\Omega}{d\mu} = -2n(\bar{x} - \mu),$$

so that the least-squares estimate of μ is $\bar{x}$. Also the absolute minimum of
Ω is

$$M = \sum_{i=1}^{n}(x_i - \bar{x})^2,$$

and we know that $E(M) = (n-1)\sigma^2$. Thus $s^2 = M/(n-1)$ is the least-squares
estimate of σ^2 and, as already stated, M/σ^2 is distributed as χ^2 with $n-1$
d.f.

Next, suppose the null hypothesis $H(\mu = \mu_0)$ is true, so that subject to
this the relative minimum of Ω is

$$\Omega_0 = \sum_{i=1}^{n}(x_i - \mu_0)^2 = M + n(\bar{x} - \mu_0)^2.$$

Now, according to the general theory for testing linear hypotheses, the sum
of squares for testing a linear null hypothesis is

S = Relative minimum of Ω with respect to the null hypothesis
    − Absolute minimum of Ω.

This is a fundamental result, and it is applicable to all kinds of linear
hypotheses. We shall see the meaning of S in a little while. By applying
the general argument to our simple case, we have the sum of squares for
testing the null hypothesis $H(\mu = \mu_0)$ as

$$S = \Omega_0 - M = n(\bar{x} - \mu_0)^2.$$

Furthermore, if the null hypothesis under test is true, then $E(x_i) = \mu_0$ and so, quite evidently, $E(\bar{x}) = \mu_0$. Hence

$$E(S) \;=\; n\,E(\bar{x} - \mu_0)^2 \;=\; n\,\mathrm{var}(\bar{x}) \;=\; \sigma^2.$$

Moreover, since $(\bar{x} - \mu_0)\sqrt{n}/\sigma$ is now a unit normal variable, it follows that S/σ^2 is distributed as χ^2 with 1 d.f.

We have thus established two results:

(i) The random variable M/σ^2 is distributed as χ^2 with $n-1$ d.f. independently of the null hypothesis $H(\mu = \mu_0)$.

(ii) If the null hypothesis $H(\mu = \mu_0)$ is true, then the random variable S/σ^2 is distributed as χ^2 with 1 d.f.

It is also known that these two χ^2's are independently distributed and hence

$$F \;=\; (n-1)S/M \tag{1}$$

has the F distribution with 1, $n-1$ d.f. This is the typical argument of the analysis of variance leading to a test of significance for the null hypothesis based on the F distribution. We shall derive, by a similar argument, tests of significance for several other linear null hypotheses.

There are two points in this analysis which deserve particular mention. Firstly, the degrees of freedom of the two χ^2's are obtained by the consideration that

$E(M) = (n-1)\sigma^2$, independently of any null hypothesis about μ; and

$E(S) = \sigma^2$, if the null hypothesis under test is true.

This argument is general and we shall use it repeatedly. Secondly, if the null hypothesis under test is not true, then $E(\bar{x}) = \mu \neq \mu_0$. Accordingly, we may now write

$$S \;=\; n(\bar{x} - \mu_0)^2$$
$$=\; n[(\bar{x} - \mu) + (\mu - \mu_0)]^2 \;=\; n[(\bar{x} - \mu)^2 + 2(\mu - \mu_0)(\bar{x} - \mu) + (\mu - \mu_0)^2],$$

so that

$$E(S) \;=\; n[E(\bar{x} - \mu)^2 + (\mu - \mu_0)^2]$$
$$=\; n[\mathrm{var}(\bar{x}) + (\mu - \mu_0)^2] \;=\; \sigma^2 + n(\mu - \mu_0)^2.$$

Thus, if the null hypothesis $H(\mu = \mu_0)$ is not true, then $E(S) > \sigma^2$. Because of this, it is usual to consider the F ratio in the form (1), so that the greater the deviations of the sample mean from the null hypothesis, the larger must be the values of F. The test of significance is carried out by finding the 100η $(0 < \eta < 1)$ percentage point of the F distribution with 1, $n-1$ d.f. Accordingly, the null hypothesis under test is rejected at the 100η per cent level of significance if the observed value of F is equal to or greater than the table value, it being accepted in the contrary case. This is a one-sided test of significance in which the upper tail of the F distribution only is used. Observed values of F which are less than unity are not tested because such values imply that the random variation between the sample observations is greater than that attributable to the difference between the sample mean $\bar{x}$ and the hypothetical value of the population mean μ.

In the conventional terminology of the analysis of variance, the quantity

M is called the *residual, error* or *within sample sum of squares* (S.S.), and S is called the *sum of squares* (S.S.) *due to the null hypothesis* $H(\mu = \mu_0)$ *under test*. The quantity M divided by its degrees of freedom, that is $M/(n-1)$, is known as the *residual, error* or *within sample mean square* (M.S.) and, of course, this is an unbiased estimate of σ^2 independently of the null hypothesis under test. Similarly, the quantity S divided by its degrees of freedom (in this case 1) is called the *mean square* (M.S.) *due to the null hypothesis under test*, and this is an unbiased estimate of σ^2 only if the null hypothesis under test is true. In this case, the two mean squares are independent estimates of σ^2, and their ratio gives the observed value of F.

The results obtained are presented in the form of an *analysis of variance table*, as proposed by Fisher.

Table 15.1   Showing the analysis of variance for testing a null hypothesis about a population mean

Source of Variation	Degrees of Freedom (d.f.)	Sum of Squares (S.S.)	Mean Square (M.S.)	F Ratio (F.R.)
Hypothesis: $\mu = \mu_0$	1	$S = n(\bar{x} - \mu_0)^2$	S	$(n-1)S/M$
Residual	$n-1$	$M = \sum_{i=1}^{n}(x_i - \bar{x})^2$	$M/(n-1)$	
Total	n	$\Omega_0 = \sum_{i=1}^{n}(x_i - \mu_0)^2$		

The above analysis of variance table reveals a characteristic feature which is common to all similar tables. If the null hypothesis $H(\mu = \mu_0)$ is true, then the total variation of the n independent sample observations is Ω_0 with n d.f. The partitioning of Ω_0 into the two components S and M with 1 and $n-1$ d.f. respectively is an algebraic identity, and the corresponding degrees of freedom are also additive. This is the underlying principle of the analysis of variance expressed in algebraic terms.

Computational notes

The numerical evaluation of M is carried out systematically as follows. Calculate

(i) $\sum_{i=1}^{n} x_i^2$, which is called the *raw* or *uncorrected* S.S. of the sample.

(ii) $\sum_{i=1}^{n} x_i$, which is simply the *grand total* of the observations.

(iii) $\left(\sum_{i=1}^{n} x_i\right)^2 / n$, which is called the *correction factor* of the sample and is usually written as C.F.(x).

(iv) $M = $ raw S.S. $-$ C.F.(x).

The calculations (i) and (ii) are carried out in one stage on a desk calculator. The grand total is accumulated in the revolutions counter and the raw S.S. in the products register.

15.5   Confidence intervals for the population mean

Since the F ratio for testing the null hypothesis $H(\mu = \mu_0)$ has $1, n-1$ d.f., it follows that

$$\sqrt{F} = t = \frac{\bar{x} - \mu_0}{s/\sqrt{n}} \quad \text{with } n - 1 \text{ d.f.,}$$

which, as expected, leads back to "Student's" test for the above null hypothesis. We can also use this t to evaluate confidence intervals for μ. Suppose, then, that no initial hypothesis is made about μ so that, in general, $E(\bar{x}) = \mu$. Then

$$t = \frac{\bar{x} - \mu}{s/\sqrt{n}}$$

still has "Student's" distribution with $n-1$ d.f. Accordingly, to calculate the $100(1 - \eta)$ per cent confidence interval for μ, we first obtain from the t table $t(\eta; n-1)$, the 100η per cent point of the t distribution with $n-1$ d.f. Therefore, since the t distribution is symmetrical about the origin, we have

$$P[|t| \geqslant t(\eta; n - 1)] = \eta,$$

or
$$P[|t| \leqslant t(\eta; n - 1)] = 1 - \eta.$$

Hence the required confidence interval for μ is

$$\bar{x} - s\,t(\eta; n - 1)/\sqrt{n} \leqslant \mu \leqslant \bar{x} + s\,t(\eta; n - 1)/\sqrt{n}.$$

It is to be noted that, unlike the percentage points of the unit normal distribution, the value of $t(\eta; n-1)$ depends upon both η and the degrees of freedom $n-1$. For any given η, the percentage point $t(\eta; n-1)$ decreases as n increases, and it achieves its limiting minimum value for $n \to \infty$ when the t distribution becomes normal. This means that for finite n, and given $\bar{x}$, s, and η, a confidence interval obtained by using the percentage point of the t distribution will always be wider than that evaluated by a normal approximation. This is intuitively reasonable, since the t distribution allows for the fact that s is a random variable and not a fixed quantity. However, as n increases, this effect of the uncertainty of s on the t distribution decreases, and in the limit as $n \to \infty$, the t distribution becomes normal.

15.6   Linear transformations and the invariance of F

There is an important property of the F ratio which quite frequently makes the computations of the analysis of variance conveniently simple. This property is the invariance of F under a linear transformation of the sample observations. We first prove this result and then illustrate its use in a numerical example.

If we make a linear transformation

$$z_i = ax_i + b, \quad \text{for } i = 1, 2, \ldots, n,$$

where a and b are constants and $a \neq 0$, then

$$E(z_i) = a E(x_i) + b = a\mu + b \equiv \lambda \quad \text{(say).}$$

Hence the null hypothesis $H(\mu = \mu_0)$ is equivalent to the null hypothesis $H(\lambda = \lambda_0)$, where $\lambda_0 \equiv a\mu_0 + b$. Furthermore,

$$\sum_{i=1}^{n} z_i = a \sum_{i=1}^{n} x_i + nb,$$

so that the mean of the z_i is $\bar{z} = a\bar{x} + b$. Therefore

$$M = \sum_{i=1}^{n} (x_i - \bar{x})^2 = \sum_{i=1}^{n} [a^{-1}(z_i - b) - a^{-1}(\bar{z} - b)]^2 = a^{-2} \sum_{i=1}^{n} (z_i - \bar{z})^2.$$

Similarly,

$$S = n(\bar{x} - \mu_0)^2 = n[a^{-1}(\bar{z} - b) - \mu_0]^2 = a^{-2} n(\bar{z} - \lambda_0)^2.$$

Therefore, in terms of the z_i, we have

$$F = (n-1)S/M = (n-1)n(\bar{x} - \mu_0)^2 \Big/ \sum_{i=1}^{n} (x_i - \bar{x})^2$$

$$= (n-1)n(\bar{z} - \lambda_0)^2 \Big/ \sum_{i=1}^{n} (z_i - \bar{z})^2.$$

Thus the F ratio is invariant under the linear transformation. This means that the structural form of F and the nature of the null hypothesis under test remain the same whether we work in terms of the original x_i or their linear transformations, the z_i. The analysis of variance can therefore be carried out as if the z_i were the original observations, and the null hypothesis considered in the equivalent form $H(\lambda = \lambda_0)$.

The $100(1-\eta)$ per cent confidence interval for λ is

$$\bar{z} - st(\eta; n-1)/\sqrt{n} \leqslant \lambda \leqslant \bar{z} + st(\eta; n-1)/\sqrt{n},$$

where, of course, now $s^2 = \sum_{i=1}^{n} (z_i - \bar{z})^2/(n-1)$. Evidently, the invariance of F ensures that of t; but since $\mu = (\lambda - b)/a$ and $\bar{x} = (\bar{z} - b)/a$, it follows that the $100(1-\eta)$ per cent confidence interval for μ is

$$a^{-1}[\bar{z} - st(\eta; n-1)/\sqrt{n} - b] \leqslant \mu \leqslant a^{-1}[\bar{z} + st(\eta; n-1)/\sqrt{n} - b].$$

This formula enables a direct evaluation to be made of the confidence interval for μ from the analysis of variance of the transformed observations.

Example 15.1: Specific gravity of lead

We now illustrate the preceding theoretical argument by an analysis of the ten measurements of the specific gravity of lead given in Table 5.1, page 65. Physical measurements of this kind are known to have a distribution which is closely represented by a normal curve. We assume that there exists an infinite conceptual population of the measurements of the specific gravity of lead such that each value is obtained from an independent experiment. This population contains all the information about the specific gravity of lead. We postulate that this population is normal with mean μ and variance σ^2, both parameters being unknown. Then μ is the true specific gravity of lead. The ten measurements actually made are assumed to be a random sample from this population, and our problem may be looked upon in three ways as follows:

    (i) to test some null hypothesis about μ, say $H(\mu = 11 \cdot 3)$;

    (ii) to obtain the best point estimate of μ; and

    (iii) to obtain a 95 per cent (say) confidence interval for μ.

It is convenient first to make the transformation $z_i = 1000(x_i - 11 \cdot 2)$, where x_i ($i = 1, 2, ..., 10$) are the original observations. Then the z_i are

96, 89, 102, 105, 99, 94, 87, 106, 97, 95.

By this transformation we have changed the given null hypothesis $H(\mu = 11{\cdot}3)$ into the equivalent one, namely, $H(\lambda = 100)$, since

$$E(z_i) = 1000[E(x_i) - 11{\cdot}2], \quad \text{or} \quad \lambda = 1000(\mu - 11{\cdot}2).$$

(i) We now have

$$\sum_{i=1}^{10} z_i^2 = 94442 \quad \text{and} \quad \sum_{i=1}^{10} z_i = 970.$$

Therefore $\quad$ C.F.$(z) = (970)^2/10 = 94090 \quad$ and $\quad \bar{z} = 97.$

Hence $\qquad M = \sum_{i=1}^{n}(z_i - \bar{z})^2 = 352 \quad$ with 9 d.f.,

and $\qquad S = n(\bar{z} - \lambda_0)^2 = 10(97 - 100)^2 = 90 \quad$ with 1 d.f.

The analysis of variance table is given below.

Table 15.2 Showing the analysis of variance of the measurements of the specific gravity of lead

Source of Variation	d.f.	S.S.	M.S.	F.R.
Hypothesis: $\lambda = 100$	1	90	90	2·30
Residual	9	352	39·11	
Total	10	442		

The five per cent value of F with 1,9 d.f. is 5·12, obtained from the table of the F distribution; and since the observed F is less than this, the null hypothesis is acceptable. The original ten observations could have arisen from a population in which $\mu = 11{\cdot}3$. In other words, the value 11·3 for the true specific gravity of lead is in reasonable agreement with the sample.

(ii) The best point estimate of μ is $\bar{x} = 11{\cdot}2 + \bar{z}/1000 = 11{\cdot}297.$

(iii) On the other hand, to obtain a 95 per cent confidence interval for μ we proceed as follows.

First, to obtain a 95 per cent confidence interval for λ, we note that the five per cent value of t with 9 d.f. is 2·262.

Next, $s = \sqrt{39{\cdot}11} = 6{\cdot}2538 \quad$ and $\quad \dfrac{s}{\sqrt{n}} = \dfrac{6{\cdot}2538}{3{\cdot}1623} = 1{\cdot}978.$

Therefore the 95 per cent confidence interval for λ is

$$97 - 2{\cdot}262 \times 1{\cdot}978 \leqslant \lambda \leqslant 97 + 2{\cdot}262 \times 1{\cdot}978,$$

or $\qquad\qquad\qquad 92{\cdot}53 \leqslant \lambda \leqslant 101{\cdot}47.$

But for the x to z transformation $a = 1000$ and $b = -11200$. Hence the required 95 per cent confidence interval for μ is

$$\frac{1}{1000}[92{\cdot}53 + 11200 \leqslant \mu \leqslant 101{\cdot}47 + 11200],$$

or $\qquad\qquad\qquad 11{\cdot}293 \leqslant \mu \leqslant 11{\cdot}301.$

15.7 Tests of significance for the means of two normal populations

It is a natural extension of the one population case to consider next tests of null hypotheses about the means of two normal populations. Suppose that $x_{i\nu}$ ($i = 1, 2$; $\nu = 1, 2, \ldots, n_i$) are random samples of n_1 and n_2 observations respectively from two normal populations with means θ_1 and θ_2 but with a common variance σ^2, the three parameters being assumed unknown. A basic point underlying the theoretical argument is that the populations have the same variance. Admittedly, this is a restriction, but the theory of least squares leading to the analysis of variance is based on this assumption. The initial interest is in linear hypotheses about the population means θ_1 and θ_2; and any general linear hypothesis about these parameters may be expressed as $H(\alpha\theta_1 + \beta\theta_2 = \delta)$, where α, β, and δ are known constants. However, we consider here only a special class of these hypotheses, namely that for which $\alpha = 1$ and $\beta = -1$. This implies that the difference between the two population means is some given quantity δ. The particular null hypothesis $H(\theta_1 - \theta_2 = 0)$ is of great importance, and we derive a test of significance for it as a special case of the test of significance for the null hypothesis $H(\theta_1 - \theta_2 = \delta)$.

Explicitly, the observations of the two samples are

$$x_{11}, x_{12}, \ldots, x_{1n_1}; \qquad x_{21}, x_{22}, \ldots, x_{2n_2}.$$

Therefore the least-squares estimates of θ_1 and θ_2 are obtained by the minimisation of Ω with respect to θ_1 and θ_2, where

$$
\begin{aligned}
\Omega &= \sum_{i=1}^{2} \sum_{\nu=1}^{n_i} (x_{i\nu} - \theta_i)^2 \\
&= \sum_{\nu=1}^{n_1} (x_{1\nu} - \theta_1)^2 + \sum_{\nu=1}^{n_2} (x_{2\nu} - \theta_2)^2 \\
&= \left[\sum_{\nu=1}^{n_1} (x_{1\nu} - \bar{x}_1)^2 + n_1(\bar{x}_1 - \theta_1)^2 \right] + \left[\sum_{\nu=1}^{n_2} (x_{2\nu} - \bar{x}_2)^2 + n_2(\bar{x}_2 - \theta_2)^2 \right] \\
&= \sum_{i=1}^{2} \sum_{\nu=1}^{n_i} (x_{i\nu} - \bar{x}_i)^2 + \sum_{i=1}^{2} n_i(\bar{x}_i - \theta_i)^2,
\end{aligned}
$$

$\bar{x}_1$ and $\bar{x}_2$ being the means of the first and second samples respectively. Differentiation with respect to θ_1 and θ_2 gives

$$\frac{\partial \Omega}{\partial \theta_1} = -2(\bar{x}_1 - \theta_1); \qquad \frac{\partial \Omega}{\partial \theta_2} = -2(\bar{x}_2 - \theta_2).$$

Hence, as expected, the least-squares estimates of θ_1 and θ_2 are $\theta_1^* = \bar{x}_1$, $\theta_2^* = \bar{x}_2$. Therefore the absolute minimum of Ω is

$$M = \sum_{i=1}^{2} \sum_{\nu=1}^{n_i} (x_{i\nu} - \bar{x}_i)^2.$$

Also,
$$E(M) = E\left[\sum_{\nu=1}^{n_1} (x_{1\nu} - \bar{x}_1)^2 + \sum_{\nu=1}^{n_2} (x_{2\nu} - \bar{x}_2)^2 \right],$$

whence by a simple extension of the argument used in the case of a single sample, we have

$$E(M) = (n_1 - 1)\sigma^2 + (n_2 - 1)\sigma^2 = (n_1 + n_2 - 2)\sigma^2.$$

Thus M/σ^2 is distributed as χ^2 with $n_1 + n_2 - 2$ d.f., and the best estimate of σ^2 is

$$s^2 = \sum_{i=1}^{2} \sum_{\nu=1}^{n_i} (x_{i\nu} - \bar{x}_i)^2 / (n_1 + n_2 - 2).$$

If we denote the sample variances by s_1^2 and s_2^2, then

$$s^2 = \frac{(n_1 - 1)s_1^2 + (n_2 - 1)s_2^2}{n_1 + n_2 - 2}.$$

This is the more usual form of writing the expression for s^2. This estimate of σ^2 is known as a *pooled estimate* since it is derived by adding together the within sample sums of squares from the two samples. This pooling is intuitively reasonable because the populations have the same variance and each sample separately gives an estimate of σ^2.

To test the null hypothesis $H(\theta_1 - \theta_2 = \delta)$, we have next to obtain the minimum of

$$\Omega = M + \sum_{i=1}^{2} n_i (\bar{x}_i - \theta_i)^2$$

relative to the hypothesis being true. A simple way of doing this is to use the equation $\theta_1 - \theta_2 = \delta$ to eliminate one of the parameters, say θ_1, from Ω and then to derive the absolute minimum of

$$\Omega_1 = M + n_1(\bar{x}_1 - \delta - \theta_2)^2 + n_2(\bar{x}_2 - \theta_2)^2$$

with respect to variation in θ_2. Thus for this absolute minimum of Ω_1, we have

$$\frac{d\Omega_1}{d\theta_2} = -2n_1(\bar{x}_1 - \delta - \theta_2) - 2n_2(\bar{x}_2 - \theta_2).$$

Hence the estimate of θ_2 obtained by this minimisation is

$$\theta_2^{**} = \bar{x} - n_1\delta/(n_1 + n_2),$$

where $\bar{x}$ is the grand mean of all the $n_1 + n_2$ observations, that is,

$$\bar{x} = \frac{n_1\bar{x}_1 + n_2\bar{x}_2}{n_1 + n_2} = \sum_{i=1}^{2} \sum_{\nu=1}^{n_i} x_{i\nu} / (n_1 + n_2).$$

It is important to note that the estimate $\theta_2^{**} \neq \theta_2^{*}$, since it has been derived on the assumption that the null hypothesis $H(\theta_1 - \theta_2 = \delta)$ is true.

Now Ω_0, the relative minimum of Ω, is the absolute minimum of Ω_1, and we have

$$\begin{aligned}
\Omega_0 &= M + n_1[\bar{x}_1 - \delta - \bar{x} + n_1\delta/(n_1+n_2)]^2 + n_2[\bar{x}_2 - \bar{x} + n_1\delta/(n_1+n_2)]^2 \\
&= M + n_1[\bar{x}_1 - \bar{x} - n_2\delta/(n_1+n_2)]^2 + n_2[\bar{x}_2 - \bar{x} + n_1\delta/(n_1+n_2)]^2 \\
&= M + n_1[n_2(\bar{x}_1 - \bar{x}_2 - \delta)/(n_1+n_2)]^2 + n_2[n_1(\bar{x}_2 - \bar{x}_1 + \delta)/(n_1+n_2)]^2 \\
&= M + n_1 n_2(\bar{x}_1 - \bar{x}_2 - \delta)^2/(n_1 + n_2).
\end{aligned}$$

Hence the sum of squares for testing the null hypothesis $H(\theta_1 - \theta_2 = \delta)$ is

$$S = \Omega_0 - M = n_1 n_2(\bar{x}_1 - \bar{x}_2 - \delta)^2/(n_1 + n_2).$$

Furthermore, if the null hypothesis $H(\theta_1 - \theta_2 = \delta)$ is true, then we have

$$E(\bar{x}_1 - \bar{x}_2 - \delta) = 0.$$

Therefore

$$E(S) = \frac{n_1 n_2}{n_1 + n_2} E(\bar{x}_1 - \bar{x}_2 - \delta)^2$$

$$= \frac{n_1 n_1}{n_1 + n_2} \, \text{var}(\bar{x}_1 - \bar{x}_2 - \delta)$$

$$= \frac{n_1 n_2}{n_1 + n_2} \, \text{var}(\bar{x}_1 - \bar{x}_2), \quad \text{as } \delta \text{ is a constant,}$$

$$= \frac{n_1 n_2}{n_1 + n_2} \, \sigma^2 \left[\frac{1}{n_1} + \frac{1}{n_2} \right], \quad \text{since the means are independent,}$$

$$= \sigma^2.$$

Thus, if the null hypothesis under test is true, then S is an unbiased estimate of σ^2; and, as previously, S/σ^2 is distributed as χ^2 with 1 d.f. The independence of the two χ^2's M/σ^2 and S/σ^2 is ensured by general theory. Hence the test of significance for the null hypothesis $H(\theta_1 - \theta_2 = \delta)$ is given by the ratio

$$F = (n_1 + n_2 - 2)S/M,$$

which has the F distribution with $1, n_1 + n_2 - 2$ d.f.

Equality of the means

For $\delta = 0$ the null hypothesis $H(\theta_1 - \theta_2 = 0)$ specifies the equality of the population means; and for testing this null hypothesis, we have

$$S = n_1 n_2 (\bar{x}_1 - \bar{x}_2)^2 / (n_1 + n_2) \quad \text{with 1 d.f.,}$$

and

$$M = \sum_{i=1}^{2} \sum_{\nu=1}^{n_i} (x_{i\nu} - \bar{x}_i)^2 \quad \text{with } n_1 + n_2 - 2 \text{ d.f.}$$

These two sums of squares do not represent the total variation of the two samples since we have still to account for the missing one degree of freedom. To do this, we first need to evaluate the sum $S + M$, which is most simply obtained indirectly by the decomposition of the sum of squares of the observations from the grand mean $\bar{x}$. Thus we have

$$\sum_{i=1}^{2} \sum_{\nu=1}^{n_i} (x_{i\nu} - \bar{x})^2 = \sum_{i=1}^{2} \sum_{\nu=1}^{n_i} [(x_{i\nu} - \bar{x}_i) + (\bar{x}_i - \bar{x})]^2$$

$$= \sum_{i=1}^{2} \sum_{\nu=1}^{n_i} [(x_{i\nu} - \bar{x}_i)^2 + (\bar{x}_i - \bar{x})^2 + 2(\bar{x}_i - \bar{x})(x_{i\nu} - \bar{x}_i)]$$

$$= \sum_{i=1}^{2} \sum_{\nu=1}^{n_i} (x_{i\nu} - \bar{x}_i)^2 + \sum_{i=1}^{2} n_i (\bar{x}_i - \bar{x})^2, \tag{1}$$

since for fixed i

$$\sum_{\nu=1}^{n_i} (x_{i\nu} - \bar{x}_i) \equiv 0.$$

Also,

$$\sum_{i=1}^{2} n_i (\bar{x}_i - \bar{x})^2 = n_1 (\bar{x}_1 - \bar{x})^2 + n_2 (\bar{x}_2 - \bar{x})^2$$

$$= n_1 [n_2 (\bar{x}_1 - \bar{x}_2)/(n_1 + n_2)]^2 + n_2 [n_1 (\bar{x}_2 - \bar{x}_1)/(n_1 + n_2)]^2$$

$$= n_1 n_2 (\bar{x}_1 - \bar{x}_2)^2 / (n_1 + n_2),$$

or

$$S = n_1 n_2 (\bar{x}_1 - \bar{x}_2)^2 / (n_1 + n_2) = \sum_{i=1}^{2} n_i (\bar{x}_i - \bar{x})^2. \tag{2}$$

The second form of the expression for S indicates that this sum of squares is proportional to the square of the deviations of the sample means from the grand mean $\bar{x}$, and so S is often also referred to as the *between samples sum of squares.*

Using (2) in (1), we immediately deduce that

$$\sum_{i=1}^{2} \sum_{\nu=1}^{n_i} (x_{i\nu} - \bar{x})^2 = S + M.$$

This sum $S + M$ is called the *total sum of squares* and it is usually denoted by T. It represents the sum of the squares of the deviations of all the observations from $\bar{x}$. Clearly, T has $n_1 + n_2 - 1$ d.f., and it does not represent the total variation of the two samples when the null hypothesis $H(\theta_1 - \theta_2 = 0)$ is true. The search for the missing degree of freedom is crucial to the understanding of the analysis of variance.

We observe that there are two independent parameters θ_1 and θ_2. The null hypothesis states that these two parameters are equal but it does not specify their common value. Accordingly, the null hypothesis can be interpreted as $H(\theta_1 = \theta_2 = \theta)$, where the common value θ is unspecified. Hence, if the null hypothesis is true, then the total variation of the two samples is

$$\sum_{i=1}^{2} \sum_{\nu=1}^{n_i} (x_{i\nu} - \theta)^2, \quad \text{and this has } n_1 + n_2 \text{ d.f.}$$

Besides,

$$\sum_{i=1}^{2} \sum_{\nu=1}^{n_i} (x_{i\nu} - \theta)^2 = \sum_{i=1}^{2} \sum_{\nu=1}^{n_i} [(x_{i\nu} - \bar{x}) + (\bar{x} - \theta)]^2$$

$$= \sum_{i=1}^{2} \sum_{\nu=1}^{n_i} [(x_{i\nu} - \bar{x})^2 + (\bar{x} - \theta)^2 + 2(\bar{x} - \theta)(x_{i\nu} - \bar{x})]$$

$$= T + (n_1 + n_2)(\bar{x} - \theta)^2,$$

since

$$\sum_{i=1}^{2} \sum_{\nu=1}^{n_i} (x_{i\nu} - \bar{x}) = \sum_{i=1}^{2} n_i(\bar{x}_i - \bar{x}) \equiv 0.$$

We thus have the fundamental equation

$$\sum_{i=1}^{2} \sum_{\nu=1}^{n_i} (x_{i\nu} - \theta)^2 - T = (n_1 + n_2)(\bar{x} - \theta)^2.$$

The quantity on the right-hand side is the sum of squares due to the missing degree of freedom. This measures the variation of the grand mean $\bar{x}$ from θ, the common but unspecified value of θ_1 and θ_2. It is conventional to ignore this single degree of freedom and its sum of squares in the analysis of variance table, as interest centres on the difference between the two means θ_1 and θ_2. However, if we wish to test any hypothesis about θ, then we have to use the sum of squares

$$(n_1 + n_2)(\bar{x} - \theta)^2 \quad \text{with } 1 \text{ d.f.}$$

We follow the standard convention of ignoring this degree of freedom and its sum of squares in the analysis of variance table.

Computational notes

The simplest (and standard) computational procedure is to evaluate T and S, and then obtain M as the difference $T - S$.

It is possible to obtain a convenient computational form for S as follows.

Table 15.3 Showing the analysis of variance for the comparison of two means

Source of Variation	d.f.	S.S.	M.S.	F.R.
Hypothesis: $\theta_1 - \theta_2 = 0$.	1	$S = \dfrac{n_1 n_2}{n_1 + n_2}(\bar{x}_1 - \bar{x}_2)^2$	S	$(n_1 + n_2 - 2)S/M$
Residual	$n_1 + n_2 - 2$	$M = \sum\limits_{i=1}^{2} \sum\limits_{\nu=1}^{n_i} (x_{i\nu} - \bar{x}_i)^2$	$M/(n_1 + n_2 - 2)$	
Total	$n_1 + n_2 - 1$	$T = \sum\limits_{i=1}^{2} \sum\limits_{\nu=1}^{n_i} (x_{i\nu} - \bar{x})^2$		

We have (Table 15.3):

$$S = \sum_{i=1}^{2} n_i(\bar{x}_i - \bar{x})^2$$

$$= \sum_{i=1}^{2} n_i(\bar{x}_i^2 - 2\bar{x}\bar{x}_i + \bar{x}^2)^2$$

$$= \sum_{i=1}^{2} n_i \bar{x}_i^2 - 2\bar{x} \sum_{i=1}^{2} n_i \bar{x}_i + (n_1 + n_2)\bar{x}^2$$

$$= \sum_{i=1}^{2} n_i \bar{x}_i^2 - (n_1 + n_2)\bar{x}^2,$$

or

$$S = \sum_{i=1}^{2} B_i^2/n_i - G^2/(n_1 + n_2); \qquad (3)$$

where

$$B_i = \sum_{\nu=1}^{n_i} x_{i\nu}$$

is the sum of the ith sample observations, $i = 1, 2$; and

$$G = \sum_{i=1}^{2} \sum_{\nu=1}^{n_i} x_{i\nu} = \sum_{i=1}^{2} B_i$$

is the grand total of the observations. The expression for S given in (3) is the appropriate form for computation. Similarly,

$$T = \sum_{i=1}^{2} \sum_{\nu=1}^{n_i} (x_{i\nu} - \bar{x})^2$$

$$= \sum_{i=1}^{2} \sum_{\nu=1}^{n_i} (x_{i\nu}^2 - 2\bar{x}x_{i\nu} + \bar{x}^2)$$

$$= \sum_{i=1}^{2} \sum_{\nu=1}^{n_i} x_{i\nu}^2 - 2\bar{x} \sum_{i=1}^{2} \sum_{\nu=1}^{n_i} x_{i\nu} + (n_1 + n_2)\bar{x}^2$$

$$= \sum_{i=1}^{2} \sum_{\nu=1}^{n_i} x_{i\nu}^2 - (n_1 + n_2)\bar{x}^2,$$

or

$$T = \text{raw S.S.} - G^2/(n_1 + n_2),$$

which is the required computational form for T.

Thus, finally, we have the following computational procedure for the analysis of variance.

Calculate

(i) the raw S.S., the sum of squares of all the $n_1 + n_2$ observations.

(ii) the B_i, the total of each sample separately, $i = 1, 2$.

(iii) G, the sum of the B_i, which is the grand total of all the $n_1 + n_2$ observations.

(iv) the correction factor for the grand mean, that is $G^2/(n_1 + n_2)$.

(v) the sum $\sum_{i=1}^{2} B_i^2/n_i$.

(vi) total S.S. (T) = raw S.S. $- G^2/(n_1 + n_2)$.

(vii) the S.S. due to the null hypothesis $H(\theta_1 = \theta_2)$ by the equation

$$S = \sum_{i=1}^{2} B_i^2/n_i - G^2/(n_1 + n_2).$$

The analysis of variance table can now be completed without any further basic calculations.

15.8   Confidence intervals for the difference of the population means

Since the F ratio for testing the null hypothesis $H(\theta_1 - \theta_2 = 0)$ has 1, $n_1 + n_2 - 2$ d.f., it follows that

$$\sqrt{F} = t = \left[\frac{n_1 n_2}{n_1 + n_2}\right]^{\frac{1}{2}} \frac{\bar{x}_1 - \bar{x}_2}{s}$$

has "Student's" distribution with $n_1 + n_2 - 2$ d.f. if the null hypothesis under test is true. This result is in conformity with the general definition of t because if the null hypothesis $H(\theta_1 - \theta_2 = 0)$ is true, then the random variable

$$\frac{\bar{x}_1 - \bar{x}_2}{\sigma\left[\dfrac{1}{n_1} + \dfrac{1}{n_2}\right]^{\frac{1}{2}}}$$

has the unit normal distribution, and $M/\sigma^2 = (n_1 + n_2 - 2)s^2/\sigma^2$ has the χ^2 distribution with $n_1 + n_2 - 2$ d.f.; and the independence of these two distributions is ensured by general theory.

Next, suppose the null hypothesis $H(\theta_1 - \theta_2 = 0)$ is not true so that $\theta_1 - \theta_2 = \delta \neq 0$. Then, irrespective of the value of δ, we have the result that

$$t = \left[\frac{n_1 n_2}{n_1 + n_2}\right]^{\frac{1}{2}} \frac{\bar{x}_1 - \bar{x}_2 - \delta}{s}$$

has "Student's" distribution with $n_1 + n_2 - 2$ d.f. Hence, if $t(\eta; n_1 + n_2 - 2)$ is the 100η $(0 < \eta < 1)$ per cent point of "Student's" distribution with $n_1 + n_2 - 2$ d.f., we must have

$$P[|t| \geqslant t(\eta; n_1 + n_2 - 2)] = \eta.$$

Therefore the $100(1 - \eta)$ per cent confidence interval for δ is

$$\bar{x}_1 - \bar{x}_2 - \left[\frac{n_1 + n_2}{n_1 n_2}\right]^{\frac{1}{2}} s\,t(\eta; n_1 + n_2 - 2) \leqslant \delta$$

$$\leqslant \bar{x}_1 - \bar{x}_2 + \left[\frac{n_1 + n_2}{n_1 n_2}\right]^{\frac{1}{2}} s\,t(\eta; n_1 + n_2 - 2).$$

Linear transformation of the observations

As in the case of one sample, it is frequently convenient to make a linear transformation of the observations $x_{i\nu}$ to simplify the computations.

Thus, if we make the transformation

$$z_{i\nu} = ax_{i\nu} + b, \quad \text{for } i = 1, 2; \quad \nu = 1, 2, \ldots, n_i,$$

where a and b are constants and $a \neq 0$, then the F ratio remains invariant. However, for $\delta \neq 0$, we have

$$\delta = \theta_1 - \theta_2 = a^{-1}[E(z_{1\nu}) - E(z_{2\nu})].$$

Therefore, in terms of the transformed observations, the $100(1 - \eta)$ per cent confidence interval for δ is

$$a^{-1}\left[\bar{z}_1 - \bar{z}_2 - \left(\frac{n_1 + n_2}{n_1 n_2}\right)^{\frac{1}{2}} s\, t(\eta;\, n_1 + n_2 - 2)\right] \leqslant \delta$$

$$\leqslant a^{-1}\left[\bar{z}_1 - \bar{z}_2 + \left(\frac{n_1 + n_2}{n_1 n_2}\right)^{\frac{1}{2}} s\, t(\eta;\, n_1 + n_2 - 2)\right],$$

where $\bar{z}_1$ and $\bar{z}_2$ are the means of the transformed observations of the two samples and

$$s^2 = \sum_{i=1}^{2} \sum_{\nu=1}^{n_i} (z_{i\nu} - \bar{z}_i)^2 / (n_1 + n_2 - 2).$$

Example 15.2: Length of cuckoo's eggs

We illustrate the preceding theory by an analysis of the data on the lengths of cuckoo's eggs found in the nests of hedge-sparrows and reed-warblers quoted in Table 5.2 (p. 66).

We first postulate an infinite population of measurements of the length of cuckoo's eggs found in the nests of hedge-sparrows, and another population of similar measurements of eggs found in the nests of reed-warblers. Next, we assume that both these populations are normal with a common variance σ^2, and that their respective means are θ_1 and θ_2. The null hypothesis of interest is $H(\theta_1 - \theta_2 = 0)$, that is, the cuckoo cannot adapt the size of its eggs according to the nest of the host in which it lays them. The alternatives to this null hypothesis are $\theta_1 - \theta_2 \neq 0$. The two samples are supposed to arise from the two populations respectively, and we use these data to test the stated null hypothesis.

The sample observations are denoted by $x_{i\nu}$, and it is convenient to make the transformation

$$z_{i\nu} = 10(x_{i\nu} - 20), \quad \text{for } i = 1, 2; \quad \nu = 1, 2, \ldots, n_i,$$

where $n_1 = 14$ and $n_2 = 10$. The transformed observations are given in the following table.

Table 15.4   Showing the transformed measurements of cuckoo's eggs found in the nests of hedge-sparrows and reed-warblers

Hedge-sparrows 1st Sample $(z_{1\nu})$ $n_1 = 14$	20, 39, 9, 38, 50, 40, 17, 38, 28, 31, 31, 35, 30, 30
Reed-warblers 2nd Sample $(z_{2\nu})$ $n_2 = 10$	32, 20, 22, 12, 16, 16, 19, 20, 29, 28

The numerical calculations for the analysis of variance are:

(i)   raw S.S. $= 19960$.

(ii) $B_1 = 436$; $B_2 = 214$. $\bar{z}_1 = 31 \cdot 14$; $\bar{z}_2 = 21 \cdot 40$.

(iii) $G = 650$.

(iv) $G^2/(n_1 + n_2) = \dfrac{422500}{24} = 17604 \cdot 17$.

(v) $\sum\limits_{i=1}^{2} B_i^2/n_i = \dfrac{190096}{14} + \dfrac{45796}{10} = 13578 \cdot 29 + 4579 \cdot 60$

$= 18\,157 \cdot 89$.

(vi) $T = 2355 \cdot 83$ with 23 d.f.

(vii) $S = 553 \cdot 72$ with 1 d.f.

Table 15.5   Showing the analysis of variance of the data of Table 15.4

Source of Variation	d.f.	S.S.	M.S.	F.R.
Hypothesis: $\theta_1 = \theta_2$	1	553·72	553·72	6·76
Residual	22	1802·11	81·91	
Total	23	2355·83		

The observed value of F is $6 \cdot 76$ with 1, 22 d.f., and the 2·5 per cent table value of F with 1, 22 d.f. is $5 \cdot 79$. Hence the observed F is significant at the 2·5 per cent level of significance, and we can therefore reject the null hypothesis. In other words, the data provide evidence that, on the average, the cuckoo has the capacity to adapt its egg length.

Since the null hypothesis has been rejected, we may assume that $\theta_1 - \theta_2 = \delta \neq 0$.

The point estimate of δ is $a^{-1}(\bar{z}_1 - \bar{z}_2) = 0 \cdot 974$. Again, in terms of the transformed observations, we have

$$s = \sqrt{81 \cdot 91} = 9 \cdot 0504; \qquad \left[\frac{n_1 + n_2}{n_1 n_2}\right]^{\frac{1}{2}} = \sqrt{0 \cdot 1714} = 0 \cdot 4140.$$

Also, the five per cent table value of t with 22 d.f. is $2 \cdot 074$. Therefore the 95 per cent confidence interval for δ is

$$\tfrac{1}{10}(9 \cdot 74 - 2 \cdot 074 \times 0 \cdot 4140 \times 9 \cdot 0504) \leqslant \delta \leqslant \tfrac{1}{10}(9 \cdot 74 + 2 \cdot 074 \times 0 \cdot 4140 \times 9 \cdot 0504),$$

or

$$0 \cdot 197 \leqslant \delta \leqslant 1 \cdot 751.$$

This confidence interval means that there is 95 per cent chance that the interval (0·197, 1·751) includes the value of δ, the true difference between lengths of cuckoo's eggs laid in the nests of hedge-sparrows and reed-warblers.

15.9   Tests of significance for the means of k ($\geqslant 2$) normal populations

The analysis of variance leading to the test of significance for the equality of the means of two normal populations is a particular case of the general analysis of variance for testing the simultaneous equality of the means of k normal populations. However, for $k > 2$ the null hypothesis of the equality of the means reveals an additional feature which the simpler case of two means does not possess. To explain this difference, suppose that the means of the k normal populations are $\theta_1, \theta_2, \ldots, \theta_k$, and that the populations have the common variance σ^2. There are now k independent samples of sizes $n_1, n_2, \ldots, n_k$, one from each population, and we set for convenience $N = \sum_{i=1}^{k} n_i$. The sample observations are $x_{i\nu}$, for $i = 1, 2, \ldots, k$; $\nu = 1, 2, \ldots, n_i$. Given these observations, we wish to derive a test of significance for the null hypothesis $H(\theta_1 = \theta_2 = \ldots = \theta_k)$, the common value of the θ_i being unspecified. As already remarked, this null hypothesis refers to the simultaneous equality of k independent parameters, and it is evidently equivalent to the statement that the θ_i satisfy simultaneously $k - 1$ independent relations of equality. According to this interpretation, we conclude that in the particular case of $k = 2$, the null hypothesis implies that the linear function $\theta_1 - \theta_2 = 0$ for all values of the parameters. On the other hand, for $k > 2$, the null hypothesis implies that, for example, the θ_i satisfy simultaneously the $k - 1$ linear relations

$$\theta_1 + \theta_2 + \ldots + \theta_{r-1} - (r - 1)\theta_r = 0, \quad \text{for } r = 2, 3, \ldots, k.$$

This is not the only set of $k - 1$ linear relations which jointly represent the null hypothesis $H(\theta_1 = \theta_2 = \ldots = \theta_k)$, but for simplicity we consider this set only. As we show, the analysis of variance now leads to an F ratio which has $k - 1, N - k$ d.f., but for $k > 2$ the test of significance based on this F cannot be expressed equivalently in terms of "Student's" distribution as is the case for $k = 2$. There are important consequences of this result, which we consider after the derivation of the appropriate F ratio.

As before, the least-squares estimates of $\theta_1, \theta_2, \ldots, \theta_k$ are obtained by the simultaneous minimisation of Ω with respect to the k parameters, where

$$
\begin{aligned}
\Omega &= \sum_{i=1}^{k} \sum_{\nu=1}^{n_i} (x_{i\nu} - \theta_i)^2 \\
&= \sum_{i=1}^{k} \sum_{\nu=1}^{n_i} [(x_{i\nu} - \bar{x}_i) + (\bar{x}_i - \theta_i)]^2, \quad \bar{x}_i \text{ being the sample means,} \\
&= \sum_{i=1}^{k} \sum_{\nu=1}^{n_i} [(x_{i\nu} - \bar{x}_i)^2 + (\bar{x}_i - \theta_i)^2 + 2(\bar{x}_i - \theta_i)(x_{i\nu} - \bar{x}_i)] \\
&= \sum_{i=1}^{k} \sum_{\nu=1}^{n_i} (x_{i\nu} - \bar{x}_i)^2 + \sum_{i=1}^{k} n_i(\bar{x}_i - \theta_i)^2,
\end{aligned}
$$

since
$$\sum_{\nu=1}^{n_i} (x_{i\nu} - \bar{x}_i) \equiv 0 \quad \text{for all } i.$$

Therefore
$$\frac{\partial \Omega}{\partial \theta_i} = -2n_i(\bar{x}_i - \theta_i), \quad \text{for } i = 1, 2, \ldots, k,$$

so that the least-squares estimates of the θ_i are $\theta_i^* = \bar{x}_i$. Hence the absolute minimum of Ω is

$$M = \sum_{i=1}^{k} \sum_{\nu=1}^{n_i} (x_{i\nu} - \bar{x}_i)^2.$$

We observe that M is simply the pooled within samples sum of squares for the k independent samples. Therefore

$$E(M) = \sum_{i=1}^{k}(n_i - 1)\sigma^2 = (N - k)\sigma^2.$$

Hence $s^2 = M/(N - k)$ is the least-squares estimate of σ^2, and M/σ^2 is distributed as χ^2 with $N - k$ d.f.

Next, to test the null hypothesis $H(\theta_1 = \theta_2 = \ldots = \theta_k = \theta)$, where θ is the unknown common value of the θ_i, we derive the relative minimum of Ω by minimising

$$\Omega_1 = M + \sum_{i=1}^{k} n_i(\bar{x}_i - \theta)^2$$

with respect to variation in θ. The estimate of θ obtained by this minimisation is

$$\theta^* = \sum_{i=1}^{k} n_i \bar{x}_i /N = \bar{x},$$

the grand mean of the N observations. Therefore the relative minimum of Ω is

$$\Omega_0 = M + \sum_{i=1}^{k} n_i(\bar{x}_i - \bar{x})^2,$$

whence the sum of squares for the null hypothesis under test is

$$S = \Omega_0 - M = \sum_{i=1}^{k} n_i(\bar{x}_i - \bar{x})^2.$$

This sum of squares is the between samples sum of squares and measures the variation of the sample means from the grand mean. We can also write this as

$$S = \sum_{i=1}^{k} n_i [(\bar{x}_i - \theta) - (\bar{x} - \theta)]^2$$
$$= \sum_{i=1}^{k} n_i(\bar{x}_i - \theta)^2 - N(\bar{x} - \theta)^2, \quad \text{as} \quad \sum_{i=1}^{k} n_i(\bar{x}_i - \theta) = N(\bar{x} - \theta).$$

Now, if the null hypothesis under test is true, we have

$$E(\bar{x}_i - \theta)^2 = \text{var}(\bar{x}_i) = \sigma^2/n_i, \quad \text{for } i = 1, 2, \ldots, k,$$

and
$$E(\bar{x} - \theta)^2 = \text{var}(\bar{x}) = \sigma^2/N.$$

Therefore
$$E(S) = \sum_{i=1}^{k} n_i \text{var}(\bar{x}_i) - N \text{var}(\bar{x}) = (k - 1)\sigma^2.$$

As expected, $S/(k - 1)$ is an unbiased estimate of σ^2 if the null hypothesis under test is true, and S/σ^2 is distributed as χ^2 with $k - 1$ d.f. The independence of the χ^2's M/σ^2 and S/σ^2 is assured by general theory, and so for testing the null hypothesis, the ratio

$$F = (N - k)S/(k - 1)M$$

has the F distribution with $k - 1$, $N - k$ d.f. The analysis of variance may now be presented.

Computational notes

The calculations for the analysis of variance in Table 15.6, overleaf, are exactly a generalisation of the two-sample case, and we indicate the outline of the procedure only. As in the two-sample case, the total sum of squares of the k samples is

Table 15.6   Showing the analysis of variance for the comparison of **k** means

Source of Variation	d.f.	M.S.	S.S.	F.R.
Hypothesis: $\theta_1 = \theta_2 = \ldots = \theta_k$	$k-1$	$S = \sum\limits_{i=1}^{k} n_i(\bar{x}_i - \bar{x})^2$	$S/(k-1)$	$(N-k)\,S/M(k-1)$
Residual	$N-k$	$M = \sum\limits_{i=1}^{k}\sum\limits_{\nu=1}^{n_i}(x_{i\nu}-\bar{x}_i)^2$	$M/(N-k)$	
Total	$N-1$	$T = \sum\limits_{i=1}^{k}\sum\limits_{\nu=1}^{n_i}(x_{i\nu}-\bar{x})^2$		

$$
\begin{aligned}
T &= \sum_{i=1}^{k}\sum_{\nu=1}^{n_i}(x_{i\nu}-\bar{x})^2 \\
&= \sum_{i=1}^{k}\sum_{\nu=1}^{n_i}[(x_{i\nu}-\bar{x}_i)+(\bar{x}_i-\bar{x})]^2 \\
&= \sum_{i=1}^{k}\sum_{\nu=1}^{n_i}(x_{i\nu}-\bar{x}_i)^2 + \sum_{i=1}^{k} n_i(\bar{x}_i-\bar{x})^2 = M + S.
\end{aligned}
$$

Thus it is simplest to evaluate T and S separately, and then to obtain the residual sum of squares M as the difference $T-S$. If we write G for the grand total of the N observations and B_i for the total of the n_i values of the ith sample, then

$$
T = \sum_{i=1}^{k}\sum_{\nu=1}^{n_i}(x_{i\nu}-\bar{x})^2 = \sum_{i=1}^{k}\sum_{\nu=1}^{n_i}x_{i\nu}^2 - G^2/N,
$$

and

$$
S = \sum_{i=1}^{k} n_i(\bar{x}_i-\bar{x})^2 = \sum_{i=1}^{k} B_i^2/n_i - G^2/N.
$$

Hence the following computational procedure.

Calculate

   (i)   the raw S.S., the sum of the squares of all the N observations.

   (ii) the B_i, the sum of each sample separately, for $i = 1, 2, \ldots, k$.

  (iii) G, the sum of all the B_i.

  (iv) the correction factor for the grand mean, that is G^2/N.

   (v)   the sum $\sum\limits_{i=1}^{k} B_i^2/n_i$.

  (vi) total S.S. (T) = raw S.S. $- G^2/N$.

 (vii) the S.S. due to the null hypothesis as $S = \sum\limits_{i=1}^{k} B_i^2/n_i - G^2/N$.

The analysis of variance table can be formed now without any further basic calculations.

Null hypothesis not acceptable

   If the null hypothesis $H(\theta_1 = \theta_2 = \ldots = \theta_k)$ is not tenable, then we may write

$$
\begin{aligned}
S &= \sum_{i=1}^{k} n_i(\bar{x}_i-\bar{x})^2 \\
&= \sum_{i=1}^{k} n_i[(\bar{x}_i-\theta_i)-(\bar{x}-\bar{\theta})+(\theta_i-\bar{\theta})]^2, \quad \text{where } \bar{\theta} \equiv \sum_{i=1}^{k} n_i\theta_i/N,
\end{aligned}
$$

$$= \sum_{i=1}^{k} n_i [(\bar{x}_i - \theta_i)^2 + (\bar{x} - \bar{\theta})^2 + (\theta_i - \bar{\theta})^2 - 2(\bar{x} - \bar{\theta})(\bar{x}_i - \theta_i) +$$

$$+ 2(\theta_i - \bar{\theta})(\bar{x}_i - \theta)],$$

the other cross-product term being zero since $\sum_{i=1}^{k} n_i (\theta_i - \bar{\theta}) \equiv 0$. Now,

$$E(\bar{x}_i) = \theta_i, \qquad E(\bar{x}) = \bar{\theta},$$

so that $\qquad E(\bar{x}_i - \theta_i)^2 = \mathrm{var}(\bar{x}_i) = \sigma^2/n_i, \qquad \text{for } i = 1, 2, \ldots, k,$

and $\qquad E(\bar{x} - \bar{\theta})^2 = \mathrm{var}(\bar{x}) = N^{-2} \sum_{i=1}^{k} n_i^2 \, \mathrm{var}(\bar{x}_i) = \sigma^2/N.$

Therefore $\qquad E(S) = (k-1)\sigma^2 + \sum_{i=1}^{k} n_i (\theta_i - \bar{\theta})^2,$

which shows that if the null hypothesis $H(\theta_1 = \theta_2 = \ldots = \theta_k)$ is not true, then $E(S) > (k-1)\sigma^2$. This implies that, on the average, the effect of the null hypothesis under test being not true is to increase the value of S and so also of the mean square $S/(k-1)$. This is a characteristic feature of the analysis of variance.

Linear hypotheses about the means

As another important conclusion, we note that when the observed F ratio is rejected at a given level of significance, it implies that, on the whole, the differences between the sample means are too large to permit the acceptance of the null hypothesis $H(\theta_1 = \theta_2 = \ldots = \theta_k)$. Alternatively, we may also, for example, infer that the $k-1$ linear relations

$$\theta_1 + \theta_2 + \ldots + \theta_{r-1} - (r-1)\theta_r = 0, \qquad \text{for } r = 2, 3, \ldots, k,$$

cannot be regarded as holding simultaneously. The F test is thus an over-all test of significance and, in general, when the observed value of F is found to be significant, it may not necessarily be true that all the above $k-1$ linear relations are untenable. Accordingly, depending upon the physical nature of the populations sampled, interest may centre upon some one linear hypothesis about the means θ_i. Now any linear relation between the θ_i may be expressed as

$$\sum_{i=1}^{k} c_i \theta_i = \delta,$$

where the c_i are known constants not all zero simultaneously, and δ is another constant not involving the θ_i or σ^2. We may thus consider a test of significance for the null hypothesis

$$H \left[\sum_{i=1}^{k} c_i \theta_i = \delta_0 \right] \quad \text{for some preassigned value } \delta_0 \text{ of } \delta.$$

A general linear hypothesis of this kind evidently includes as special cases null hypotheses such as $H(\theta_1 = \delta_0)$ or $H(\theta_2 - \theta_3 = \delta_0)$ pertaining to a specific mean or to the difference between two specified means respectively. On the other hand, we may also look upon δ as an unknown parameter defined as a known linear function of the θ_i, and we may wish to obtain a $100(1-\eta)$ per cent confidence interval for δ. Both these problems can be solved by a slight extension of "Student's" argument based on a simple but important result in the theory of least squares.

Since the sample means $\bar{x}_i$ are the least-squares estimates of the θ_i, it is known that the least-squares estimate of δ is

$$\delta^* = \sum_{i=1}^{k} c_i \bar{x}_i.$$

This means that the least-squares estimate δ^* is obtained by substituting for the θ_i their least-squares estimates in the linear expression for δ. This result is intuitively reasonable, but we do not give its proof here. Next, we observe that δ^* is a linear function of the independent and normally distributed sample means, the $\bar{x}_i$, such that

$$E(\delta^*) = E\left[\sum_{i=1}^{k} c_i \bar{x}_i\right] = \sum_{i=1}^{k} c_i \theta_i = \delta,$$

and $\qquad \text{var}(\delta^*) = \text{var}\left[\sum_{i=1}^{k} c_i \bar{x}_i\right] = \sum_{i=1}^{k} c_i^2 \, \text{var}(\bar{x}_i) = \sigma^2 \sum_{i=1}^{k} c_i^2/n_i.$

Therefore $\qquad\qquad \text{s.e.}(\delta^*) = s\left[\sum_{i=1}^{k} c_i^2/n_i\right]^{\frac{1}{2}}.$

It is also known that δ^* is normally distributed with the above mean and variance, and that this distribution is independent of the χ^2 distribution of M/σ^2 with $N-k$ d.f. Hence, whatever be the true value of δ, the random variable

$$t = \frac{\delta^* - \delta}{\text{s.e.}(\delta^*)} \qquad (1)$$

has "Student's" distribution with $N-k$ d.f. This distribution may be used to test any null hypothesis about δ, and also to obtain confidence intervals for δ. This test is a particular example of the "generalised" t test which we consider in a subsequent section of this chapter. The important point about the random variable t defined in (1) is that it uses the residual mean square of the analysis of variance as an estimate of σ^2.

Example 15.3: Comparative animal feeding experiment

We now illustrate the preceding theory by analysing the data of Crampton and Hopkins obtained from a comparative feeding trial with pigs and quoted in Table 5.8 (page 70). Here $k = 5$ and all $n_i = 10$, so that $N = 50$. The equality of the sample sizes slightly simplifies the numerical calculations. However, before we carry out the computations, it is important to understand how the theoretical model is applicable to the data. The five lots correspond to different treatments which may be regarded as different types of diets. We postulate an infinite population of gains in weight in a physical population of pigs given any one of the five diets. There are thus five distinct populations of gains in weight, and these are assumed to be normal with means $\theta_1, \theta_2, \theta_3, \theta_4, \theta_5$ respectively, but with the same unknown variance σ^2. The practical assumption in such experiments is that the animals are distributed randomly amongst the different treatments. Consequently, a natural extension of this assumption is that the pigs in the postulated five populations are comparable, and so the differences between their gains in weight are, apart from intrinsic chance variability, due to the diets. The randomisation of the pigs in the finite experiment and the comparability of the five populations are fundamental principles which justify respectively the independence of the observations and the theoretical comparability of the θ_i. In the present context, the comparability of the populations implies that they have the same variance, and without this assumption the analysis of variance is not applicable.

In order to simplify the computations of the analysis of variance, it is possible to use a linear transformation of the observations such as

$$z_{i\nu} = x_{i\nu} - 100, \quad \text{or} \quad z_{i\nu} = x_{i\nu} - 130.$$

However, this is not particularly necessary in this rather simple example, and we analyse the untransformed observations. The computations are as follows:

(i)   raw S.S. = 1451305.

(ii) $B_1 = 1566; \quad B_2 = 1735; \quad B_3 = 1619; \quad B_4 = 1802; \quad B_5 = 1755.$

(iii) $G = 8477.$

(iv) $G^2/N = (8477)^2/50 = 1437190 \cdot 58.$

(v) $\sum\limits_{i=1}^{k} B_i^2/n_i = 1441097 \cdot 1.$

(vi) $T = 14114 \cdot 42$ with 49 d.f.

(vii) $S = 3906 \cdot 52$ with 4 d.f.

The analysis of variance table deduced from the above calculations is given below.

Table 15.7 Showing the analysis of variance of the gains of weight in pigs in a comparative feeding trial

Source of Variation	d.f.	S.S.	M.S.	F.R.
Hypothesis: $\theta_1 = \theta_2 = \theta_3 = \theta_4 = \theta_5$	4	3906·52	976·63	4·31
Residual	45	10207·90	226·84	
Total	49	14114·42		

The observed value of F is highly significant since the one per cent table value of F with 4, 45 d.f. is 3·77. The null hypothesis under test may be safely rejected, and we conclude that it is unlikely that the gains in weight observed in the five lots of pigs are in conformity with the hypothesis that, on the average, the five treatments do not give different gains in weight.

As an illustration, the overall inequality of the θ_i may be further analysed as follows. The five lot means are

$$\bar{x}_1 = 156 \cdot 6; \quad \bar{x}_2 = 173 \cdot 5; \quad \bar{x}_3 = 161 \cdot 9; \quad \bar{x}_4 = 180 \cdot 2; \quad \bar{x}_5 = 175 \cdot 5.$$

Also, each $\bar{x}_i$ is normally distributed with mean θ_i and variance σ^2/n_i. Suppose, for example, that interest centres on the average gain in weight recorded by pigs given the first treatment. A confidence interval for θ_1 can be derived by using the fact that

$$t = \frac{\bar{x}_1 - \theta_1}{s/\sqrt{n_1}}$$

has "Student's" distribution with $N - k$ d.f.

The one per cent point of "Student's" distribution with 45 d.f. is $2 \cdot 688$. Also, with our data, we have

$$s = \sqrt{226 \cdot 84} = 15 \cdot 0612; \qquad \frac{s}{\sqrt{n_1}} = \frac{15 \cdot 0612}{3 \cdot 1623} = 4 \cdot 7627.$$

Hence the 99 per cent confidence interval for θ_1 is

$$156 \cdot 6 - 2 \cdot 688 \times 4 \cdot 7627 \leqslant \theta_1 \leqslant 156 \cdot 6 + 2 \cdot 688 \times 4 \cdot 7627,$$

or

$$143 \cdot 80 \leqslant \theta_1 \leqslant 169 \cdot 40.$$

Similarly, a confidence interval for the difference $\theta_2 - \theta_3 = \delta$ is obtained by using the fact that

$$t = \frac{\overline{x}_2 - \overline{x}_3 - \delta}{s \left[\dfrac{1}{n_2} + \dfrac{1}{n_3} \right]^{\frac{1}{2}}}$$

has "Student's" distribution with $N - k$ d.f.

With our data, we have

$$s[1/n_2 + 1/n_3]^{\frac{1}{2}} = 15 \cdot 0612 \times 0 \cdot 4472 = 6 \cdot 7354.$$

Therefore the 99 per cent confidence interval for δ is

$$11 \cdot 6 - 2 \cdot 688 \times 6 \cdot 7354 \leqslant \delta \leqslant 11 \cdot 6 + 2 \cdot 688 \times 6 \cdot 7354,$$

or

$$-6 \cdot 50 \leqslant \delta \leqslant 29 \cdot 70.$$

This confidence interval reveals an interesting feature of the analysis of variance. Although the null hypothesis $H(\theta_1 = \theta_2 = \theta_3 = \theta_4 = \theta_5)$ is significant at the one per cent level, the above 99 per cent confidence interval includes the point $\delta = 0$, which implies that the null hypothesis $H(\theta_2 = \theta_3)$ is acceptable at the one per cent level of significance. Tests of this kind often give a deeper understanding of the differences between the population means, which are usually of greater interest than the simple F test for the null hypothesis of the equality of all the means simultaneously.

15.10   Tests of significance for the variance of a normal population

The mean and variance are the two independent parameters ot a normal distribution, but we have seen that a test of significance of any null hypothesis about the mean is mathematically dependent upon the magnitude of σ^2 or of its sample estimate. This role of the variance extends generally to the tests of significance for linear hypotheses about the means of several normal populations, and it is of fundamental importance in statistical analysis. Besides, the variance of a normal population has a physical interpretation which is of interest quite apart from the magnitude of the population mean. A simple example brings out this point clearly. Suppose an automatic machine is set to produce screws of length μ; but we know that, however accurately the machine is set, the screws actually produced will deviate from the specified length. These deviations are due partly to the impossibility of measuring length without an approximation error and partly to the intrinsic random variations of the production process. Thus, if x_i is the observed length of a screw produced, we postulate that

$$x_i = \mu + \epsilon_i, \quad \text{or} \quad x_i - \mu = \epsilon_i,$$

HYATT REGENCY ✿ SEOUL

REGENCY BAR

SERVER 012		CASHIER 106	2 GUESTS
PAGE 01	Cover F: 0 B: 2		POS 07
011 22:59	10/01/88		TABLE 0004

011 22:59	2 1264 CARLSB BEER		6400
	1 1381 L-ADE		2500
	1 1271 COCA COK		1800
	1 1116 BACA RUM		3300
011 22:59	NET SALES		14000
	SERVICE CHARGE		1400
	V.A.T		1540
	TOTAL		16940
011 22:59	CASH		16940

Above tax (V.A.T) include value added tax,
special consumption tax and defense tax.
上記税金には特別消費税と防衛税が含めています。

CHECK NO.

№ 009763

Name (Block Letters)	Room/Card No.
Tax Exemption No.	서울특별시용산구한남동747 – 7
	서울미라마관광주식회사
	대표이사 회장 閔丙鎬
	대표이사 사장 重愉明
Signature	106 – 81 – 07782
	797 – 1234
	부가가치세법
	제32조 제1항규정에
	의한 영수증 RECEIPT

where ϵ_i is the sum of the effects of the approximation error and the random variation of the production process. It is clear that, in view of the uncertainty of machine calibration, the value of μ is unknown even though the production is aimed at a specific value for the length of the screws produced. Thus both μ and ϵ_i are unknown, and only their sum can be observed. Accordingly, the assumption that the x_i are normally distributed with mean μ and variance σ^2 strictly refers to the distributional behaviour of the errors ϵ_i; and an equivalent way of describing this normality is that the ϵ_i are normally distributed random variables with zero mean and variance σ^2. We thus infer that σ^2 is a measure of the random deviations of the observations from expectation, and its magnitude is of intrinsic importance quite independently of the mean μ. Thus, if a modification in the automatic machine leads to a reduction in the value of σ^2, it is evident that such a change would be regarded as an improvement in the production process because it implies a decrease in the variation in the lengths of the screws produced.

We thus conclude that there are practical reasons for viewing the population variance as a distinct parameter of the population, which are apart from its theoretical importance in the distribution of the sample average. Accordingly, as in the case of the population mean, we consider tests of significance for null hypotheses about σ^2 and also methods of the determination of confidence intervals for it.

Suppose $x_1, x_2, \ldots, x_n$ are n independent observations from a normal population with mean μ and variance σ^2. Then it is known that the sample variance s^2 is the least-squares estimate of σ^2, and the random variable $(n-1)s^2/\sigma^2$ has the χ^2 distribution with $n-1$ d.f. This result may be used to test the null hypothesis $H(\sigma = \sigma_0)$, where σ_0 is some preassigned value of σ.

(i) If the alternatives to the null hypothesis are $\sigma > \sigma_0$, then we use the upper tail of the χ^2 distribution as the region of rejection. The required test of significance is obtained by using $(n-1)s^2/\sigma_0^2$ as a χ^2 variable with $n-1$ d.f. If the observed value of this variable is greater than the 100η per cent point of the χ^2 distribution with $n-1$ d.f., then the null hypothesis is rejected at the 100η per cent level of significance, it being accepted in the contrary event.

(ii) Similarly, if the alternatives to the null hypothesis under test are $\sigma < \sigma_0$, then we use the lower tail of the χ^2 distribution with $n-1$ d.f. as the region of rejection. The null hypothesis is now rejected at the 100η per cent level of significance if $(n-1)s^2/\sigma_0^2$ is less than the lower 100η per cent point of the χ^2 distribution with $n-1$ d.f., the null hypothesis being accepted in the opposite event.

(iii) If the alternatives to the null hypothesis are $\sigma \neq \sigma_0$, then we have to use a two-sided test of significance. Thus, if χ_1^2 and χ_2^2 are the lower and upper percentage points of the χ^2 distribution with $n-1$ d.f. such that

$$P(\chi^2 \leqslant \chi_1^2) = P(\chi^2 \geqslant \chi_2^2) = \tfrac{1}{2}\eta, \qquad (1)$$

then the null hypothesis is rejected at the 100η per cent level of significance if $(n-1)s^2/\sigma_0^2$ lies outside the interval (χ_1^2, χ_2^2), it being accepted in the opposite event.

As an example, suppose $n = 12$. Then the table for the χ^2 distribution with 11 d.f. gives

$$P(\chi^2 \geqslant 19\cdot 68) \;=\; P(\chi^2 \leqslant 4\cdot 57) \;=\; 0\cdot 05.$$

Thus $19\cdot 68$ and $4\cdot 57$ are the upper and lower five per cent points of the χ^2 distribution with 11 d.f., and for a two-sided test their use would give a ten per cent level of significance. It is important to remember that the range of the χ^2 variable is $\chi^2 \geqslant 0$, and the distribution is asymmetrical for any finite number of degrees of freedom. Accordingly, the upper and lower percentage points are unequal and do not possess the symmetry found in the normal and "Student's" distributions. This implies that we have to use a somewhat different argument to determine a "symmetrical" $100(1-\eta)$ per cent confidence interval for σ^2.

It follows from (1) that

$$P(\chi_1^2 \leqslant \chi^2 \leqslant \chi_2^2) \;=\; 1-\eta,$$

so that for the $100(1-\eta)$ per cent confidence interval for σ^2 we have

$$P[\chi_1^2 \leqslant (n-1)s^2/\sigma^2 \leqslant \chi_2^2] \;=\; 1-\eta.$$

Therefore the $100(1-\eta)$ per cent confidence interval for $1/\sigma^2$ is

$$\chi_1^2/(n-1)s^2 \;\leqslant\; 1/\sigma^2 \;\leqslant\; \chi_2^2/(n-1)s^2,$$

whence, by reversing the inequality, the $100(1-\eta)$ per cent confidence interval for σ^2 is

$$(n-1)s^2/\chi_2^2 \;\leqslant\; \sigma^2 \;\leqslant\; (n-1)s^2/\chi_1^2.$$

This is the $100(1-\eta)$ per cent *symmetrical* confidence interval for σ^2, since the percentage points χ_1^2 and χ_2^2 are determined by considering equal areas in both tails of the χ^2 distribution. This method is intuitively reasonable and simple, but it does not give the "shortest" $100(1-\eta)$ per cent confidence interval for σ^2, as defined by Neyman and Pearson.

To determine this "shortest" confidence interval for σ^2, suppose A and B are two positive numbers such that

$$P[A \leqslant (n-1)s^2/\sigma^2 \leqslant B] \;=\; 1-\eta.$$

Then by transforming the inequality, a $100(1-\eta)$ per cent confidence interval for σ^2 is obtained as

$$(n-1)s^2/B \leqslant \sigma^2 \leqslant (n-1)s^2/A.$$

Now this confidence interval will be the "shortest" for σ^2 if A and B are so chosen that

$$(n-1)s^2\left[\frac{1}{A} - \frac{1}{B}\right]$$

is a minimum, subject to the condition that

$$P(A \leqslant \chi^2 \leqslant B) \;=\; 1-\eta,$$

where χ^2 has the χ^2 distribution with $n-1$ d.f.

It is somewhat awkward to determine A and B in this way, and we shall concern ourselves only with symmetrical confidence intervals which can be readily evaluated by using the available tables of the χ^2 distribution. It is also known that these symmetrical confidence intervals are the "shortest" when the distributions are symmetrical, as is the case with the normal and "Student's" distributions. However, this is not true of skew distributions.

Example 15.4: Heights of men

For a random sample of 15 men accepted for life assurance on ordinary terms, it was found that their average height was 68·57 inches and the sample variance was 5·76 square inches. if height may be assumed to be a normal variable with mean μ and variance σ^2, test, using a five per cent level of significance, that σ^2 is

(i)	less than 3·2 ;	(ii)	more than 3·2;
(iii)	equal to 3·2;	(iv)	less than 13·0 ;
(v)	more than 13·0 ;	(vi)	equal to 13·0.

Also, determine the symmetrical 90 per cent confidence interval for σ^2.

Here $(n-1)s^2 = 80·64$ and $(n-1)s^2/\sigma^2$ is distributed as χ^2 with 14 d.f. Also, for 14 d.f. the table of the χ^2 distribution gives

$$P(\chi^2 \leqslant 6·57) \quad = \quad P(\chi^2 \geqslant 23·68) \quad = \quad 0·05;$$
$$P(\chi^2 \leqslant 5·63) \quad = \quad P(\chi^2 \geqslant 26·12) \quad = \quad 0·025.$$

(i)   The test of significance for the null hypothesis $H(\sigma^2 = 3·2)$ with the alternatives $\sigma^2 < 3·2$ is obtained by using $(n-1)s^2/\sigma^2 = 25·20$ as χ^2 with 14 d.f. We have to consider a one-sided test, and the lower five per cent point of the χ^2 distribution with 14 d.f. is 6·57. Therefore the null hypothesis $H(\sigma^2 = 3·2)$ is acceptable, since values of $\chi^2 \leqslant 6·57$ only are significant.

(ii)   To test the null hypothesis $H(\sigma^2 = 3·2)$ with the alternatives $\sigma^2 > 3·2$, we again use a one-sided test, but the region of rejection at the five per cent level of significance is $\chi^2 \geqslant 23·68$. The observed value 25·20 is now significant.

(iii)   To test the null hypothesis $H(\sigma^2 = 3·2)$ with the alternatives $\sigma^2 \neq 3·2$, we must use a two-sided test of significance, and the corresponding five per cent region of rejection is now defined by the two tails $\chi^2 \leqslant 5·63$ and $\chi^2 \geqslant 26·12$. The observed value of 25·20 is therefore not significant at the five per cent level.

The above three tests have assessed the credibility of the same null hypothesis $H(\sigma^2 = 3·2)$ but in comparison with different sets of alternative hypotheses. Consequently, we have used different regions of rejection, and this explains the difference between the results of the three tests. The test of significance (i) implies that if we know in advance that the possible values of σ^2 are $\leqslant 3·2$, then the observed value of the sample variance $s^2 = 5·76$ provides reasonable evidence for accepting the null hypothesis $H(\sigma^2 = 3·2)$. On the other hand, if the possible values of σ^2 are known to be $\geqslant 3·2$, then the same observed result suggests that the null hypothesis $H(\sigma^2 = 3·2)$ is not tenable. This is the implication of the second test of significance. Furthermore, if there is complete uncertainty about the possible alternatives to the null hypothesis $H(\sigma^2 = 3 \cdot 2)$, then the third test of significance suggests that the observed value $s^2 = 5 \cdot 76$ is consistent with the null hypothesis under test.

To test the null hypothesis $H(\sigma^2 = 13 \cdot 0)$, the observed value of $(n-1)s^2/\sigma^2$ is $6 \cdot 20$.

(iv) Thus if the alternatives to the null hypothesis $H(\sigma^2 = 13 \cdot 0)$ are $\sigma^2 < 13 \cdot 0$, then the one-sided test of significance based on the lower tail of the χ^2 distribution with 14 d.f. gives a significant result at the five per cent level of significance.

(v) Again, if the alternatives to the null hypothesis $H(\sigma^2 = 13 \cdot 0)$ are $\sigma^2 > 13 \cdot 0$, then the one-sided test based on the upper tail of the χ^2 distribution with 14 d.f. indicates that the null hypothesis under test is acceptable at the five per cent level of significance.

(vi) Finally, if the alternatives to the null hypothesis $H(\sigma^2 = 13 \cdot 0)$ are $\sigma^2 \neq 13 \cdot 0$, then the two-sided test based on both the tails of the χ^2 distribution with 14 d.f. again shows that the null hypothesis under test is acceptable at the five per cent level of significance.

We thus conclude that if there is no prior knowledge about the possible alternatives to the null hypothesis, then the same observed value of $s^2 = 5 \cdot 76$ is found to be consistent with the null hypotheses $H(\sigma^2 = 3 \cdot 2)$ and $H(\sigma^2 = 13 \cdot 0)$. But, once the alternatives to the null hypothesis are known, then the appropriate one-sided tests give varying results, and these, as a rule, discriminate more sensitively between the null hypothesis and its alternatives.

Confidence interval for σ^2

To obtain the 90 per cent symmetrical confidence interval for σ^2, we observe that the lower and upper five per cent points of the χ^2 distribution with 14 d.f. are $6 \cdot 57$ and $23 \cdot 68$ respectively. Hence, according to the notation used earlier, $\chi^2_{1} = 6 \cdot 57$ and $\chi^2_{2} = 23 \cdot 68$. Therefore the required confidence interval for σ^2 is

$$\frac{80 \cdot 64}{23 \cdot 68} \leqslant \sigma^2 \leqslant \frac{80 \cdot 64}{6 \cdot 57},$$

or

$$3 \cdot 405 \leqslant \sigma^2 \leqslant 12 \cdot 274.$$

15.11 Comparison of the variances of two normal populations

We have shown that "Student's" distribution or the equivalent analysis of variance provides an exact test of significance for comparing the means of two normal populations having the same unknown variance. In general, therefore, it is theoretically necessary to test the validity of the assumption regarding the equality of the population variances before using the exact method for testing any null hypothesis about the population means. This is one good reason for comparing the variances of the populations. Moreover, since the variance of a population of measurements is a theoretical measure of their random variations, it is, at times, of interest to compare two such population measures. For example, the accuracies of two technicians carrying out similar measurements (such as the determination of the extensibility of yarn) may be compared by testing the equality of the variances. In such a situation, the model postulates two independent normal populations of conceptual measurements made by the two technicians. If these populations have variances σ_1^2 and σ_2^2, then these are the true measures of the accuracies of the two technicians.

The equality of the variances of two normal populations is a null hypothesis of particular interest, but we consider the problem of comparing the two variances from a slightly more general point of view. Suppose, then, that the two normal populations have means θ_1, θ_2 and variances σ_1^2, σ_2^2 respectively. Given random samples of n_1 and n_2 observations respectively from the populations, it is known that the corresponding sample variances are the least-squares estimates of σ_1^2 and σ_2^2 irrespective of the magnitude of the population means θ_1 and θ_2. We denote the two sample variances by s_1^2 and s_2^2. Then it is also known that $(n_1-1)s_1^2/\sigma_1^2$ and $(n_2-1)s_2^2/\sigma_2^2$ are independent χ^2 variables with n_1-1 and n_2-1 d.f. respectively. Hence it follows that

$$F = \frac{s_1^2}{\sigma_1^2}\bigg/\frac{s_2^2}{\sigma_2^2} = s_1^2/\lambda s_2^2, \quad \text{where } \lambda \equiv \sigma_1^2/\sigma_2^2,$$

has the F distribution with n_1-1, n_2-1 d.f. By the symmetry of F, we know that the reciprocal

$$F^{-1} = \lambda s_2^2/s_1^2$$

also has the F distribution but with n_2-1, n_1-1 d.f.

These results may be used to test null hypotheses about the population variance ratio λ. As already mentioned, a null hypothesis of special importance is $H(\lambda = 1)$; but before this can be tested, its alternatives have to be specified. We consider two sets of alternatives, namely, that $\lambda > 1$ or that $\lambda \neq 1$. By the symmetry of the F distribution the case $\lambda < 1$ is similar to that of $\lambda > 1$ and so may be ignored.

(i) We first consider the test of significance of the null hypothesis $H(\lambda = 1)$ with the alternatives $\lambda > 1$. If this null hypothesis is true, then the observed ratio $F_0 = s_1^2/s_2^2$ has the F distribution with n_1-1, n_2-1 d.f. The significance of F_0 is to be assessed by a one-sided test, the region of rejection being in the upper tail of the F distribution. Thus the null hypothesis $H(\lambda = 1)$ is rejected at the 100η per cent level of significance if F_0 is greater than the upper 100η per cent table value of the F distribution with n_1-1, n_2-1 d.f., the null hypothesis being accepted in the opposite event. It is to be noted that the parameters of the F tables are the degrees of freedom, that is n_1-1 and n_2-1, and *not* the sample sizes n_1 and n_2.

An observed value $F_0 > 1$ is tested according to the above rule, since the values given in the F tables are the upper percentage points which correspond to the instruction for calculating the ratio of the larger mean square to the smaller one. Nevertheless, despite this usual rule, it is important to remember that the appropriate ratio for testing the null hypothesis $H(\lambda = 1)$ with the alternatives $\lambda > 1$ is F_0, as defined above, whether $F_0 >$ or <1. However, an observed value $F_0 < 1$ invariably indicates acceptance of the null hypothesis $H(\lambda = 1)$ in preference to the alternatives $\lambda > 1$ for any level of significance $\eta < 0{\cdot}5$. This follows from the fact that for $\eta < 0{\cdot}5$, the upper percentage points of the F distributions lie in the region $F > 1$.

(ii) If the alternatives to the null hypothesis $H(\lambda = 1)$ are $\lambda \neq 1$, then a two-sided test of significance is used. Thus, suppose F_1 and F_2 are the lower and upper percentage points of the F distribution with $n_1 - 1$, $n_2 - 1$ d.f. such that

$$P(F \leqslant F_1) \;=\; P(F \geqslant F_2) \;=\; \tfrac{1}{2}\eta.$$

Then the null hypothesis under test is rejected at the 100η per cent level of significance if F_0 lies outside the interval (F_1, F_2), it being accepted in the opposite event.

Now F_2 is the upper 50η per cent point of the F distribution with $n_1 - 1$, $n_2 - 1$ d.f. and is obtained directly from the F tables. The value F_1 is the lower 50η per cent point of the F distribution with $n_1 - 1$, $n_2 - 1$ d.f. and it has to be obtained from the F tables by using the symmetry of the F distribution. We know that if $F' = 1/F$, then F' has the F distribution with $n_2 - 1$, $n_1 - 1$ d.f. Therefore

$$P(F \leqslant F_1) \;=\; P(F' \geqslant 1/F_1) \;=\; \tfrac{1}{2}\eta.$$

Hence $1/F_1$ is the upper 50η per cent point of F', and it is thus obtained directly from the F tables.

Example 15.5

As an illustration, suppose $n_1 = 11$, $n_2 = 13$, and $\eta = 0{\cdot}05$. Then from the F tables we have

$$F_2 \;=\; 3{\cdot}37 \quad \text{and} \quad 1/F_1 \;=\; 3{\cdot}62, \quad \text{or} \quad F_1 \;=\; 0{\cdot}276.$$

This means that for the F distribution with 10, 12 d.f.,

$$P(F \leqslant 0{\cdot}276) \;=\; P(F \geqslant 3{\cdot}37) \;=\; 0{\cdot}025.$$

Confidence interval for the variance ratio λ

For finite degrees of freedom the F distribution is asymmetrical, and we shall not determine the "shortest" confidence interval for λ when the null hypothesis $H(\lambda = 1)$ is not tenable. To obtain the $100(1 - \eta)$ per cent symmetrical confidence interval for $\lambda \neq 1$, we use the fact that, for F_1 and F_2 as specified in (ii) above,

$$P(F_1 \leqslant F_0/\lambda \leqslant F_2) \;=\; 1 - \eta.$$

Hence the required confidence interval for λ is

$$F_0/F_2 \;\leqslant\; \lambda \;\leqslant\; F_0/F_1.$$

To illustrate this numerically, suppose $F_0 = 7{\cdot}65$ in Example 15.5. Then the 95 per cent confidence interval for λ is

$$7{\cdot}65/3{\cdot}37 \;\leqslant\; \lambda \;\leqslant\; 7{\cdot}65 \times 3{\cdot}62,$$

or
$$2{\cdot}270 \;\leqslant\; \lambda \;\leqslant\; 27{\cdot}693.$$

We could also conclude from this confidence interval that the observed value $F_0 = 7{\cdot}65$ is not consistent with the null hypothesis $H(\lambda = 1)$ at the five per cent level of significance. Of course, this is a two-sided test of significance and it is only applicable when the alternatives to the null hypothesis are $\lambda \neq 1$.

Note

It seems natural to extend the test of significance for comparing the variances of two normal populations to that for testing the simultaneous equality of k (> 2) population variances. Unfortunately, no exact test is available for this general comparison. There is, however, an approximate test of significance due to M.S. Bartlett, but we do not consider it here, as its derivation is based on an approach which does not lie within the framework of the analysis of variance.

15.12 Regression analysis

The term "regression" was originally used by Francis Galton (1822 – 1911) in a statistical examination of human inheritance to denote a certain hereditary relationship. Of the "law of universal regression" Galton wrote, "Each peculiarity in a man is shared by his kinsman, but *on the average* in a less degree." For example, though tall fathers tend to have tall sons, yet the average height of sons of a group of tall fathers is less than their fathers' height. Thus there is a regression, or going back, of sons' heights towards the average height of all men. From Galton's work the word regression has passed into standard statistical terminology. In its widest sense, regression analysis may be defined as that branch of statistical theory which is concerned with an objective analysis of the association between two or more random variables. This idea of association between random variables is a new and fundamentally important concept which extends the scope of statistical analysis considerably. The general aim of regression analysis is to derive a mathematical description of the relationship between the variables together with an assessment of the inevitable uncertainty of the relationship found. This leads to a variety of approaches for attacking the problem, and our concern here is only with an exposition of the simplest, but nonetheless very useful, techniques for the analysis of statistical association. More specifically, we are largely interested in the relationships of dependence of one random variable y on another non-random variable x. The latter variable is usually called an *independent* or *explanatory* variable. We shall see the reason for this usage a little later.

Now, in general, the dependence of y on x means that the distribution of y depends upon x and so, in particular, the moments of y are functions of x. We may therefore write

$$E(y) = \psi_1(x; \alpha, \beta, \gamma, \ldots),$$
$$\text{var}(y) = \psi_2(x; \alpha, \beta, \gamma, \ldots),$$

and so on for the higher moments of y, where $\psi_1, \psi_2, \ldots$ are particular functions of x and of the unknown parameters $\alpha, \beta, \gamma, \ldots$. However, in regression analysis the functional form of ψ_1 is assumed to be known, and estimates of the unknown parameters $\alpha, \beta, \gamma, \ldots$ are determined from the data together with an estimate of $\text{var}(y)$.

15.13 Linear regression

In the simplest specification of the regression problem, it is assumed that

(i) $E(y) = \alpha + \beta x$;

(ii) $\mathrm{var}(y) = \sigma^2$, an independent and unknown parameter;   and

(iii) the probability distribution of y is normal with the specified mean and variance.

We observe that $E(y)$ is a linear function of x, and this relationship is known as the *linear regression* of the random variable y on the explanatory variable x. This case is of particular importance and it is also mathematically the simplest. More generally, when $E(y)$ is a non-linear function of x, then the regression is said to be *non-linear*. However, in this book, we shall not be concerned with non-linear regression, though this too has much practical and theoretical interest. The condition (ii) regarding $\mathrm{var}(y)$ is also restrictive, but it is the usual one which enables an application of the theory of least squares to be made for the estimation of the three parameters α, β, and σ^2. The parameter β is known as the *coefficient of linear regression of y on x*. Finally, the assumption of the normality of the distribution of y provides the basis for a mathematical justification of the analysis of variance for testing null hypotheses about the parameters α, β, and σ^2.

Physical interpretation of linear regression

In order to understand the physical meaning of this linear regression problem and how the sample data are related to the model, it is appropriate to consider a simple experimental situation. Suppose a farmer is interested in determining the effect of the quantity of fertilizer on the yield of potatoes. We assume that the farmer has n experimental field plots which are equal in area and of comparable fertility. On each of these n plots, the farmer uses a quantity of fertilizer, the amounts used being $x_1, x_2, \ldots, x_n$; and we assume for simplicity that $x_1 < x_2 < \ldots < x_n$. If, at harvest time, the potato yields from the n plots are $y_1, y_2, \ldots, y_n$, then we have a sample of n pairs of observations (x_ν, y_ν), for $\nu = 1, 2, \ldots, n$. A typical pair (x_ν, y_ν) means that the potato yield from the plot which received x_ν amount of fertilizer is y_ν. Clearly, y_ν is a random observation, but x_ν is a specific amount of fertilizer determined by the farmer. In fact, to be exact, it is also assumed that the error of measurement in x_ν is negligible, and for present purposes x_ν is to be regarded as a known exact quantity. The randomness lies wholly in y_ν, and the assumption is made that the y_ν are independent normal variables such that

$$E(y_\nu) = \alpha + \beta x_\nu ; \quad \text{and}$$
$$\mathrm{var}(y_\nu) = \sigma^2 \quad \text{for } \nu = 1, 2, \ldots, n.$$

It is to be noted that the y_ν are random observations but not from one normal population. Indeed, the exact conceptual specification of the y_ν is a subtle idea which lies at the heart of the regression problem. The argument is as follows. Consider any x_ν, that is, any one of the n specific quantities of fertilizer used by the farmer per plot. We now visualise an infinity of plots of the same area and of comparable fertility such that on each of these plots an amount x_ν of fertilizer is used. Next we denote the conceptual potato yields from these plots as $y_{\nu_1}, y_{\nu_2}, y_{\nu_3}, \ldots$. This infinite sequence of conceptual observations constitutes a statistical population which is assumed to be normal with mean $\alpha + \beta x_\nu$ and variance σ^2. The actual observation y_ν is regarded as a single chance observation from this population. Hence the specification of the mean and variance of y_ν.

Now, each of the n actual observations $y_1, y_2, \ldots, y_n$ is to be interpreted in exactly the same way. In other words, we visualise n normal populations with differing means $\alpha + \beta x_1, \alpha + \beta x_2, \ldots, \alpha + \beta x_n$, but with the same variance σ^2; and from each of these populations we have obtained an actual observation. Accordingly, $y_1, y_2, \ldots, y_n$ are independent observations from n *different* normal populations. With this understanding, the conditions

$$E(y_\nu) = \alpha + \beta x_\nu, \quad \text{for } \nu = 1, 2, \ldots, n,$$

are thus seen to imply that the n means of the populations lie on some straight line whose equation may be generally written as

$$E(y) = \alpha + \beta x.$$

We may also infer that, from the experimental point of view, the expected yield per plot from comparable plots dressed with x_ν amount of fertilizer is the sum $\alpha + \beta x_\nu$. The parameter α represents the average yield without the fertilizer, and βx_ν is the additive contribution to the yield of the amount of fertilizer used. Admittedly, this is a rigid assumption and it may well not be strictly true. Nevertheless, this is the experimental meaning of the linear regression model. It is known that such a model holds reasonably well in many experimental situations, and this gives importance to the methods of statistical analysis derived from it. On the assumption that the stated conditions are valid, we have given physical meaning to the parameters α and β. The estimation of the regression equation

$$E(y) = \alpha + \beta x$$

can now be interpreted as follows. If it may be assumed that, on the average, potato yield varies linearly with the amount of fertilizer used per plot, then we wish to estimate this relationship over the range of variation of the fertilizer amounts $x_1, x_2, \ldots, x_n$.

Estimation of linear regression

For mathematical convenience in estimation, it is usual to change the specification of α and to write

$$E(y_\nu) = \alpha + \beta(x_\nu - \bar{x}),$$

where $\bar{x}$ is the average of the x_ν. This means simply that we have replaced the parameter α by $\alpha + \beta\bar{x}$, and the change is equivalent to measuring the x_ν from their average instead of from an arbitrary origin. In all future reference to the linear regression equation we shall use this modified form in conformity with general practice.

Given the n pairs of observations (x_ν, y_ν), for $\nu = 1, 2, \ldots, n$, we derive the least-squares estimates of α and β by minimising

$$\Omega = \sum_{\nu=1}^{n} [\, y_\nu - a - \beta(x_\nu - \bar{x})]^2$$

with respect to variations in a and β. Since

$$\frac{\partial \Omega}{\partial a} = -2 \sum_{\nu=1}^{n} [y_\nu - a - \beta(x_\nu - \bar{x})],$$

and

$$\frac{\partial \Omega}{\partial \beta} = -2 \sum_{\nu=1}^{n} (x_\nu - \bar{x})[y_\nu - a - \beta(x_\nu - \bar{x})],$$

the least-squares estimates a^* and β^* are obtained from the equations

$$\sum_{\nu=1}^{n} [y_\nu - a^* - \beta^*(x_\nu - \bar{x})] = 0, \tag{1}$$

and

$$\sum_{\nu=1}^{n} (x_\nu - \bar{x})[y_\nu - a^* - \beta^*(x_\nu - \bar{x})] = 0. \tag{2}$$

It is convenient at this stage to introduce the compact notation

$$X = \sum_{\nu=1}^{n} (x_\nu - \bar{x})^2; \quad Y = \sum_{\nu=1}^{n} (y_\nu - \bar{y})^2; \quad Z = \sum_{\nu=1}^{n} (x_\nu - \bar{x})(y_\nu - \bar{y}),$$

$\bar{y}$ being the average of the y_ν.

Now, since $\sum_{\nu=1}^{n} (x_\nu - \bar{x}) \equiv 0$, therefore we have from (1)

$$\sum_{\nu=1}^{n} y_\nu - na^* = 0, \quad \text{or} \quad a^* = \bar{y}.$$

Then from (2) we have

$$Z - X\beta^* = 0, \quad \text{or} \quad \beta^* = Z/X.$$

Hence the absolute minimum of Ω is

$$\begin{aligned}
M &= \sum_{\nu=1}^{n} [y_\nu - a^* - \beta^*(x_\nu - \bar{x})]^2 \\
&= \sum_{\nu=1}^{n} [(y_\nu - \bar{y}) - \beta^*(x_\nu - \bar{x})]^2 \\
&= \sum_{\nu=1}^{n} [(y_\nu - \bar{y})^2 - 2\beta^*(x_\nu - \bar{x})(y_\nu - \bar{y}) + \beta^{*2}(x_\nu - \bar{x})^2] \\
&= Y - 2\beta^* Z + \beta^{*2} X = Y - \beta^{*2} X = Y - Z^2/X.
\end{aligned}$$

This is the sum of the squared residuals, and we shall show that this is an unbiased estimate of $(n-2)\sigma^2$. Thus $s^2 = M/(n-2)$ is an unbiased estimate of σ^2. It also follows from general theory that M/σ^2 is distributed as χ^2 with $n - 2$ d.f. This distribution can be used to test any null hypothesis about the variance σ^2. We observe that the number of degrees of freedom for M is in conformity with the general principle, since two parameters a and β are estimated from the n random observations.

The expected value of M is conveniently proved by first evaluating the means and variances of the estimates a^* and β^*.

Properties of the estimates a^* and β^*

We have

$$E(a^*) = \frac{1}{n} E\left[\sum_{\nu=1}^{n} y_\nu\right]$$

$$= \frac{1}{n} \sum_{\nu=1}^{n} [a + \beta(x_\nu - \bar{x})] = a, \quad \text{since } \sum_{\nu=1}^{n} (x_\nu - \bar{x}) \equiv 0.$$

Similarly, we have

$$E(\beta^*) = \frac{1}{X} E\left[\sum_{\nu=1}^{n} (y_\nu - \bar{y})(x_\nu - \bar{x}) \right]$$

$$= \frac{1}{X} E\left[\sum_{\nu=1}^{n} (x_\nu - \bar{x}) y_\nu \right], \quad \text{since } \bar{y} \sum_{\nu=1}^{n} (x_\nu - \bar{x}) \equiv 0,$$

$$= \frac{1}{X} \sum_{\nu=1}^{n} (x_\nu - \bar{x})[a + \beta(x_\nu - \bar{x})]$$

$$= \frac{1}{X} \beta \sum_{\nu=1}^{n} (x_\nu - \bar{x})^2 = \beta.$$

Thus a^* and β^* are verified to be unbiased estimates of a and β respectively. Of course, this result was expected in view of the general property of unbiasedness of all least-squares estimates.

It is also important to note that algebraically

$$\beta^* = \frac{Z}{X} = \frac{1}{X} \sum_{\nu=1}^{n} (y_\nu - \bar{y})(x_\nu - \bar{x}) = \sum_{\nu=1}^{n} \frac{(x_\nu - \bar{x})}{X} y_\nu.$$

Thus β^* is a linear function of the random observations, and this form of the expression for β^* will be used repeatedly.

The variances of a^* and β^* are easily obtained. In fact, we have

$$\text{var}(a^*) = \text{var}(\bar{y}) = \sigma^2/n,$$

and

$$\text{var}(\beta^*) = \text{var}\left[\sum_{\nu=1}^{n} \frac{1}{X} (x_\nu - \bar{x}) y_\nu \right] = \sum_{\nu=1}^{n} \frac{1}{X^2} (x_\nu - \bar{x})^2 \text{var}(y_\nu) = \sigma^2/X.$$

Hence

$$E[n(a^* - a)^2] = n \text{ var}(a^*) = \sigma^2,$$

and

$$E[X(\beta^* - \beta)^2] = X \text{ var}(\beta^*) = \sigma^2.$$

We now use these expectations in deriving the expected value of M.

Expectation of M

Finally, to evaluate the expected value of M, we first establish the identity

$$\sum_{\nu=1}^{n} [y_\nu - a - \beta(x_\nu - \bar{x})]^2 = n(a^* - a)^2 + X(\beta^* - \beta)^2 + M.$$

To prove this, we note that

$$y_\nu - a - \beta(x_\nu - \bar{x}) = [y_\nu - a^* - \beta^*(x_\nu - \bar{x})] + (a^* - a) + \\ + (\beta^* - \beta)(x_\nu - \bar{x}),$$

whence, on squaring both sides and summing over for $1 \leqslant \nu \leqslant n$, we obtain

$$\sum_{\nu=1}^{n} [y_\nu - a - \beta(x_\nu - \bar{x})]^2 = \sum_{\nu=1}^{n} [\{y_\nu - a^* - \beta^*(x_\nu - \bar{x})\} + (a^* - a) + \\ + (\beta^* - \beta)(x_\nu - \bar{x})]^2 \\ = M + n(a^* - a)^2 + X(\beta^* - \beta)^2,$$

the three cross-product terms on the right-hand side being zero as

(a) $$2(a^* - a) \sum_{\nu=1}^{n} [y_\nu - a^* - \beta^*(x_\nu - \bar{x})] = 0, \quad \text{by (1)};$$

(b) $\qquad 2(\alpha^* - \alpha)(\beta^* - \beta) \sum_{\nu=1}^{n} (x_\nu - \bar{x}) \equiv 0;$ and

(c) $\qquad 2(\beta^* - \beta) \sum_{\nu=1}^{n} (x_\nu - \bar{x})[y_\nu - \alpha^* - \beta^*(x_\nu - \bar{x})] = 0,$ by (2).

Therefore

$$M = \sum_{\nu=1}^{n} [y_\nu - \alpha - \beta(x_\nu - \bar{x})]^2 - n(\alpha^* - \alpha)^2 - X(\beta^* - \beta)^2.$$

Hence, taking expectations on both sides, we have

$$E(M) = \sum_{\nu=1}^{n} E[y_\nu - \alpha - \beta(x_\nu - \bar{x})]^2 - E[n(\alpha^* - \alpha)^2] - E[X(\beta^* - \beta)^2]$$

$$= \sum_{\nu=1}^{n} \text{var}(y_\nu) - n\,\text{var}(\alpha^*) - X\,\text{var}(\beta^*) = (n - 2)\sigma^2.$$

This completes the estimation problem of linear regression, and the next stage in the analysis is to derive the appropriate sums of squares for testing null hypotheses about α and β.

Tests of significance for null hypotheses about α and β

To test the null hypothesis $H(\alpha = \alpha_0)$, where α_0 is some preassigned value of α, we have to obtain the relative minimum of

$$\Omega = \sum_{\nu=1}^{n} [y_\nu - \alpha - \beta(x_\nu - \bar{x})]^2 = M + n(\alpha^* - \alpha)^2 + X(\beta^* - \beta)^2,$$

where M is independent of both α and β. Thus, if the null hypothesis $H(\alpha = \alpha_0)$ is true, then the relative minimum of Ω is obtained for $\beta = \beta^*$. Therefore the relative minimum of Ω is

$$M + n(\alpha^* - \alpha_0)^2.$$

Hence the sum of squares for testing the null hypothesis $H(\alpha = \alpha_0)$ is

$$S_1 = n(\alpha^* - \alpha_0)^2,$$

and $\qquad\qquad\qquad E(S_1) = \sigma^2,$

if the null hypothesis under test is true. It is also known that S_1/σ^2 is distributed as χ^2 with 1 d.f. if the null hypothesis $H(\alpha = \alpha_0)$ is true.

Similarly, to test the null hypothesis $H(\beta = \beta_0)$, where β_0 is some pre-assigned value of β, we have to obtain the relative minimum of

$$\Omega = M + n(\alpha^* - \alpha)^2 + X(\beta^* - \beta)^2.$$

Accordingly, if the null hypothesis $H(\beta = \beta_0)$ is true, then the relative minimum of Ω is

$$M + X(\beta^* - \beta_0)^2,$$

which is obtained for $\alpha = \alpha^*$. Therefore the sum of squares for testing the null hypothesis $H(\beta = \beta_0)$ is

$$S_2 = X(\beta^* - \beta_0)^2,$$

and $\qquad\qquad\qquad E(S_2) = \sigma^2,$

if the null hypothesis under test is true. It is also known that S_2/σ^2 is distributed as χ^2 with 1 d.f. if the null hypothesis $H(\beta = \beta_0)$ is true.

We have thus deduced three results:

(i) M/σ^2 is distributed as χ^2 with $n - 2$ d.f. independently of any null hypothesis about α or β.

(ii) S_1/σ^2 is distributed as χ^2 with 1 d.f. if the null hypothesis $H(\alpha = \alpha_0)$ is true.

(iii) S_2/σ^2 is distributed as χ^2 with 1 d.f. if the null hypothesis $H(\beta = \beta_0)$ is true.

These three χ^2's are also known to be independently distributed. Hence the test of significance for the null hypothesis $H(\alpha = \alpha_0)$ is obtained by using the fact that if the null hypothesis under test is true, then the ratio

$$F = (n-2)S_1/M = n(n-2)(\alpha^* - \alpha_0)^2/M$$

has the F distribution with 1, $n-2$ d.f. In the same way, the test of significance for the null hypothesis $H(\beta = \beta_0)$ is obtained by using the fact that if the null hypothesis under test is true, then the ratio

$$F = (n-2)S_2/M = X(n-2)(\beta^* - \beta_0)^2/M$$

has the F distribution with 1, $n-2$ d.f.

The above two tests of significance are, in fact, independent, and so the two null hypotheses $H(\alpha = \alpha_0)$ and $H(\beta = \beta_0)$ can be tested by the same analysis of variance. It is useful to remember that this convenient independence is mathematically possible only because we have used the modified regression equation

$$E(y_\nu) = \alpha + \beta(x_\nu - \bar{x}).$$

Indeed, this is the main reason for the modification of the regression equation. Moreover, when the above two null hypotheses are true, the total sum of squares with n d.f. is

$$\sum_{\nu=1}^{n} [y_\nu - \alpha_0 - \beta_0(x_\nu - \bar{x})]^2 = M + S_1 + S_2.$$

This gives the complete partitioning of the total variation and leads to the following analysis of variance.

Table 15.8 Showing the analysis of variance for linear regression

Source of Variation	d.f.	S.S.	M.S.	F.R.
Hypothesis: $\alpha = \alpha_0$	1	$S_1 = n(\alpha^* - \alpha_0)^2$	S_1	$(n-2)S_1/M$
Hypothesis: $\beta = \beta_0$	1	$S_2 = X(\beta^* - \beta_0)^2$	S_2	$(n-2)S_2/M$
Residual	$n-2$	$M = Y - Z^2/X$	$M/(n-2)$	
Total	n	$\sum_{\nu=1}^{n} [y_\nu - \alpha_0 - \beta_0(x_\nu - \bar{x})]^2$		

The above table gives the complete analysis of variance for the linear regression model, but we have to look into this analysis more closely to understand its further implications.

Geometrical interpretation of the analysis

Suppose the n pairs of observations (x_ν, y_ν), for $\nu = 1, 2, ..., n$, are regarded as co-ordinates of points in the (x, y) plane. Fig 15.1, overleaf, gives a typical representation of these points denoted by $A_1, A_2, ..., A_n$.

The co-ordinates of any point A_ν are (x_ν, y_ν), which means that $OH_\nu = x_\nu$ and $A_\nu H_\nu = y_\nu$, where $A_\nu H_\nu$ is the perpendicular from A_ν on to the

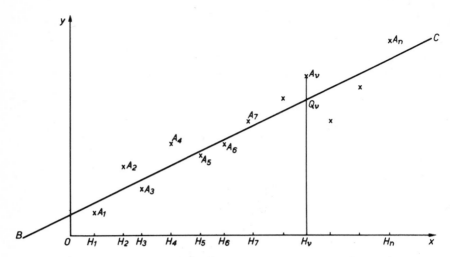

Fig 15.1   Representation of a linear regression and the observations

x axis. Now the postulated model of linear regression implies that, apart from random variations, the points A_ν lie on a straight line

$$y_e = \alpha + \beta(x - \bar{x}),$$

where $y_e \equiv E(y)$ corresponding to any value of x. Of course, y_e is unknown since α and β are unknown; and in using the least-squares procedure for estimating α and β, we have, in fact, minimised the sum of the squares of the differences between the observed y_ν and their expected values

$$y_{e\nu} \equiv E(y_\nu) = \alpha + \beta(x_\nu - \bar{x}).$$

We have thus determined the "best" fitting line to the given observations. Accordingly, if the estimated value of $y_{e\nu}$ is denoted by

$$y_{e\nu}^* \equiv \alpha^* + \beta^*(x_\nu - \bar{x}),$$

then the estimation procedure implies that

$$\sum_{\nu=1}^{n} (y_\nu - y_{e\nu}^*)^2 \quad \text{is a minimum.}$$

This sum of squares is, in fact, the residual S. S. (M) of the analysis of variance, and the equivalence of the two forms of M is readily verified. The value $y_{e\nu}^*$ is the estimated or "fitted" ordinate corresponding to x_ν, as determined from the estimated regression line

$$y_e^* = \alpha^* + \beta^*(x - \bar{x})$$

shown in Fig. 15.1 as BC. Therefore $Q_\nu H_\nu = y_{e\nu}^*$ and $A_\nu Q_\nu = y_\nu - y_{e\nu}^*$. Thus $A_\nu Q_\nu$, for $\nu = 1, 2, \ldots, n$, are the unexplained or random variations of the observations y_ν from the estimated regression line, and it is intuitively acceptable that these differences should provide an estimate of the variance σ^2. This is the geometrical meaning of the residual S.S. (M).

Furthermore, we have shown that

$$M = Y - Z^2/X = Y - \beta^{*2}X,$$

which clearly indicates that M is the difference between two positive quantities. The first quantity (Y) is simply the sum of squares of the deviations of the observations y_ν from their own mean $\bar{y}$; this would be the usual residual S.S. if β were, in fact, zero so that $E(y_\nu) = a$, a constant. The quantity $\beta^{*2}X$ may now be interpreted as the reduction in the residual S.S. of the random variable y by its postulated linear relationship with the non-random variable x. This reduction of the residual S.S. with every additional parameter introduced in the expectation of the random variable y is a principle of complete generality. We have exemplified it in the present case of the two parameters a and β.

Another way of observing the statistical similarity between the parameters a and β is to note that

$$M = \sum_{\nu=1}^{n} y_\nu^2 - na^{*2} - X\beta^{*2},$$

and we recall that the sums of squares for testing the null hypotheses $H(a = 0)$ and $H(\beta = 0)$ are na^{*2} and $X\beta^{*2}$ respectively. Thus M is the difference between the raw S.S. of the observations and the sums of squares due to the null hypotheses $H(a = 0)$ and $H(\beta = 0)$.

Confidence intervals for a and β

We have seen that the estimates a^* and β^* are linear functions of the normally distributed random variables $-y_\nu$. Hence both a^* and β^* are also normally distributed. Furthermore, since

$$\text{var}(a^*) = \sigma^2/n \quad \text{and} \quad \text{var}(\beta^*) = \sigma^2/X,$$

it follows that $\quad$ s.e.$(a^*) = s/\sqrt{n} \quad$ and $\quad$ s.e.$(\beta^*) = s/\sqrt{X},$

s^2 being the residual mean square obtained from the analysis of variance. Hence, to test the null hypothesis $H(a = a_0)$, we use the fact that if the null hypothesis under test is true, then the random variable

$$t = \frac{a^* - a_0}{s/\sqrt{n}}$$

has "Student's" distribution with $n - 2$ d.f. The square of this t is easily seen to be $(n - 2)S_1/M$, the appropriate F ratio in the analysis of variance. In the same way, to test the null hypothesis $H(\beta = \beta_0)$, we use the fact that if the null hypothesis under test is true, then the random variable

$$t = \frac{\beta^* - \beta_0}{s/\sqrt{X}}$$

has "Student's" distribution with $n - 2$ d.f. The square of this t is the F ratio $(n - 2)S_2/M$ in the analysis of variance.

We can thus use "Student's" distribution to obtain any required confidence intervals for the parameters a and β in exactly the same way as we obtained a confidence interval for μ, the mean of a normal population. It is useful to remember that although a and β are very different parameters geometrically, their estimates are linear functions of the random variables y_ν. This explains the distributional similarity of the tests of significance of null hypotheses about the two parameters and also the manner in which confidence intervals may be obtained for them. The geometrical meaning of α and β is important, but the best theoretical approach is to view them as two parameters occurring linearly in the basic equations of expectations, namely,

$$E(y_\nu) = \alpha + \beta(x_\nu - \bar{x}), \quad \text{for} \quad \nu = 1, 2, \ldots, n.$$

Estimation of a y value from the regression equation

Suppose x_0 is a value of the explanatory variable x such that $x_0 \neq x_\nu$, for $\nu = 1, 2, \ldots, n$. Then the estimated regression line can be used to determine an estimated value of the random variable y corresponding to x_0. In fact, the required estimate is simply

$$y^*_{e0} = \alpha^* + \beta^*(x_0 - \bar{x}).$$

To determine $\operatorname{var}(y^*_{e0})$, it is convenient to express the estimate as an explicit linear function of the y_ν. Using the expressions for α^* and β^*, we have

$$y^*_{e0} = \frac{1}{n}\sum_{\nu=1}^{n} y_\nu + (x_0 - \bar{x})\sum_{\nu=1}^{n}\left[\frac{1}{X}(x_\nu - \bar{x})\right]y_\nu$$

$$= \sum_{\nu=1}^{n}\left[\frac{1}{n} + \frac{1}{X}(x_0 - \bar{x})(x_\nu - \bar{x})\right]y_\nu.$$

Hence

$$\operatorname{var}(y^*_{e0}) = \sum_{\nu=1}^{n}\left[\frac{1}{n} + \frac{1}{X}(x_0 - \bar{x})(x_\nu - \bar{x})\right]^2\sigma^2$$

$$= \sum_{\nu=1}^{n}\left[\frac{1}{n^2} + \frac{2}{nX}(x_0 - \bar{x})(x_\nu - \bar{x}) + \frac{1}{X^2}(x_0 - \bar{x})^2(x_\nu - \bar{x})^2\right]\sigma^2$$

$$= \sigma^2\left[\frac{1}{n} + \frac{(x_0 - \bar{x})^2}{X}\right], \quad \text{since} \quad \sum_{\nu=1}^{n}(x_\nu - \bar{x}) \equiv 0.$$

Therefore

$$\text{s.e.}(y^*_{e0}) = s\left[\frac{1}{n} + \frac{(x_0 - \bar{x})^2}{X}\right]^{\frac{1}{2}}.$$

Furthermore, if $y_{e0} \equiv \alpha + \beta(x_0 - \bar{x})$ is the value of the random variable y corresponding to $x = x_0$ obtained from the true but unknown regression line, then evidently $E(y^*_{e0}) \equiv y_{e0}$; and since y^*_{e0} is a linear function of the y_ν, we infer that

$$t = \frac{y^*_{e0} - y_{e0}}{\text{s.e.}(y^*_{e0})}$$

has "Student's" distribution with $n-2$ d.f. If $t(\eta; n-2)$ is the 100η per cent point of "Student's" distribution with $n-2$ d.f., then the $100(1-\eta)$ per cent confidence interval for y_{e0} is

$$y^*_{e0} - t(\eta; n-2) \times \text{s.e.}(y^*_{e0}) \leqslant y_{e0} \leqslant y^*_{e0} + t(\eta; n-2) \times \text{s.e.}(y^*_{e0}).$$

This method of estimating a y value corresponding to any value x_0 of x is legitimate, provided it can be assumed that the linearity of regression holds over the range of values of x which includes x_0. However, if x_0 is appreciably outside this range, then there is a possibility that the regression may not be linear much beyond the range covered by the initial values of x. Thus, for example, it is obvious that an estimated linear relationship between potato yield and amount of fertilizer used could not hold beyond a certain value of x. In any statistical relationship of this kind, linearity of regression would, in general, hold only over a limited range of the explanatory variable. We mention this as a caution against an automatic use of the estimated regression for evaluating y values corresponding to values of the explanatory variable not included in the sample observations.

Linear transformations in linear regression analysis

We have so far postulated a clear distinction between the variables x and y of the linear regression model by assuming that x is a predetermined non-random variable whilst y is a random variable subject to the chance errors of experimentation. This model is appropriate in many situations where the x values can be chosen in advance by the experimenter. On the other hand, there are instances when, in principle, both x and y are random variables. For example, x and y may correspond to the height and weight measurements of adults in a country. The mathematical derivation of the appropriate regression analysis is now somewhat different, but it happily turns out that the randomness of x does not affect the analysis of the regression of y on x. Consequently, the same tests of significance can be used irrespective of whether x is a non-random or a random variable. There is, however, another point which arises when both x and y are random variables. Formally, one could now think of the regression of y on x, and also that of x on y. In general, these two regressions will be different, and this duality leads to certain consequences which we shall consider in the next chapter. For the present, we confine our attention to the single regression equation of y on x, where x may or may not be a random variable.

When x is a predetermined variable, it occasionally happens that the values of x are equispaced. For example, in the experiment to determine the effect of quantity of fertilizer on the yield of potatoes, the quantity of fertilizer used on each of the n experimental plots could be chosen to vary by equal amounts. A linear transformation of the x values would now lead to considerable simplification in the computations required for statistical analysis. More generally, it is possible to extend this idea and use two different linear transformations on the x and y values to simplify the calculations for the analysis of variance. We next show how the analysis of variance is modified by such linear transformation of the observations.

Suppose, then, that (x_ν, y_ν), for $\nu = 1, 2, \ldots, n$, are the original sample observations such that

$$E(y_\nu) = \alpha + \beta(x_\nu - \bar{x}).$$

We now assume that these observations are transformed to the pairs (u_ν, v_ν) by the linear transformations

$$x_\nu = a_1 + b_1 u_\nu \quad \text{and} \quad y_\nu = a_2 + b_2 v_\nu.$$

Here $a_1, a_2, b_1,$ and b_2 are known constants, but both b_1 and b_2 are restricted to non-zero values. With these transformations, the sample means are transformed as

$$\bar{x} = a_1 + b_1 \bar{u}, \quad \bar{y} = a_2 + b_2 \bar{v},$$

where $\bar{u}$ and $\bar{v}$ are the means of the u_ν and v_ν respectively.

We now have

$$E(v_\nu) = \frac{1}{b_2} E(y_\nu - a_2)$$

$$= \frac{1}{b_2} [a + \beta(x_\nu - \bar{x}) - a_2]$$

$$= \frac{a - a_2}{b_2} + \frac{\beta b_1}{b_2}(u_\nu - \bar{u})$$

$$= \gamma + \delta(u_\nu - \bar{u}),$$

where we set $\delta \equiv \beta b_1/b_2$ and $\gamma \equiv (a - a_2)/b_2$. Thus the regression of v on u is also linear but with parameters γ and δ. Hence the original null hypotheses $H(a = a_0)$ and $H(\beta = \beta_0)$ are equivalent to the null hypotheses $H(\gamma = \gamma_0)$ and $H(\delta = \delta_0)$, where $\gamma_0 \equiv (a_0 - a_2)/b_2$ and $\delta_0 \equiv b_1\beta_0/b_2$.

We show next that the F ratios for testing the original null hypotheses are transformed appropriately into F ratios for testing the equivalent null hypotheses about γ and δ. We have

$$X = \sum_{\nu=1}^{n} (x_\nu - \bar{x})^2 = \sum_{\nu=1}^{n} [(a_1 + b_1 u_\nu) - (a_1 + b_1\bar{u})]^2$$

$$= b_1^2 \sum_{\nu=1}^{n} (u_\nu - \bar{u})^2 = b_1^2 U, \quad \text{say.}$$

Similarly, $\quad Y = b_2^2 \sum_{\nu=1}^{n} (v_\nu - \bar{v})^2 = b_2^2 V, \quad \text{say;} \quad \text{and}$

$$Z = b_1 b_2 \sum_{\nu=1}^{n} (u_\nu - \bar{u})(v_\nu - \bar{v}) = b_1 b_2 W, \quad \text{say.}$$

Hence $\quad M = Y - Z^2/X = b_2^2(V - W^2/U); \quad \text{and}$

$$\beta^* = Z/X = b_2 W/b_1 U = b_2\delta^*/b_1,$$

where $\delta^* = W/U$ is the least-squares estimate of δ. Also, the least-squares estimate of γ is $\gamma^* = \bar{v}$. Accordingly, we have

$$n(a^* - a_0)^2 = n[(a_2 + b_2\bar{v}) - (a_2 + b_2\gamma_0)]^2$$

$$= b_2^2 n(\bar{v} - \gamma_0)^2 = b_2^2 n(\gamma^* - \gamma_0)^2.$$

Therefore the F ratio for testing the null hypothesis $H(a = a_0)$ is

$$\frac{n(n - 2)(a^* - a_0)^2}{Y - Z^2/X} = \frac{n(n - 2)(\gamma^* - \gamma_0)^2}{V - W^2/U},$$

which is the F ratio for testing the equivalent null hypothesis $H(\gamma = \gamma_0)$. Similarly, the F ratio for testing the null hypothesis $H(\beta = \beta_0)$ is

$$\frac{(n - 2) X(\beta^* - \beta_0)^2}{Y - Z^2/X} = \frac{(n - 2)U(\delta^* - \delta_0)^2}{V - W^2/U},$$

which is the F ratio for testing the equivalent null hypothesis $H(\delta = \delta_0)$. Of course, both these results are intuitively obvious because of the invariance of F under linear transformation of the observations.

Next, the estimated regression equation of v on u is

$$v_{e\nu}^* = \gamma^* + \delta^*(u_\nu - \bar{u}), \quad \text{where we write } E(v_\nu) \equiv v_{e\nu},$$

$$= \left[\frac{1}{b_2}(a^* - a_2) + \frac{b_1\beta^*}{b_2}(x_\nu - \bar{x})\frac{1}{b_1} \right]$$

$$= [\{a^* + \beta^*(x_\nu - \bar{x})\} - a_2]/b_2 = (y_{e\nu}^* - a_2)/b_2,$$

so that
$$y_{e\nu}^* = a_2 + b_2 v_{e\nu}^*.$$

This means that the estimated y values can be obtained from the corresponding estimated v values.

In the same way, confidence intervals for α and β can be evaluated simply from those obtained for γ and δ. In fact, if $t(\eta; n-2)$ is the 100η per cent point of "Student's" distribution with $n-2$ d.f., then

(i) the $100(1-\eta)$ per cent confidence interval for γ is
$$\gamma^* - t(\eta; n-2) \times \text{s.e.}(\gamma^*) \leqslant \gamma \leqslant \gamma^* + t(\eta; n-2) \times \text{s.e.}(\gamma^*);$$
and

(ii) the $100(1-\eta)$ per cent confidence interval for δ is
$$\delta^* - t(\eta; n-2) \times \text{s.e.}(\delta^*) \leqslant \delta \leqslant \delta^* + t(\eta; n-2) \times \text{s.e.}(\delta^*).$$

But $\alpha = a_2 + b_2\gamma$ and $\beta = b_2\delta/b_1$. Hence the corresponding confidence intervals for α and β are

(i)'
$$a_2 + b_2[\gamma^* - t(\eta; n-2) \times \text{s.e.}(\gamma^*)] \leqslant \alpha$$
$$\leqslant a_2 + b_2[\gamma^* + t(\eta; n-2) \times \text{s.e.}(\gamma^*)];$$
and

(ii)'
$$\frac{b_2}{b_1}[\delta^* - t(\eta; n-2) \times \text{s.e.}(\delta^*)] \leqslant \beta \leqslant \frac{b_2}{b_1}[\delta^* + t(\eta; n-2) \times \text{s.e.}(\delta^*)].$$

The interest of (i)' and (ii)' is that the analysis of the transformed data leads directly to these confidence intervals for α and β.

Finally, we extend the analysis of the transformed data to the estimation of a y value y_{eo} corresponding to the x value x_0. If $x_0 = a_1 + b_1 u_0$ and $v_{eo} = \gamma + \delta(u_0 - \bar{u})$, then the estimate of v_{eo} is
$$v_{eo}^* = \gamma^* + \delta^*(u_0 - \bar{u}),$$
and
$$\text{s.e.}(v_{eo}^*) = s\left[\frac{1}{n} + \frac{(u_0 - \bar{u})^2}{U}\right]^{\frac{1}{2}},$$

where s^2 is now the residual mean square obtained from the analysis of variance of the transformed observations. The $100(1-\eta)$ per cent confidence interval for v_{eo} is

(iii) $v_{eo}^* - t(\eta; n-2) \times \text{s.e.}(v_{eo}^*) \leqslant v_{eo} \leqslant v_{eo}^* + t(\eta; n-2) \times \text{s.e.}(v_{eo}^*).$

But $y_{eo} = a_2 + b_2 v_{eo}$, so that the point estimate of y_{eo} is $y_{eo}^* = a_2 + b_2 v_{eo}^*$, and the $100(1-\eta)$ per cent confidence interval for y_{eo} is

(iii)' $a_2 + b_2[v_{eo}^* - t(\eta; n-2) \times \text{s.e.}(v_{eo}^*)] \leqslant y_{eo}$
$$\leqslant a_2 + b_2[v_{eo}^* + t(\eta; n-2) \times \text{s.e.}(v_{eo}^*)].$$

Computational notes

If the analysis of variance is carried out in terms of the original paired observations (x_ν, y_ν), then the three basic sums X, Y, and Z are evaluated as follows. We have

$$X = \sum_{\nu=1}^{n}(x_\nu - \bar{x})^2 = \sum_{\nu=1}^{n} x_\nu^2 - \left[\sum_{\nu=1}^{n} x_\nu\right]^2 \bigg/ n = \text{raw S.S.}(x) - \text{C.F.}(x).$$

Similarly,

$$Y = \sum_{\nu=1}^{n}(y_\nu - \bar{y})^2 = \sum_{\nu=1}^{n} y_\nu^2 - \left[\sum_{\nu=1}^{n} y_\nu\right]^2 \Big/ n = \text{raw S.S.}(y) - \text{C.F.}(y).$$

Lastly,
$$Z = \sum_{\nu=1}^{n}(x_\nu - \bar{x})(y_\nu - \bar{y}) = \sum_{\nu=1}^{n} x_\nu y_\nu - \bar{x}\sum_{\nu=1}^{n} y_\nu$$

$$= \sum_{\nu=1}^{n} x_\nu y_\nu - \left[\sum_{\nu=1}^{n} x_\nu\right]\left[\sum_{\nu=1}^{n} y_\nu\right]\Big/ n.$$

The sum $\sum_{\nu=1}^{n} x_\nu y_\nu$ is known as the *raw sum of products of x and y*, and is denoted briefly as raw S. P.(x,y). Also,

$$\left[\sum_{\nu=1}^{n} x_\nu\right]\left[\sum_{\nu=1}^{n} y_\nu\right]\Big/ n$$

is referred to as the *correction factor for the sum of products*, and is denoted as C.F.(x,y). Therefore, in conformity with the computational forms for X and Y, we can also write

$$Z = \text{raw S.P.}(x, y) - \text{C.F.}(x, y).$$

The computational expressions for U, V, and W derived from the transformed observations are evidently analogous to those for X, Y, and Z respectively.

We conclude by an example showing the computational procedure in the analysis of data.

Example 15.6: Speed records in motor racing

Table 15.9 gives the speed records (in miles per hour) attained in the Indianapolis Memorial Day car races during the years 1911–1939, there having been no racing in 1917 and 1918. We assume that the speed (y) increased linearly over the years (x), and then estimate the parameters α and β of the linear regression equation

$$E(y) = \alpha + \beta(x - \bar{x}).$$

The fact that there was no racing in 1917 and 1918 means that no observations are available for the two years, and it would be completely wrong to take zero for the missing y values. The correct procedure is to estimate the regression from the records available for the other 27 years. Next, we use the fitted regression to estimate the speeds for the years 1917 and 1918, and also to find the 95 per cent confidence intervals for these speeds as determined from the true regression. The 95 per cent confidence intervals for α and β are also evaluated.

We could analyse the given x and y observations, but it is simpler to make the transformations

$$u_\nu = x_\nu - 1925 \quad \text{and} \quad v_\nu = y_\nu - 70.$$

The u and v values are given in the last two columns of Table 15.9, and we work with these transformed observations.

In Table 15.9,

$$n = 27; \quad \sum_{\nu=1}^{n} u_\nu = 15; \quad \sum_{\nu=1}^{n} v_\nu = 700 \cdot 6; \quad \bar{u} = 0 \cdot 5556; \quad \bar{v} = 25 \cdot 95.$$

Raw S.S.$(u) = 1917$;   Raw S.S.$(v) = 21738 \cdot 56$;   Raw S.P.$(u,v) = 2889 \cdot 8$.
  C.F.$(u) = 8 \cdot 33$;   C.F.$(v) = 18179 \cdot 27$;   C.F.$(u, v) = 389 \cdot 22$.

Table 15.9   Showing the speed records (in miles per hour) attained in the Indianapolis Memorial Day car races during the years 1911 – 1939

Year (x)	Speed (y)	$u = x - 1925$	$v = y - 70$
1911	74.7	-14	4·7
1912	78.7	-13	8·7
1913	75.9	-12	5·9
1914	82.5	-11	12·5
1915	89.8	-10	19·8
1916	83.3	-9	13·3
1917	no races		
1918	no races		
1919	81.1	-6	11·1
1920	88.5	-5	18·5
1921	89.6	-4	19·6
1922	94.5	-3	24·5
1923	91.0	-2	21·0
1924	98.2	-1	28·2
1925	101.1	0	31·1
1926	95.9	1	25·9
1927	97.5	2	27·5
1928	99.5	3	29·5
1929	97.6	4	27·6
1930	100.4	5	30·4
1931	96.6	6	26·6
1932	104.1	7	34·1
1933	104.1	8	34·1
1934	104.9	9	34·9
1935	106.2	10	36·2
1936	109.1	11	39·1
1937	113.6	12	43·6
1938	117.2	13	47·2
1939	115.0	14	45·0

Hence $U = 1908·67;$ $V = 3559·29;$ $W = 2500·58.$

Therefore $\gamma^* = 25·95;$ $\delta^* = 1·3101;$

and the estimated regression equation of v on u is

$$v_e^* = 25·95 + 1·3101(u - 0·5556)$$
$$= 25·22 + 1·3101u.$$

Residual S.S. (M) = $3559·29 - 3275·97 = 283·32$ with 25 d.f.

Residual M.S. is $s^2 = 11·3328;$ $s = 3·3664.$

If required, the analysis of variance table could now be formed to test any null hypotheses about γ and δ and so implicitly about α and β.

To estimate the missing speed values, we observe that the years 1917 and 1918 correspond to $u = -8$ and $u = -7$. For $u = -7$, the estimated v value is

$$v_e^*(-7) = 25·22 - 9·17 = 16·05,$$

and that for $u = -8$

$$v_e^*(-8) = 25·22 - 10·48 = 14·74.$$

For any u, the standard error of the estimated v value is

$$\text{s.e.}(v_e^*) = s\left[\frac{1}{n} + \frac{(u - \bar{u})^2}{U}\right]^{\frac{1}{2}}.$$

Therefore $\text{s.e.}[v_e^*(-7)] = 3\cdot3664\left[\frac{1}{27} + \frac{(7\cdot5556)^2}{1908\cdot67}\right]^{\frac{1}{2}}$

$$= 3\cdot3664 \times 0\cdot258740 = 0\cdot8710;$$

and

$$\text{s.e.}[v_e^*(-8)] = 3\cdot3664\left[\frac{1}{27} + \frac{(8\cdot5556)^2}{1908\cdot67}\right]^{\frac{1}{2}}$$

$$= 3\cdot3664 \times 0\cdot274567 = 0\cdot9243.$$

The five per cent value of "Student's" distribution with 25 d.f. is 2·060. Therefore the 95 per cent confidence interval for $v_e(-7)$ is

$$16\cdot05 - 2\cdot060 \times 0\cdot8710 \leqslant v_e(-7) \leqslant 16\cdot05 + 2\cdot060 \times 0\cdot8710,$$

or $14\cdot26 \leqslant v_e(-7) \leqslant 17\cdot84.$

Similarly, the 95 per cent confidence interval for $v_e(-8)$ is

$$14\cdot74 - 2\cdot060 \times 0\cdot9243 \leqslant v_e(-8) \leqslant 14\cdot74 + 2\cdot060 \times 0\cdot9243,$$

or $12\cdot84 \leqslant v_e(-8) \leqslant 16\cdot64.$

Transforming to the original speed scale, we infer that the estimated speeds for the years 1917 and 1918 are 84·74 and 86·05 respectively. Furthermore, if $y_e(17)$ and $y_e(18)$ are the true speeds for 1917 and 1918 as given by the assumed true linear regression, then their 95 per cent confidence intervals are

$$82\cdot84 \leqslant y_e(17) \leqslant 86\cdot64,$$

and $84\cdot26 \leqslant y_e(18) \leqslant 87\cdot84.$

Since $b_1 = b_2 = 1$, we have $\delta^* = \beta^* = 1\cdot3101$;

and $\text{s.e.}(\beta^*) = \left[\frac{11\cdot3328}{1908\cdot67}\right]^{\frac{1}{2}} = \sqrt{0\cdot00593754} = 0\cdot077055.$

Therefore the 95 per cent confidence interval for β is

$$1\cdot3101 - 2\cdot060 \times 0\cdot077055 \leqslant \beta \leqslant 1\cdot3101 + 2\cdot060 \times 0\cdot077055,$$

or $1\cdot1514 \leqslant \beta \leqslant 1\cdot4688.$

Again, since $b_2 = 1$ and $a_2 = 70$, we have $\alpha^* = \delta^* + 70 = 95\cdot95$;

and $\text{s.e.}(\alpha^*) = \left[\frac{11\cdot3328}{27}\right]^{\frac{1}{2}} = \sqrt{0\cdot419733} = 0\cdot647868.$

Therefore the 95 per cent confidence interval for α is

$$95\cdot95 - 2\cdot060 \times 0\cdot647868 \leqslant \alpha \leqslant 95\cdot95 + 2\cdot060 \times 0\cdot647868,$$

or $94\cdot62 \leqslant \alpha \leqslant 97\cdot28.$

15.14   Comparison of two linear regressions

As we have seen in the preceding section, a linear regression measures, on the average, the variation in a random variable over a predetermined range of values of another explanatory variable. Accordingly, if there are two comparable but independent experimental situations, then it is possible to arrive at two different linear regressions. In such a case, it would be natural to derive methods for comparing the two regressions, and this is the object of our investigation in this section. However, it is useful to view this problem in a meaningful physical context, and this can be done quite conveniently by extending the argument of the farmer's problem of determining the effect of quantity of a fertilizer on the yield of potatoes. Suppose, then, the farmer is faced with a choice of one of two fertilizers A and B for improving his potato harvest returns. He knows that, in general, the use of either fertilizer, within limits, will tend to increase yield, and, quite obviously, his interest lies in determining which of A and B is likely to give him, on the average, bigger harvest returns. It is thus clear that a pertinent comparison is between the two linear rates of increase of yield obtained by using A and B respectively.

Suppose, then, that the farmer has $n_1 + n_2$ similar experimental plots. He uses fertilizer A in quantities $x_{11}, x_{12}, \ldots, x_{1n_1}$ on n_1 plots and obtains potato yields $y_{11}, y_{12}, \ldots, y_{1n_1}$ respectively. Similarly, he uses fertilizer B in quantities $x_{21}, x_{22}, \ldots, x_{2n_2}$ on n_2 plots and has the returns $y_{21}, y_{22}, \ldots, y_{2n_2}$. This is a general formulation of the two experiments, since it is quite legitimate to plan the experimentation in such a way that $n_1 = n_2$ and the quantities of the fertilizers used are also the same. However, we shall consider the general case. We have thus obtained two independent sets of paired observations, which may be compactly written as

$$(x_{i\nu}, y_{i\nu}), \quad \text{for } i = 1, 2; \ \nu = 1, 2, \ldots, n_i.$$

The first suffix i refers to the fertilizers and the second suffix ν to the plots in each group. We now postulate that the $y_{i\nu}$ are normally distributed with

$$E(y_{i\nu}) = \alpha_i + \beta_i(x_{i\nu} - \bar{x}_i) \quad \text{and} \quad \text{var}(y_{i\nu}) = \sigma^2,$$

where $\alpha_i, \beta_i, \sigma^2$ are unknown parameters and $\bar{x}_i$ is the mean of the $x_{i\nu}$, for $\nu = 1, 2, \ldots, n_i$.

It follows from an easy extension of our previous argument that the $y_{i\nu}$ are independent observations from $N \equiv n_1 + n_2$ normal populations. Furthermore, the means of the n_1 populations corresponding to the quantities used of fertilizer A are assumed to lie on one straight line. Similarly, the means of the n_2 populations corresponding to the quantities used of fertilizer B lie on another straight line. These are the two regression lines, and our purpose is to test null hypotheses about α_1, β_1 and α_2, β_2, the parameters of these lines. The tests to be derived are exact, and equality of the variances of the N populations sampled is an essential assumption in the derivation of these tests.

Estimation of the parameters

The first stage in the analysis is to determine the least-squares estimates of the five parameters α_i, β_i $(i = 1, 2)$, and σ^2. We therefore minimise

$$\Omega = \sum_{i=1}^{2} \sum_{\nu=1}^{n_i} [y_{i\nu} - \alpha_i - \beta_i(x_{i\nu} - \bar{x}_i)]^2$$

with respect to α_i and β_i. Differentiation with respect to α_i and β_i gives

$$\frac{\partial \Omega}{\partial \alpha_i} = -2 \sum_{\nu=1}^{n_i} [y_{i\nu} - \alpha_i - \beta_i(x_{i\nu} - \bar{x}_i)], \tag{1}$$

and
$$\frac{\partial \Omega}{\partial \beta_i} = -2 \sum_{\nu=1}^{n_i} [y_{i\nu} - a_i - \beta_i(x_{i\nu} - \bar{x}_i)](x_{i\nu} - \bar{x}_i). \tag{2}$$

It is convenient to introduce the following notation:

$$X_i = \sum_{\nu=1}^{n_i}(x_{i\nu} - \bar{x}_i)^2; \qquad Y_i = \sum_{\nu=1}^{n_i}(y_{i\nu} - \bar{y}_i)^2;$$

$$Z_i = \sum_{\nu=1}^{n_i}(x_{i\nu} - \bar{x}_i)(y_{i\nu} - \bar{y}_i),$$

where $\bar{y}_i$ is the average of the $y_{i\nu}$; and

$$X = \sum_{i=1}^{2} X_i; \qquad Y = \sum_{i=1}^{2} Y_i; \qquad Z = \sum_{i=1}^{2} Z_i.$$

It now follows immediately from (1) and (2) that the estimates of a_i and β_i are

$$a_i^* = \bar{y}_i \qquad \text{and} \qquad \beta_i^* = Z_i/X_i, \quad \text{for } i = 1, 2.$$

These are, in fact, the usual estimates which would be obtained by separate minimisation of the two independent sets of deviations of the observations from their respective expectations. Furthermore, the absolute minimum of Ω is

$$M = \sum_{i=1}^{2} \sum_{\nu=1}^{n_i} [y_{i\nu} - \bar{y}_i - \beta_i^*(x_{i\nu} - \bar{x}_i)]^2 = \sum_{i=1}^{2}(Y_i - Z_i^2/X_i),$$

which is equal to the pooled residual sums of squares obtained from the two independent samples of n_1 and n_2 observations respectively. Accordingly,

$$E(M) = \sum_{i=1}^{2}(n_i - 2)\sigma^2 = (N - 4)\sigma^2.$$

Hence $s^2 = M/(N - 4)$ is the least-squares estimate of σ^2, and M/σ^2 has the χ^2 distribution with $N - 4$ d.f.

If we set $M_i = Y_i - Z_i^2/X_i$, the residual sum of squares from the ith sample, then by a simple extension of the one-sample case we can write

$$\Omega = \sum_{i=1}^{2}[M_i + n_i(a_i^* - a_i)^2 + X_i(\beta_i^* - \beta_i)^2]$$

$$= M + \sum_{i=1}^{2}[n_i(a_i^* - a_i)^2 + X_i(\beta_i^* - \beta_i)^2].$$

We shall use this expression for Ω in deriving the appropriate sums of squares for the null hypotheses to be tested.

Tests of significance for null hypotheses about a_1, a_2 *and* β_1, β_2

To test the null hypothesis $H(a_1 - a_2 = a_0)$, where a_0 is some pre-assigned quantity, we have first to derive the minimum of Ω relative to the null hypothesis. The relative minimum is the absolute minimum of

$$\Omega_1 = M + \sum_{i=1}^{2} X_i(\beta_i^* - \beta_i)^2 + n_1(a_1^* - a_2 - a_0)^2 + n_2(a_2^* - a_2)^2$$

for variations in β_1, β_2, and a_2. Clearly, the required minimum is obtained for $\beta_i = \beta_i^*$ and $a_2 = a_2^{**}$, where a_2^{**} is determined from the equation

$$n_1(a_1^* - a_2^{**} - a_0) + n_2(a_2^* - a_2^{**}) = 0.$$

Thus
$$a_2^{**} = (n_1 a_1^* + n_2 a_2^* - n_1 a_0)/N = \bar{y} - n_1 a_0/N,$$

where $\bar{y} = (n_1 a_1^* + n_2 a_2^*)/N$, the grand mean. Hence the relative minimum of Ω is

$$M + n_1(\alpha_1^* - \alpha_2^{**} - \alpha_0)^2 + n_2(\alpha_2^* - \alpha_2^{**})^2,$$

and so the sum of squares for testing the null hypothesis $H(\alpha_1 - \alpha_2 = \alpha_0)$ is

$$
\begin{aligned}
S_1 &= n_1(\alpha_1^* - \alpha_2^{**} - \alpha_0)^2 + n_2(\alpha_2^* - \alpha_2^{**})^2 \\
&= \frac{n_1}{N^2}[n_2(\alpha_1^* - \alpha_2^*) - n_2\alpha_0]^2 + \frac{n_2}{N^2}[-n_1(\alpha_1^* - \alpha_2^*) + n_1\alpha_0]^2 \\
&= \frac{n_1 n_2}{N}(\alpha_1^* - \alpha_2^* - \alpha_0)^2.
\end{aligned}
$$

If the null hypothesis $H(\alpha_1 - \alpha_2 = \alpha_0)$ is true, then

$$E(\alpha_1^*) = \alpha_2 + \alpha_0 \quad \text{and} \quad E(\alpha_2^*) = \alpha_2.$$

Also, since α_1^* and α_2^* are obtained from independent samples, we have

$$
\begin{aligned}
E(S_1) &= \frac{n_1 n_2}{N} E[(\alpha_1^* - \alpha_2 - \alpha_0) - (\alpha_2^* - \alpha_2)]^2 \\
&= \frac{n_1 n_2}{N}[\text{var}(\alpha_1^*) + \text{var}(\alpha_2^*)] = \frac{n_1 n_2}{N}\left[\frac{\sigma^2}{n_1} + \frac{\sigma^2}{n_2}\right] = \sigma^2.
\end{aligned}
$$

Hence S_1 is an unbiased estimate of σ^2 if the null hypothesis $H(\alpha_1 - \alpha_2 = \alpha_0)$ is true, and S_1/σ^2 is then distributed as χ^2 with 1 d.f.

In the same way, to test the null hypothesis $H(\beta_1 - \beta_2 = \beta_0)$, where β_0 is some pre-assigned quantity, we have first to derive the minimum of Ω relative to the null hypothesis. This relative minimum is the absolute minimum of

$$\Omega_2 = M + \sum_{i=1}^{2} n_i(\alpha_i^* - \alpha_i)^2 + X_1(\beta_1^* - \beta_2 - \beta_0)^2 + X_2(\beta_2^* - \beta_2)^2$$

for variations in α_1, α_2, and β_2. This minimum is obtained for $\alpha_i = \alpha_i^*$, and $\beta_2 = \beta_2^{**}$, where β_2^{**} is determined from the equation

$$X_1(\beta_1^* - \beta_2^{**} - \beta_0) + X_2(\beta_2^* - \beta_2^{**}) = 0.$$

Therefore $\beta_2^{**} = (X_1\beta_1^* + X_2\beta_2^* - X_1\beta_0)/X = \bar{\beta}^* - X_1\beta_0/X,$

where $\bar{\beta}^* = (X_1\beta_1^* + X_2\beta_2^*)/X$. Hence the relative minimum of Ω is

$$M + X_1(\beta_1^* - \beta_2^{**} - \beta_0)^2 + X_2(\beta_2^* - \beta_2^{**})^2.$$

Therefore the sum of squares for testing the null hypothesis $H(\beta_1 - \beta_2 = \beta_0)$ is

$$
\begin{aligned}
S_2 &= X_1(\beta_1^* - \beta_2^{**} - \beta_0)^2 + X_2(\beta_2^* - \beta_2^{**})^2 \\
&= \frac{X_1}{X^2}[X_2(\beta_1^* - \beta_2^* - \beta_0)]^2 + \frac{X_2}{X^2}[-X_1(\beta_1^* - \beta_2^* - \beta_0)]^2 \\
&= \frac{X_1 X_2}{X}(\beta_1^* - \beta_2^* - \beta_0)^2.
\end{aligned}
$$

Again, if the null hypothesis $H(\beta_1 - \beta_2 = \beta_0)$ is true, then

$$E(\beta_1^*) = \beta_2 + \beta_0 \quad \text{and} \quad E(\beta_2^*) = \beta_2.$$

Hence, since β_1^* and β_2^* are estimates obtained from independent samples, we have

$$E(S_2) = \frac{X_1 X_2}{X} E[(\beta_1^* - \beta_2 - \beta_0) - (\beta_2^* - \beta_2)]^2$$

$$= \frac{X_1 X_2}{X}[\text{var}(\beta_1^*) + \text{var}(\beta_2^*)]^2 = \frac{X_1 X_2}{X}\left[\frac{\sigma^2}{X_1} + \frac{\sigma^2}{X_2}\right] = \sigma^2.$$

Therefore S_2 is an unbiased estimate of σ^2 if the null hypothesis $H(\beta_1 - \beta_2 = \beta_0)$ is true, and then S_2/σ^2 is distributed as χ^2 with 1 d.f.

We have thus deduced three main results:

(i)   If the null hypothesis $H(\alpha_1 - \alpha_2 = \alpha_0)$ is true, then S_1/σ^2 is distributed as χ^2 with 1 d.f.

(ii)   If the null hypothesis $H(\beta_1 - \beta_2 = \beta_0)$ is true, then S_2/σ^2 is distributed as χ^2 with 1 d.f.

(iii) M/σ^2 is distributed as χ^2 with $N - 4$ d.f. independently of any hypothesis about α_i and β_i.

We also know from general theory that these three χ^2's are independent.

Hence the test of significance for the null hypothesis $H(\alpha_1 - \alpha_2 = \alpha_0)$ is obtained by using the fact that if the null hypothesis is true, then the ratio

$$F = (N - 4)S_1/M$$

has the F distribution with $1, N - 4$ d.f.

Similarly, the test of significance for the null hypothesis $H(\beta_1 - \beta_2 = \beta_0)$ is obtained by using the fact that if the null hypothesis is true, then the ratio

$$F = (N - 4)S_2/M$$

has the F distribution with $1, N - 4$ d.f.

Confidence intervals for $\alpha_1 - \alpha_2$ and $\beta_1 - \beta_2$

It is clear that since each of the above two F ratios has $1, N - 4$ d.f., the respective null hypotheses can also be tested by using tests based on the equivalent "Student's" distribution. In fact, to test the null hypothesis $H(\alpha_1 - \alpha_2 = \alpha_0)$, we have

$$t = \sqrt{F} = \frac{\alpha_1^* - \alpha_2^* - \alpha_0}{s\left[\frac{1}{n_1} + \frac{1}{n_2}\right]^{\frac{1}{2}}},$$

where t has "Student's" distribution with $N - 4$ d.f. Similarly, to test the null hypothesis $H(\beta_1 - \beta_2 = \beta_0)$, we have

$$t = \sqrt{F} = \frac{\beta_1^* - \beta_2^* - \beta_0}{s\left[\frac{1}{X_1} + \frac{1}{X_2}\right]^{\frac{1}{2}}},$$

where this t also has "Student's" distribution with $N - 4$ d.f.

Furthermore, we can use "Student's" distribution for obtaining confidence intervals for the parametric differences $\alpha_1 - \alpha_2 = \alpha \neq 0$ and $\beta_1 - \beta_2 = \beta \neq 0$. Thus, if $t(\eta; N - 4)$ is the 100η per cent point of "Student's" distribution with $N - 4$ d.f., then the $100(1 - \eta)$ per cent confidence intervals for α and β respectively are

$$\alpha_1^* - \alpha_2^* - s\left[\frac{1}{n_1} + \frac{1}{n_2}\right]^{\frac{1}{2}} t(\eta; N-4) \leqslant \alpha$$

$$\leqslant \alpha_1^* - \alpha_2^* + s\left[\frac{1}{n_1} + \frac{1}{n_2}\right]^{\frac{1}{2}} t(\eta; N-4);$$

and
$$\beta_1^* - \beta_2^* - s\left[\frac{1}{X_1} + \frac{1}{X_2}\right]^{\frac{1}{2}} t(\eta; N-4) \leqslant \beta$$

$$\leqslant \beta_1^* - \beta_2^* + s\left[\frac{1}{X_1} + \frac{1}{X_2}\right]^{\frac{1}{2}} t(\eta; N-4).$$

Tests of significance for the null hypotheses $H(\alpha_1 - \alpha_2 = 0)$ *and*
  $H(\beta_1 - \beta_2 = 0)$

The null hypotheses $H(\alpha_1 - \alpha_2 = 0)$ and $H(\beta_1 - \beta_2 = 0)$ are of special interest in statistical analysis; and taken jointly, these hypotheses imply that the two regression lines have the same parameters. The sums of squares for testing these null hypotheses are

$$S_1 = n_1 n_2 (\alpha_1^* - \alpha_2^*)^2 / N \quad \text{and} \quad S_2 = X_1 X_2 (\beta_1^* - \beta_2^*)^2 / X \quad \text{respectively.}$$

The residual sum of squares is, as before,

$$M = Y - \sum_{i=1}^{2} Z_i^2 / X_i = Y - \sum_{i=1}^{2} X_i \beta_i^{*2} \quad \text{with } N-4 \text{ d.f.}$$

Hence the total sum of squares in the analysis of variance is

$$T = S_1 + S_2 + M = \frac{n_1 n_2}{N}(\alpha_1^* - \alpha_2^*)^2 + \frac{X_1 X_2}{X}(\beta_1^* - \beta_2^*)^2 + Y - \sum_{i=1}^{2} X_i \beta_i^{*2}$$

$$= \left[\sum_{i=1}^{2} \sum_{\nu=1}^{n_i} (y_{i\nu} - \bar{y}_i)^2 + \frac{n_1 n_2}{N}(\bar{y}_1 - \bar{y}_2)^2 \right] +$$

$$+ \left[\frac{X_1 X_2}{X}(\beta_1^* - \beta_2^*)^2 - (X_1 \beta_1^{*2} + X_2 \beta_2^{*2}) \right]$$

$$= \sum_{i=1}^{2} \sum_{\nu=1}^{n_i} (y_{i\nu} - \bar{y})^2 - (X_1 \beta_1^* + X_2 \beta_2^*)^2 / X$$

$$= \sum_{i=1}^{2} \sum_{\nu=1}^{n_i} (y_{i\nu} - \bar{y})^2 - X\bar{\beta}^{*2} \quad \text{with } N-2 \text{ d.f.}$$

This now leads to the analysis of variance given in Table 15.10. We observe that two degrees of freedom are unaccounted for since we are only testing the equality of the parameters α_1, α_2, and β_1, β_2.

Table 15.10 Showing the analysis of variance for the comparison of two linear regressions

Source of Variation	d.f.	S.S.	M.S.	F.R.
Hypothesis: $\alpha_1 - \alpha_2 = 0$	1	$S_1 = n_1 n_2 (\alpha_1^* - \alpha_2^*)^2 / N$	S_1	$(N-4)S_1/M$
Hypothesis: $\beta_1 - \beta_2 = 0$	1	$S_2 = X_1 X_2 (\beta_1^* - \beta_2^*)^2 / X$	S_2	$(N-4)S_2/M$
Residual	$N-4$	$M = Y - \sum_{i=1}^{2} X_i \beta_i^{*2}$	$M/(N-4)$	
Total	$N-2$	$T = \sum_{i=1}^{2} \sum_{\nu=1}^{n_i} (y_{i\nu} - \bar{y})^2 - X\bar{\beta}^{*2}$		

Null hypotheses tenable

Suppose the null hypothesis $H(\alpha_1 - \alpha_2 = 0)$ is true. Then the least-squares estimate of $\bar{a}$, the common value of a_1 and a_2, is

$$\bar{a}^* \;=\; \bar{y} \;=\; (n_1 a_1^* + n_2 a_2^*)/N.$$

This estimate is, in fact, the same as a_2^{**} obtained above with $a_0 = 0$. Clearly, $\mathrm{var}(\bar{a}^*) = \sigma^2/N$, and since $\bar{a}^*$ is a linear function of the random observations, it follows that

$$t \;=\; \frac{\bar{a}^* - \bar{a}}{s/\sqrt{N}}$$

has "Student's" distribution with $N - 4$ d.f. Hence the $100(1 - \eta)$ per cent confidence interval for $\bar{a}$ is

$$\bar{a}^* - \frac{s}{\sqrt{N}}\, t(\eta;\, N - 4) \;\leqslant\; \bar{a} \;\leqslant\; \bar{a}^* + \frac{s}{\sqrt{N}}\, t(\eta;\, N - 4).$$

Similarly, suppose the null hypothesis $H(\beta_1 - \beta_2 = 0)$ is true. Then the least-squares estimate $\bar{\beta}$, the common value of β_1 and β_2, is

$$\bar{\beta}^* \;=\; (X_1 \beta_1^* + X_2 \beta_2^*)/X \;=\; (Z_1 + Z_2)/X \;=\; Z/X.$$

This estimate is the same as β_2^{**} obtained above with $\beta_0 = 0$, and it is known as the *pooled estimate of the common regression coefficient* $\bar{\beta}$.

Also,

$$\mathrm{var}(\bar{\beta}^*) \;=\; \mathrm{var}\!\left[\frac{1}{X}(X_1\beta_1^* + X_2\beta_2^*)\right]$$

$$=\; \frac{1}{X^2}\left[X_1^2\, \mathrm{var}(\beta_1^*) + X_2^2\, \mathrm{var}(\beta_2^*)\right] \;=\; \frac{\sigma^2}{X};$$

and as $\bar{\beta}^*$ is a linear function of the random observations, we conclude that

$$t \;=\; \frac{\bar{\beta}^* - \bar{\beta}}{s/\sqrt{X}}$$

has "Student's" distribution with $N - 4$ d.f. Hence the $100(1 - \eta)$ per cent confidence interval for $\bar{\beta}$ is

$$\bar{\beta}^* - \frac{s}{\sqrt{X}}\, t(\eta;\, N - 4) \;\leqslant\; \bar{\beta} \;\leqslant\; \bar{\beta}^* + \frac{s}{\sqrt{X}}\, t(\eta;\, N - 4).$$

Comparison of variances

Throughout the above analysis it was assumed that the $n_1 + n_2$ observations $y_{i\nu}$ ($i = 1, 2;\; \nu = 1, 2, ..., n_i$) were obtained from different normal populations having a common variance σ^2, and the tests of significance considered were for various null hypotheses about the parameters occurring in the expectations of the $y_{i\nu}$. This is the usual interest in regression analysis, but we now need to look more closely at the assumption of the equality of the variances. From the experimental point of view, the $y_{i\nu}$ constitute two groups of n_1 and n_2 independent observations respectively, arising from two separate experiments. For instance – to continue with the farmer's problem of choosing between fertilizers A and B – the first set of n_1 observations $(y_{1\nu})$ are obtained by varying the quantity per plot of fertilizer A, whilst the remaining n_2 observations arise similarly from the use of fertilizer B. Now it is conceivable that the two fertilizers may give, on the average, not only different linear rates of increase of yield, but the yields may also be of different intrinsic variability. A situation of this kind could, for example, occur when one fertilizer gives, on the average, a low but consistent yield, whereas the second fertilizer tends to give better returns but only under favourable conditions such as a certain amount of rainfall or sunshine. We may thus postulate that

$$E(y_{1\nu}) = a_1 + \beta_1(x_{1\nu} - \bar{x}_1) \quad \text{and} \quad \text{var}(y_{1\nu}) = \sigma_1^2, \quad \text{for } \nu = 1, 2, \ldots, n_1;$$
$$E(y_{2\nu}) = a_2 + \beta_2(x_{2\nu} - \bar{x}_2) \quad \text{and} \quad \text{var}(y_{2\nu}) = \sigma_2^2, \quad \text{for } \nu = 1, 2, \ldots, n_2.$$

With such a model $(\sigma_1 \neq \sigma_2)$, we cannot test exactly any linear null hypotheses about the a's and β's. However,

$$s_1^2 = M_1/(n_1 - 2) = (Y_1 - Z_1^2/X_1)/(n_1 - 2),$$
and
$$s_2^2 = M_2/(n_2 - 2) = (Y_2 - Z_2^2/X_2)/(n_2 - 2)$$

are still independent least-squares estimates of σ_1^2 and σ_2^2 respectively. Hence we can use the ratio of s_1^2 and s_2^2 to test the null hypothesis $H(\sigma_1^2 = \sigma_2^2)$. Thus if the null hypothesis is true then the observed $F = s_1^2/s_2^2$ has the F distribution with $n_1 - 2$, $n_2 - 2$ d.f.

This test of significance is not always carried out in regression analysis, but it has a dual purpose. Firstly, a non-significant value of F would give assurance that the data are consistent with a basic assumption of regression analysis. Secondly, as we have indicated, the intrinsic variability of the observations in the two experiments may also have practical relevance.

Example 15.7: Absorption of salts by plant cells

We conclude this section by showing how the preceding theory can be used to answer several questions based on the data of Steward and Harrison, already quoted in Table 5.3 (page 67). Here $n_1 = n_2 = 5$. We also note that the time of immersion of the potato slices (x) is the explanatory variable, and its values are the same for the two sets, each of five observations, pertaining to the uptake of Rb and Br ions. The corresponding y values are $y_{1\nu}$ and $y_{2\nu}$, for $\nu = 1, 2, \ldots, 5$. Using the notation introduced in the theoretical analysis, we have the following computations:

$$\sum_{\nu=1}^{5} x_\nu = 320 \cdot 4 \quad \text{and} \quad \bar{x}_1 = \bar{x}_2 = 64 \cdot 08.$$

$$\sum_{\nu=1}^{5} x_\nu^2 = 24332 \cdot 18 \quad \text{and} \quad \text{C.F.}(x) = 20531 \cdot 232.$$

Therefore
$$X_1 = X_2 = 3800 \cdot 948.$$

$$\sum_{\nu=1}^{5} y_{1\nu} = 71 \cdot 9 \quad \text{and} \quad \bar{y}_1 = a_1^* = 14 \cdot 38;$$

$$\sum_{\nu=1}^{5} y_{1\nu}^2 = 1148 \cdot 25 \quad \text{, and} \quad \text{C.F.}(y_1) = 1033 \cdot 922.$$

Therefore
$$Y_1 = 114 \cdot 328.$$

$$\sum_{\nu=1}^{5} x_\nu y_{1\nu} = 5264 \cdot 86 \quad \text{and} \quad \text{C.F.}(x, y_1) = 4607 \cdot 352.$$

Therefore
$$Z_1 = 657 \cdot 508 \quad \text{and} \quad \beta_1^* = 0 \cdot 172985,$$
and
$$M_1 = 114 \cdot 328 - 113 \cdot 7390 = 0 \cdot 5890 \quad \text{with 3 d.f.}$$

$$\sum_{\nu=1}^{5} y_{2\nu} = 45 \cdot 6 \quad \text{and} \quad \bar{y}_2 = a_2^* = 9 \cdot 12;$$

$$\sum_{\nu=1}^{5} y_{2\nu}^2 = 552 \cdot 94 \quad \text{and} \quad \text{C.F.}(y_2) = 415 \cdot 872.$$

Therefore $$Y_2 = 137 \cdot 068.$$

$$\sum_{\nu=1}^{5} x_{\nu} y_{2\nu} = 3636 \cdot 35 \quad \text{and} \quad \text{C.F.}(x, y_2) = 2922 \cdot 048.$$

Therefore $\quad Z_2 = 714 \cdot 302 \quad$ and $\quad \beta_2^* = 0 \cdot 187927,$

and $\quad\quad M_2 = 137 \cdot 068 - 134 \cdot 2366 = 2 \cdot 8314 \quad$ with 3 d.f.

We can use these calculations to make the following inferences:

(i) To test the null hypothesis $H(\sigma_1^2 = \sigma_2^2)$, we have

$$F = \frac{2 \cdot 8314}{0 \cdot 5890} = 4 \cdot 81 \quad \text{with 3, 3 d.f.}$$

The five per cent table value of F with 3, 3 d.f. is $9 \cdot 28$, so that the null hypothesis under test is acceptable. We are thus justified in carrying out the regression analysis under the assumption that $y_{1\nu}$ and $y_{2\nu}$ have the same variance.

(ii) The pooled estimate of this common population variance σ^2 is

$$s^2 = \frac{3 \cdot 4204}{6} = 0 \cdot 570067 \quad \text{with 6 d.f.}$$

(iii) To test the null hypothesis $H(\beta_1 = \beta_2)$, we have

$$\text{s.e.}(\beta_1^* - \beta_2^*) = \left[\frac{1 \cdot 140134}{3800 \cdot 948} \right]^{\frac{1}{2}} = [10^{-4} \times 2 \cdot 999604]^{\frac{1}{2}} = 10^{-2} \times 1 \cdot 731936,$$

whence $\quad\quad t = \dfrac{0 \cdot 014942}{0 \cdot 017319} = 0 \cdot 863 \quad$ with 6 d.f.

The five per cent value of "Student's" distribution with 6 d.f. is $2 \cdot 447$, so that the observed value of t is not significant. Thus the hypothesis of the equality of β_1 and β_2 is acceptable. In other words, over the time-interval in which the observations were made, there is an equal rate of uptake of Rb and Br ions by the potato slices.

(iv) Since $X_1 = X_2$, the pooled estimate of $\bar{\beta}$, the common value of β_1 and β_2 is

$$\bar{\beta}^* = (\beta_1^* + \beta_2^*)/2 = 0 \cdot 180456,$$

and
$$\text{s.e.}(\bar{\beta}^*) = \left[\frac{0 \cdot 570067}{7601 \cdot 896} \right]^{\frac{1}{2}} = [10^{-4} \times 0 \cdot 749901]^{\frac{1}{2}} = 10^{-2} \times 0 \cdot 865968.$$

Therefore the 95 per cent confidence interval for $\bar{\beta}$ is

$$0 \cdot 180456 - 10^{-2} \times 0 \cdot 865968 \times 2 \cdot 447 \leqslant \bar{\beta}$$
$$\leqslant 0 \cdot 180456 + 10^{-2} \times 0 \cdot 865968 \times 2 \cdot 447,$$

or $$0 \cdot 159266 \leqslant \bar{\beta} \leqslant 0 \cdot 201646.$$

(v) To test the null hypothesis $H(a_1 = a_2)$, we have

$$\text{s.e.}(a_1^* - a_2^*) = \left[\frac{1 \cdot 140134}{5} \right]^{\frac{1}{2}} = (0 \cdot 228027)^{\frac{1}{2}} = 0 \cdot 477522,$$

whence $\qquad t = \dfrac{5 \cdot 26}{0 \cdot 477522} = 11 \cdot 015 \quad \text{with } 6 \, \text{d.f.}$

This is a highly significant result, and so we can confidently reject the null hypothesis $H(a_1 = a_2)$. Furthermore, if $a_1 - a_2 = a \neq 0$, then the 95 per cent confidence interval for a is

$$5 \cdot 26 - 0 \cdot 477522 \times 2 \cdot 447 \leqslant a \leqslant 5 \cdot 26 + 0 \cdot 477522 \times 2 \cdot 447,$$

or $\qquad\qquad\qquad 4 \cdot 092 \leqslant a \leqslant 6 \cdot 428.$

(vi) The estimated regression line for the uptake of Rb ions is

$$\begin{aligned} y_{1e}^* &= 14 \cdot 38 + 0 \cdot 172985(x - 64 \cdot 08) \\ &= 3 \cdot 30 + 0 \cdot 172985\, x. \end{aligned}$$

Similarly, the estimated regression line for the uptake of Br ions is

$$\begin{aligned} y_{2e}^* &= 9 \cdot 12 + 0 \cdot 187927(x - 64 \cdot 08) \\ &= -2 \cdot 92 + 0 \cdot 187927\, x. \end{aligned}$$

Fig. 15.2, overleaf, shows the observations made and the two estimated regression lines.

Now, since the null hypothesis $H(\beta_1 = \beta_2)$ is accepted and the null hypothesis $H(a_1 = a_2)$ is rejected, we conclude that the two regression lines are parallel but non-coincident. This implies that, on the experimental evidence obtained, the uptake rates are the same for Rb and Br ions, but the average amounts are different. We therefore infer that the uptake rates must have been different prior to the period of observation. This is an important conclusion of our analysis. To investigate the differences in the uptake rates of the two kinds of ions, another experiment is required in which the uptake measurements would be made at frequent intervals in the immersion period up to 20 hours. This positive indication of how future experimentation might be carried out is one of the most important functions of statistical analysis in scientific research.

15.15 Comparison of k $(\geqslant 2)$ linear regressions

It is relatively a simple matter to extend the analysis for two linear regressions to the general case of comparing k linear regressions. Indeed, the preceding analysis for $k = 2$ is a particular case of the general analysis, and we have considered it separately partly because of simplicity and partly because the two-sample situation is of special practical use. The general analysis would be appropriate if, for example, the farmer of our previous illustration were interested in making a choice from k fertilizers $A_1, A_2, \ldots, A_k$ to improve his yield of potatoes. In such a situation, we assume that the farmer has $N = \sum\limits_{i=1}^{k} n_i$ similar experimental plots, and that he uses fertilizer A_i in quantities $x_{i1}, x_{i2}, \ldots, x_{in_i}$ on n_i plots to obtain the corresponding yields $y_{i1}, y_{i2}, \ldots, y_{in_i}$, for $i = 1, 2, \ldots, k$. There are thus k independent sets of paired observations which may be compactly denoted as

$$(x_{i\nu}, y_{i\nu}), \quad \text{where} \quad i = 1, 2, \ldots, k; \quad \nu = 1, 2, \ldots, n_i.$$

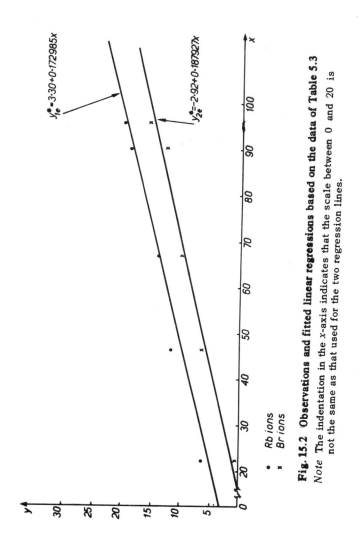

$y^*_{1e} = 3.30 + 0.172985x$

$y^*_{2e} = -2.92 + 0.187927x$

• Rb ions
× Br ions

Fig. 15.2 Observations and fitted linear regressions based on the data of Table 5.3

Note The indentation in the x-axis indicates that the scale between 0 and 20 is not the same as that used for the two regression lines.

We now postulate that the $y_{i\nu}$ are normally distributed random variables such that

$$E(y_{i\nu}) = \alpha_i + \beta_i(x_{i\nu} - \bar{x}_i); \qquad \text{var}(y_{i\nu}) = \sigma^2,$$

where α_i, β_i, σ^2 are unknown but independent parameters and $\bar{x}_i$ is the average of the $x_{i\nu}$, for $\nu = 1, 2, \ldots, n_i$. Furthermore, in consonance with our earlier explanation, the $y_{i\nu}$ are independent observations from N normal populations. The means of n_i of these populations corresponding to the quantities used of fertilizer A_i are assumed to lie on one straight line, which is the regression of yield (y_i) on the non-random variable x_i denoting the quantity used of fertilizer A_i. We are interested in comparing these k regression lines obtained for $i = 1, 2, \ldots, k$. In particular, we first derive tests of significance for the two null hypotheses $H(\alpha_1 = \alpha_2 = \ldots = \alpha_k)$ and $H(\beta_1 = \beta_2 = \ldots = \beta_k)$.

Estimation of the parameters

As usual, the first stage in this regression analysis is the derivation of the estimates of the parameters by the minimisation of

$$\Omega = \sum_{i=1}^{k} \sum_{\nu=1}^{n_i} [y_{i\nu} - a_i - \beta_i (x_{i\nu} - \bar{x}_i)]^2$$

with respect to variations in a_i and β_i. Differentiation with respect to a_i and β_i gives

$$\frac{\partial \Omega}{\partial a_i} = -2 \sum_{\nu=1}^{n_i} [y_{i\nu} - a_i - \beta_i (x_{i\nu} - \bar{x}_i)],$$

and

$$\frac{\partial \Omega}{\partial \beta_i} = -2 \sum_{\nu=1}^{n_i} (x_{i\nu} - \bar{x}_i)[y_{i\nu} - a_i - \beta_i (x_{i\nu} - \bar{x}_i)],$$

for $i = 1, 2, \ldots, k$.

We again use the notation

$$X_i = \sum_{\nu=1}^{n_i} (x_{i\nu} - \bar{x}_i)^2, \qquad Y_i = \sum_{\nu=1}^{n_i} (y_{i\nu} - \bar{y}_i)^2,$$

$$Z_i = \sum_{\nu=1}^{n_i} (x_{i\nu} - \bar{x}_i)(y_{i\nu} - \bar{y}_i),$$

where $\bar{y}_i$ is the average of the $y_{i\nu}$, and

$$X = \sum_{i=1}^{k} X_i, \qquad Y = \sum_{i=1}^{k} Y_i, \qquad Z = \sum_{i=1}^{k} Z_i.$$

The least-squares estimates of a_i and β_i are obtained from the equations

$$\frac{\partial \Omega}{\partial a_i} = \frac{\partial \Omega}{\partial \beta_i} = 0, \quad \text{for } i = 1, 2, \ldots, k.$$

Solution of these equations immediately gives the estimates as

$$a_i^* = \bar{y}_i \quad \text{and} \quad \beta_i^* = Z_i / X_i.$$

Therefore the absolute minimum of Ω is

$$M = \sum_{i=1}^{k} \sum_{\nu=1}^{n_i} [y_{i\nu} - \bar{y}_i - \beta_i^*(x_{i\nu} - \bar{x}_i)]^2$$

$$= \sum_{i=1}^{k} [Y_i - Z_i^2/X_i] = Y - \sum_{i=1}^{k} X_i \beta_i^{*2}.$$

Also, since the k samples are independent, we have

$$E(M) = \sum_{i=1}^{k} E[Y_i - Z_i^2/X_i] = \sum_{i=1}^{k} (n_i - 2)\sigma^2 = (N - 2k)\sigma^2.$$

Hence $s^2 = M/(N - 2k)$ is the pooled least-squares estimate of σ^2, and M/σ^2 has the χ^2 distribution with $N - 2k$ d.f.

As previously, we set $M_i = Y_i - Z_i^2/X_i$, the residual sum of squares from the ith sample. Then

$$\Omega = \sum_{i=1}^{k} [M_i + n_i(a_i^* - a_i)^2 + X_i(\beta_i^* - \beta_i)^2]$$

$$= M + \sum_{i=1}^{k} [n_i(a_i^* - a_i)^2 + X_i(\beta_i^* - \beta_i)^2].$$

This identity is obtained simply by the method used in the case of a single sample, and we use this second expression for Ω to derive the sums of squares for testing the null hypotheses $H(\alpha_1 = \alpha_2 = \ldots = \alpha_k)$ and $H(\beta_1 = \beta_2 = \ldots = \beta_k)$.

Tests of significance for the null hypotheses $H(\alpha_1 = \alpha_2 = \ldots = \alpha_k)$ *and*
$$H(\beta_1 = \beta_2 = \ldots = \beta_k)$$

If the null hypothesis $H(\alpha_1 = \alpha_2 = \ldots = \alpha_k)$ is true, then we assume that the common but unknown value of the α_i is $\bar{\alpha}$. Hence the relative minimum of Ω is the absolute minimum of

$$\Omega_1 = M + \sum_{i=1}^{k} [n_i(\alpha_i^* - \bar{\alpha})^2 + X_i(\beta_i^* - \beta_i)^2]$$

for variations in $\bar{\alpha}$ and $\beta_1, \beta_2, \ldots, \beta_k$. For this minimisation, we have

$$\frac{\partial \Omega_1}{\partial \bar{\alpha}} = -2 \sum_{i=1}^{k} n_i(\alpha_i^* - \bar{\alpha}),$$

and

$$\frac{\partial \Omega_1}{\partial \beta_i} = -2X_i(\beta_i^* - \beta_i), \quad \text{for } i = 1, 2, \ldots, k.$$

Therefore the absolute minimum of Ω_1 is obtained for $\beta_i = \beta_i^*$ and

$$\bar{\alpha} = \bar{\alpha}^* = \sum_{i=1}^{k} n_i \alpha_i^* / N = \bar{y}, \quad \text{the grand mean of the } y_{i\nu}.$$

Thus the relative minimum of Ω is

$$M + \sum_{i=1}^{k} n_i(\alpha_i^* - \bar{\alpha}^*)^2,$$

whence the required sum of squares for testing the null hypothesis $H(\alpha_1 = \alpha_2 = \ldots = \alpha_k)$ is

$$S_1 = \sum_{i=1}^{k} n_i(\alpha_i^* - \bar{\alpha}^*)^2.$$

Furthermore, if the null hypothesis under test is true, then $E(\alpha_i^*) = E(\bar{\alpha}^*) = \bar{\alpha}$. Hence, using this expectation, we have

$$\begin{aligned}
E(S_1) &= E\left[\sum_{i=1}^{k} n_i(\alpha_i^* - \bar{\alpha}^*)^2 \right] \\
&= E\left[\sum_{i=1}^{k} n_i \{(\alpha_i^* - \bar{\alpha}) - (\bar{\alpha}^* - \bar{\alpha})\}^2 \right] \\
&= E\left[\sum_{i=1}^{k} n_i(\alpha_i^* - \bar{\alpha})^2 - N(\bar{\alpha}^* - \bar{\alpha})^2 \right], \quad \text{since } \sum_{i=1}^{k} n_i(\alpha_i^* - \bar{\alpha}) = N(\bar{\alpha}^* - \bar{\alpha}), \\
&= \sum_{i=1}^{k} n_i \operatorname{var}(\alpha_i^*) - N \operatorname{var}(\bar{\alpha}^*) = (k-1)\sigma^2.
\end{aligned}$$

Hence, if the null hypothesis under test is true, then $S_1/(k-1)$ is an unbiased estimate of σ^2, and S_1/σ^2 is distributed as χ^2 with $k-1$ d.f.

Similarly, if the null hypothesis $H(\beta_1 = \beta_2 = \ldots = \beta_k)$ is true, then we assume that the common but unknown value of the β_i is $\bar{\beta}$. Hence, to test this null hypothesis, the relative minimum of Ω is the absolute minimum of

$$\Omega_2 = M + \sum_{i=1}^{k} [n_i(\alpha_i^* - \alpha_i)^2 + X_i(\beta_i^* - \bar{\beta})^2]$$

for variations in $\alpha_1, \alpha_2, \ldots, \alpha_k$ and $\bar{\beta}$. For this minimisation, we have

$$\frac{\partial \Omega_2}{\partial \alpha_i} = -2n_i(\alpha_i^* - \alpha_i), \quad \text{for } i = 1, 2, \ldots, k,$$

and

$$\frac{\partial \Omega_2}{\partial \bar{\beta}} = -2 \sum_{i=1}^{k} X_i(\beta_i^* - \bar{\beta}).$$

Hence the absolute minimum of Ω_2 is obtained for $\alpha_i = \alpha_i^*$ and

$$\bar{\beta} = \bar{\beta}^* = \sum_{i=1}^{k} X_i \beta_i^* / X.$$

Therefore the relative minimum of Ω is

$$M + \sum_{i=1}^{k} X_i(\beta_i^* - \bar{\beta}^*)^2,$$

whence the required sum of squares for testing the null hypothesis $H(\beta_1 = \beta_2 = \ldots = \beta_k)$ is

$$S_2 = \sum_{i=1}^{k} X_i(\beta_i^* - \bar{\beta}^*)^2.$$

To obtain the expectation of S_2, we observe that if the null hypothesis under test is true, then $E(\beta_i^*) = E(\bar{\beta}^*) = \bar{\beta}$. Hence

$$
\begin{aligned}
E(S_2) &= E\left[\sum_{i=1}^{k} X_i(\beta_i^* - \bar{\beta}^*)^2 \right] \\
&= E\left[\sum_{i=1}^{k} X_i \{(\beta_i^* - \bar{\beta}) - (\bar{\beta}^* - \bar{\beta})\}^2 \right] \\
&= E\left[\sum_{i=1}^{k} X_i(\beta_i^* - \bar{\beta})^2 - X(\bar{\beta}^* - \bar{\beta})^2 \right], \quad \text{since } \sum_{i=1}^{k} X_i(\beta_i^* - \bar{\beta}) = X(\bar{\beta}^* - \bar{\beta}), \\
&= \sum_{i=1}^{k} X_i \text{ var}(\beta_i^*) - X \text{ var}(\bar{\beta}^*) = (k-1)\sigma^2.
\end{aligned}
$$

Hence, if the null hypothesis under test is true, then $S_2/(k-1)$ is an unbiased estimate of σ^2, and S_2/σ^2 is distributed as χ^2 with $k-1$ d.f.

We have thus deduced three main results:

(i) If the null hypothesis $H(\alpha_1 = \alpha_2 = \ldots = \alpha_k)$ is true, then S_1/σ^2 is distributed as χ^2 with $k-1$ d.f.

(ii) If the null hypothesis $H(\beta_1 = \beta_2 = \ldots = \beta_k)$ is true, then S_2/σ^2 is distributed as χ^2 with $k-1$ d.f.

(iii) M/σ^2 is distributed as χ^2 with $N - 2k$ d.f. irrespective of any hypotheses about α_i and β_i.

We also know from general theory that these three χ^2's are independent.

Hence, to test the null hypothesis $H(\alpha_1 = \alpha_2 = \ldots = \alpha_k)$, we use the fact that if this hypothesis is true, then the ratio

$$F = (N - 2k) S_1 / (k - 1)M$$

has the F distribution with $k - 1, N - 2k$ d.f.

Similarly, to test the null hypothesis $H(\beta_1 = \beta_2 = \ldots = \beta_k)$, we use the fact that if this hypothesis is true, then the ratio

$$F = (N - 2k) S_2 / (k - 1)M$$

has the F distribution with $k - 1, N - 2k$ d.f.

Finally, to complete the analysis of variance, we note that the total sum of squares is

$$
\begin{aligned}
T &= S_1 + S_2 + M \\
&= \sum_{i=1}^{k} n_i(\alpha_i^* - \bar{\alpha}^*)^2 + \sum_{i=1}^{k} X_i(\beta_i^* - \bar{\beta}^*)^2 + Y - \sum_{i=1}^{k} X_i\beta_i^{*2} \\
&= \left[\sum_{i=1}^{k} \sum_{\nu=1}^{n_i} (y_{i\nu} - \bar{y}_i)^2 + \sum_{i=1}^{k} n_i(\bar{y}_i - \bar{y})^2 \right] + \sum_{i=1}^{k} X_i(\beta_i^* - \bar{\beta}^*)^2 - \sum_{i=1}^{k} X_i\beta_i^{*2} \\
&= \sum_{i=1}^{k} \sum_{\nu=1}^{n_i} (y_{i\nu} - \bar{y})^2 - X\bar{\beta}^{*2}.
\end{aligned}
$$

This sum of squares has $N - 2$ d.f., and we have not accounted for two degrees of freedom since we are testing only the equality of the α_i and the β_i respectively without specifying their common values $\bar{\alpha}$ and $\bar{\beta}$. Hence the following analysis of variance table.

Table 15.11   Showing the analysis of variance for the comparison of $k \ (\geqslant 2)$ linear regressions

Source of Variation	d.f.	S. S.	M. S.	F. R.
Hypothesis: $a_1 = a_2 = \ldots = a_k$	$k-1$	$S_1 = \sum_{i=1}^{k} n_i(\alpha_i^* - \bar{\alpha}^*)^2$	$S_1/(k-1)$	$(N-2k)S_1/M(k-1)$
Hypothesis: $\beta_1 = \beta_2 = \ldots = \beta_k$	$k-1$	$S_2 = \sum_{i=1}^{k} X_i(\beta_i^* - \bar{\beta}^*)^2$	$S_2/(k-1)$	$(N-2k)S_2/M(k-1)$
Residual	$N-2k$	$M = Y - \sum_{i=1}^{k} X_i\beta_i^{*2}$	$M/(n-2k)$	
Total	$N-2$	$T = \sum_{i=1}^{k} \sum_{\nu=1}^{n_i} (y_{i\nu} - \bar{y})^2 - X\bar{\beta}^{*2}$		

Computational notes

We have

$$
\begin{aligned}
S_1 &= \sum_{i=1}^{k} n_i(\alpha_i^* - \bar{\alpha}^*)^2 \\
&= \sum_{i=1}^{k} n_i(\bar{y}_i - \bar{y})^2 = \sum_{i=1}^{k} n_i\bar{y}_i^2 - N\bar{y}^2 = \sum_{i=1}^{k} T_i^2/n_i - G^2/N,
\end{aligned}
$$

where $T_i = \sum_{i=1}^{n_i} y_{i\nu}$ and $G = \sum_{i=1}^{k} T_i$, the grand total of the $y_{i\nu}$. The last expression for S_1 is a good computational form.

Again,

$$
\begin{aligned}
S_2 &= \sum_{i=1}^{k} X_i(\beta_i^* - \bar{\beta}^*)^2 \\
&= \sum_{i=1}^{k} X_i\beta_i^{*2} - X\bar{\beta}^{*2} = \sum_{i=1}^{k} Z_i^2/X_i - Z^2/X.
\end{aligned}
$$

The last expression is the best computational form for S_2.

Finally,

$$
T = \sum_{i=1}^{k} \sum_{\nu=1}^{n_i} (y_{i\nu} - \bar{y})^2 - X\bar{\beta}^{*2} = \sum_{i=1}^{k} \sum_{\nu=1}^{n_i} y_{i\nu}^2 - G^2/N - Z^2/X.
$$

Thus a convenient computational procedure is to evaluate T, S_1, S_2, and then to obtain M as the difference: $M = T - S_1 - S_2$.

Alternatively, M may be directly computed from the formula

$$M = Y - \sum_{i=1}^{k} X_i \beta_i^{*2} = \sum_{i=1}^{k} \left[Y_i - Z_i^2 / X_i \right].$$

In this case, S_1 is obtained as the difference: $S_1 = T - M - S_2$.

Confidence intervals for α_i and β_i

If the null hypothesis $H(\alpha_1 = \alpha_2 = \ldots = \alpha_k)$ is rejected, then this means that, on the whole, the k α's cannot be regarded as equal. We may then obtain confidence intervals for any α_i by noting that

$$t = \frac{\alpha_i^* - \alpha_i}{s/\sqrt{n_i}}$$

has "Student's" distribution with $N - 2k$ d.f. Hence, if $t(\eta; N - 2k)$ is the 100η per cent point of "Student's" distribution with $N - 2k$ d.f., then the $100(1 - \eta)$ per cent confidence interval for α_i is

$$\alpha_i^* - \frac{s}{\sqrt{n_i}} t(\eta; N - 2k) \leqslant \alpha_i \leqslant \alpha_i^* + \frac{s}{\sqrt{n_i}} t(\eta; N - 2k).$$

Similarly, if the null hypothesis $H(\beta_1 = \beta_2 = \ldots = \beta_k)$ is rejected, then the k β's cannot be regarded as being equal on the whole. In this case, a confidence interval for any β_i is obtained by using the fact that

$$t = \frac{\beta_i^* - \beta_i}{s/\sqrt{X_i}}$$

has "Student's" distribution with $N - 2k$ d.f. Hence the $100(1 - \eta)$ per cent confidence interval for β_i is

$$\beta_i^* - \frac{s}{\sqrt{X_i}} t(\eta; N - 2k) \leqslant \beta_i \leqslant \beta_i^* + \frac{s}{\sqrt{X_i}} t(\eta; N - 2k).$$

Confidence intervals for $\bar{\alpha}$ and $\bar{\beta}$

On the other hand, if the null hypothesis $H(\alpha_1 = \alpha_2 = \ldots = \alpha_k)$ is accepted, then the least-squares estimate of $\bar{\alpha}$, the common value of α_i, is

$$\bar{\alpha}^* = \bar{y}, \quad \text{and} \quad \text{var}(\bar{\alpha}^*) = \sigma^2/N.$$

Since $\bar{\alpha}^*$ is a linear function of the observations, it also follows that

$$t = \frac{\bar{\alpha}^* - \bar{\alpha}}{s/\sqrt{N}}$$

has "Student's" distribution with $N - 2k$ d.f. Hence the $100(1 - \eta)$ per cent confidence interval for $\bar{\alpha}$ is

$$\bar{\alpha}^* - \frac{s}{\sqrt{N}} t(\eta; N - 2k) \leqslant \bar{\alpha} \leqslant \bar{\alpha}^* + \frac{s}{\sqrt{N}} t(\eta; N - 2k).$$

Similarly, if the null hypothesis $H(\beta_1 = \beta_2 = \ldots = \beta_k)$ is accepted, then the least-squares estimate of the $\bar{\beta}$; the common value of the β_i, is

$$\bar{\beta}^* = \sum_{i=1}^{k} X_i \beta_i^*/X, \quad \text{and} \quad \text{var}(\bar{\beta}^*) = \sigma^2/X.$$

Therefore it may be inferred that

$$t = \frac{\bar{\beta}^* - \bar{\beta}}{s/\sqrt{X}}$$

has "Student's" distribution with $N - 2k$ d.f. Hence the $100(1 - \eta)$ per cent confidence interval for $\bar{\beta}$ is

$$\bar{\beta}^* - \frac{s}{\sqrt{X}} t(\eta; N - 2k) \leqslant \bar{\beta} \leqslant \bar{\beta}^* + \frac{s}{\sqrt{X}} t(\eta; N - 2k).$$

Estimated linear regressions

The estimated linear regression for the ith group of populations is

$$y_{ie}^* = \bar{y}_i + \beta_i^*(x_i - \bar{x}_i), \quad \text{for } i = 1, 2, ..., k.$$

These are the k best-fitting lines obtained by a least-squares analysis of the sample observations.

Example 15.8: Inheritance of shell breadth in snails

We illustrate the use of the above theory by an analysis of the measurements of the breadth of shells of the snail *Arianta arbustorum* (L) taken from the Charles Oldham collection in the British Museum, London. The data pertain to three samples of wild-caught parental snails and their progeny from the United Kingdom, Haute-Savoie, and the rest of France respectively. These snails are cross-fertilizing hermaphrodites so that the "mother" and "father" cannot be specified unless the brood is seen being laid. In view of this difficulty, the explanatory variable (x) is the midparent breadth, that is, the average of the breadth of the shells of both parents. From the progeny of each parental pair, the average breadth of shells of five randomly selected offspring provides the y values. The data are presented in Table 15.12, the measurements being correct to the nearest tenth of a millimetre. The regression of average offspring breadth on midparent breadth gives a valid estimate of the heritability of the characteristic shell breadth. Accordingly, the purpose of statistical analysis is to compare the linear regressions in the three regions and, if no significant difference is found, to determine a pooled estimate of the common value of the three regression coefficients.

Here $\quad n_1 = 22, \quad n_2 = 6, \quad n_3 = 5, \quad$ and $\quad N = 33.$

$$\sum_{\nu=1}^{22} x_{1\nu} = 408 \cdot 9, \quad \sum_{\nu=1}^{22} y_{1\nu} = 423 \cdot 5;$$

$$\sum_{\nu=1}^{22} x_{1\nu}^2 = 7663 \cdot 13, \quad \sum_{\nu=1}^{22} y_{1\nu}^2 = 8200 \cdot 95, \quad \sum_{\nu=1}^{22} x_{1\nu} y_{1\nu} = 7911 \cdot 33;$$

$$\text{C.F.}(x_1) = \frac{167199 \cdot 21}{22} = 7599 \cdot 964, \quad \text{C.F.}(y_1) = \frac{179352 \cdot 25}{22} = 8152 \cdot 375,$$

$$\text{C.F.}(x_1, y_1) = \frac{173169 \cdot 15}{22} = 7871 \cdot 325.$$

Therefore

$$X_1 = 63 \cdot 166, \quad Y_1 = 48 \cdot 575, \quad Z_1 = 40 \cdot 005;$$
$$\beta_1^* = 0 \cdot 633331, \quad \bar{x}_1 = 18 \cdot 5864, \quad \bar{y}_1 = 19 \cdot 2500;$$

and

$$M_1 = 48 \cdot 575 - 25 \cdot 336 = 23 \cdot 239 \quad \text{with 20 d.f.}$$

Table 15.12 Showing the measurements of the breadths of shells of the snail *Arianta arbustorum* (L)

United Kingdom		Haute-Savoie		Rest of France	
Midparent breadth (x_1)	Average breadth of 5 offspring (y_1)	Midparent breadth (x_2)	Average breadth of 5 offspring (y_2)	Midparent breadth (x_3)	Average breadth of 5 offspring (y_3)
17·3	16·6	24·5	24·9	20·0	19·9
21·1	20·8	23·3	23·1	18·0	19·2
20·1	20·5	24·4	23·7	18·2	19·3
21·0	21·5	24·2	23·6	17·4	18·0
20·6	21·2	28·1	26·4	19·3	19·2
20·8	21·1	25·2	24·2		
16·5	19·6				
16·2	19·5				
18·6	20·5				
20·2	19·5				
17·0	16·6				
18·4	17·8				
18·4	19·5				
19·8	18·4				
19·1	19·3				
19·7	19·4				
20·3	21·4				
16·6	17·0				
16·8	17·9				
16·4	18·7				
17·1	18·6				
16·9	18·1				

Source: Private communication from Dr. L.M. Cook, University of Manchester.

$$\sum_{\nu=1}^{6} x_{2\nu} = 149\cdot7, \quad \sum_{\nu=1}^{6} y_{2\nu} = 145\cdot9;$$

$$\sum_{\nu=1}^{6} x_{2\nu}^2 = 3748\cdot79, \quad \sum_{\nu=1}^{6} y_{2\nu}^2 = 3554\cdot87, \quad \sum_{\nu=1}^{6} x_{2\nu}y_{2\nu} = 3649\cdot36;$$

$$\text{C.F.}(x_2) = \frac{22410\cdot09}{6} = 3735\cdot015, \quad \text{C.F.}(y_2) = \frac{21286\cdot81}{6} = 3547\cdot802,$$

$$\text{C.F.}(x_2, y_2) = \frac{21341\cdot23}{6} = 3640\cdot205.$$

Therefore
$$X_2 = 13\cdot775, \quad Y_2 = 7\cdot068, \quad Z_2 = 9\cdot155;$$
$$\beta_2^* = 0\cdot664610, \quad \bar{x}_2 = 24\cdot9500, \quad \bar{y}_2 = 24\cdot3167;$$

and
$$M_2 = 7\cdot068 - 6\cdot085 = 0\cdot983 \quad \text{with 4 d.f.}$$

$$\sum_{\nu=1}^{5} x_{3\nu} = 92\cdot9, \quad \sum_{\nu=1}^{5} y_{3\nu} = 95\cdot6;$$

$$\sum_{\nu=1}^{5} x_{3\nu}^2 = 1730\cdot49, \quad \sum_{\nu=1}^{5} y_{3\nu}^2 = 1829\cdot78, \quad \sum_{\nu=1}^{5} x_{3\nu}y_{3\nu} = 1778\cdot62;$$

$$\text{C.F.}(x_3) \;=\; \frac{8630 \cdot 41}{5} \;=\; 1726 \cdot 082, \quad \text{C.F.}(y_3) \;=\; \frac{9139 \cdot 36}{5} \;=\; 1827 \cdot 872,$$

$$\text{C.F.}(x_3,\, y_3) \;=\; \frac{8881 \cdot 24}{5} \;=\; 1776 \cdot 248.$$

Therefore
$$X_3 \;=\; 4 \cdot 408, \qquad Y_3 \;=\; 1 \cdot 908, \qquad Z_3 \;=\; 2 \cdot 372;$$
$$\beta_3^* \;=\; 0 \cdot 538113, \qquad \bar{x}_3 \;=\; 18 \cdot 5800, \qquad \bar{y}_3 \;=\; 19 \cdot 1200;$$

and
$$M_3 \;=\; 1 \cdot 908 \,-\, 1 \cdot 276 \;=\; 0 \cdot 632 \quad \text{with 3 d.f.}$$

Hence
$$X \;=\; 81 \cdot 349, \qquad Y \;=\; 57 \cdot 551, \qquad Z \;=\; 51 \cdot 532,$$

and the pooled residual sum of squares is $M = 24 \cdot 854$ with 27 d.f. Also,

$$\bar{\beta}^* \;=\; 0 \cdot 633468 \qquad \text{and} \qquad X\bar{\beta}^{*2} \;=\; 32 \cdot 644.$$

Therefore the sum of squares for testing the equality of the β_i is

$$S_2 \;=\; 32 \cdot 697 \,-\, 32 \cdot 644 \;=\; 0 \cdot 053 \quad \text{with 2 d.f.}$$

Finally,

$$\sum_{i=1}^{3} \sum_{\nu=1}^{n_i} y_{i\nu}^2 \;=\; 13585 \cdot 60 \quad \text{and} \quad G^2/N \;=\; \frac{442225 \cdot 00}{33} \;=\; 13400 \cdot 758.$$

Therefore $\quad T \;=\; 184 \cdot 842 \,-\, 32 \cdot 644 \;=\; 152 \cdot 198 \quad$ with 31 d.f.

Hence the sum of squares for testing the equality of the α_i is

$$S_1 \;=\; T \,-\, M \,-\, S_2 \;=\; 127 \cdot 291 \quad \text{with 2 d.f.}$$

The analysis of variance is as follows.

Table 15.13   Showing the analysis of variance of the data of Table 15.12

Source of Variation	d.f.	S.S.	M.S.	F.R.
Hypothesis: $\alpha_1 = \alpha_2 = \alpha_3$	2	127·291	63·646	69·14
Hypothesis: $\beta_1 = \beta_2 = \beta_3$	2	0·053	0·0265	< 1
Residual	27	24·854	0·9205	
Total	31	152·198		

There is no doubt that the α_i are significantly different, but the β_i can be taken to be equal.

The pooled estimate of the error variance is $s^2 = 0 \cdot 920519$. Therefore

$$\text{s.e.}(\bar{\beta}^*) \;=\; \frac{s}{\sqrt{X}} \;=\; \frac{0 \cdot 959437}{9 \cdot 019368} \;=\; 0 \cdot 106375.$$

The five per cent point of "Student's" distribution with 27 d.f. is $2 \cdot 052$. Hence the 95 per cent confidence interval for $\bar{\beta}$ is

$$0 \cdot 633468 \,-\, 0 \cdot 106375 \times 2 \cdot 052 \;\leqslant\; \bar{\beta} \;\leqslant\; 0 \cdot 633468 \,+\, 0 \cdot 106375 \times 2 \cdot 052,$$
or
$$0 \cdot 4152 \;\leqslant\; \bar{\beta} \;\leqslant\; 0 \cdot 8518.$$

The three separate estimated regression lines are:

$$y_{1e}^* = 19 \cdot 2500 + 0 \cdot 633331(x_1 - 18 \cdot 5684)$$
$$= 7 \cdot 4787 + 0 \cdot 633331 x_1,$$
$$y_{2e}^* = 24 \cdot 3167 + 0 \cdot 664610(x_2 - 24 \cdot 9500)$$
$$= 7 \cdot 7347 + 0 \cdot 664610 x_2,$$

and

$$y_{3e}^* = 19 \cdot 1200 + 0 \cdot 538113(x_3 - 18 \cdot 5800)$$
$$= 9 \cdot 1219 + 0 \cdot 538113 x_3.$$

15.16 The generalised t test

All the null hypotheses about means and regression coefficients that we have considered in this chapter are special cases of general linear hypotheses and, under the stated assumptions, these hypotheses can be tested by the analysis of variance based on the method of least squares. The associated tests of significance have been shown to be dependent upon the F distribution; but in the simple case of a single linear parametric hypothesis, the F test is equivalent to the one based on "Student's" distribution. As we have seen, the derivation of the F ratio for testing a null hypothesis requires two separate minimisations of Ω, the sum of squares of the deviations of the observations from their postulated expectations. The first minimisation gives the least-squares estimates of the parameters occurring in the model, whilst the second minimisation is relative to the null hypothesis under test. The difference between this relative minimum and the first (absolute) minimum of Ω gives the sum of squares due to the null hypothesis under test. However, when the null hypothesis refers to a single linear parametric function, then a simpler but mathematically equivalent approach leads to a test of significance based on "Student's" distribution. The simplicity is due to the fact that it is now not necessary to carry out the relative minimisation of Ω to arrive at the appropriate t test. This test is known as the *generalised t test* since, as the name implies, it is based on a generalisation of the model leading to the simple case of testing some null hypothesis about the mean of a normal population.

To explain the method, suppose $x_1, x_2, \ldots, x_n$ are n independent and normally distributed observations such that

$$E(x_i) = \sum_{j=1}^{k} c_{ij} \theta_j \quad \text{and} \quad \text{var}(x_i) = \sigma^2, \quad \text{for } i = 1, 2, \ldots, n,$$

where

    (i) c_{ij} are known constants;

    (ii) θ_j are $k < n$ independent unknown parameters; and

    (iii) σ^2 is another independent unknown parameter.

Given these observations, we wish to test the null hypothesis

$$H\left[\mu \equiv \sum_{j=1}^{k} a_j \theta_j = \mu_0\right],$$

where a_j are known constants such that they are not all simultaneously zero and μ_0 is a preassigned constant.

The first stage in the derivation of the test for the null hypothesis is to estimate the θ_j and σ^2 by the minimisation of

$$\Omega = \sum_{i=1}^{n} \left[x_i - \sum_{j=1}^{k} c_{ij} \, \theta_j \right]^2$$

with respect to variations in the θ_j. The equations for the least-squares estimates θ_j^* are

$$\left[\frac{\partial \Omega}{\partial \theta_j} \right]_{\theta\text{'s}=\theta^*\text{'s}} = 0, \quad \text{for } j = 1, 2, \ldots, k.$$

Then the absolute minimum of Ω is

$$M = \sum_{i=1}^{n} \left[x_i - \sum_{j=1}^{k} c_{ij} \, \theta_j^* \right]^2,$$

and it is known from general theory that $E(M) = (n - k)\sigma^2$. Therefore the least-squares estimate of σ^2 is $s^2 = M/(n-k)$, and M/σ^2 has the χ^2 distribution with $n - k$ d.f.

Next, in order to test the null hypothesis $H(\mu = \mu_0)$, it is not necessary to minimise Ω relative to the null hypothesis. Alternatively, we observe that the least-squares estimate of μ is

$$\mu^* = \sum_{j=1}^{k} a_j \theta_j^*.$$

Also, we have $\text{var}(\mu^*) = \text{var}(\text{a linear function of the } x_i)$,
since the θ_j^* are linear functions of the observations,

$$= \sigma^2 \times (\text{a function of the } a_j, c_{ij}, n, k)$$
$$= \sigma^2 \times g(a_j, c_{ij}, n, k), \quad \text{say.}$$

Therefore $\quad$ s.e.$(\mu^*) = s\sqrt{g(a_j, c_{ij}, n, k)}$.

Hence the generalised t for testing the null hypothesis $H(\mu = \mu_0)$ is

$$t = \frac{\mu^* - \mu_0}{\text{s.e.}(\mu^*)},$$

where t has "Student's" distribution with $n - k$ d.f.

As a simple check, it is not difficult to verify that in the elementary case when $E(x_i) = \mu$, for $i = 1, 2, \ldots, n$, the generalised t test for the null hypothesis $H(\mu = \mu_0)$ reduces to the familiar "Student's" t. The convenience of the generalised t test is better revealed by the following somewhat more elaborate problem.

Example 15.9

Each of the $3n$ independent normally distributed observations $x_1, x_2, \ldots, x_n$; $y_1, y_2, \ldots, y_n$; $z_1, z_2, \ldots, z_n$ has the same unknown variance σ^2, and

$$E(x_i) = \theta_1; \quad E(y_i) = \theta_2; \quad E(z_i) = \theta_1 - \theta_2, \quad \text{for } i = 1, 2, \ldots, n.$$

From these data, we derive the generalised t to test the null hypothesis $H(\theta_1 - c\theta_2 = 0)$, where c is a known constant.

Let $\bar{x}, \bar{y}$, and $\bar{z}$ be the averages of the x, y, and z observations respectively. To estimate θ_1 and θ_2, we have

$$\Omega = \sum_{i=1}^{n}[(x_i - \theta_1)^2 + (y_i - \theta_2)^2 + (z_i - \theta_1 + \theta_2)^2]$$

$$= \sum_{i=1}^{n}[\{(x_i - \bar{x}) + (\bar{x} - \theta_1)\}^2 + \{(y_i - \bar{y}) + (\bar{y} - \theta_2)\}^2 +$$

$$+ \{(z_i - \bar{z}) + (\bar{z} - \theta_1 + \theta_2)\}^2]$$

$$= \sum_{i=1}^{n}[(x_i - \bar{x})^2 + (y_i - \bar{y})^2 + (z_i - \bar{z})^2] + n[(\bar{x} - \theta_1)^2 + (\bar{y} - \theta_2)^2 +$$

$$+ (\bar{z} - \theta_1 + \theta_2)^2]$$

$$= M_0 + n[(\bar{x} - \theta_1)^2 + (\bar{y} - \theta_2)^2 + (\bar{z} - \theta_1 + \theta_2)^2], \quad \text{say,}$$

where M_0 is a function of the observations only. This is the most convenient form of Ω to carry out the minimisation. Differentiation with respect to θ_1 and θ_2 gives

$$\frac{\partial \Omega}{\partial \theta_1} = -2n[(\bar{x} - \theta_1) + (\bar{z} - \theta_1 + \theta_2)],$$

$$\frac{\partial \Omega}{\partial \theta_2} = -2n[(\bar{y} - \theta_2) - (\bar{z} - \theta_1 + \theta_2)].$$

Therefore the equations for the estimates θ_1^* and θ_2^* are

$$(\bar{x} - \theta_1^*) + (\bar{z} - \theta_1^* + \theta_2^*) = 0, \quad \text{or} \quad 2\theta_1^* - \theta_2^* = \bar{x} + \bar{z},$$

and

$$(\bar{y} - \theta_2^*) - (\bar{z} - \theta_1^* + \theta_2^*) = 0, \quad \text{or} \quad -\theta_1^* + 2\theta_2^* = \bar{y} - \bar{z}.$$

The solution of these simultaneous equations is

$$\theta_1^* = (2\bar{x} + \bar{y} + \bar{z})/3; \qquad \theta_2^* = (\bar{x} + 2\bar{y} - \bar{z})/3.$$

Hence the absolute minimum of Ω is

$$M = M_0 + n[(\bar{x} - \theta_1^*)^2 + (\bar{y} - \theta_2^*)^2 + (\bar{z} - \theta_1^* + \theta_2^*)^2].$$

But, since

$$\bar{x} - \theta_1^* = -(\bar{y} - \theta_2^*) = -(\bar{z} - \theta_1^* + \theta_2^*) = (\bar{x} - \bar{y} - \bar{z})/3,$$

therefore

$$M = M_0 + \frac{n}{3}(\bar{x} - \bar{y} - \bar{z})^2.$$

Next, to determine the degrees of freedom of M, we have

$$E(M) = E(M_0) + \frac{n}{3}E(\bar{x} - \bar{y} - \bar{z})^2$$

$$= 3(n - 1)\sigma^2 + \frac{n}{3}E[(\bar{x} - \theta_1) - (\bar{y} - \theta_2) - (\bar{z} - \theta_1 + \theta_2)]^2$$

$$= 3(n - 1)\sigma^2 + \frac{n}{3}E[(\bar{x} - \theta_1)^2 + (\bar{y} - \theta_2)^2 + (\bar{z} - \theta_1 + \theta_2)^2],$$

since the expectations of the cross-product terms are zero,

$$= 3(n - 1)\sigma^2 + \frac{n}{3}[\text{var}(\bar{x}) + \text{var}(\bar{y}) + \text{var}(\bar{z})]$$

$$= (3n - 2)\sigma^2, \quad \text{as expected.}$$

Thus $s^2 = M/(3n - 2)$ is the least-squares estimate of σ^2, and M/σ^2 has the χ^2 distribution with $3n - 2$ d.f.

Finally, to test the null hypothesis $H(\theta_1 - c\theta_2 = 0)$, we observe that the least-squares estimate of $\theta_1 - c\theta_2$ is

$$\theta_1^* - c\theta_2^* = \frac{2\bar{x} + \bar{y} + \bar{z}}{3} - \frac{c(\bar{x} + 2\bar{y} - \bar{z})}{3} = \frac{(2-c)\bar{x} + (1-2c)\bar{y} + (1+c)\bar{z}}{3};$$

and

$$\text{var}(\theta_1^* - c\theta_2^*) = \frac{1}{9}[(2-c)^2 + (1-2c)^2 + (1+c)^2]\frac{\sigma^2}{n} = \frac{2(c^2 - c + 1)\sigma^2}{3n}.$$

Therefore

$$\text{s.e.}(\theta_1^* - c\theta_2^*) = s\left[\frac{2(c^2 - c + 1)}{3n}\right]^{\frac{1}{2}}.$$

Also, if the null hypothesis $H(\theta_1 - c\theta_2 = 0)$ is true, then $E(\theta_1^* - c\theta_2^*) = 0$. Hence the appropriate generalised t test for testing the significance of this null hypothesis is

$$t = \frac{\theta_1^* - c\theta_2^*}{s\left[\frac{2(c^2 - c + 1)}{3n}\right]^{\frac{1}{2}}},$$

where t has "Student's" distribution with $3n - 2$ d.f.

15.17   Increasing the sensitivity of the t test

The generalised t test is one way of extending the applicability of the classical result of "Student". This approach makes possible a wider use of "Student's" distribution for testing meaningful linear hypotheses about parameters occurring in the equations defining the expectations of the observations. There is, however, another kind of extension of "Student's" distribution, which increases the *sensitivity* of the tests of significance for assessing the credibility of a null hypothesis in the light of the sample observations. This notion of sensitivity can best be explained by an elementary illustration.

Suppose a random sample of n observations from a normal population with mean μ and variance σ^2 gives the sample average $\bar{x}$ and the sample variance s^2. Then we have seen that a test of significance for the null hypothesis $H(\mu = \mu_0)$, μ_0 preassigned, is obtained by using

$$z = \frac{\bar{x} - \mu_0}{\sigma/\sqrt{n}} \tag{1}$$

as a unit normal variable. Strictly, this is a valid method only if σ is known. When, as is usually the case, σ is unknown, then we have indicated that the above null hypothesis can be tested by using

$$t = \frac{\bar{x} - \mu_0}{s/\sqrt{n}} \tag{2}$$

as a random variable having "Student's" distribution with $n - 1$ d.f. The fundamental difference between z and t is that σ, an unknown but fixed parameter, in z is replaced by s, a known random variable, to obtain t. Now, the greater the number of the degrees of freedom for t, the more closely does its distribution approach the unit normal distribution of z, assuming, of course, that the null hypothesis under test is true. This implies that for increasing degrees of freedom, the uncertainty in the denominator of t tends to decrease, and this correspondingly increases the discriminatory power of the test of significance based on the use of "Student's" distribution. In other words, the t test becomes increasingly sensitive to the departures from the null hypothesis as the degrees of freedom increase.

It is known that $(n - 1)s^2/\sigma^2$ has the χ^2 distribution with $n - 1$ d.f., and that the degrees of freedom of t in (2) are the same as those of this χ^2. Next, suppose s_0^2 is another independent unbiased estimate of σ^2 such that $\nu s_0^2/\sigma^2$ has the χ^2 distribution with ν d.f. Clearly, both these χ^2's are independent and

$$s_1^2 = \frac{(n - 1)s^2 + \nu s_0^2}{n + \nu - 1}$$

is obviously the pooled unbiased estimate of σ^2. It is also known that $(n + \nu - 1)s_1^2/\sigma^2$ is distributed as χ^2 with $n + \nu - 1$ d.f., and this χ^2 and $\bar{x}$ are independent random variables. Hence we may also test the null hypothesis $H(\mu = \mu_0)$ by using

$$t = \frac{\bar{x} - \mu_0}{s_1/\sqrt{n}} \tag{3}$$

as a random variable having "Student's" distribution with $n + \nu - 1$ d.f. We have thus developed another test for the null hypothesis $H(\mu = \mu_0)$ by using the additional independent information about σ^2 provided by the estimate s_0^2. This test of significance has a larger number of degrees of freedom than the one based on the random variable (2). Accordingly, (3) provides a more sensitive test than that obtained by using (2) for the null hypothesis under test.

It is also possible to obtain confidence intervals for the test parameter μ by using the pooled estimate of the population variance with the total available degrees of freedom. Thus, for example, if $t(\eta; n - 1)$ and $t(\eta; n + \nu - 1)$ are the 100η per cent points of "Student's" distribution with $n - 1$ and $n + \nu - 1$ d.f. respectively, then the corresponding $100(1 - \eta)$ per cent confidence intervals for μ are

$$\bar{x} - \frac{s}{\sqrt{n}} t(\eta; n - 1) \leqslant \mu \leqslant \bar{x} + \frac{s}{\sqrt{n}} t(\eta; n - 1),$$

and $\qquad \bar{x} - \frac{s_1}{\sqrt{n}} t(\eta; n + \nu - 1) \leqslant \mu \leqslant \bar{x} + \frac{s_1}{\sqrt{n}} t(\eta; n + \nu - 1).$

The second is the better confidence interval for μ because it is based on a more stable estimate of the population variance.

This principle of incorporating additional information about σ^2 to obtain a more stable estimate can be applied to any number of independent estimates to derive a pooled estimate. The use of the latter estimate increases the sensitivity of the t test for the null hypothesis $H(\mu = \mu_0)$. Furthermore, the argument can also be applied to the generalised t test for any linear hypothesis involving the means of a number of normal populations with a common unknown variance σ^2. Indeed, the tests for the comparison of two or more population means use this principle, since the least-squares procedure ensures that the pooled estimate of the population is the weighted average of the sample variances. This averaging also occurs in estimating the population variance in regression analysis.

15.18   Null hypotheses and the assumptions underlying tests of significance

In common usage, there is considerable overlap in the meaning of the words "hypothesis" and "assumption", but in statistical theory a clear and important distinction is drawn between them. With the broad background knowledge of statistically testable linear hypotheses, it is our purpose now to clarify the difference between the assumptions underlying a test of significance and the null hypotheses which can be tested by its use.

The simplest example to illustrate the difference is, again, that of testing the null hypothesis $H(\mu = \mu_0)$, μ_0 preassigned, pertaining to the mean of a normal population. That the sample observations are drawn from a normal population with mean μ and variance σ^2 is an assumption which leads mathematically to "Student's" test for the null hypothesis $H(\mu = \mu_0)$. The statistical inference from this test of significance refers only to this null hypothesis. Thus if the observed difference between the sample mean and μ_0 is found to be significant (not significant) at a certain level of significance, then the null hypothesis under test is rejected (accepted) at that level. However, whatever be the result of the test, it does not provide a basis for inferring anything about, for example, the normality of the population sampled. Indeed, it may well be that the rejection or acceptance of the null hypothesis $H(\mu = \mu_0)$ in any given case is due to considerable non-normality of the population sampled, but even so the t test does not provide a basis for any inference about this. The proper inference from a significant (non-significant) result of the t test is that if the population sampled is normal, then the data do not (do) provide credible evidence, at the chosen level of significance, for the acceptance of the null hypothesis under test. On the other hand, if there is doubt about the normality of the population sampled, then it is possible to use another test of significance (to be considered in Chapter 17) to assess whether or not the sampled population may reasonably be regarded as normal. In other words, an assumption underlying one test of significance for a null hypothesis can itself be used as another null hypothesis which may be tested by a different test of significance. But it is to be noted that, in general, the sample information adequate for the test of significance of one null hypothesis may be inadequate for another test of significance for a different null hypothesis.

Similarly, we tested null hypotheses about the means of two populations on the assumption that the sampled populations are normal and have the same variance. As we have seen, a different test of significance is required to test the equality of the population variances. Moreover, we show in Chapter 17 that the normality of the sampled populations can also be tested by a different method. This argument also applies to tests of significance for more general linear hypotheses about the means of $k > 2$ normal populations with the same unknown variance. Furthermore, in linear regression analysis, the normality of the populations sampled, the equality of their variances, and the linearity of regressions are all assumptions underlying the tests of significance for the regression parameters.

It is also to be noted that for large samples these linear parametric hypotheses can be tested approximately by relaxing the strict assumption of the normality of the populations sampled; and for this reason, this assumption is often not explicitly stated in practical statistical analysis. Nevertheless a correct use of these tests of significance requires that their underlying assumptions should at all times be carefully borne in mind, and also viewed as clearly distinct from the specific null hypotheses under test. Failure to observe this simple but completely general rule can lead to patently erroneous inferences. Indeed, even the most general inferential procedures have their applicational limitations. This is a fundamental caution in all statistical analysis.

CHAPTER 16

CORRELATION ANALYSIS

16.1   The duality of linear regression

In linear regression analysis, the study of the association between two variables is, in principle, the analysis of the random variation of one variable for predetermined or fixed values of the other variable. Thus, given the n pairs of observations (x_ν, y_ν), for $\nu = 1, 2, ..., n$, we postulated that for the specified values of x, the observed y values are independent random variables such that

$$E(y_\nu) = \alpha + \beta(x_\nu - \bar{x}), \quad \bar{x} \text{ being the average of the } x_\nu,$$

and $\mathrm{var}(y_\nu) = \sigma^2$.

In many situations, this non-randomness of the x values is physically meaningful and provides the correct basis for statistical analysis. As an example, in the data of the speeds attained in the annual Indianapolis Memorial Day car races considered in the previous chapter, the years are evidently the non-random variable. The linear regression of speed on time gives the linear annual advance in speed. On the other hand, as already pointed out, in many instances there is no clear distinction between the random and the non-random variables, and the choice has to be made in terms of the objectives of the analysis. For example, if the observations (x_ν, y_ν) refer to the height and weight measurements of n randomly selected adults in a country, then, quite clearly, both x_ν and y_ν are random variables. Accordingly, it is possible to define two distinct models for linear regression analysis. We now indicate how these models are arrived at and then derive the least-squares solutions for the two cases. It is helpful to continue with our height — weight terminology to keep the physical interpretation in view.

Firstly, suppose we are interested in determining how weight (y) depends upon the height (x) of the adults. We therefore postulate that corresponding to each of the n observed values of height (x_ν), there exists a population of adults with a normal weight distribution, and y_ν is a chance observation from this population such that

$$E(y_\nu) = \alpha_y + \beta_y(x_\nu - \bar{x}) \equiv y_{\nu e}, \quad \text{say,}$$

and $\mathrm{var}(y_\nu) = \sigma_y^2, \quad \text{for } \nu = 1, 2, ..., n.$

The suffix y in the parameters α_y, β_y, and σ_y^2 is introduced to indicate that they refer to the distributions of the y_ν, otherwise the above is the usual statement of the linear regression model.

Suppose $\bar{x}$ and $\bar{y}$ are the averages of the x_ν and y_ν. As before, we introduce the notation

$$X = \sum_{\nu=1}^{n}(x_\nu - \bar{x})^2; \quad Y = \sum_{\nu=1}^{n}(y_\nu - \bar{y})^2; \quad Z = \sum_{\nu=1}^{n}(x_\nu - \bar{x})(y_\nu - \bar{y}).$$

Now the standard least-squares estimation procedure of minimising

$$\Omega_y = \sum_{\nu=1}^{n}[y_\nu - E(y_\nu)]^2 = \sum_{\nu=1}^{n}[y_\nu - a_y - \beta_y(x_\nu - \bar{x})]^2$$

with respect to variations in a_y and β_y gives the estimates

$$a_y^* = \bar{y}, \quad \beta_y^* = Z/X, \quad \text{and}$$

$$\sigma_y^{2*} = \sum_{\nu=1}^{n}[y_\nu - a_y^* - \beta_y^*(x_\nu - \bar{x})]^2/(n-2) = (Y - Z^2/X)/(n-2).$$

Hence the estimated linear regression of y on x is

$$y_e^* = a_y^* + \beta_y^*(x - \bar{x}). \tag{1}$$

Secondly, suppose we now interchange the roles of x and y. This implies that we wish to consider a model for the variation of height (x) for the observed values of the weight (y). Thus, we postulate that corresponding to each of the n observed values of the weights (y_ν), there exists a population of adults with a normal height distribution. Furthermore, we assume that for any y_ν, the random variable x_ν is normally distributed with

$$E(x_\nu) = a_x + \beta_x(y_\nu - \bar{y}) \equiv x_{\nu e}, \quad \text{say,}$$

and $$\text{var}(x_\nu) = \sigma_x^2, \quad \text{for } \nu = 1, 2, ..., n.$$

This is a different model with different parameters, and its application leads to the second estimated linear regression. More specifically, the least-squares estimation is now done by minimising

$$\Omega_x = \sum_{\nu=1}^{n}[x_\nu - E(x_\nu)]^2 = \sum_{\nu=1}^{n}[x_\nu - a_x - \beta_x(y_\nu - \bar{y})]^2$$

with respect to variations in a_x and β_x. This gives the estimates

$$a_x^* = \bar{x}, \quad \beta_x^* = Z/Y, \quad \text{and}$$

$$\sigma_x^{2*} = \sum_{\nu=1}^{n}[x_\nu - a_x^* - \beta_x^*(y_\nu - \bar{y})]^2/(n-2) = (X - Z^2/Y)/(n-2).$$

Hence the estimated linear regression of x on y is

$$x_e^* = a_x^* + \beta_x^*(y - \bar{y}). \tag{2}$$

It is important to understand clearly the difference in the assumptions leading to the two estimated linear regressions. Equation (1) is the linear regression of the random variable y for fixed values of x, and the value y_e^* denotes the estimated weight of a randomly selected person whose height is known to be x. On the other hand, equation (2) is the linear regression of the random variable x for fixed values of y, and the value x_e^* is the estimated height of a randomly selected adult whose weight is known to be y. In general, these two regressions are distinct and so also is their role in prediction. In fact, (1) should always be used to predict the weight of an adult of known height, and (2) to predict the height of an adult of known weight. The parameter β_y is the linear regression coefficient of y on x, and its least-squares estimate β_y^* is obtained by minimising the random deviations of y_ν from their expectations. In contrast, β_x is the linear regression coefficient of x on y, and its least-squares estimate β_x^* is obtained by minimising the random deviations of x_ν from their expectations.

16.2   The product-moment correlation coefficient

By definition, the variances of the x and y observations are

$$s_x^2 = X/(n-1) \quad \text{and} \quad s_y^2 = Y/(n-1) \quad \text{respectively.}$$

We extend these definitions to define the *covariance* of the x and y observations, or the *sample product-moment* as

$$s_{xy} = Z/(n-1).$$

Then the *sample product-moment correlation coefficient* is defined as

$$r = Z/(XY)^{\frac{1}{2}} = s_{xy}/s_x s_y.$$

This definition of the product-moment correlation coefficient is due to Karl Pearson. It should be noted that a non-zero value of r can be either positive or negative, and its sign is the same as that of Z.

Now, in this new notation, the estimate of the regression coefficient of y on x is

$$\beta_y^* = Z/X = r(XY)^{\frac{1}{2}}/X = r s_y/s_x.$$

Similarly, the estimate of the regression coefficient of x on y is

$$\beta_x^* = Z/Y = r(XY)^{\frac{1}{2}}/Y = r s_x/s_y.$$

Hence

$$\beta_x^* \beta_y^* = r^2,$$

so that the square of the product-moment correlation coefficient between the x and y observations is the product of the least-squares estimates of the two linear regression coefficients β_x and β_y. This is a fundamental result, and it is also convenient for the computation of r.

The two estimated linear regressions (1) and (2) may now be written alternatively as

$$y_e^* - \bar{y} = \frac{r s_y}{s_x}(x - \bar{x}),$$

or

$$\frac{y_e^* - \bar{y}}{s_y} = r\left[\frac{x - \bar{x}}{s_x}\right]; \tag{1'}$$

and

$$x_e^* - \bar{x} = \frac{r s_x}{s_y}(y - \bar{y}),$$

or

$$\frac{x_e^* - \bar{x}}{s_x} = r\left[\frac{y - \bar{y}}{s_y}\right]. \tag{2'}$$

Also, since the estimated variances σ_x^{2*} and σ_y^{2*} are necessarily non-negative, it follows that

$$XY - Z^2 \geqslant 0, \quad \text{or} \quad XY \geqslant Z^2,$$

so that

$$r^2 \leqslant 1, \quad \text{or} \quad -1 \leqslant r \leqslant 1.$$

We have thus established another important property of r.

The product-moment correlation coefficient is one statistical measure of the strength of association between the x and y sample observations.

If $r = \pm 1$, then $XY - Z^2 = 0$, and hence each separate residual

$$y_\nu - \bar{y} - \beta_y^*(x_\nu - \bar{x}) \quad \text{and} \quad x_\nu - \bar{x} - \beta_x^*(y_\nu - \bar{y})$$

must be identically zero. Hence all the n pairs of observations (x_ν, y_ν) lie on a straight line, and the two linear regressions coincide with this line. For $r = +1$, the two regressions coincide with

either $\quad (y_e^* - \bar{y})/s_y = (x - \bar{x})/s_x, \quad$ or $\quad (x_e^* - \bar{x})/s_x = (y - \bar{y})/s_y,$

and the correlation between the x and y observations is said to be *perfect*. On the other hand, for $r = -1$, the two regressions coincide with

either $(y_e^* - \bar{y})/s_y = -(x - \bar{x})/s_x, \quad$ or $\quad (x_e^* - \bar{x})/s_x = -(y - \bar{y})/s_y,$

and the observations are said to be *perfectly negatively correlated*. If $r = 0$, then nothing can be deduced connecting the two series of observations. The two regressions now are

$$y_e^* = \bar{y} \quad \text{and} \quad x_e^* = \bar{x},$$

which means that, on the average, y does not alter with x and vice versa. The x and y observations are now said to be *uncorrelated*. This can be caused by a general scatter of the points (x_ν, y_ν) when plotted on the (x, y) plane, or by a strong *curvilinear association* between them, that is, when the points lie rather closely on a curve. The essential point of the product-moment correlation coefficient is that it is a proper measure of the association between the x and y observations when there is, apart from errors of sampling, a linear relation between the sample values. We shall return to a further consideration of this point a little later.

Invariance of r under linear transformations of the observations

Regarded simply as a formal measure of association between the x and y observations, the product-moment correlation coefficient r is a numerical characteristic of the two series, which is invariant under linear transformations of the x and y values. This is a useful property of the correlation coefficient, and one which often considerably simplifies its computation.

To prove the invariance, suppose we make the linear transformations

$$x_\nu = a_1 + b_1 u_\nu \quad \text{and} \quad y_\nu = a_2 + b_2 v_\nu,$$

where a_1, a_2, b_1, b_2 are known constants and the b's are assumed to be non-zero. If now $\bar{u}$ and $\bar{v}$ denote the means of the u_ν and v_ν respectively, then

$$\bar{x} = a_1 + b_1 \bar{u} \quad \text{and} \quad \bar{y} = a_2 + b_2 \bar{v}.$$

Therefore $\quad x_\nu - \bar{x} = b_1(u_\nu - \bar{u}); \quad y_\nu - \bar{y} = b_2(v_\nu - \bar{v}),$

for $\nu = 1, 2, ..., n$. Hence, on addition, we have

$$X = \sum_{\nu=1}^{n}(x_\nu - \bar{x})^2 = b_1^2 \sum_{\nu=1}^{n}(u_\nu - \bar{u})^2 = b_1^2 U, \quad \text{say,}$$

and $\quad {}'Y = \sum_{\nu=1}^{n}(y_\nu - \bar{y})^2 = b_2^2 \sum_{\nu=1}^{n}(v_\nu - \bar{v})^2 = b_2^2 V, \quad \text{say.}$

It is also easily seen that

$$Z = \sum_{\nu=1}^{n}(x_\nu - \bar{x})(y_\nu - \bar{y}) = b_1 b_2 \sum_{\nu=1'}^{n}(u_\nu - \bar{u})(v_\nu - \bar{v}) = b_1 b_2 W, \quad \text{say.}$$

Hence the square of the product-moment correlation coefficient between the x and y observations is

$$r_{xy}^2 \;=\; \frac{Z^2}{XY} \;=\; \frac{W^2}{UV} \;=\; r_{uv}^2,$$

the square of the product-moment correlation coefficient between the transformed u and v observations. Hence the invariance of the product-moment correlation coefficient under general linear transformations.

Example 16.1: Measurement of the intelligence of schoolchildren

We next use the data of Roberts and Griffiths, quoted in Table 5.6, page 65, pertaining to the scores obtained by 65 schoolchildren on the Binet and Otis tests to illustrate the calculation of the two linear regressions and the product-moment correlation coefficient. We assume that the Binet and Otis test scores of a child are the x and y observations of our theory. It is not particularly worth while to transform the data and we carry out the computations with the given values.

Here $\qquad n \;=\; 65, \quad \sum_{\nu=1}^{65} x_\nu \;=\; 6366, \quad \sum_{\nu=1}^{65} y_\nu \;=\; 6739;$

$$\sum_{\nu=1}^{65} x_\nu^2 \;=\; 638650, \quad \sum_{\nu=1}^{65} y_\nu^2 \;=\; 781685, \quad \sum_{\nu=1}^{65} x_\nu y_\nu \;=\; 691449;$$

$\text{C.F.}(x) \;=\; \dfrac{(6366)^2}{65} \;=\; 623476{\cdot}2462, \; \text{C.F.}(y) \;=\; \dfrac{(6739)^2}{65} \;=\; 698678{\cdot}7846,$

and $\qquad\qquad \text{C.F.}(x,y) \;=\; \dfrac{6366 \times 6739}{65} \;=\; 660007{\cdot}2923.$

Therefore $\qquad\qquad \bar{x} \;=\; 97{\cdot}94, \quad \bar{y} \;=\; 103{\cdot}68;$

$\qquad X \;=\; 15173{\cdot}7538, \quad Y \;=\; 83006{\cdot}2154, \quad Z \;=\; 31441{\cdot}7077.$

Hence $\qquad\qquad \beta_y^* \;=\; 2{\cdot}072111, \quad \beta_x^* \;=\; 0{\cdot}378787,$

and $\qquad\qquad r \;=\; (2{\cdot}072111 \times 0{\cdot}378787)^{\frac{1}{2}} \;=\; 0{\cdot}885940.$

The estimated linear regression of y on x is

$$y_e^* \;=\; 103{\cdot}68 + 2{\cdot}072111(x - 97{\cdot}94)$$
$$= \; -99{\cdot}26 + 2{\cdot}072111\,x;$$

and the estimated linear regression of x on y is

$$x_e^* \;=\; 97{\cdot}94 + 0{\cdot}378787(y - 103{\cdot}68)$$
$$= \; 58{\cdot}67 + 0{\cdot}378787\,y.$$

Jointly, these two estimated linear regressions give more information about the relationship between the Binet and Otis test scores than the product-moment correlation coefficient. The estimate $\beta_y^* = 2{\cdot}072111$ means that for every unit advance in the Binet score (x), the Otis score (y_e^*) should, on the average, increase by $2{\cdot}072111$. On the other hand, the estimate $\beta_x^* = 0{\cdot}378787$ indicates that for every unit advance in the Otis score (y), the Binet score (x_e^*) should, on the average, increase by $0{\cdot}378787$. Decreases in the scores on one test corresponding to decreases on the other can be similarly interpreted. Such prediction is not possible with the product-moment correlation coefficient. In general, therefore, regression analysis is to be preferred to correlation analysis, although there is obviously a close computational relationship between them. Nevertheless, the product-moment correlation does give an overall measure of the association between the x and y observations, and we consider next the meaning of this measure of association.

16.3   The population product-moment correlation coefficient

As we have seen, the product-moment correlation coefficient r is a statistical measure of the strength of association between n pairs of sample observations (x_ν, y_ν), for $\nu = 1, 2, \ldots, n$. In the context of our first example, the observations x_ν and y_ν are the height and weight measurements of a randomly selected adult, and this is the basis for the pairing of the observations. Of course, each pair is a random set and so it is independently distributed with respect to any other similar pair. Now, as r is a function of the observations it can be regarded as an estimate, and we naturally wish to find out what r actually estimates. To answer this question, we have to first build up a model of the population sampled. We therefore postulate that there exists a conceptually infinite physical population of adults, and each adult in it provides the two measurements of height and weight. Accordingly, we can next visualise an infinite statistical population of paired height and weight measurements, each pair of observations pertaining to one adult. This totality of paired measurements constitutes an infinite *bivariate population* which may be formally regarded as an extension of the idea of a univariate population. Suppose this population has the probability density function $f(x,y)$ and, for simplicity, we assume that the limits of variation of x and y are $-\infty < x, y < \infty$. Then

(i) $f(x,y) \geqslant 0$, for all possible pairs (x,y);

(ii) $\int\limits_{-\infty}^{\infty} \int\limits_{-\infty}^{\infty} f(x,y)\,dx\,dy = 1$; and

(iii) the probability that a randomly selected adult has height and weight measurements in the infinitesimal region $(x,\ x+dx;\ y,\ y+dy)$ is

$$f(x,y)\,dx\,dy.$$

The observed sample of n pairs is assumed to be a random sample of paired observations from this population.

We now postulate that in such a bivariate population there is a fixed but unknown measure of the strength of association between the random variables x and y. This is the *population product-moment correlation coefficient* and it is usually denoted by ρ. To give an algebraic expression for ρ, we have to introduce the following formal definitions of the moments of the bivariate probability distribution having the density function $f(x,y)$.

$$E(x) = \int\limits_{-\infty}^{\infty} \int\limits_{-\infty}^{\infty} x\, f(x,y)\,dx\,dy, \quad E(y)= \int\limits_{-\infty}^{\infty} \int\limits_{-\infty}^{\infty} y\, f(x,y)\,dx\,dy,$$

$$\operatorname{var}(x) = E[x - E(x)]^2 = \int\limits_{-\infty}^{\infty} \int\limits_{-\infty}^{\infty} [x - E(x)]^2\, f(x,y)\,dx\,dy,$$

$$\operatorname{var}(y) = E[y - E(y)]^2 = \int\limits_{-\infty}^{\infty} \int\limits_{-\infty}^{\infty} [y - E(y)]^2\, f(x,y)\,dx\,dy,$$

and the *population covariance* between x and y is defined as

$$\text{cov}(x, y) = E[\{x - E(x)\}\{y - E(y)\}]$$

$$= \int_{-\infty}^{\infty} \int_{-\infty}^{\infty} [x - E(x)][y - E(y)]\, f(x,y)\,dx\,dy.$$

It is also to be observed that

$$\text{cov}(x, y) = E[\{x - E(x)\}\{y - E(x)\}]$$

$$= E[xy - xE(y) - yE(x) + E(x)E(y)] = E(xy) - E(x)E(y).$$

This result is in conformity with the corresponding expressions

$$\text{var}(x) = E(x^2) - E^2(x) \quad \text{and} \quad \text{var}(y) = E(y^2) - E^2(y).$$

We shall not consider here how the above defining integrals can be evaluated in any particular case when the form of $f(x,y)$ is specified. It is enough to note these formal definitions, and then by an obvious extension of Pearson's formula we have

$$\rho \equiv \text{corr}(x,y) = \frac{\text{cov}(x, y)}{[\text{var}(x)\,\text{var}(y)]^{\frac{1}{2}}}.$$

The sample value r is an estimate of the parameter ρ.

Limits of variation of ρ

To determine the permissible limits of variation of ρ, suppose

$$E(x) = \theta_1, \qquad E(y) = \theta_2;$$

$$\text{var}(x) = \sigma_1^2, \quad \text{var}(y) = \sigma_2^2, \quad \text{and} \quad \text{cov}(x,y) = \rho\sigma_1\sigma_2.$$

If we now define the standardised random variables

$$u = (x - \theta_1)/\sigma_1 \quad \text{and} \quad v = (y - \theta_2)/\sigma_2,$$

then, quite clearly, $E(u) = E(v) = 0$; $\text{var}(u) = \text{var}(v) = 1$; and

$$\text{cov}(u, v) = E(uv) = \text{cov}(x,y)/\sigma_1\sigma_2 = \rho.$$

Next, consider two new random variables w and z defined by the relations

$$w = u + v \quad \text{and} \quad z = u - v.$$

Then $E(w) = E(z) = 0.$

Hence $\text{var}(w) = E(w^2) = E(u + v)^2$

$$= E(u^2) + E(v^2) + 2E(uv) = 2(1 + \rho).$$

Similarly, $\text{var}(z) = E(z^2) = E(u - v)^2$

$$= E(u^2) + E(v^2) - 2E(uv) = 2(1 - \rho).$$

But since both var(w) and var(z) must always be non-negative, we must have

$$1 \pm \rho \geqslant 0, \quad \text{so that} \quad -1 \leqslant \rho \leqslant 1.$$

We thus observe that the limits of r are in conformity with those of ρ.

The bivariate normal distribution

In general, a bivariate population can have a variety of forms in much the same way as a univariate one; but, as in the case of a single random variable, our interest centres largely on the bivariate normal distribution, which is a straightforward generalisation of the univariate normal distribution. The probability density function of the bivariate normal distribution is

$$f(x, y) = \frac{1}{2\pi\sigma_1\sigma_2\sqrt{1 - \rho^2}} \exp\left[-\frac{1}{2(1 - \rho^2)}\left\{\left(\frac{x - \theta_1}{\sigma_1}\right)^2 + \right.\right.$$
$$\left.\left. + \left(\frac{y - \theta_2}{\sigma_2}\right)^2 - 2\rho\left(\frac{x - \theta_1}{\sigma_1}\right)\left(\frac{y - \theta_2}{\sigma_2}\right)\right\}\right],$$

for $-\infty < x, \ y < \infty$. The five parameters θ_1, θ_2, σ_1, σ_2, and ρ have the following interpretation, though we do not prove these results:

(i) $E(x) = \theta_1$, $E(y) = \theta_2$;

(ii) $\text{var}(x) = \sigma_1^2$, $\text{var}(y) = \sigma_2^2$; and

(iii) $\text{cov}(x, y) = \rho\sigma_1\sigma_2$, $\text{corr}(x, y) = \rho$.

We consider the relationship between r and ρ only in such a population. Accordingly, if the sample observations are from the above bivariate normal population, then it is known that r is the maximum-likelihood estimate of ρ. Also, for large n, we have

$$E(r) \sim \rho\left[1 - \frac{1 - \rho^2}{2(n - 1)}\right]; \quad \text{var}(r) \sim \frac{(1 - \rho^2)^2}{n - 1}.$$

The exact values of $E(r)$ and $\text{var}(r)$ are very complicated functions of ρ and n, but the above simple approximations are adequate for our present purpose. It is clear that r is an unbiased estimate of ρ only if $\rho = 0$, that is, when the random variables x and y are uncorrelated in the population. Here we have another example of a maximum-likelihood estimate of a parameter which is not fully unbiased. Of course, r is a consistent estimate of ρ since $E(r) \to \rho$ and $\text{var}(r) \to 0$, as $n \to \infty$.

It is a fundamental property of the bivariate normal distribution that if the two random variables x and y have this joint distribution, then the regression of x on y and that of y on x are both linear with the regression coefficients $\beta_1 = \rho\sigma_1/\sigma_2$ and $\beta_2 = \rho\sigma_2/\sigma_1$, respectively, whence $\beta_1\beta_2 = \rho^2$. Thus the population product-moment correlation coefficient ρ and the true linear regression coefficients β_1 and β_2 satisfy a relation analogous to the sample relationship.

We also state, without proof, that if x and y have the bivariate normal distribution then

(i) for any *fixed* y the distribution of x is normal with

$$E(x) = \theta_1 + \beta_1(y - \theta_2); \quad \text{var}(x) = \sigma_1^2(1 - \rho^2); \quad \text{and}$$

(ii) for any *fixed* x the distribution of y is normal with

$$E(y) = \theta_2 + \beta_2(x - \theta_1); \quad \text{var}(y) = \sigma_2^2(1 - \rho^2).$$

In fact, the expectations in (i) and (ii) are the linear regressions of x on y, and of y on x respectively.

16.4   Tests of significance for null hypotheses about ρ

The usual interest in correlation analysis is to test whether the observed correlation coefficient r is small enough to infer that the random variables x and y can be taken to be uncorrelated in the population. The test of significance for this important null hypothesis $H(\rho = 0)$ is equivalent to that for the null hypothesis $H(\beta_1 = 0)$ or $H(\beta_2 = 0)$. It now follows that if x and y have a joint bivariate normal distribution, then the test for the null hypothesis $H(\rho = 0)$ is obtained by using the fact that if the null hypothesis under test is true, then

$$F = \frac{(n-2)Z^2}{XY - Z^2} = \frac{(n-2)r^2}{1 - r^2}$$

has the F distribution with $1, n-2$ d.f. An equivalent test of significance for the null hypothesis is obtained by using the fact that if the null hypothesis is true, then

$$t = \frac{r\sqrt{n-2}}{\sqrt{1 - r^2}}$$

has "Student's" distribution with $n - 2$ d.f.

For any non-zero null hypothesis about ρ there is no parallelism between the correlation coefficient ρ and the regression coefficients β_1 and β_2. In fact, no exact test of significance is available for testing readily non-zero null hypotheses about ρ. Fisher has given an approximate method for such null hypotheses, but we do not consider this here.

16.5   Calculation of the product-moment correlation coefficient from grouped data

The method for calculating the sample product-moment correlation coefficient that we have considered is generally used only when the sample size n is small. For large n, it is usual first to group the paired observations to form a *two-way frequency table*. This table is simply a bivariate extension of the frequency table used for presenting the observed distribution of a large number of observations of a single random variable. The principle for calculating the correlation coefficient from a two-way table is the same as that of its evaluation from a number of ungrouped paired observations, but we have to take account of the groupings of the two random variables in evaluating their sums of squares and the sum of products. We now show how these modifications are carried out. The theoretical exposition is illustrated by a numerical example.

Suppose the observed grouping of x has ν_1 class-intervals each of length h_1, and let the midpoints of these class-intervals be $x_1, x_2, ...,$ x_{ν_1}. Similarly, we assume that the observed grouping of y has ν_2 class-intervals each of length h_2, and the midpoints of the class-intervals are $y_1, y_2, ..., y_{\nu_2}$. The combination of any x and y class-intervals constitutes a *cell* of the two-way table and, quite evidently, there are $\nu_1 \nu_2$ such cells in all. Next, to describe the observed two-way frequency distribution, we introduce the notation that there are f_{ij} pairs in the sample whose x values are in the ith x class-interval with class-limits $x_i \pm h_1/2$, and whose y values are in the jth y class-interval with class-limits $y_j \pm h_2/2$. By summing over all the $\nu_1 \nu_2$ cell frequencies for the possible combinations (i, j), we obtain the total frequency of the sample, say N, so that

$$\sum_{i=1}^{\nu_1} \sum_{j=1}^{\nu_2} f_{ij} = N.$$

It is also convenient at this stage to introduce the *marginal frequencies*

$$f_{i.} = \sum_{j=1}^{\nu_2} f_{ij} \quad \text{and} \quad f_{.j} = \sum_{i=1}^{\nu_1} f_{ij}$$

Then it is obvious that

$$\sum_{i=1}^{\nu_1} f_{i.} = \sum_{j=1}^{\nu_2} f_{.j} = N.$$

We also observe that $f_{i.}$ is the total observed frequency of pairs whose x values lie in the ith x class-interval $x_i \pm h_1/2$ irrespective of their y values. Similarly, $f_{.j}$ is the total observed frequency of pairs whose y values lie in the jth y class-interval $y_j \pm h_2/2$ irrespective of their x values. Considered separately, $f_{i.}$ and $f_{.j}$ are the usual observed frequencies in the class-intervals of the univariate distributions of x and y. Hence, by an obvious extension of earlier definitions, we define the sample means of the x and y observations as

$$\bar{x} = \sum_{i=1}^{\nu_1} f_{i.} x_i /N \quad \text{and} \quad \bar{y} = \sum_{j=1}^{\nu_2} f_{.j} y_j /N.$$

Also,
$$X = \sum_{i=1}^{\nu_1} f_{i.}(x_i - \bar{x})^2, \quad Y = \sum_{j=1}^{\nu_2} f_{.j}(y_j - \bar{y})^2,$$

and
$$Z = \sum_{i=1}^{\nu_1} \sum_{j=1}^{\nu_2} f_{ij}(x_i - \bar{x})(y_j - \bar{y}).$$

Then the sample product-moment correlation coefficient r obtained from the two-way frequency table is again defined as

$$r = Z/(XY)^{\frac{1}{2}}.$$

It is easily seen that if $f_{ij} = 1$ for $i = j$ and zero otherwise, then the above definition of r reduces to the simpler one in the case of ungrouped data, the sample size now being the smaller of ν_1 and ν_2.

The two-way frequency table may be presented schematically as in Table 16.1. The pairs $(x_i, f_{i.})$, for $i = 1, 2, ..., \nu_1$, define formally the

Table 16.1   Showing schematically a bivariate frequency distribution

Midpoints of x intervals	y_1	y_2	y_3	y_4	.	.	.	y_{ν_2}	Total
x_1	f_{11}	f_{12}	f_{13}	f_{14}	.	.	.	$f_{1\nu_2}$	$f_{1.}$
x_2	f_{21}	f_{22}	f_{23}	f_{24}	.	.	.	$f_{2\nu_2}$	$f_{2.}$
x_3	f_{31}	f_{32}	f_{33}	f_{34}	.	.	.	$f_{3\nu_2}$	$f_{3.}$
x_4	f_{41}	f_{42}	f_{43}	f_{44}	.	.	.	$f_{4\nu_2}$	$f_{4.}$
.	.	.	.	.	.	.	.	.	.
.	.	.	.	..	.	.	.	.	.
.	.	.	.	.	.	.	.	.	.
x_{ν_1}	$f_{\nu_1 1}$	$f_{\nu_1 2}$	$f_{\nu_1 3}$	$f_{\nu_1 4}$	.	.	.	$f_{\nu_1 \nu_2}$	$f_{\nu_1.}$
Total	$f_{.1}$	$f_{.2}$	$f_{.3}$	$f_{.4}$	.	.	.	$f_{.\nu_2}$	N

(Midpoints of y intervals)

observed frequency distribution of x irrespective of the y values, and the pairs $(y_j, f_{.j})$, for $j = 1, 2, ..., \nu_2$, similarly give the observed frequency distribution of y regardless of the x values. These distributions are usually referred to as the *marginal distributions* of x and y respectively because they are obtained from the margins of the two-way frequency table. In fact, these are the usual univariate frequency distributions obtained from the observed values of x and y.

Computational notes

It is to be noted that for any given two-way frequency distribution, many cell frequencies (f_{ij}) will be zero, and so the summation over the $\nu_1\nu_2$ cells of the table is, in general, simpler than that suggested by the formal scheme presented in Table 16.1. Given such a frequency table, the calculation of $\bar{x}$ and $\bar{y}$ is straightforward. Also,

$$X = \sum_{i=1}^{\nu_1} f_{i.}(x_i - \bar{x})^2 = \sum_{i=1}^{\nu_1} f_{i.}(x_i^2 - 2x_i\bar{x} + \bar{x}^2)$$

$$= \sum_{i=1}^{\nu_1} f_{i.}x_i^2 - N\bar{x}^2 = \sum_{i=1}^{\nu_1} f_{i.}x_i^2 - \left[\sum_{i=1}^{\nu_1} f_{i.}x_i\right]^2 \Big/ N.$$

Similarly, $\quad Y = \sum_{j=1}^{\nu_2} f_{.j}y_j^2 - \left[\sum_{j=1}^{\nu_2} f_{.j}y_j\right]^2 \Big/ N.$

Thus X and Y are obtained from the two marginal distributions. However, the evaluation of Z requires some care. We observe that

$$Z = \sum_{i=1}^{\nu_1}\sum_{j=1}^{\nu_2} f_{ij}(x_i - \bar{x})(y_j - \bar{y})$$

$$= \sum_{i=1}^{\nu_1}\sum_{j=1}^{\nu_2} f_{ij}(x_i - \bar{x})y_j - \bar{y}\sum_{i=1}^{\nu_1}\sum_{j=1}^{\nu_2} f_{ij}(x_i - \bar{x})$$

$$= \sum_{i=1}^{\nu_1}\sum_{j=1}^{\nu_2} f_{ij}(x_i - \bar{x})y_j - \bar{y}\sum_{i=1}^{\nu_1} f_{i.}(x_i - \bar{x})$$

$$= \sum_{i=1}^{\nu_1}\sum_{j=1}^{\nu_2} f_{ij}\,x_i y_j - \bar{x}\sum_{i=1}^{\nu_1}\sum_{j=1}^{\nu_2} f_{ij}y_j, \quad \text{since} \quad \sum_{i=1}^{\nu_1} f_{i.}(x_i - \bar{x}) \equiv 0,$$

$$= \sum_{i=1}^{\nu_1}\sum_{j=1}^{\nu_2} f_{ij}\,x_i y_j - \bar{x}\sum_{j=1}^{\nu_2} f_{.j}\,y_j$$

$$= \sum_{i=1}^{\nu_1}\sum_{j=1}^{\nu_2} f_{ij}\,x_i y_j - \left[\sum_{i=1}^{\nu_1} f_{i.}x_i\right]\sum_{j=1}^{\nu_2} f_{.j}\,y_j \Big/ N.$$

This expression for Z is in conformity with the earlier formula for the sample sum of products. The computations are done in two parts. Firstly, the correction factor is evaluated from the marginal totals obtained for determining the means $\bar{x}$ and $\bar{y}$. Secondly, the double sum

$$\sum_{i=1,j=1}^{\nu_1\,\nu_2} f_{ij}x_iy_j = \sum_{i=1}^{\nu_1}\left[x_i \times \sum_{j=1}^{\nu_2} f_{ij}y_j\right],$$

so that this part of the computations is carried out in two stages. The first stage gives for each *fixed* i the sum

$$\sum_{j=1}^{\nu_2} f_{ij}y_j,$$

which is then multiplied by x_i and the product summed for $1 \leqslant i \leqslant \nu_1$.

This computational procedure can be further simplified by using linear transformations on the x_i and the y_j. This is equivalent to a choice of working means for the x and y variables. Thus, if we set

$$x_i = a_1 + b_1 u_i \quad \text{and} \quad y_j = a_2 + b_2 v_j,$$

where a_1, a_2, b_1, b_2 are constants and the b's are non-zero, then for the means we have

$$\bar{x} = a_1 + b_1 \bar{u} \quad \text{and} \quad \bar{y} = a_2 + b_2 \bar{v}.$$

Hence $X = \sum_{i=1}^{\nu_1} f_{i.}(x_i - \bar{x})^2 = b_1^2 \sum_{i=1}^{\nu_1} f_{i.}(u_i - \bar{u})^2 = b_1^2 U$, say,

$Y = \sum_{j=1}^{\nu_2} f_{.j}(y_j - \bar{y})^2 = b_2^2 \sum_{j=1}^{\nu_2} f_{.j}(v_j - \bar{v})^2 = b_2^2 V$, say,

and $Z = \sum_{i=1}^{\nu_1} \sum_{j=1}^{\nu_2} f_{ij}(x_i - \bar{x})(y_j - \bar{y})$

$= b_1 b_2 \sum_{i=1}^{\nu_1} \sum_{j=1}^{\nu_2} f_{ij}(u_i - \bar{u})(v_j - \bar{v}) = b_1 b_2 W$,   say.

Therefore $r^2 = Z^2/XY = W^2/UV,$

which again proves the invariance of r when calculated from a two-way frequency table. Of course, the computation of U, V, and W is done in exactly the same way as that of the corresponding quantities X, Y, and Z.

Example 16.2: Correlation of men's sitting height and stature

We illustrate the computation of the product-moment correlation coefficient r from a two-way frequency table of 6,999 observations on the sitting height (x) and stature (y) in inches of men aged 40·5 years. The data and the basic calculations are presented in Table 16.2, overleaf. Here

$$\nu_1 = 14, \quad \nu_2 = 12, \quad \text{and} \quad N = 6999.$$

The transformations from x to u and from y to v are made by choosing the working means at the midpoints of the x class-interval 35– and the y class-interval 66– respectively. Hence

$$U = 15131 - \frac{(1217)^2}{6999} = 15131 - 211·61 = 14919·39,$$

$$V = 13552 - \frac{(2328)^2}{6999} = 13552 - 774·34 = 12777·66,$$

and $W = 10789 - \dfrac{1217 \times 2328}{6999} = 10789 - 404·80 = 10384·20.$

Therefore $r^2 = \dfrac{107831610}{190634893} = 0·565645,$

so that $r = 0·752094.$

The check used in Table 16.2 is based on the identity

$$\sum_{j=1}^{\nu_2} f_{.j} v_j \equiv \sum_{i=1}^{\nu_1} \left[\sum_{j=1}^{\nu_2} f_{ij} v_j \right],$$

and this should always be used.

Table 16.2 Showing the calculation of the correlation coefficient between sitting height and stature from a frequency table

Classmarks of sitting height intervals \ Classmarks of stature intervals	56–	58–	60–	62–	64–	66–	68–	70–	72–	74–	76–	78–	Total $(f_{i.})$	u_i	$f_{i.}u_i$	$f_{i.}u_i^2$	$\sum_{j=1}^{v_2} f_{ij}v_j$	$u_i \times \sum_{j=1}^{v_2} f_{ij}v_j$
28–	1			1	1								3	−7	−21	147	−8	56
29–	2						1						3	−6	−18	108	−9	54
30–		2	3		1		1						7	−5	−35	175	−17	85
31–		8	18	23	4	6	1	1					56	−4	−224	896	−133	532
32–		5	27	59	33	107	3						133	−3	−399	1197	−249	747
33–	1	2	28	194	274	529	14	10	2				622	−2	−1244	2488	−734	1468
34–		1	6	142	533	864	122	91	4	1			1347	−1	−1347	1347	−691	691
35–			1	31	364	426	604	307	42	3	1		1960	0	0	0	381	0
36–			1	3	63	67	802	377	128	8		1	1647	1	1647	1647	1482	1482
37–			1	2	2		294	114	101	20	2		878	2	1756	3512	1465	2930
38–			1	1	1	1	37	9	23	17	1		277	3	831	2493	652	1956
39–			1		1		7		1	3	1		60	4	240	960	163	652
40–	1												5	5	25	125	20	100
41–												1	1	6	6	36	6	36
Total $(f_{.j})$	4	18	86	454	1277	2001	1886	912	301	52	6	2	6999	Total	1217	15131	2328	10789
v_j	−5	−4	−3	−2	−1	0	1	2	3	4	5	6		Total				
$f_{.j}v_j$	−20	−72	−258	−908	−1277	0	1886	1824	903	208	30	12		2328				
$f_{.j}v_j^2$	100	288	774	1816	1277	0	1886	3648	2709	832	150	72		13552				

CHECK

Source: H. A. Ruger (1932 – 33), *Annals of Eugenics*, Vol. 5, p. 95.

16.6   Linear functions of correlated random variables

So far as practical work is concerned, the sample product-moment correlation coefficient r is of little use since, in general, linear regression analysis offers a better way for analysing the observed association between the measurements of two random variables. However, the concept of correlated variables is a fundamental idea in statistical theory and it leads to many important statistical methods. We now consider some theoretical results concerning linear functions of two or more correlated random variables. These results are of intrinsic importance and they also provide a deeper understanding of the concept of correlation.

Suppose x_1 and x_2 are two correlated random variables such that

$$E(x_1) = \theta_1, \quad E(x_2) = \theta_2,$$

$$\operatorname{var}(x_1) = \sigma_1^2, \quad \operatorname{var}(x_2) = \sigma_2^2, \quad \text{and} \quad \operatorname{cov}(x_1, x_2) = \rho_{12}\sigma_1\sigma_2,$$

where we now use the notation ρ_{12} to denote the product-moment correlation coefficient between x_1 and x_2. It is important to note that for the present we make no assumption about the form of the joint distribution of x_1 and x_2 apart from the existence of the moments stated above.

(i) Consider first the new random variable

$$u = x_1 + x_2.$$

Then
$$E(u) = E(x_1 + x_2) = E(x_1) + E(x_2) = \theta_1 + \theta_2.$$

This shows that the correlation of x_1 and x_2 does not affect the expectation of their sum. However, by definition, we have

$$
\begin{aligned}
\operatorname{var}(u) &= E[u - E(u)]^2 = E[(x_1 - \theta_1) + (x_2 - \theta_2)]^2 \\
&= E[(x_1 - \theta_1)^2 + (x_2 - \theta_2)^2 + 2(x_1 - \theta_1)(x_2 - \theta_2)] \\
&= \operatorname{var}(x_1) + \operatorname{var}(x_2) + 2\operatorname{cov}(x_1, x_2) \\
&= \sigma_1^2 + \sigma_2^2 + 2\rho_{12}\sigma_1\sigma_2.
\end{aligned}
$$

(ii) In exactly the same manner, we deduce that for the random variable

$$v = x_1 - x_2$$

$$E(v) = \theta_1 - \theta_2 \quad \text{and} \quad \operatorname{var}(v) = \sigma_1^2 + \sigma_2^2 - 2\rho_{12}\sigma_1\sigma_2.$$

(iii) Now u and v are two random variables which are linear functions of two other random variables x_1 and x_2 with given moments. Considered jointly, these random variables are, in general, correlated. Indeed, we have

$$
\begin{aligned}
\operatorname{cov}(u, v) &= E[\{u - E(u)\}\{v - E(v)\}] \\
&= E[\{(x_1 - \theta_1) + (x_2 - \theta_2)\}\{(x_1 - \theta_1) - (x_2 - \theta_2)\}] \\
&= E[(x_1 - \theta_1)^2 - (x_2 - \theta_2)^2] \\
&= \operatorname{var}(x_1) - \operatorname{var}(x_2) = \sigma_1^2 - \sigma_2^2.
\end{aligned}
$$

Hence we observe that u and v will be uncorrelated random variables only if x_1 and x_2 have equal variances. This result is of considerable importance and we shall have occasion to use it later.

(iv) The results derived in (i) to (iii) are particular cases of general ones in the distribution theory of linear functions of correlated random variables. We next consider some of these general results. Suppose $x_1, x_2, \ldots, x_n$ are n correlated random variables such that

$$E(x_i) = \theta_i, \quad \text{var}(x_i) = \sigma_i^2, \quad \text{for } i = 1, 2, \ldots, n,$$

and
$$\text{cov}(x_i, x_j) = \rho_{ij}\sigma_i\sigma_j, \quad \text{for } i \neq j = 1, 2, \ldots, n.$$

If we now consider a linear function

$$z_1 = \sum_{i=1}^{n} a_i x_i,$$

where the a_i are constants not all zero simultaneously, then z_1 is also a random variable. Furthermore, we have

$$E(z_1) = E\left[\sum_{i=1}^{n} a_i x_i\right] = \sum_{i=1}^{n} a_i E(x_i) = \sum_{i=1}^{n} a_i \theta_i.$$

Again, by definition, we have

$$\text{var}(z_1) = E[z_1 - E(z_1)]^2 = E\left[\sum_{i=1}^{n} a_i(x_i - \theta_i)\right]^2$$

$$= E\left[\sum_{i=1}^{n} a_i^2(x_i - \theta_i)^2 + \sum_{i=1}^{n}\sum_{j\neq i} a_i a_j(x_i - \theta_i)(x_j - \theta_j)\right]$$

$$= \sum_{i=1}^{n} a_i^2 \text{var}(x_i) + \sum_{i=1}^{n}\sum_{j\neq i} a_i a_j \text{cov}(x_i, x_j)$$

$$= \sum_{i=1}^{n} a_i^2 \sigma_i^2 + \sum_{i=1}^{n}\sum_{j\neq i} a_i a_j \rho_{ij}\sigma_i\sigma_j.$$

The formulae for $E(z_1)$ and $\text{var}(z_1)$ are of fundamental importance.

(v) As a particular case of these formulae, suppose $\sigma_i = \sigma$ and $a_i = a \neq 0$ for all i. Then we have

$$\text{var}(z_1) = na^2\sigma^2 + a^2\sigma^2 \sum_{i=1}^{n}\sum_{j\neq i} \rho_{ij}$$

$$= n(n - 1)a^2\sigma^2\left[\frac{1}{n - 1} + \sum_{i=1}^{n}\sum_{j\neq i} \frac{\rho_{ij}}{n(n - 1)}\right].$$

Now there are n variables and $\binom{n}{2}$ possible correlations ρ_{ij} amongst them. If we set $\bar{\rho}$ as the average of all these correlations, then

$$\sum_{i=1}^{n}\sum_{j\neq i} \rho_{ij}/n(n - 1) = \sum_{i=1}^{n}\sum_{j<i} \rho_{ij}/\binom{n}{2} = \bar{\rho}.$$

Hence
$$\text{var}(z_1) = n(n - 1)a^2\sigma^2\left[\frac{1}{n - 1} + \bar{\rho}\right];$$

and since $\text{var}(z_1) \geqslant 0$, we must have $\bar{\rho} \geqslant -1/(n - 1)$. This is a basic restriction on the correlations ρ_{ij}. It means that although any individual ρ_{ij} has the possible limits $-1 \leqslant \rho_{ij} \leqslant 1$, the $\binom{n}{2}$ correlations jointly satisfy the condition that their average $\bar{\rho} \geqslant -1/(n - 1)$.

If we further set $\rho_{ij} = \rho$, a constant, and $a_i = 1/n$, then the n random variables x_i become equicorrelated and have the same variance σ^2. Also, we now have $z_1 = \bar{x}$. Accordingly,

$$\text{var}(\bar{x}) = \sigma^2\left[n\frac{1}{n^2} + \rho n(n - 1)\frac{1}{n^2}\right] = \frac{\sigma^2}{n}\left[1 + (n - 1)\rho\right].$$

This result is of great importance as it shows the effect of correlation on the variance of the average of n equicorrelated random variables with a constant variance. Moreover, since $\text{var}(\bar{x}) \geqslant 0$, it follows that $\rho \geqslant -1/(n-1)$. This means that if n random variables have the same variance, and each pair of random variables has the same correlation ρ, then $\rho \geqslant -1/(n-1)$. This is a surprising but important result.

(vi) Another important general result is obtained by considering the covariance of two linear functions of the observations $x_1, x_2, \ldots, x_n$ as specified in (iv) above. Suppose z_1, as defined in (iv), is one linear function and let the other linear function be

$$z_2 = \sum_{i=1}^{n} b_i x_i,$$

where the b's are constants not all zero simultaneously. Evidently,

$$E(z_2) = \sum_{i=1}^{n} b_i \theta_i \quad \text{and} \quad \text{var}(z_2) = \sum_{i=1}^{n} b_i^2 \sigma_i^2 + \sum_{i=1}^{n} \sum_{j \neq i} b_i b_j \rho_{ij} \sigma_i \sigma_j.$$

Furthermore,

$$\begin{aligned}
\text{cov}(z_1, z_2) &= E[\{z_1 - E(z_1)\}\{z_2 - E(z_2)\}] \\
&= E\left[\left\{\sum_{i=1}^{n} a_i(x_i - \theta_i)\right\}\left\{\sum_{i=1}^{n} b_i(x_i - \theta_i)\right\}\right] \\
&= E\left[\sum_{i=1}^{n} a_i b_i(x_i - \theta_i)^2 + \sum_{i=1}^{n} \sum_{j \neq i} a_i b_j(x_i - \theta_i)(x_j - \theta_j)\right] \\
&= \sum_{i=1}^{n} a_i b_i \sigma_i^2 + \sum_{i=1}^{n} \sum_{j \neq i} a_i b_j \rho_{ij} \sigma_i \sigma_j.
\end{aligned}$$

Certain useful particular results may be derived from this general formula.

(vii) If $\sigma_i^2 = \sigma^2$ and $\rho_{ij} = \rho$, so that the x_i are equicorrelated random variables with a constant variance, then

$$\text{cov}(z_1, z_2) = \sigma^2\left[\sum_{i=1}^{n} a_i b_i + \rho \sum_{i=1}^{n} \sum_{j \neq i} a_i b_j\right].$$

(viii) Furthermore, suppose $\rho = 0$ so that the x_i are uncorrelated random variables with the same variance. Then

$$\text{cov}(z_1, z_2) = \sigma^2 \sum_{i=1}^{n} a_i b_i.$$

Also, $\qquad \text{var}(z_1) = \sigma^2 \sum_{i=1}^{n} a_i^2 \quad \text{and} \quad \text{var}(z_2) = \sigma^2 \sum_{i=1}^{n} b_i^2;$

these results are in agreement with our earlier formula for the variance of a linear function of n independent observations with equal variance. This is in consonance with the fact that independent observations are necessarily uncorrelated.

If the x_i are uncorrelated observations with equal variance, then the linear functions z_1 and z_2 will also be uncorrelated if

$$\sum_{i=1}^{n} a_i b_i = 0. \tag{1}$$

This is a fundamental result.

(ix) If, in addition to (1), the constants a_i and b_i are so chosen that they also satisfy the conditions

$$\sum_{i=1}^{n} a_i^2 = \sum_{i=1}^{n} b_i^2 = 1, \tag{2}$$

then $\text{var}(z_1) = \text{var}(z_2) = \sigma^2$ and $\text{cov}(z_1, z_2) = 0.$

In this case, z_1 and z_2 are known as *orthogonal linear functions* of the random variables $x_1, x_2, \ldots, x_n$.

(x) Finally, if the a_i and b_i satisfy (1), (2), and

$$\sum_{i=1}^{n} a_i = \sum_{i=1}^{n} b_i = 0, \tag{3}$$

then the linear functions z_1 and z_2 are known as *orthogonal linear contrasts*. As an example, suppose $n = 3$. Then it is not difficult to verify that

$$z_1 = \frac{x_1 + x_2 + x_3}{\sqrt{3}} \quad \text{and} \quad z_2 = \frac{x_1 + x_2 - 2x_3}{\sqrt{6}}$$

are orthogonal linear functions, though z_2 is also a contrast.

Some applications

We consider next a few examples to illustrate various points of the preceding theory.

Example 16.3

Suppose $x_1, x_2,$ and x_3 are uncorrelated random variables, each with zero mean and variance σ^2. Consider first the set of three linear functions

$$w_1 = x_1 + x_2 + x_3, \quad w_2 = x_1 - x_2, \quad \text{and} \quad w_3 = x_1 + x_2 - 2x_3.$$

Then, quite clearly, $E(w_1) = E(w_2) = E(w_3) = 0.$

Therefore

$$\begin{aligned}
\text{var}(w_1) &= E(x_1 + x_2 + x_3)^2 \\
&= E(x_1^2 + x_2^2 + x_3^2), \text{ as the } x_i \text{ are uncorrelated,} \\
&= \text{var}(x_1) + \text{var}(x_2) + \text{var}(x_3), \text{ as the } x_i \text{ have zero means,} \\
&= 3\sigma^2.
\end{aligned}$$

Similarly, $\text{var}(w_2) = \text{var}(x_1) + \text{var}(x_2) = 2\sigma^2,$

and $\text{var}(w_3) = \text{var}(x_1) + \text{var}(x_2) + 4\,\text{var}(x_3) = 6\sigma^2.$

Also, $\text{cov}(w_1, w_2) = E[(x_1 - x_2)(x_1 + x_2 + x_3)]$

$$= E[x_1^2 - x_2^2 + x_3(x_1 - x_2)] = \text{var}(x_1) - \text{var}(x_2) = 0.$$

Similarly,

$$\begin{aligned}
\text{cov}(w_2, w_3) &= E[(x_1 - x_2)(x_1 + x_2 - 2x_3)] \\
&= E(x_1^2 - x_2^2) = \text{var}(x_1) - \text{var}(x_2) = 0,
\end{aligned}$$

and

$$\begin{aligned}
\text{cov}(w_1, w_3) &= E[(x_1 + x_2 + x_3)(x_1 + x_2 - 2x_3)] \\
&= E(x_1^2 + x_2^2 - 2x_3^2) = \text{var}(x_1) + \text{var}(x_2) - 2\,\text{var}(x_3) \\
&= 0.
\end{aligned}$$

Thus $w_1, w_2,$ and w_3 form a set of three mutually uncorrelated linear functions.

Next, consider the linear functions

$$y_1 = x_1 + x_2, \quad y_2 = x_1 - x_2 + x_3, \quad \text{and} \quad y_3 = x_2 + x_3.$$

Here, again, $E(y_1) = E(y_2) = E(y_3) = 0;$

and it is also easily seen that

$$\text{var}(y_1) = 2\sigma^2, \quad \text{var}(y_2) = 3\sigma^2, \quad \text{and} \quad \text{var}(y_3) = 2\sigma^2.$$

Also,
$$\text{cov}(y_1, y_2) = E(x_1^2 - x_2^2) = 0,$$
$$\text{cov}(y_2, y_3) = E(x_3^2 - x_2^2) = 0,$$
but
$$\text{cov}(y_1, y_3) = E(x_2^2) = \sigma^2.$$

Thus, although y_1 and y_2 are uncorrelated, and so also are y_2 and y_3, nevertheless y_1 and y_3 are correlated. Therefore y_1, y_2, and y_3 do not constitute a set of mutually uncorrelated linear functions. This is an important point to note.

Example 16.4

As a somewhat more general illustration, suppose $x_1, x_2, \ldots, x_n$ are independent observations such that

$$E(x_i) = \theta_i \quad \text{and} \quad \text{var}(x_i) = \sigma^2, \quad \text{for} \quad i = 1, 2, \ldots, n.$$

Let $\bar{x}$ be the average of the n observations, and for any fixed j and k such that $1 \leqslant j, \ k \leqslant n$, we define two other linear functions $x_j - \bar{x}$ and $x_k - \bar{x}$.

Explicitly,
$$\bar{x} = \frac{1}{n} \sum_{i=1}^{n} x_i,$$

$$x_j - \bar{x} = -\frac{1}{n} \sum_{i \neq j \neq k} x_i + (1 - 1/n)x_j - \frac{1}{n}x_k,$$

and
$$x_k - \bar{x} = -\frac{1}{n} \sum_{i \neq j \neq k} x_i + (1 - 1/n)x_k - \frac{1}{n}x_j.$$

Hence, applying the results obtained in (viii) above, we have

$$\text{cov}(\bar{x}, x_j - \bar{x}) = \sigma^2(-1/n)(1/n)(n-1) + (1/n)(1-1/n)] = 0.$$

Therefore $\bar{x}$ and $x_j - \bar{x}$ are uncorrelated linear functions. This implies that $\bar{x}$ and $x_k - \bar{x}$ are also uncorrelated linear functions. However,

$$\text{cov}(x_j - \bar{x}, x_k - \bar{x}) = \sigma^2[(-1/n)^2(n-2) + 2(1-1/n)(-1/n)]$$
$$= -\sigma^2/n.$$

Also, $\text{var}(x_j - \bar{x}) = \text{var}(x_k - \bar{x})$
$$= \sigma^2[(-1/n)^2(n-1) + (1-1/n)^2] = \sigma^2(1-1/n).$$

Hence $\text{corr}(x_j - \bar{x}, x_k - \bar{x}) = -(\sigma^2/n)/\sigma^2(1-1/n) = -1/(n-1)$.
Thus the deviations from the sample mean of any two random observations are negatively correlated.

Example 16.5

Suppose x is a unit normal variable, and we define another random variable y by the relation
$$y = \alpha + \beta x + \gamma x^2,$$

where α, β, and γ are constants. Then

$$E(y) = \alpha + \beta E(x) + \gamma E(x^2) = \alpha + \gamma,$$

since x is unit normal. Again,

$$E(y - \alpha)^2 = E[x^2(\beta + \gamma x)^2]$$
$$= \beta^2 E(x^2) + \gamma^2 E(x^4), \quad \text{since} \quad E(x^3) = 0,$$
$$= \beta^2 + 3\gamma^2,$$

since $E(x^4) = 3$ for a unit normal variable. Therefore
$$\mathrm{var}(y) = \mathrm{var}(y - a), \quad \text{since } a \text{ is a constant,}$$
$$= E(y - a)^2 - E^2(y - a) = (\beta^2 + 3\gamma^2) - \gamma^2 = \beta^2 + 2\gamma^2.$$
Also,
$$\mathrm{cov}(x, y) = E(xy), \quad \text{since } E(x) = 0,$$
$$= E[x(a + \beta x + \gamma x^2)] = \beta.$$
Hence
$$\mathrm{corr}(x, y) = \beta/(\beta^2 + 2\gamma^2)^{\frac{1}{2}}.$$

Thus, for $\beta = 0$, we have $\mathrm{corr}(x, y) = 0$, whereas for $\gamma = 0$, $\mathrm{corr}(x, y) = \pm 1$ according as $\beta >$ or < 0.

Now, for $\beta = 0$, $y = a + \gamma x^2$ is an exact quadratic relationship between the random variables x and y, though $\mathrm{corr}(x, y) = 0$. On the other hand, for $\gamma = 0$, $y = a + \beta x$ is an exact linear relationship between the random variables x and y. This is appropriately reflected in the values $\mathrm{corr}(x, y) = \pm 1$ according as $\beta >$ or < 0. Thus the product-moment correlation coefficient is a meaningful measure of association between two random variables if there is a linear association between them. This coefficient is of no reliable use when strong non-linearity exists in the relation between the random variables, as is the case when $\beta = 0$. This reaffirms our earlier caution about the use of the product-moment correlation coefficient.

Example 16.6

For our concluding example, we go back to the evaluation of the expectation of the residual sum of squares in Example 11.7 of Chapter 11 (p. 296).

We recall that in this example
$$E(x_\nu) = \theta_1 + \nu\theta_2, \quad \mathrm{var}(x_\nu) = \sigma^2, \quad \text{for } \nu = 1, 2, ..., n.$$
The least-squares estimates θ_1^* and θ_2^* were shown to be the linear functions
$$\theta_1^* = \frac{2}{n(n-1)} \sum_{\nu=1}^{n} (2n + 1 - 3\nu) x_\nu, \qquad \theta_2^* = \frac{6}{n(n^2 - 1)} \sum_{\nu=1}^{n} (2\nu - n - 1) x_\nu,$$
with
$$\mathrm{var}(\theta_1^*) = \frac{2(2n + 1)\sigma^2}{n(n-1)} \quad \text{and} \quad \mathrm{var}(\theta_2^*) = \frac{12\sigma^2}{n(n^2 - 1)}.$$

Also, the residual sum of squares was
$$M = \sum_{\nu=1}^{n} (x_\nu - \theta_1^* - \nu\theta_2^*)^2.$$

To derive the expectation of M, it is convenient to obtain a different expression for it. To do this, we observe that the sum of the squares of the deviations of the observations from their expectation is
$$\sum_{\nu=1}^{n} (x_\nu - \theta_1 - \nu\theta_2)^2 = \sum_{\nu=1}^{n} [(x_\nu - \theta_1^* - \nu\theta_2^*) - (\theta_1^* - \theta_1) - \nu(\theta_2^* - \theta_2)]^2$$
$$= M + n(\theta_1^* - \theta_1)^2 + (\theta_2^* - \theta_2)^2 \sum_{\nu=1}^{n} \nu^2 +$$
$$+ 2(\theta_1^* - \theta_1)(\theta_2^* - \theta_2) \sum_{\nu=1}^{n} \nu.$$

the other two cross-product terms being zero because of the least-squares equations for θ_1^* and θ_2^*,

$$= M + n(\theta_1^* - \theta_1)^2 + \frac{n(n + 1)(2n + 1)}{6} (\theta_2^* - \theta_2)^2 +$$

$$+ n(n + 1)(\theta_1^* - \theta_1)(\theta_2^* - \theta_2),$$

whence

$$M = \sum_{\nu=1}^{n} (x_\nu - \theta_1 - \nu\theta_2)^2 - n(\theta_1^* - \theta_1)^2 - \frac{n(n + 1)(2n + 1)}{6} (\theta_2^* - \theta_2)^2 -$$

$$- n(n + 1)(\theta_1^* - \theta_1)(\theta_2^* - \theta_2).$$

Therefore

$$E(M) = \sum_{\nu=1}^{n} E(x_\nu - \theta_1 - \nu\theta_2)^2 - nE(\theta_1^* - \theta_1)^2 - \frac{n(n + 1)(2n + 1)}{6} E(\theta_2^* - \theta_2)^2 -$$

$$- n(n + 1) E[(\dot{\theta}_1^* - \theta_1)(\theta_2^* - \theta_2)]$$

$$= n\sigma^2 - n \operatorname{var}(\theta_1^*) - \frac{n(n + 1)(2n + 1)}{6} \operatorname{var}(\theta_2^*) - n(n + 1) \operatorname{cov}(\theta_1^*, \theta_2^*),$$

since

$$E(\theta_1^*) = \theta_1, \quad E(\theta_2^*) = \theta_2, \quad E(x_\nu) = \theta_1 + \nu\theta_2, \quad \text{and} \quad \operatorname{var}(x_\nu) = \sigma^2.$$

Next, substituting the values of the variances of θ_1^* and θ_2^*, we deduce after some simplification that

$$E(M) = \sigma^2 \left[n - \frac{4(2n + 1)}{n - 1} \right] - n(n + 1) \operatorname{cov}(\theta_1^*, \theta_2^*). \tag{1}$$

The main problem here is to determine $\operatorname{cov}(\theta_1^*, \theta_2^*)$. Now θ_1^* and θ_2^* are linear functions of independent random variables with equal variance. Hence, using (viii) above, we have

$$\operatorname{cov}(\theta_1^*, \theta_2^*) = \frac{12\sigma^2}{n^2 (n - 1)^2(n + 1)} \sum_{\nu=1}^{n} (2\nu - n - 1)(2n + 1 - 3\nu)$$

$$= \frac{12\sigma^2}{n^2(n - 1)^2(n + 1)} \sum_{\nu=1}^{n} [- (n + 1)(2n + 1) - 6\nu^2 + (7n + 5)\nu]$$

$$= \frac{12\sigma^2}{n^2(n - 1)^2(n + 1)} \left[- n(n + 1)(2n + 1) - 6 \frac{n(n + 1)(2n + 1)}{6} + \right.$$

$$\left. + (7n + 5)\frac{n(n + 1)}{2} \right]$$

$$= - 6\sigma^2 /n(n - 1), \text{ on reduction.}$$

Hence, from (1), we finally have

$$E(M) = \sigma^2 \left[n - \frac{4(2n + 1)}{n - 1} + \frac{6(n + 1)}{n - 1} \right] = (n - 2)\sigma^2,$$

the required result.

The main difficulty in the determination of this expectation is due to the fact that θ_1^* and θ_2^* are *correlated* least-squares estimates; and, indeed, it was to avoid a similar difficulty that we chose to define the linear regression of y on x as $E(y_\nu) = \alpha + \beta(x_\nu - \bar{x})$ and not as $E(y_\nu) = \alpha + \beta x_\nu$, where the symbols used here have their standard meaning in regression analysis. With the accepted specification of linear regression, we have obtained

$$\alpha^* = \frac{1}{n}\sum_{\nu=1}^{n} y_\nu; \quad \beta^* = \frac{Z}{X} = \sum_{\nu=1}^{n} [(x_\nu - \bar{x})/X]y_\nu.$$

Therefore, since the y_ν are independent observations with variance σ^2, we have from (viii) above

$$\text{cov}(a^*, \beta^*) \;=\; \sigma^2 \sum_{\nu=1}^{n} (1/n)\,[(x_\nu - \bar{x})/X] \;\equiv\; 0.$$

Thus a^* and β^* are uncorrelated linear functions of the y_ν, and under the assumption of normality of the y_ν this also ensures the independence of the estimates. This implies that it is possible to test null hypotheses about a and β independently within the same analysis of variance, as already indicated.

16.7 Tests of significance for null hypotheses about the means and variances of a bivariate normal population

The tests of significance for the comparison of the means and variances of two independent normal populations are of frequent application in the analysis of data. Thus, for example, we may wish to compare the average lengths of screws produced by two different machines, or to compare the variances of the lengths as measures of the consistency of the production performance of the machines. The fact that the screws are produced by two different machines would, in general, ensure the independence of the model normal populations. Occasionally, cases arise when this independence cannot reasonably be assumed and then the tests of significance cannot be carried out in the usual way. As an example, we may cite the data of Roberts and Griffiths pertaining to the Binet (x) and Otis (y) test scores obtained by 65 children, and quoted in Example 16.1. Here the two sets of scores are obviously related since each pair (x, y) refers to the achievement of one child on the two tests. Accordingly, we cannot postulate that these paired scores are from two independent populations. However, we may postulate legitimately an infinite bivariate population of such paired (x, y) observations from which the sample of 65 pairs has been obtained. It is then usual to assume that this population is bivariate normal such that

$$E(x) = \theta_1, \qquad E(y) = \theta_2;$$
$$\text{var}(x) = \sigma_1^2, \qquad \text{var}(y) = \sigma_2^2, \qquad \text{and} \qquad \text{cov}(x, y) = \rho\sigma_1\sigma_2.$$

With such a model, the two null hypotheses of particular interest are that $H(\theta_1 - \theta_2 = 0)$ and $H(\sigma_1^2 - \sigma_2^2 = 0)$. In the context of our example, the interpretation of the first null hypothesis is that in the bivariate population of the Binet and Otis test scores, the average scores of the children are the same. Similarly, the meaning of the second null hypothesis is that the variabilities of the children's performances on the two tests are the same. We now show how these two null hypotheses can be tested. The mathematical justification of these tests is based on two general results in bivariate normal distribution theory, which we first state without proof.

Consider the two random variables

$$u = x - y \qquad \text{and} \qquad v = x + y.$$

Then we know that $E(u) = \theta_1 - \theta_2, \qquad E(v) = \theta_1 + \theta_2,$

$$\text{var}(u) = \sigma_1^2 + \sigma_2^2 - 2\rho\sigma_1\sigma_2, \qquad \text{var}(v) = \sigma_1^2 + \sigma_2^2 + 2\rho\sigma_1\sigma_2,$$

and

$$\text{cov}(u, v) = \sigma_1^2 - \sigma_2^2.$$

These results hold generally, but if x and y have a bivariate normal distribution, then it is also known that

(i) u and v have a bivariate normal distribution with the stated moments; and

(ii) u has a normal distribution with the stated mean and variance.

Test of significance for the null hypothesis $H(\sigma_1^2 - \sigma_2^2 = 0)$

Since $\text{cov}(u, v) = \sigma_1^2 - \sigma_2^2$, it follows that the null hypothesis $H(\sigma_1^2 - \sigma_2^2 = 0)$ is equivalent to the null hypothesis $H[\text{corr}(u, v) = 0]$, where u and v have a bivariate normal distribution by (i) above. Next, suppose (x_ν, y_ν), for $\nu = 1, 2, ..., n$, is a random sample of n paired observations from the bivariate normal population of the random variables x and y. Then the pairs (u_ν, v_ν), where

$$u_\nu = x_\nu - y_\nu \quad \text{and} \quad v_\nu = x_\nu + y_\nu, \quad \text{for } \nu = 1, 2, ..., n,$$

may be interpreted as a random sample of n paired observations from the bivariate normal population of the random variables u and v. In standard notation, the sample means are

$$\bar{u} = \bar{x} - \bar{y} \quad \text{and} \quad \bar{v} = \bar{x} + \bar{y}.$$

Also, the sample sums of squares of the u_ν and v_ν are

$$U = \sum_{\nu=1}^{n} (u_\nu - \bar{u})^2, \quad V = \sum_{\nu=1}^{n} (v_\nu - \bar{v})^2,$$

and the sample sum of products is

$$W = \sum_{\nu=1}^{n} (u_\nu - \bar{u})(v_\nu - \bar{v}).$$

Hence the sample product-moment correlation coefficient between the u and v observations is

$$r_{uv} = W/(UV)^{\frac{1}{2}}.$$

The test for the null hypothesis $H[\text{corr}(u, v) = 0]$ is now obtained by using the fact that if the null hypothesis under test is true, then

$$F = (n-2)r_{uv}^2/(1 - r_{uv}^2)$$

has the F distribution with $1, n-2$ d.f. This test, due to E.J. Pitman, is interesting because it shows how a test for the equality of the variances can be made to depend upon a test for a population correlation to be zero.

For computation, we observe that

$$U = \sum_{\nu=1}^{n} (u_\nu - \bar{u})^2$$

$$= \sum_{\nu=1}^{n} [(x_\nu - \bar{x}) - (y_\nu - \bar{y})]^2$$

$$= \sum_{\nu=1}^{n} [(x_\nu - \bar{x})^2 + (y_\nu - \bar{y})^2 - 2(x_\nu - \bar{x})(y_\nu - \bar{y})]$$

$$= X + Y - 2Z,$$

where X, Y, and Z have their earlier meaning. Similarly,

$$V = X + Y + 2Z \quad \text{and} \quad W = X - Y.$$

Therefore, in terms of the original (x, y) observations, we have

$$r_{uv}^2 = \frac{(X - Y)^2}{(X + Y - 2Z)(X + Y + 2Z)}$$

$$= \frac{(X - Y)^2}{(X + Y)^2 - 4Z^2}$$

$$= \frac{(X - Y)^2}{(X - Y)^2 + 4(XY - Z^2)} = \frac{1}{1 + 4(XY - Z^2)/(X - Y)^2},$$

whence the square root gives r_{uv}. This is a convenient computational form for r_{uv}, and the sign of this correlation coefficient is the same as that of $W = X - Y$.

Test of significance for the null hypothesis $H(\theta_1 - \theta_2 = \theta_0)$
Since the random variable $u = x - y$ is a normal variable with mean $\theta_1 - \theta_2$ and variance $\sigma_1^2 + \sigma_2^2 - 2\rho\sigma_1\sigma_2$, therefore the least-squares estimate of $\theta_1 - \theta_2$ is $\bar{u}$, and that of var (u) is

$$s_u^2 = \sum_{\nu=1}^{n} (u_\nu - \bar{u})^2/(n-1) = (X + Y - 2Z)/(n-1).$$

Hence the test of significance for the null hypothesis $H(\theta_1 - \theta_2 = \theta_0)$, θ_0 preassigned, is obtained by using the fact that if the null hypothesis under test is true, then

$$t = \frac{\bar{u} - \theta_0}{s_u/\sqrt{n}}$$

has "Student's" distribution with $n - 1$ d.f. This t test is known as "Student's" *paired* t because it is based on the differences of the paired (x, y) observations. It is also to be noted that because the observations are paired, we have obtained an exact test of significance for the comparison of the means θ_1 and θ_2 even though the variances of x and y are not necessarily equal. The use of the t test in this way was originally presented by "Student" himself.

It is simpler computationally to evaluate s_u^2 from the differences $u_\nu = x_\nu - y_\nu$ if only the null hypothesis $H(\theta_1 - \theta_2 = \theta_0)$ is to be tested. However, if both the null hypotheses $H(\theta_1 - \theta_2 = \theta_0)$ and $H(\sigma_1^2 - \sigma_2^2 = 0)$ are to be tested, then it is somewhat more convenient to work in terms of the original observations and evaluate X, Y, and Z.

Example 16.7: Measurement of the intelligence of children
We illustrate the use of the above two tests of significance by an analysis of the data of Example 16.1. As our population model, the Binet (x) and Otis (y) test scores are assumed to have a bivariate normal distribution with the five parameters $\theta_1, \theta_2, \sigma_1^2, \sigma_2^2$, and ρ in the preceding notation.
For the observed sample of scores from this population, we have

$$\bar{x} = 97\cdot94, \qquad \bar{y} = 103\cdot68, \qquad n = 65;$$
$$X = 15173\cdot7538, \qquad Y = 83006\cdot2154, \qquad Z = 31441\cdot7077.$$

Therefore $\qquad 4(XY - Z^2)/(X - Y)^2 = 0\cdot2355$,
so that $\qquad r_{uv}^2 = 1/1\cdot2355 = 0\cdot8094$.

Hence, to test the null hypothesis $H(\sigma_1^2 - \sigma_2^2 = 0)$, we obtain

$$F = \frac{50 \cdot 9922}{0 \cdot 1906} = 267 \cdot 5 \quad \text{with 1, 63 d.f.}$$

The significance of this observed F is beyond question, and we conclude that the variances of the Binet and Otis test scores in the bivariate population are almost certainly not equal.

Again, $\bar{u} = -5 \cdot 74$, and since $U = 35296 \cdot 5538$, we deduce that

$$s_u^2 = 551 \cdot 5087, \quad \text{whence} \quad s_u = 23 \cdot 4842.$$

Hence, to test the null hypothesis $H(\theta_1 - \theta_2 = 0)$, we have

$$|t| = \frac{46 \cdot 2776}{23 \cdot 4842} = 1 \cdot 97 \quad \text{with 64 d.f.}$$

This value of t is not significant at the five per cent level, and so we infer that the means of the Binet and Otis scores in the bivariate population are not different.

In view of these two tests of significance, our conclusion is that though, on the average, the performances of the children on the two tests are the same, the performance on the Otis test is much more variable than that on the Binet test.

16.8 Linear functions of frequencies

The methods of correlation analysis can be extended to the case of correlated discrete random variables, and our purpose now is to consider this extension in the context of one general and useful model for correlated discrete random variables. As is usual, we present the derivation of the model distribution from a finite sampling experiment. Suppose a finite number of N identical balls are distributed randomly amongst k distinct cells in such a way that the probability of a ball being put in the ith cell is p_i, for $i = 1, 2, \ldots, k$, where $0 \leqslant p_i \leqslant 1$ and $\sum_{i=1}^{k} p_i = 1$. This implies that each ball is necessarily placed in some one of the k cells. We assume that the observed distribution of the N balls is $n_1, n_2, \ldots, n_k$ so that any n_i satisfies the inequality $0 \leqslant n_i \leqslant N$ subject to the condition that $\sum_{i=1}^{k} n_i = N$. Then, quite clearly, the n_i are k random variables having a multinomial distribution, and the probability of the observed distribution of the N balls is

$$N! \prod_{i=1}^{k} p_i^{n_i} / n_i! \tag{1}$$

Of course, the n_i are not independent random variables, since their sum is necessarily a constant N. We wish to determine first

$$E(n_i), \quad \text{var}(n_i), \quad \text{and} \quad \text{cov}(n_i, n_j), \quad \text{for} \quad i \neq j \quad \text{and} \quad 1 \leqslant i, j \leqslant k.$$

These moments can be evaluated from the joint distribution (1) of the n_i. However, we shall determine them by a simple and powerful indirect method which depends upon the use of what are known as *pseudo-random variables*.

Suppose $x_1, x_2, \ldots, x_N$ are N random variables associated with the N balls. For any fixed i, let x_r be a variable which can take the values 1 and 0 with probabilities p_i and $1 - p_i$ respectively. If the N balls are numbered serially and x_r is assumed associated with the rth ball, then the definition of x_r implies that it takes the value 1 if the rth ball is in the ith cell and the value 0 otherwise. It therefore follows that

$$n_i = \sum_{r=1}^{n} x_r, \tag{2}$$

as it is known that exactly n_i balls were observed in the ith cell so that n_i of the x_r must be unity. These x_r are the pseudo-random variables which we have introduced simply to express n_i as their sum. As we now show, it is easy to evaluate the moments of the x_r and thus to determine the moments of n_i. It follows immediately from (2) that

$$E(n_i) = E\left[\sum_{r=1}^{N} x_r\right] = \sum_{r=1}^{N} E(x_r).$$

Now, by the definition of the expectation of a discrete random variable, we have
$$E(x_r) = 1.p_i + 0.(1 - p_i) = p_i, \quad \text{so that} \quad E(n_i) = Np_i.$$
Again,

$$\text{var}(n_i) = \text{var}\left[\sum_{r=1}^{N} x_r\right] = \sum_{r=1}^{N} \text{var}(x_r) + \sum_{r=1}^{N} \sum_{s \neq r} \text{cov}(x_r, x_s). \tag{3}$$

Hence we need only evaluate $\text{var}(x_r)$ and $\text{cov}(x_r, x_s)$ to determine $\text{var}(n_i)$. But
$$\text{var}(x_r) = E(x_r^2) - E^2(x_r) = E(x_r^2) - p_i^2.$$
Also, by using the definition of the expectation of a function of a discrete random variable, we have

$$E(x_r^2) = 1^2.p_i + 0^2.(1 - p_i) = p_i,$$
so that $\quad\quad \text{var}(x_r) = p_i - p_i^2 = p_i(1 - p_i).$

Similarly, $\text{cov}(x_r, x_s) = E(x_r x_s) - E(x_r)E(x_s) = E(x_r x_s) - p_i^2.$

But the evaluation of $E(x_r x_s)$ requires some care. We observe that x_r and x_s are two random variables associated with the rth and sth balls respectively. By definition $x_r = 1$ if the rth ball is in the ith cell and zero otherwise; and, in the same way, $x_s = 1$ if the sth ball is in the ith cell and zero otherwise. Furthermore, as the distribution of the balls is random, the product $x_r x_s$ can have four possible values depending upon whether the rth and sth balls are or are not in the specified cells. Accordingly, we have

$$E(x_r x_s) = 1.1.p_i^2 + 1.0.p_i(1 - p_i) + 0.1.(1 - p_i)p_i + 0.0.(1 - p_i)^2$$
$$= p_i^2,$$
so that $\quad\quad\quad\quad \text{cov}(x_r, x_s) = 0.$

Hence, from (3), we have

$$\text{var}(n_i) = \sum_{r=1}^{N} \text{var}(x_r) = Np_i(1 - p_i).$$

As a check, we observe that for $k = 2$, $E(n_i)$ and $\text{var}(n_i)$ both reduce to the corresponding moments of a binomial distribution.

To evaluate $\text{cov}(n_i, n_j)$ for fixed but unequal values of i and j, we have to think in terms of the random number of balls in two specified cells. We extend the preceding argument and associate the pseudo-random variables

$y_1, y_2, \ldots, y_N$ with the N balls such that any y_m takes the values 1 and 0 with probabilities p_j and $1 - p_j$ respectively. Hence

$$n_j = \sum_{m=1}^{N} y_m, \quad E(n_j) = Np_j, \quad \text{and} \quad \text{var}(n_j) = Np_j(1 - p_j).$$

Furthermore,

$$\begin{aligned}
\text{cov}(n_i, n_j) &= E(n_i n_j) - E(n_i) E(n_j) \\
&= E\left[\left\{\sum_{r=1}^{N} x_r\right\}\left\{\sum_{m=1}^{N} y_m\right\}\right] - N^2 p_i p_j \\
&= E\left[\sum_{r=1}^{N} x_r y_r + \sum_{r=1}^{N} \sum_{m \neq r} x_r y_m\right] - N^2 p_i p_j \\
&= \sum_{r=1}^{N} E(x_r y_r) + \sum_{r=1}^{N} \sum_{m \neq r} E(x_r y_m) - N^2 p_i p_j.
\end{aligned}$$
(4)

But

$$E(x_r y_r) = 1.0.p_i(1 - p_j) + 0.1.(1 - p_i)p_j + 0.0.(1 - p_i)(1 - p_j) = 0,$$

as the rth ball cannot be simultaneously in the ith and jth cells. On the other hand, for $r \neq m$, x_r and y_m refer to two different balls in relation to different cells, and so

$$\begin{aligned}
E(x_r y_m) &= 1.1.p_i p_j + 1.0.p_i(1 - p_j) + 0.1.(1 - p_i)p_j + 0.0.(1 - p_i)(1 - p_j) \\
&= p_i p_j.
\end{aligned}$$

Hence, from (4), we have

$$\text{cov}(n_i, n_j) = \sum_{r=1}^{N} \sum_{m \neq r} p_i p_j - N^2 p_i p_j = N(N - 1) p_i p_j - N^2 p_i p_j = -Np_i p_j.$$

We have thus established the three fundamental results for the random variables n_i, namely,

$$E(n_i) = Np_i; \quad \text{var}(n_i) = Np_i(1 - p_i); \quad \text{cov}(n_i, n_j) = -Np_i p_j,$$

for $i \neq j$. It is intuitively reasonable that the correlation between n_i and n_j should be negative because, as N is fixed, an increase in the number of balls in any one cell will. on the average, tend to decrease the number found in any other cell. and vice versa.

Finally, it is useful for later application to extend the above results to linear functions of the correlated random variables n_i. Suppose, then, that

$$L_1 = \sum_{i=1}^{k} a_i n_i \quad \text{and} \quad L_2 = \sum_{i=1}^{k} b_i n_i$$

are two linear functions of the n_i such that the a_i and b_i are fixed sets of constants not simultaneously zero. Then, using the earlier general results for linear functions of correlated random variables, we have

$$E(L_1) = \sum_{i=1}^{k} a_i E(n_i) = N \sum_{i=1}^{k} a_i p_i;$$

and

$$\begin{aligned}
\text{var}(L_1) &= \text{var}\left[\sum_{i=1}^{k} a_i n_i\right] \\
&= \sum_{i=1}^{k} a_i^2 \text{var}(n_i) + \sum_{i=1}^{k} \sum_{j \neq i} a_i a_j \text{cov}(n_i, n_j) \\
&= \sum_{i=1}^{k} a_i^2 Np_i(1 - p_i) - \sum_{i=1}^{k} \sum_{j \neq i} a_i a_j Np_i p_j
\end{aligned}$$

$$= N\left[\sum_{i=1}^{k} a_i^2 p_i - \left\{\sum_{i=1}^{k} a_i^2 p_i^2 + \sum_{i=1}^{k}\sum_{j\neq i} a_i a_j p_i p_j\right\}\right]$$

$$= N\left[\sum_{i=1}^{k} a_i^2 p_i - \left\{\sum_{i=1}^{k} a_i p_i\right\}^2\right]$$

Similarly,

$$E(L_2) = N\sum_{i=1}^{k} b_i p_i; \quad \text{var}(L_2) = N\left[\sum_{i=1}^{k} b_i^2 p_i - \left\{\sum_{i=1}^{k} b_i p_i\right\}^2\right].$$

Also,

$$\text{cov}(L_1, L_2) = \sum_{i=1}^{k} a_i b_i \,\text{var}(n_i) + \sum_{i=1}^{k}\sum_{j\neq i} a_i b_j \,\text{cov}(n_i, n_j)$$

$$= N\left[\sum_{i=1}^{k} a_i b_i p_i(1 - p_i) - \sum_{i=1}^{k}\sum_{j\neq i} a_i b_j p_i p_j\right]$$

$$= N\left[\sum_{i=1}^{k} a_i b_i p_i - \left\{\sum_{i=1}^{k} a_i b_i p_i^2 + \sum_{i=1}^{k}\sum_{j\neq i} a_i b_j p_i p_j\right\}\right]$$

$$= N\left[\sum_{i=1}^{k} a_i b_i p_i - \left\{\sum_{i=1}^{k} a_i p_i\right\}\left\{\sum_{i=1}^{k} b_i p_i\right\}\right].$$

In particular, if $E(L_1) = E(L_2) = 0$, then

$$\text{var}(L_1) = N\sum_{i=1}^{k} a_i^2 p_i; \quad \text{var}(L_2) = N\sum_{i=1}^{k} b_i^2 p_i;$$

and

$$\text{cov}(L_1, L_2) = N\sum_{i=1}^{k} a_i b_i p_i.$$

These are important results and we shall use them in the next chapter.

16.9 Rank correlation

We have used the sample product-moment correlation coefficient r as a measure of the strength of association between the paired observations (x_ν, y_ν), for $\nu = 1, 2, \ldots, n$. In general, these observations are assumed to be measurements of the correlated random variables x and y which have a bivariate normal distribution. It is known from empirical evidence that this distributional assumption is reasonable when we are dealing with measurable characteristics such as height, weight, age, etc., that is, quantitative variables which have a natural scale of measurement. However, there are cases when it is either not possible or not considered worth while to measure certain variables since it is simpler to arrange in order the individuals possessing the characteristics. For example, qualitative variables such as sociability, honesty, and human intelligence are difficult to measure satisfactorily in the way height or age can be measured. Nevertheless, it is usually possible to arrange a group of n individuals in an order with respect to a qualitative characteristic without assigning any specific measurements. In the same way, it is also possible to order a group of n individuals with respect to a quantitative variable without measuring the variable. Thus n persons can be ordered simply with respect to height without measuring their heights.

Formally, an ordered arrangement of n objects according to some characteristic is called a *ranking*, and the order given to an object is called its *rank*. Thus with n objects arranged in order there will be one rank corresponding to each of the n integers from 1 to n, provided that no two or more objects are given the same rank. When two or more objects are given the same rank, there is said to be a *tie*. We first consider the case when there are no ties so that each object in the ranking has a unique ordinal position.

The ranking of objects with respect to some characteristic is a simple concept which precedes the notion of measurement. For example, it is relatively straightforward to arrange a number of shades of red in an order from the darkest to the lightest, though the measurement of redness is more difficult. In the same way, a classic method of assessing the hardness of minerals is by a rank order. A mineral X is said to be harder than another mineral Y if a piece of X scratches a piece of Y when the two pieces are rubbed together. If X scratches Y and Y scratches another mineral Z, then the proper order of hardness of the minerals is X, Y, and Z, which could be assigned the ordinal ranking $1, 2, 3$. This simple ordering does not give any measure of the hardness of the three minerals or even of their true relative differences in hardness; nevertheless the ranking does convey a certain amount of information about the hardness of the minerals. Now, in so far as it is possible to measure intensity of colour or hardness of minerals, the corresponding idea of ranking colour shades or mineral hardness seems of little consequence. However, many social and psychological traits cannot yet be measured in a satisfactory manner because they have no natural scale of measurement. Moreover, all subjective assessment of traits such as beauty, palatability, and pleasure suffers from the same limitation; and it is therefore natural to think of ranking such traits in preference to measuring them by using an imposed artificial scale of measurement.

In general, if the same n objects are ranked in two ways, say by different judges or according to different characteristics, then the degree of agreement or *concordance* between the two rankings may be of interest. For example, two judges may be asked to rank in order of preference n paintings in an exhibition, and the measure of concordance between the two rankings would then indicate how far the two judges were in agreement in their personal assessments. Similarly, a group of n children may be ranked with respect to musical and mathematical abilities, and the measure of concordance between the two rankings would indicate the degree of association between the two traits. A particular measure of concordance based on two rankings is known as the *coefficient of rank correlation*. This was proposed by the psychometrician C. Spearman in 1904.

Spearman's coefficient of rank correlation

Suppose a set of n objects A_1, A_2, ..., A_n is given the rankings x_1, x_2, ..., x_n by one criterion and y_1, y_2, ..., y_n by another, there being no tied ranks. Since there are no tied ranks, each ranking is some ordering of the first n natural numbers. Therefore

$$\sum_{\nu=1}^{n} x_\nu = \sum_{\nu=1}^{n} y_\nu = n(n + 1)/2; \quad \text{and}$$

$$\sum_{\nu=1}^{n} x_\nu^2 = \sum_{\nu=1}^{n} y_\nu^2 = n(n + 1)(2n + 1)/6.$$

If we denote the difference of the ranks given to A_ν as $d_\nu = x_\nu - y_\nu$, then

Spearman's coefficient of rank correlation for untied ranks is defined as

$$R = 1 - 6 \sum_{\nu=1}^{n} d_{\nu}^2 / n(n^2 - 1).$$

We shall see the reason for this definition a little later, but we first show that R satisfies the obvious requirement for a correlation coefficient, that is, $-1 \leqslant R \leqslant 1$. Clearly, if there is perfect concordance between the two rankings, each d_{ν} must be zero so that $R = 1$. On the other hand, for maximum discordance between the two rankings $R = -1$. To prove this, we observe that $\sum_{\nu=1}^{n} d_{\nu}^2$ will be a maximum when one ranking is in its natural order and the other is in the reverse order. Accordingly, for maximum discordance, the rank ν of the first ranking corresponds to the rank $n - \nu + 1$ of the second ranking. We thus have

$$\max \sum_{\nu=1}^{n} d_{\nu}^2 = \sum_{\nu=1}^{n} [\nu - (n - \nu + 1)]^2$$

$$= \sum_{\nu=1}^{n} [4\nu^2 - 4(n + 1)\nu + (n + 1)^2]$$

$$= 4 \frac{n(n + 1)(2n + 1)}{6} - 4(n + 1) \frac{n(n + 1)}{2} + n(n + 1)^2 = \frac{n(n^2 - 1)}{3}.$$

Hence for maximum discordance $R = -1$. In general, we have $-1 < R < 1$, so that R does satisfy the first requirement for a correlation coefficient.

A further insight into the definition of R is obtained by noting that Spearman's coefficient is, in fact, the product-moment correlation coefficient between the pairs (x_{ν}, y_{ν}), assuming that these ranks were ordinary measurements. To prove this, we observe that the means of the x_{ν} and y_{ν} are

$$\bar{x} = \bar{y} = (n + 1)/2,$$

and the product-moment correlation coefficient between the two rankings is

$$r = \frac{\sum_{\nu=1}^{n} (x_{\nu} - \bar{x})(y_{\nu} - \bar{y})}{\left[\sum_{\nu=1}^{n} (x_{\nu} - \bar{x})^2 \sum_{\nu=1}^{n} (y_{\nu} - \bar{y})^2\right]^{\frac{1}{2}}}.$$

But, since the rankings are permutations of the first n natural numbers, we have

$$\sum_{\nu=1}^{n} (y_{\nu} - \bar{y})^2 = \sum_{\nu=1}^{n} (x_{\nu} - \bar{x})^2$$

$$= \sum_{\nu=1}^{n} x_{\nu}^2 - \left[\sum_{\nu=1}^{n} x_{\nu}\right]^2 \bigg/ n$$

$$= \frac{n(n + 1)(2n + 1)}{6} - \frac{n(n + 1)^2}{4} = \frac{n(n^2 - 1)}{12}.$$

Also,

$$\sum_{\nu=1}^{n} d_{\nu}^2 = \sum_{\nu=1}^{n} (x_{\nu} - y_{\nu})^2$$

$$= \sum_{\nu=1}^{n} (x_{\nu}^2 + y_{\nu}^2 - 2x_{\nu}y_{\nu}) = 2 \frac{n(n + 1)(2n + 1)}{6} - 2 \sum_{\nu=}^{n} x_{\nu}y_{\nu},$$

so that
$$\sum_{\nu=1}^{n} x_\nu y_\nu = \frac{n(n+1)(2n+1)}{3} - \frac{1}{2}\sum_{\nu=1}^{n} d_\nu^2 . \tag{1}$$

Hence

$$\sum_{\nu=1}^{n}(x_\nu - \bar{x})(y_\nu - \bar{y}) = \sum_{\nu=1}^{n} x_\nu y_\nu - \left[\sum_{\nu=1}^{n} x_\nu\right]\left[\sum_{\nu=1}^{n} y_\nu\right]\bigg/n$$

$$= \frac{n(n+1)(2n+1)}{3} - \frac{1}{2}\sum_{\nu=1}^{n} d_\nu^2 - \frac{n(n+1)^2}{4}, \text{ by (1)}$$

$$= \frac{n(n^2-1)}{12} - \frac{1}{2}\sum_{\nu=1}^{n} d_\nu^2 .$$

Substitution for the expressions in the numerator and denominator of the defining formula for r now immediately shows that it is equal to R, as stated. This numerical equivalence of r and R should not blur the essential difference between the two coefficients of correlation. The important point is that no distributional assumption is made regarding the x_ν and the y_ν when they are correctly interpreted as ranks. In contrast, the proper use of the product-moment correlation coefficient r rests on the assumption that the pairs (x_ν, y_ν) are random observations from a bivariate normal population. Indeed, it is this freedom from any such population assumption that gives to the rank correlation coefficient R its importance in statistical theory.

Tied ranks

We can use the algebraic equivalence of the product-moment and rank correlation coefficients to define the latter for the case when there are tied ranks in one or both rankings. In general, if k objects receive the same rank in a ranking, then the k objects are said to be tied. Thus, if the mth, $(m+1)$th, ..., $(m+k-1)$th objects are tied, then each of the k objects is given the average rank

$$\sum_{i=0}^{k-1}(m+i)/k.$$

One or more ties, each involving two or more objects, can occur in a ranking; and our purpose is to indicate next how the rank correlation coefficient may be calculated in such cases. It is clear that when there are tied ranks, the ranking is no longer an ordering of the first n natural numbers and, consequently, Spearman's simple formula is inapplicable. However, we can calculate the rank correlation coefficient between two rankings of which one or both have tied ranks by using the product-moment formula with the suitably adjusted tied ranks. It is to be noted that the adjustment for tied ranks ensures that the sum of the ranks in each of the two rankings remains $n(n+1)/2$; but, in general,

$$\sum_{\nu=1}^{n}(x_\nu - \bar{x})^2 \neq \sum_{\nu=1}^{n}(y_\nu - \bar{y})^2 \neq n(n^2-1)/12,$$

if x_ν and y_ν, for $\nu = 1, 2, ..., n$, now denote the ranks adjusted for ties in the two rankings. The rank correlation coefficient is calculated simply by using the formula for the product-moment correlation coefficient with the adjusted ranks.

Significance of R

Whether there are ties in the rankings or not, the rank correlation co-efficient R is a measure of the concordance between the two rankings. As in the case of the product-moment correlation coefficient r, the main reason for evaluating R is to determine whether there is any genuine concordance between the two rankings. Thus, if R measures the observed concordance between the rankings given to n paintings by two judges, then we may wish to test whether or not R is sufficiently small to indicate that there is, in fact, no agreement between the assessments of the paintings by the judges.

It is known that for $n > 10$, the rank correlation coefficient R has a distribution closely approximating that of r when the population product-moment correlation coefficient ρ is zero. Accordingly, an approximate but useful test of significance for the null hypothesis that there is no concordance between the two rankings is obtained by using the fact that if the null hypothesis under test is true, then

$$F = (n - 2)R^2/(1 - R^2)$$

has the F distribution with 1, $n - 2$ d.f.

Example 16.8: Assessment of sociability in children

Two schoolmasters A and B gave the following rankings to ten children $X_1, X_2, ..., X_{10}$ for sociability in the school.

Table 16.3 Showing the rankings of ten children for sociability by two schoolmasters

Children	X_1	X_2	X_3	X_4	X_5	X_6	X_7	X_8	X_9	X_{10}
Ranking of A	1	2·5	2·5	4	5	6	8	8	8	10
Ranking of B	2	3	1	6	7	5	4	8	10	9

We observe that A has tied X_2 and X_3, and also separately X_7, X_8, and X_9. The ranking of B has no ties. If we denote the ranks of A by x_ν and those of B by y_ν ($\nu = 1, 2, ..., 10$), then

$$\sum_{\nu=1}^{10} x_\nu = \sum_{\nu=1}^{10} y_\nu = 55; \quad \sum_{\nu=1}^{10} x_\nu^2 = 382\cdot5; \quad \sum_{\nu=1}^{10} y_\nu^2 = 385; \quad \sum_{\nu=1}^{10} x_\nu y_\nu = 367.$$

Also, since $\left[\sum_{\nu=1}^{10} x_\nu\right]^2/10 = 302\cdot5$, we have

$$\sum_{\nu=1}^{10} (x_\nu - \bar{x})^2 = 80\cdot0; \quad \sum_{\nu=1}^{10} (y_\nu - \bar{y})^2 = 82\cdot5; \quad \sum_{\nu=1}^{10} (x_\nu - \bar{x})(y_\nu - \bar{y}) = 64\cdot5.$$

Hence
$$R = \frac{64\cdot5}{\sqrt{6600}} = \frac{64\cdot5}{81\cdot24} = 0\cdot7939.$$

To test the significance of R, we have

$$F = \frac{5\cdot0422}{0\cdot3697} = 13\cdot64 \quad \text{with } 1, 8 \text{ d.f.}$$

The one per cent table value of F with 1, 8 d.f. is 11·26, and so there is no doubt about the high significance of R. We therefore conclude that there is a large measure of agreement between A and B regarding the ranking of the children for sociability. This agreement does not imply that the school-masters are right in their rankings in the sense that we still do not know the correct ordering of the children for sociability.

Example 16.9: Measurement of colour discrimination

We can extend this measurement of concordance to the case when there is a known objective order. Suppose, in a test of colour discrimination, 20 discs of varying shades of red were presented to a subject who made the following ranking.

Table 16.4 Showing the true order and the ranking by a subject of 20 discs of varying shades of red

True order (x_ν)	1	2	3	4	5	6	7	8	9	10	11	12	13	14	15	16	17	18	19	20
Subject's ranking (y_ν)	2	4	1	16	7	3	10	11	19	13	18	20	5	6	12	17	14	15	9	8
d_ν	−1	−2	2	−12	−2	3	−3	−3	−10	−3	−7	−8	8	8	3	−1	3	3	10	12

Here there are no ties in either of the rankings, and we have

$$\sum_{\nu=1}^{20} d_\nu^2 = 806.$$

Therefore, by Spearman's formula, we have

$$R = 1 - \frac{6 \times 806}{7980} = \frac{3144}{7980} = 0 \cdot 3940.$$

To test the significance of R, we have

$$F = \frac{2 \cdot 7942}{0 \cdot 8448} = 3 \cdot 31 \quad \text{with } 1, 18 \text{ d.f.}$$

The five per cent table value of F with $1, 18$ d.f. is $4 \cdot 41$ so that the observed F is not significant. We therefore conclude that the experimental evidence does not suggest that the subject has any marked sense of colour discrimination.

Example 16.10: Ranking of paintings

The ranking given by a judge to N paintings $A_1, A_2, ..., A_N$ is $1, 2, ..., N$. A second judge is in general agreement with this ordering, but he divides his ranking into k tied groups of $n_1, n_2, ..., n_k$ paintings, where

$$\sum_{i=1}^{k} n_i = N.$$

Thus the paintings $A_1, A_2, ..., A_{n_1}$ receive the same rank; the next A_{n_1+1}, $A_{n_1+2}, ..., A_{n_1+n_2}$ are tied; and so on till the kth group of n_k paintings have the same rank. These k ranks are in an increasing numerical order.

If $\sum_{i=1}^{s} n_i = t_s$, for $1 \leqslant s \leqslant k$, prove that the coefficient of rank correlation between the two rankings is

$$R = \left[3 \sum_{i=2}^{k} n_i\, t_i\, t_{i-1} \Big/ N(N^2 - 1) \right]^{\frac{1}{2}}.$$

Verify that for $N = 10$, and $n_1 = 2,\ n_2 = 4,\ n_3 = 3,\ n_4 = 1$,

$$R = (10/11)^{\frac{1}{2}} = 0{\cdot}9535.$$

Since the first ranking is $1, 2, \ldots, N$, therefore the sum of its ranks is $N(N + 1)/2$. The sum of the squares of these ranks is

$$\frac{N(N + 1)(2N + 1)}{6} - \frac{N(N + 1)^2}{4} = \frac{N(N^2 - 1)}{12}.$$

The sum of the ranks of the second ranking is again $N(N + 1)/2$. Also, the first n_1 paintings are tied with rank $(t_1 + 1)/2$, since $t_1 = n_1$. The next n_2 paintings are tied with rank

$$\sum_{x=n_1+1}^{n_1+n_2} x/n_2 = \frac{1}{2}(n_1 + n_2)(n_1 + n_2 + 1) - \frac{1}{2}n_1(n_1 + 1)$$

$$= \frac{1}{2}[(n_1 + n_2) + n_1 + 1] = \frac{1}{2}(t_2 + t_1 + 1).$$

Similarly, the next n_3 paintings are tied with rank $\frac{1}{2}(t_3 + t_2 + 1)$, and so on till the last n_k paintings are tied with rank $\frac{1}{2}(t_k + t_{k-1} + 1)$. Hence the sum of the squares of the ranks for the second ranking is

$$= \frac{1}{4}[t_1(t_1 + 1)^2 + (t_2 - t_1)(t_2 + t_1 + 1)^2 + \ldots + (t_k - t_{k-1})(t_k + t_{k-1} + 1)^2 - $$

$$- N(N + 1)^2]$$

$$= \frac{1}{4}[(t_1^3 + 2t_1^2 + t_1) + \{(t_2^3 - t_1^3) + 2(t_2^2 - t_1^2) + (t_2 - t_1) + t_2 t_1(t_2 - t_1)\} + \ldots +$$

$$+ \{(t_k^3 - t_{k-1}^3) + 2(t_k^2 - t_{k-1}^2) + (t_k - t_{k-1}) + t_k t_{k-1}(t_k - t_{k-1})\} - N(N + 1)^2]$$

$$= \frac{1}{4}\left[(t_k^3 + 2t_k^2 + t_k) + \sum_{j=1}^{k-1} t_j\, t_{j+1}(t_{j+1} - t_j) - N(N + 1)^2\right]$$

$$= \frac{1}{4}\sum_{j=1}^{k-1} t_j\, t_{j+1}\, n_{j+1} = \frac{1}{4}\sum_{i=2}^{k} n_i\, t_i\, t_{i-1}, \quad \text{since } t_k = N.$$

Again, it is easily verified that this is also the sum of the products of the two rankings. Therefore, using the product-moment formula, we have

$$R = \frac{1}{4}\sum_{i=2}^{k} n_i\, t_i\, t_{i-1} \Big/ \left[\frac{N(N^2 - 1)}{12} \frac{1}{4}\sum_{i=2}^{k} n_i\, t_i\, t_{i-1}\right]^{\frac{1}{2}}$$

$$= \left[3 \sum_{i=2}^{k} n_i\, t_i\, t_{i-1} \Big/ N(N^2 - 1)\right]^{\frac{1}{2}}.$$

For $N = 10$, $n_1 = 2$, $n_2 = 4$, $n_3 = 3$, $n_4 = 1$, we have
$$t_1 = 2, \ t_2 = 6, \ t_3 = 9, \ t_4 = 10.$$

Therefore
$$t_1 t_2 n_2 + t_2 t_3 n_3 + t_3 t_4 n_4 = 48 + 162 + 90 = 300,$$

so that
$$R = \left[\frac{3 \times 300}{990}\right]^{\frac{1}{2}} = \left[\frac{10}{11}\right]^{\frac{1}{2}} = \sqrt{0 \cdot 9091} = 0 \cdot 9535.$$

CHAPTER 17

APPLICATIONS OF THE χ^2 DISTRIBUTION

17.1 Goodness of fit

In this final chapter, we consider first a general statistical problem whose approximate solution is based on the χ^2 distribution. This problem is concerned with an empirical justification of the theoretical model which, in any particular case, provides the basis for the statistical analysis of the data. In the preceding chapters, we have often assumed that a given random sample of observations was from a population which could be adequately represented by some probability distribution such as the binomial, normal, or Poisson. From a purely mathematical point of view, an assumption of any theoretical model is adequate; but, in order to develop a practically useful methodology, it is essential to justify the model as reasonably representative of the observable random phenomenon under study. Indeed, the tests of significance and the associated techniques of statistical analysis are only valid for practical work if this correspondence between the mathematically defined model and the data of finite experience can be established. However, in view of the finiteness of all sample information, this correspondence can never be complete, and we have to make allowance for the random variations of the sample observations themselves. We are thus led intuitively to the notion of a test of significance for assessing the agreement between the sample data and their expectation based on some postulated model. Such a test of significance is known as a *test of goodness of fit*, and we shall see that under certain mild restrictions it can be carried out by using the χ^2 distribution. Such tests were first proposed by Karl Pearson.

17.2 Goodness of fit of a binomial distribution

A simple approach to goodness of fit is obtained by the reconsideration of an elementary problem of coin tossing. Suppose a penny is tossed 200 times and it turns up heads 150 times. If the penny were, in fact, unbiased, then we would expect an equal number of heads and tails in the 200 trials. Of course, we know that this expectation would hardly ever be realised even if the penny were, indeed, unbiased. The statistical problem is to test whether the difference between observation and expectation can be reasonably attributed to chance, or whether it indicates that the penny might be biased. To formulate a test of significance, we assume that the 200 trials are Bernoullian in character with a constant probability p of obtain-

482

ing a head at any one trial. Then the maximum-likelihood estimate of p is

$$\hat{p} = \frac{150}{200} = 0.75,$$

and $\qquad$ s.d.$(\hat{p}) = \left[\frac{p(1-p)}{200}\right]^{\frac{1}{2}} = \frac{1}{\sqrt{800}} = \frac{1}{28.2843} = 0.03536,$

if the null hypothesis $H(p = 0.5)$ is true. Hence the usual large-sample test of significance for this null hypothesis is obtained by using

$$z = \frac{\hat{p} - 0.5}{\text{s.d.}(\hat{p})}$$

as a unit normal variable; and for the specified values of $\hat{p}$ and s.d.$(\hat{p})$ we have

$$z = 0.25 \times 28.2843 = 7.071.$$

This value of z is so large that the null hypothesis under test can be rejected quite decisively. In other words, the observed relative frequency of heads almost certainly indicates that the penny could not be unbiased.

Since z is a unit normal variable, an equivalent method for carrying out this test of significance is to use z^2 as χ^2 with 1 d.f., that is,

$$\chi^2 = (\hat{p} - 0.5)^2/\text{var}(\hat{p}) \quad \text{with 1 d.f.}$$

If the null hypothesis $H(p = 0.5)$ is true, then the observed value of χ^2 is easily seen to be 50. For a single degree of freedom this value of χ^2 is highly significant, and the null hypothesis under test is again rejected decisively. The interest of this χ^2 test lies in the fact that it can be derived in another symmetrical way, which leads to an easy but important generalisation. To obtain this alternative derivation of χ^2, it is necessary first to express the data in the form of the following table and then to use the discrepancies between the observed and expected frequencies for heads and tails separately.

Table 17.1 Showing the observed and expected frequencies of heads and tails in 200 trials with an unbiased penny

Frequency	Heads	Tails
Observed	150	50
Expected	100	100

In general, if n_1 and n_2 are the observed, and Np and $N(1-p)$ the expected frequencies of heads and tails respectively such that $n_1 + n_2 = N$, the total frequency, then the alternative form of the χ^2 for testing the null hypothesis $H(p = p_0)$, p_0 preassigned, is

$$
\begin{aligned}
\chi^2 &= \frac{[n_1 - Np_0]^2}{Np_0} + \frac{[n_2 - N(1-p_0)]^2}{N(1-p_0)} \\
&= \frac{(1-p_0)(n_1 - Np_0)^2 + p_0(n_2 - N + Np_0)^2}{Np_0(1-p_0)}
\end{aligned}
$$

$$= \frac{(n_1 - Np_0)^2}{Np_0(1 - p_0)} = \frac{(\hat{p} - p_0)^2}{p_0(1 - p_0)/N}, \quad \text{since } \hat{p} = n_1/N.$$

This χ^2 is just z^2 for testing the null hypothesis $H(p = p_0)$, and it clearly has one degree of freedom. For the particular set of observations given in Table 17.1 and $p_0 = 0\cdot5$, we have

$$\chi^2 = \frac{(150 - 100)^2}{100} + \frac{(50 - 100)^2}{100} = 50 \quad \text{with 1 d.f., as before.}$$

The structural form of this χ^2 is now clear. It is evaluated by summing the squares of the differences between the observed and expected frequencies divided by their respective expected frequencies for each of the two distinct classes of heads and tails. Symbolically, we can write

$$\chi^2 = \Sigma \, (\text{Observed} - \text{Expected})^2 / \text{Expected} \quad \text{with 1 d.f.,}$$

the summation being over the two classes of the binomial distribution. This is the simplest illustration of the goodness of fit χ^2, and it tests the agreement between the observed and expected forms of the binomial distribution.

17.3   General form of the goodness of fit χ^2

More generally, if the empirical frequency distribution has k distinct classes ($k \geqslant 2$) with observed frequencies $n_1, n_2, \ldots, n_k$, and the corresponding expected frequencies on some null hypothesis are $\nu_1, \nu_2, \ldots, \nu_k$, where

$$\sum_{i=1}^{k} n_i = \sum_{i=1}^{k} \nu_i = N, \text{ say,}$$

then the goodness of fit χ^2 for testing the agreement between the observed and expected frequency distributions is

$$\chi^2 = \sum_{i=1}^{k} (n_i - \nu_i)^2 / \nu_i \quad \text{with } k-1 \text{ d.f.} \tag{1}$$

This is the general form of the goodness of fit χ^2. Like the normal approximation for the binomial distribution, this χ^2 is also a large-sample approximation. It is adequate for the valid application of this test that all the expected frequencies (ν_i) should be $\geqslant 5$. This number 5 is conventional, but it is necessary to ensure that no ν_i is too small, for otherwise the χ^2 approximation tends to break down. We shall consider later the modifications needed for the application of the test when some of the ν_i are small. It is also to be noted that the degrees of freedom for χ^2 are $k-1$ and *not* k. An intuitively acceptable reason for this is that the sum of the expected frequencies must necessarily be N, a fixed number, so that a null hypothesis can assign only any $k-1$ of the ν_i, the remaining expected frequency being determined by the equation

$$\sum_{i=1}^{k} \nu_i = N.$$

It is not usual to evaluate the goodness of fit χ^2 by a direct substitution of the n_i and ν_i in the formal expression (1). A simpler form is obtained by noting that

$$\chi^2 = \sum_{i=1}^{k} (n_i^2 - 2\nu_i n_i + \nu_i^2)/\nu_i$$

$$= \sum_{i=1}^{k} (n_i^2/\nu_i - 2n_i + \nu_i)$$

$$= \sum_{i=1}^{k} n_i^2/\nu_i - N, \tag{2}$$

since the sum of the observed and also that of the expected frequencies is N. This last expression for χ^2 is generally a convenient computational form.

Example 17.1: Goodness of fit of a multinomial distribution

As an illustration of the use of the general goodness of fit χ^2, we consider the following experiment which leads to a multinomial distribution. Suppose the experiment consists of tossing a penny twice. There are four mutually exclusive and exhaustive ordered outcomes of the experiment, namely, two heads (*HH*), a head followed by a tail (*HT*), a tail followed by a head (*TH*), and two tails (*TT*). If the probability of obtaining a head (*H*) at a trial is p and that of a tail (*T*) $1 - p$, then the probabilities of the four ordered outcomes of the experiment are:

$$P(HH) = p^2, \quad P(HT) = p(1-p), \quad P(TH) = (1-p)p, \quad P(TT) = (1-p)^2.$$

Next, suppose this experiment is repeated 560 times and the observed distribution of the four outcomes is found to be as given in Table 17.2.

Table 17.2 Showing the observed distribution of the ordered outcomes of 560 experiments of tossing a penny twice

Outcome	HH	HT	TH	TT	Total
Frequency	207	146	121	86	560

Whatever be the value of p in its permissible range $(0 < p < 1)$, the probability of the observed frequency distribution is given by the multinomial distribution as

$$\frac{560!}{207!\,146!\,121!\,86!} [p^2]^{207}[p(1-p)]^{267}[(1-p)^2]^{86}$$

$$= \frac{560!}{207!\,146!\,121!\,86!} p^{681}(1-p)^{439}.$$

This probability can be used to determine the maximum-likelihood estimate $\hat{p}$ of p and then, for example, the null hypothesis $H(p = 0{\cdot}5)$ can be tested by using the large-sample normal approximation for the probability distribution of $\hat{p}$. Alternatively, we may compare the observed and expected frequencies in the four classes by a goodness of fit χ^2. Thus, if the null hypothesis $H(p = 0{\cdot}5)$ is true, then the expected frequency in each of the four classes is 140. Hence the goodness of fit χ^2 is

$$= \frac{1}{140}[207^2 + 146^2 + 121^2 + 86^2] - 560$$

$$= \frac{86202}{140} - 560 = 55\cdot73 \qquad \text{with 3 d.f.}$$

This observed value of χ^2 is highly significant since the $0\cdot1$ per cent point of the χ^2 distribution with 3 d.f. is $16\cdot27$. We therefore confidently reject the null hypothesis that the penny is unbiased. The important point is that the χ^2 test is based on the comparison of the observed frequency distribution with a frequency distribution expected if the null hypothesis under test is true.

As an extension of the above procedure, suppose we assume that the probability $P(H) = 5/8$ and $P(T) = 3/8$. Then the null hypothesis is $H(p = 0\cdot625)$. Hence the expected frequencies in the four classes are:

Class (HH): $560 \times (0\cdot625)^2$ $= 218\cdot75$,

Class (HT): $560 \times (0\cdot625)(0\cdot375) = 131\cdot25$,

Class (TH): $560 \times (0\cdot375)(0\cdot625) = 131\cdot25$,

Class (TT): $560 \times (0\cdot375)^2$ $= 78\cdot75$.

Therefore the goodness of fit χ^2 is

$$= \frac{207^2}{218\cdot75} + \frac{146^2}{131\cdot25} + \frac{121^2}{131\cdot25} + \frac{86^2}{78\cdot75} - 560$$

$$= 195\cdot88 + 162\cdot41 + 111\cdot55 + 93\cdot92 - 560 = 3\cdot76 \quad \text{with 3 d.f.}$$

This value of χ^2 is clearly not significant since the 20 per cent point of the χ^2 distribution with 3 d.f. is $4\cdot64$. Accordingly, the null hypothesis $H(p = 0\cdot625)$ can be accepted.

The two tests of significance based on the χ^2 distribution have compared the same observed frequency distribution with two different expected distributions obtained on the null hypotheses $H(p = 0\cdot5)$ and $H(p = 0\cdot625)$ respectively. The degrees of freedom for both χ^2's are three since in either case the only restriction on the expected frequencies in the four classes is that their sum must be 560.

Example 17.2: Analysis of Pharbitis data

As another instructive illustration, we reconsider the analysis of Imai's *Pharbitis* data already quoted in Chapter 9. The observed frequencies and expected proportions in the AB, Ab, aB, ab classes are reproduced in Table 17.3, θ being an unknown parameter.

Table 17.3   Showing the observed frequencies and expected proportions in Imai's *Pharbitis* example.

Class	AB	Ab	aB	ab	Total
Observed frequency	187	35	37	31	290
Expected proportion	$\frac{1}{4}(2+\theta)$	$\frac{1}{4}(1-\theta)$	$\frac{1}{4}(1-\theta)$	$\frac{1}{4}\theta$	1

Because of certain genetical considerations, a null hypothesis of particular interest is $H(\theta = 0\cdot25)$; and if this null hypothesis is true, then the expected frequencies in the four classes are:

Class AB: $290 \times 9/16 = 163\cdot125,$
Class Ab: $290 \times 3/16 = 54\cdot375,$
Class aB: $290 \times 3/16 = 54\cdot375,$
Class ab : $290 \times 1/16 = 18\cdot125.$

Therefore the χ^2 for testing the agreement between the observed and expected frequency distributions is

$$= \frac{187^2}{163\cdot125} + \frac{35^2}{54\cdot375} + \frac{37^2}{54\cdot375} + \frac{31^2}{18\cdot125} - 290$$

$$= 214\cdot37 + 22\cdot53 + 25\cdot18 + 53\cdot02 - 290 = 25\cdot10 \quad \text{with 3 d.f.}$$

This value of χ^2 is highly significant since the 0·1 per cent point of the χ^2 distribution with 3 d.f. is 16·27. We therefore infer that the data are almost certainly not consistent with the null hypothesis $H(\theta = 0\cdot25)$.

We showed in Chapter 11 that the maximum-likelihood estimate of θ is

$$\hat{\theta} = 0\cdot4835 \quad \text{and} \quad \text{s.e.}(\hat{\theta}) = 0\cdot04663.$$

Therefore an alternative method for testing the null hypothesis $H(\theta = 0\cdot25)$ is to use

$$z = \frac{\hat{\theta} - 0\cdot25}{\text{s.e.}(\hat{\theta})}$$

as a unit normal variable. For the given observations, we have

$$z = 0\cdot2335/0\cdot04663 = 5\cdot01.$$

This is a large value of the unit normal variable and so the null hypothesis under test is again rejected decisively. In this extreme case the two tests are in good agreement, but the χ^2 test is the better method because it is based on a comparison of the complete observed and expected distributions.

Next, suppose we wish to test that the observed frequency distribution is in agreement with an expected frequency distribution with $\theta = 0\cdot4835$, that is, θ has, in fact, the value obtained as the maximum-likelihood estimate. If this null hypothesis is true, then the expected frequencies in the four classes are:

Class AB: $290 \times 2\cdot4835 \times 0\cdot25 = 180\cdot05,$
Class Ab: $290 \times 0\cdot5165 \times 0\cdot25 = 37\cdot45,$
Class aB: $290 \times 0\cdot5165 \times 0\cdot25 = 37\cdot45,$
Class ab : $290 \times 0\cdot4835 \times 0\cdot25 = 35\cdot05.$

Hence the goodness of fit χ^2 for testing the null hypothesis $H(\theta = 0\cdot4835)$ is

$$= \frac{187^2}{180\cdot05} + \frac{35^2}{37\cdot45} + \frac{37^2}{37\cdot45} + \frac{31^2}{35\cdot05} - 290$$

$$= 194\cdot22 + 32\cdot71 + 36\cdot56 + 27\cdot42 - 290 = 0\cdot91.$$

This χ^2 has only 2 d.f., and this is an important point. There are four classes of the observed frequency distribution, but we have so evaluated the expected frequencies as to satisfy two conditions. Firstly, the sum of the expected frequencies is 290, the same as the observed total. Secondly, we have used the value of the estimate $\hat{\theta}$ for evaluating the expected frequencies and thus ensured that the observed and expected frequency distributions have the same value of θ. These are the two restrictions on the expected frequencies, and for large samples these imply a reduction in the degrees of freedom for the goodness of fit χ^2 to two. The observed χ^2 is thus $0 \cdot 91$ with 2 d.f., and this is obviously not significant. We therefore conclude that the observed frequency distribution is in agreement with a multinomial frequency distribution for which $\theta = 0 \cdot 4835$.

This reduction in the degrees of freedom of the goodness of fit χ^2 by one due to the estimation of a parameter from the data is a general principle. Thus, if there are k classes in all and c parameters are estimated from the observations, then the resulting goodness of fit χ^2 has only $k - c - 1$ d.f.

17.4   Combination of binomial estimates

We have seen how the χ^2 distribution can be used to test null hypotheses about a single binomial parameter. An important extension of this method can be made to combine the information obtained from different independent estimates. In general, suppose there are k binomial populations with parameters $p_1, p_2, ..., p_k$ respectively. We assume that we are given samples of $n_1, n_2, ..., n_k$ observations respectively from these populations, and the corresponding maximum-likelihood estimates of the parameters are $\hat{p}_1, \hat{p}_2, ..., \hat{p}_k$.

Now, for fixed i and large n_i, we can test the null hypothesis $H(p_i = p_0)$, p_0 preassigned, by using

$$z_i = (\hat{p}_i - p_0)/[p_0(1-p_0)/n_i]^{\frac{1}{2}}$$

as a unit normal variable, or z_i^2 as χ^2 with 1 d.f. Clearly, this is true for all $i = 1, 2, ..., k$, provided the sample sizes are assumed to be sufficiently large. Furthermore, since the k samples are independent, it follows that the z_i are also independent. Hence $\sum_{i=1}^{k} z_i^2$ is distributed as χ^2 with k d.f. This χ^2 tests jointly the null hypotheses $H(p_i = p_0)$, for $i = 1, 2, ..., k$.

Next, suppose the null hypothesis $H(p_1 = p_2 = ... = p_k)$ is true irrespective of the common value p (say) of the p_i. Then the estimate of p is

$$\hat{p} = \sum_{i=1}^{k} n_i \hat{p}_i/N, \quad \text{where} \quad N = \sum_{i=1}^{k} n_i.$$

This estimate $\hat{p}$ is simply the maximum-likelihood estimate of p when it is assumed that the k binomial populations have the same unknown parameter. Hence, if this assumption is true, a test for the null hypothesis $H(p = p_0)$ is obtained by using

$$z = (\hat{p} - p_0)/[p_0(1 - p_0)/N]^{\frac{1}{2}}$$

as a unit normal variable or, equivalently, z^2 as χ^2 with 1 d.f.

Finally, consider the difference

$$\sum_{i=1}^{k} z_i^2 - z^2 = \frac{1}{p_0(1-p_0)} \left[\sum_{i=1}^{k} n_i(\hat{p}_i - p_0)^2 - N(\hat{p} - p_0)^2 \right]$$

$$= \frac{1}{P_0(1-P_0)}\left[\sum_{i=1}^{k} n_i\{(\hat{p}_i - p_0)^2 - (\hat{p} - p_0)^2\}\right]$$

$$= \frac{1}{P_0(1-P_0)}\left[\sum_{i=1}^{k} n_i\{\hat{p}_i^2 - 2p_0\hat{p}_i - \hat{p}^2 + 2p_0\hat{p}\}\right]$$

$$= \frac{1}{P_0(1-P_0)}\left[\sum_{i=1}^{k} n_i\hat{p}_i^2 - N\hat{p}^2\right]$$

$$= \frac{1}{P_0(1-P_0)}\sum_{i=1}^{k} n_i(\hat{p}_i - \hat{p})^2 .$$

Thus the difference $\sum_{i=1}^{k} z_i^2 - z^2$ is always non-negative, and it is also known that, for large samples, this difference is distributed as χ^2 with $k-1$ d.f. This is known as the *heterogeneity* χ^2, and it provides a valid test for the null hypothesis $H(p_1 = p_2 = \dots = p_k)$, that is, of the simultaneous equality of the k binomial parameters. If this null hypothesis is acceptable, then $\hat{p}$ is a valid pooled estimate of the common parameter p. Hence $\hat{p}$ may be used to test the null hypothesis $H(p = p_0)$. This method is inappropriate if the heterogeneity χ^2 is significant, for it then indicates that the k binomial populations cannot be regarded as having the same parameter. The parameter p may now be interpreted as the weighted average of the k binomial parameters, and with this understanding the null hypothesis $H(p = p_0)$ has meaning. A test for this null hypothesis is obtained by using the fact that

$$F = (k-1)z^2\Big/\left[\sum_{i=1}^{k} z_i^2 - z^2\right] = (k-1)N(\hat{p} - p_0)^2\Big/\sum_{i=1}^{k} n_i(\hat{p}_i - \hat{p})^2$$

has the F distribution with $1, k-1$ d.f. It is to be observed that this partitioning of the total χ^2 with k d.f. into two component χ^2's — the heterogeneity χ^2 with $k-1$ d.f. and z^2 with 1 d.f. — is analogous to the analysis of variance for the comparison of the means of several normal populations.

Example 17.3: Comparison of Mendelian proportions
As an application of the above methods, we consider the analysis of the data of the pea breeding experiments of Mendel and six other experimenters quoted in Table 5.5 of Chapter 5. The seven experimenters carried out independent but similar experiments to determine the proportion of yellow peas in the offspring of self-fertilised hybrid yellow peas. Suppose $p_1, p_2, \dots, p_7$ denote the true proportions of yellow peas in the populations investigated by the experimenters. Then, on the basis of the data, we have

$$\hat{p}_1 = 0.7505, \quad n_1 = 8023; \quad \hat{p}_2 = 0.7547, \quad n_2 = 1847;$$
$$\hat{p}_3 = 0.7505, \quad n_3 = 4770; \quad \hat{p}_4 = 0.7464, \quad n_4 = 1755;$$
$$\hat{p}_5 = 0.7530, \quad n_5 = 15806; \quad \hat{p}_6 = 0.7367, \quad n_6 = 1952;$$
$$\hat{p}_7 = 0.7509, \quad n_7 = 145246.$$

According to Mendel's theory, the true proportion of yellow offspring should be 0.75. Hence the appropriate χ^2's for testing the null hypotheses

$H(p_i = 0.75)$, for $i = 1, 2, ..., 7$, are as follows:

$$z_1^2 = \frac{8023\,(0.7505 - 0.75)^2}{0.1875} = \frac{0.0020}{0.1875} = 0.0107;$$

$$z_2^2 = \frac{1847\,(0.7547 - 0.75)^2}{0.1875} = \frac{0.0408}{0.1875} = 0.2176;$$

$$z_3^2 = \frac{4770\,(0.7505 - 0.75)^2}{0.1875} = \frac{0.0012}{0.1875} = 0.0064;$$

$$z_4^2 = \frac{1755\,(0.7464 - 0.75)^2}{0.1875} = \frac{0.0227}{0.1875} = 0.1211;$$

$$z_5^2 = \frac{15806\,(0.7530 - 0.75)^2}{0.1875} = \frac{0.1423}{0.1875} = 0.7589;$$

$$z_6^2 = \frac{1952\,(0.7367 - 0.75)^2}{0.1875} = \frac{0.3453}{0.1875} = 1.8416;$$

$$z_7^2 = \frac{145246(0.7509 - 0.75)^2}{0.1875} = \frac{0.1176}{0.1875} = 0.6272.$$

Hence the total χ^2 is

$$\sum_{i=1}^{7} z_i^2 = 3.584 \quad \text{with 7 d.f.}$$

Next, by combining the data from the seven experiments, we obtain

$$\hat{p} = \frac{134710 \cdot 1872}{179399} = 0.7509, \quad \text{since here } N = 179,399.$$

Therefore the χ^2 for testing the null hypothesis $H(p = 0.75)$ is

$$z^2 = \frac{179399\,(0.7509 - 0.75)^2}{0.1875} = \frac{0.1453}{0.1875} = 0.775 \quad \text{with 1 d.f.}$$

Hence the heterogeneity χ^2 for testing the null hypothesis $H(p_1 = p_2 = ... = p_7)$ is

$$\sum_{i=1}^{7} z_i^2 - z^2 = 2.809 \quad \text{with 6 d.f.}$$

Since the five per cent point of the χ^2 with 6 d.f. is 12·59, the observed χ^2 is not significant, and we therefore conclude that the null hypothesis under test is acceptable. Thus $\hat{p} = 0.7509$ is a valid estimate of p obtained by combining the information from the seven experiments. Also, as $z^2 = 0.775$ has the χ^2 distribution with 1 d.f., it is obviously not significant. It thus follows that the pooled estimate $\hat{p}$ is in agreement with Mendel's theory which indicates that $p = 0.75$.

Finally, to obtain the 95 per cent confidence interval for p, we use $(\hat{p} - p)/\text{s.e.}(\hat{p})$ as a unit normal variable. But

$$\text{s.e.}(\hat{p}) = \left[\frac{0.7509 \times 0.2491}{179398}\right]^{\frac{1}{2}} = \frac{0.432492}{423.5540} = 0.001021.$$

Hence the 95 per cent confidence interval for p is

$$0{\cdot}7509 - 1{\cdot}96 \times 0{\cdot}001021 \leqslant p \leqslant 0{\cdot}7509 + 1{\cdot}96 \times 0{\cdot}001021,$$

or

$$0{\cdot}7489 \leqslant p \leqslant 0{\cdot}7529.$$

The extreme closeness of the confidence limits for p (they are the same correct to two significant figures) is, of course, due to the fact that the total number of observations in the seven experiments is very large ($N = 179{,}399$). The statistical evidence in support of Mendel's theory is overwhelming. The genetical interest of these historic experiments lies in the fact that they provide jointly an empirical verification of Mendel's theory which postulates a particular quantitative way for the inheritance of the colour of peas in successive generations.

17.5 Goodness of fit of a normal distribution

The majority of the methods of statistical analysis that we have discussed in this book are based on the assumption that the populations sampled are normal. We are now in a position to consider how this assumption could be tested in any particular case by using a goodness of fit χ^2. A normal distribution has two parameters — the mean and the variance — and it is, in principle, possible to compare an observed frequency distribution with a theoretical normal distribution having any specified values for its mean and variance. However, such a comparison is usually not of much interest. Indeed, the particular comparison of practical value is that between the observed frequency distribution and a normal distribution having the same mean and variance as the observed distribution. This implies that the theoretical frequencies corresponding to the observed class-intervals are subject to two further restrictions, apart from the general one that their sum is the same as the total observed frequency. Besides, it generally happens that the expected frequencies in some of the class-intervals in the two tails of the fitted normal distribution are small, and therefore we have to modify these class-intervals for a valid application of the goodness of fit χ^2. This procedure is explained best by the following example.

Example 17.4: Distribution of the weights of ears of maize

We consider the data of Lindstrom pertaining to the frequency distribution of the weight in grams of 327 ears of maize. This distribution is given in Example 7.1 of Chapter 7 together with the expected frequencies of the fitted normal distribution. The observed distribution extends from 44·5 to 324·5, whereas the range of the theoretical distribution is from $-\infty$ to ∞. Accordingly, our first modification is to consider that the observed frequency in the first class-interval (44·5 – 64·5) corresponds to the expected frequency below 64·5. Similarly, the second modification is that the observed frequency in the last class-interval (304·5 – 324·5) corresponds to the expected frequency beyond 304·5. We have already indicated that these modifications do not cause appreciable error since the areas under the fitted normal curve below 44·5 and above 324·5 are both negligibly small. The observed and expected frequencies in the class-intervals are given in Table 17.4 overleaf.

Table 17.4   Showing the observed and expected frequency distribution of
the weights of 327 ears of maize

Class-interval	Observed frequency (n_i)	Expected frequency (ν_i)	n_i^2/ν_i
$-\infty - 64 \cdot 5$	2 ⎤	0·785 ⎤	
$64 \cdot 5 - 84 \cdot 5$	4 ⎪ 12*	1·929 ⎪ 7·913*	18·198
$84 \cdot 5 - 104 \cdot 5$	6 ⎦	5·199 ⎦	
$104 \cdot 5 - 124 \cdot 5$	12	11·903	12·098
$124 \cdot 5 - 144 \cdot 5$	14	22·726	8·624
$144 \cdot 5 - 164 \cdot 5$	33	36·395	29·922
$164 \cdot 5 - 184 \cdot 5$	49	48·756	49·245
$184 \cdot 5 - 204 \cdot 5$	61	54·772	67·936
$204 \cdot 5 - 224 \cdot 5$	51	51·568	50·438
$224 \cdot 5 - 244 \cdot 5$	43	40·679	45·453
$244 \cdot 5 - 264 \cdot 5$	30	26·814	33·565
$264 \cdot 5 - 284 \cdot 5$	15	14·813	15·189
$284 \cdot 5 - 304 \cdot 5$	5 ⎤ 7*	6·900 ⎤ 10·726*	4·568
$304 \cdot 5 - \infty$	2 ⎦	3·826 ⎦	
Total	327	327·065	335·236

*Pooled frequencies

Next, we observe that there are in all 14 class-intervals in Table 17.4,
and of these the first two and the last class-interval have expected frequencies which are less than the conventional number 5. We therefore
cannot apply the goodness of fit test with these class-intervals. The proper
procedure here is to combine the first three class-intervals into one class-
interval extending from $-\infty$ to 104·5, and, similarly, to combine the last
two class-intervals to form one class-interval extending from 284·5 to ∞.
The corresponding observed and expected frequencies are also pooled. This
reduces the number of class-intervals to 11, and in each of them the expected frequency is now greater than 5. It is to be noted that the inequality
of the lengths of the class-intervals does not vitiate the goodness of fit
test so long as the class-intervals are distinct and have expected frequencies greater than 5.

The χ^2 test can now be applied to compare the observed and expected
frequencies in the 11 class-intervals in Table 17.4. The last column of
this table gives the values of n_i^2/ν_i. Hence

$$\chi^2 = 335 \cdot 236 - 327 = 8 \cdot 236.$$

There are 11 class-intervals, and since two parameters (the mean and the
variance of the fitted normal distribution) have been estimated from the
data to determine the expected frequencies, the degrees of freedom for the
calculated χ^2 are $11 - 2 - 1 = 8$. But for 8 d.f. the five per cent point of
the χ^2 distribution is 15·51. Thus the observed value of χ^2 is not significant, and we infer that the observed distribution could reasonably have
arisen from a normal population with the same mean and variance.

In conclusion, two general points are worth comment. Firstly, we cannot lay down in advance how many, if any, class-intervals in the tails of a

distribution have to be combined to ensure a valid comparison between the observed and expected distributions by the χ^2 test. The appropriate combination of the class-intervals is indicated by the magnitude of their expected frequencies, and this will vary for different distributions. Besides, there is some flexibility in the choice of the class-intervals which may be combined in a given distribution. For example, in the above illustration, we could have combined only the first two and not the first three class-intervals. In general, such small variations will not cause any appreciable difference in the χ^2 test. Secondly, the χ^2 test is a large-sample procedure, and it should not be used unless the sample size is reasonably large. This restriction is intuitively obvious since a few random observations from a population cannot convey much information about its possible form. In this respect, the goodness of fit χ^2 differs from the tests of null hypotheses about the mean and variance of a normal population because these latter tests can be used legitimately with small samples. This difference exemplifies an earlier remark that the same data do not necessarily answer different questions about the population sampled. Accordingly, the correct procedure is to collect data in the light of the questions to which answers are sought. This becomes even more important when the sampling procedure is elaborate and expensive.

17.6 Goodness of fit of a Poisson distribution

As another example, we consider the goodness of fit of a Poisson distribution. In Chapter 9, we cited a number of observed frequency distributions which could all be represented by Poisson models with appropriately estimated means. To illustrate the details of the numerical calculations leading to the χ^2 test for the goodness of fit, we consider here the data of Rutherford *et al.* on the emission of α-particles from a radioactive source. The upper tail of the observed frequency distribution was pooled in the earlier reference to these data in Example 9.8 of Chapter 9. The complete frequency distribution is given overleaf in Table 17.5. The mean of the observed frequency distribution is

$$\sum_{w=0}^{\infty} wn_w \bigg/ \sum_{w=0}^{\infty} n_w = \frac{10094}{2608} = 3 \cdot 870,$$

and the expected frequencies are evaluated from a Poisson distribution with mean $3 \cdot 870$. The expected frequencies in the classes beyond $w = 10$ are all less than 5, and so we pool these frequencies and the corresponding observed ones for $w \geqslant 11$. Accordingly, we now have 12 classes in all, and in each of them the expected frequency is greater than 5. The goodness of fit χ^2 can now be evaluated. The last column of Table 17.5 gives the values of n_w^2 / ν_w, whence $\chi^2 = 2620 \cdot 973 - 2608 = 12 \cdot 973$ with 10 d.f. since only one parameter is estimated from the data. The five per cent point of χ^2 with 10 d.f. is $18 \cdot 31$, and so the observed value of χ^2 is not significant. We therefore infer that the observed frequency distribution of the emission of α-particles is in agreement with a Poisson distribution having the mean $3 \cdot 870$.

Table 17.5   Showing the observed and expected distributions of the number of α-particles in 2,608 intervals of 7·5 seconds each

Number of particles emitted in a time-interval of 7·5 seconds (w)	Observed frequency (n_w)	Expected frequency (ν_w)	n_w^2/ν_w
0	57	54·40	59·724
1	203	210·52	195·749
2	383	407·36	360·097
3	525	525·50	524·500
4	532	508·42	556·674
5	408	393·52	423·013
6	273	253·82	293·629
7	139	140·32	137·692
8	45	67·88	29·832
9	27	29·19	24·974
10	10	11·30	8·850
11	4 ⎤	3·97 ⎤	
12	2 ⎟ 6*	1·28 ⎟ 5·77*	6·239
$\geqslant 13$	0 ⎦	0·52 ⎦	
Total	2608	2608·00	2620·973

* Pooled frequencies

17.7   Small samples from a Poisson distribution

The preceding goodness of fit χ^2 can always be used to test agreement between an observed and an expected distribution, provided the sample size is reasonably large. However, in the case of a Poisson distribution, a small sample test is also possible. In principle, this procedure depends upon the fact that the mean and variance are equal for a Poisson distribution. Suppose $x_1, x_2, \ldots, x_n$ are random observations, where n is not necessarily large, and we wish to test whether these could have been obtained from a Poisson distribution with some mean μ. The observations x_i may, for example, denote the number of α-particles emitted by a radioactive source in successive time-intervals of length T (fixed), or the number of telephone calls arriving at an exchange in successive half-hourly periods. The essential point is that although the events (emission of an α-particle or the arrival of a telephone call) happen in a temporal sequence, they occur randomly. This implies that the occurrence of an event in no way affects the occurrence of a subsequent event.

It can be shown that under the above conditions the x_i have a Poisson distribution, and the goodness of fit can be tested as for the α-particle data if n is large. For n not necessarily large, a test that the x_i are random observations from a Poisson distribution may be deduced as follows. If x_i has a Poisson distribution with mean μ, then

$$E(x_i) = \text{var}(x_i) = \mu, \quad \text{for } i = 1, 2, \ldots, n.$$

Also, the sample mean $\bar{x}$ is the maximum-likelihood estimate of μ. Besides it is known that as a first approximation

$$z_i = \frac{x_i - \bar{x}}{\sqrt{\bar{x}}}$$

is a unit normal variable. Hence it follows that

$$\sum_{i=1}^{n} z_i^2 = \sum_{i=1}^{n} (x_i - \bar{x})^2 / \bar{x}$$

is distributed approximately as χ^2 with $n - 1$ d.f. The degrees of freedom are $n - 1$ since the z_i are subject to the condition that

$$\sum_{i=1}^{n} z_i \equiv 0.$$

This χ^2 provides a valid method for testing the null hypothesis that the sample observations are from a Poisson distribution with mean $\bar{x}$.

Example 17.5: Arrival of telephone calls

As an example of this χ^2 test, we consider the analysis of the data presented in Table 17.6 below.

Table **17.6** Showing the frequency of telephone calls arriving at an exchange in 120 successive half-hourly intervals

3	0	4	1	1	2	3	4	3	2	0	4	2	3	3	2	1	3	4	0
0	4	0	3	4	1	0	3	1	4	2	4	3	4	4	4	4	2	3	3
0	2	3	4	3	1	3	2	4	3	2	3	2	2	1	4	2	0	1	4
0	0	1	1	1	1	1	4	3	1	2	0	4	4	3	3	4	1	1	0
2	2	0	4	1	0	0	3	1	4	1	3	4	0	4	2	2	0	1	4
1	1	4	0	2	1	1	4	0	2	4	0	3	3	0	4	4	0	4	0

The observations pertain to the frequency of telephone calls which arrived at an exchange in 120 successive half-hourly periods. If we denote these numbers by x_i, then we have

$$\sum_{i=1}^{120} x_i = 255 \quad \text{and} \quad \sum_{i=1}^{120} x_i^2 = 803.$$

Therefore $\qquad \bar{x} = 2 \cdot 125, \quad \text{C.F.}(x) = \dfrac{65025}{120} = 541 \cdot 875,$

and $\qquad \displaystyle\sum_{i=1}^{120} (x_i - \bar{x})^2 = 261 \cdot 125.$

Hence $\qquad \chi^2 = \dfrac{261 \cdot 125}{2 \cdot 125} = 122 \cdot 88 \quad$ with 119 d.f.

We cannot immediately test the significance of this observed value of χ^2 because the usual tables of the χ^2 distribution do not go much beyond 100 d.f. However, we know that if the degrees of freedom (ν) of χ^2 are large, then $(\chi^2 - \nu)/\sqrt{2\nu}$ is approximately a unit normal variable. It is to be observed that since the mean and variance of χ^2 are ν and 2ν respectively, the random variable $(\chi^2 - \nu)/\sqrt{2\nu}$ has zero mean and unit variance for all values of ν. The normal approximation holds for large ν. In our example, $\nu = 119$ and so

$$(\chi^2 - \nu)/\sqrt{2\nu} = \frac{3 \cdot 88}{15 \cdot 427} = 0 \cdot 252,$$

which is obviously not a significant value for a unit normal variable. We therefore conclude that the observed frequency distribution of telephone calls conforms well with a Poisson distribution with mean $2 \cdot 125$.

It is to be noted that the sample size $n = 120$ is rather small for a goodness of fit χ^2, but it is adequately large for the normal approximation for the χ^2 distribution to be applicable. A somewhat better approximation is the following one due to Fisher.

17.8   Fisher's approximation for χ^2

If χ^2 has the χ^2 distribution with ν d.f., then for large ν

$$z = \sqrt{2\chi^2} - \sqrt{2\nu - 1}$$

is approximately a unit normal variable. This approximation can be used confidently for values of $\nu \geqslant 100$ to test the significance of observed χ^2 values. In the above example,

$$\chi^2 = 122 \cdot 88 \quad \text{and} \quad \nu = 119.$$

Therefore $z = \sqrt{245 \cdot 76} - \sqrt{237} = 15 \cdot 677 - 15 \cdot 395 = 0 \cdot 282.$

This value of z is slightly larger than the value obtained by using the standardised χ^2 as a unit normal variable, though our main conclusion about the sample observations remains unaffected.

Note

Though the sample size is relatively small, we could still have tested the conformity of the telephone data with a Poisson distribution by forming a frequency distribution of the 120 observations and comparing it with the expected distribution derived from a Poisson model with mean $2 \cdot 125$. However, this goodness of fit test is known to be less powerful (in the sense of the Neyman – Pearson theory) than the χ^2 test we have actually used.

17.9   Analysis of contingency tables

There is another general problem in statistical analysis whose approximate large-sample solution is also dependent upon the χ^2 distribution. This problem is concerned with the testing of the independence of two qualitatively distinct characteristics or attributes which cannot be measured accurately and according to which it is not even possible to rank the individuals of a sample. However, such an attribute has a certain finite number of distinct classes, and it is possible to distribute the sample individuals amongst them. Such attributes arise quite frequently in different sciences such as medicine, biometry, psychometry, sociometry, econometrics, etc. As a simple example, a sample of women may be classified by the colour of their hair as "fair-haired", "red-haired", "brown-haired" or "black-haired". In the same way, a sample of undergraduates may be classified according to their first preferences for different team games such as cricket, football, hockey, etc. With such attributes, it is always possible to distribute a sample of individuals into the qualitatively distinct classes and thus obtain a frequency distribution. However, it is worth mentioning here that a distribution of this kind can also be determined if the sample observations are classified into distinct groups with respect to a quantitative characteristic. Thus, for example, a sample of schools may be clas ified into those with not more than 300 students, more than 300 but not more than 500 students, more than 500 but not more than 1,000 students, and more than 1,000 students.

A frequency table in which a sample is classified according to the distinct classes of two different attributes A and B (quantitative or qualitative) is called a *contingency table*. It looks rather like a correlation table, except that the rows and columns do not necessarily correspond to any numerical values of the attributes A and B. In general, suppose the attribute A has α distinct classes denoted by A_1, A_2, ..., A_α, and the attribute B has the β distinct classes B_1, B_2, ..., B_β. Then there are in all $\alpha\beta$ distinct classes or cells in the contingency table, a typical cell $A_i B_j$ corresponding to the ith A-class and the jth B-class. Such a table is called an $\alpha \times \beta$ *contingency table*.

Suppose, next, that we have distributed a random sample of N objects in the $\alpha\beta$ cells and that the observed frequency in the cell $A_i B_j$ is n_{ij}. Clearly, we have

$$\sum_{i=1}^{\alpha} \sum_{j=1}^{\beta} n_{ij} = N.$$

It is also convenient to introduce the notation

$$\sum_{j=1}^{\beta} n_{ij} = n_{i.}, \quad \text{for } i = 1, 2, ..., \alpha; \quad \text{and} \quad \sum_{i=1}^{\alpha} n_{ij} = n_{.j}, \quad \text{for } j = 1, 2, ..., \beta$$

Then

$$\sum_{i=1}^{\alpha} n_{i.} = \sum_{j=1}^{\beta} n_{.j} = N.$$

If the A_i correspond to rows and the B_j to columns, then the general $\alpha \times \beta$ contingency table has the form shown in Table 17.7.

Table 17.7 Showing a schematic representation of an $\alpha \times \beta$ contingency table

A-class \ B-class	B_1	B_2	B_3	B_4	·	·	·	·	B_β	Total
A_1	n_{11}	n_{12}	n_{13}	n_{14}	·	·	·	·	$n_{1\beta}$	$n_{1.}$
A_2	n_{21}	n_{22}	n_{23}	n_{24}	·	·	·	·	$n_{2\beta}$	$n_{2.}$
A_3	n_{31}	n_{32}	n_{33}	n_{34}	·	·	·	·	$n_{3\beta}$	$n_{3.}$
A_4	n_{41}	n_{42}	n_{43}	n_{44}	·	·	·	·	$n_{4\beta}$	$n_{4.}$
·	·	·	·	·	·	·	·	·	·	·
A_α	$n_{\alpha 1}$	$n_{\alpha 2}$	$n_{\alpha 3}$	$n_{\alpha 4}$	·	·	·	·	$n_{\alpha\beta}$	$n_{\alpha.}$
Total	$n_{.1}$	$n_{.2}$	$n_{.3}$	$n_{.4}$	·	·	·	·	$n_{.\beta}$	N

We wish to test that in the population from which the sample is drawn, the attributes A and B are independent. Suppose, then, that in this population the probability is p_{ij} that an individual random observation occurs in the $A_i B_j$ class, where

$$\sum_{i=1}^{\alpha} \sum_{j=1}^{\beta} p_{ij} = 1.$$

We further define that

$$p_{i.} = \sum_{j=1}^{\beta} p_{ij}, \quad \text{for } i = 1, 2, \ldots, \alpha;$$

and

$$p_{.j} = \sum_{i=1}^{\alpha} p_{ij}, \quad \text{for } j = 1, 2, \ldots, \beta.$$

Then

$$\sum_{i=1}^{\alpha} p_{i.} = \sum_{j=1}^{\beta} p_{.j} = \sum_{i=1}^{\alpha} \sum_{j=1}^{\beta} p_{ij} = 1.$$

It is clear that $p_{i.}$ is the probability of a random observation from the population being in A_i irrespective of its B-class. Similarly, $p_{.j}$ is the probability of the random observation being in B_j regardless of its A-class. Hence, if the attributes A and B are independent in the population sampled, we have

$$p_{ij} = p_{i.}p_{.j}, \quad \text{for all } i \text{ and } j.$$

Therefore, given the N observations, the expected frequency in the cell $A_i B_j$ is

$$N p_{ij} = N p_{i.} p_{.j},$$

if the attributes A and B are independent. But these expected frequencies are unknown since the p_{ij} are unknown population parameters. However, it can be shown (we do not prove this here) that the maximum-likelihood estimates of $p_{i.}$ and $p_{.j}$ are

$$\hat{p}_{i.} = n_{i.}/N \quad \text{and} \quad \hat{p}_{.j} = n_{.j}/N \quad \text{respectively.}$$

Hence, if the attributes A and B are independent, the maximum-likelihood estimates of the expected frequencies $N p_{ij}$ are

$$N \hat{p}_{ij} = N \hat{p}_{i.} \hat{p}_{.j} = n_{i.} \times n_{.j}/N, \quad \text{for all } i \text{ and } j.$$

We therefore conclude that, under the null hypothesis of the independence of the attributes A and B, the deviation of the observed frequencies in the contingency table from their respective expectations is measured by the goodness of fit χ^2

$$= \sum_{i=1}^{\alpha} \sum_{j=1}^{\beta} \left[n_{ij} - N\hat{p}_{ij} \right]^2 / N\hat{p}_{ij}$$

$$= \sum_{i=1}^{\alpha} \sum_{j=1}^{\beta} \left[n_{ij} - \frac{n_{i.} \times n_{.j}}{N} \right]^2 \bigg/ \left[\frac{n_{i.} \times n_{.j}}{N} \right]$$

$$= N \sum_{i=1}^{\alpha} \sum_{j=1}^{\beta} \left[\frac{n_{ij}^2}{n_{i.} \times n_{.j}} - 2 \frac{n_{ij}}{N} + \frac{n_{i.} \times n_{.j}}{N^2} \right]$$

$$= N \left[\sum_{i=1}^{\alpha} \sum_{j=1}^{\beta} n_{ij}^2 / (n_{i.} \times n_{.j}) - 2 + 1 \right],$$

so that

$$\chi^2 = N \left[\sum_{i=1}^{\alpha} \sum_{j=1}^{\beta} n_{ij}^2 / (n_{i.} \times n_{.j}) - 1 \right]. \tag{1}$$

Next, to determine the degrees of freedom for this χ^2, we observe that there are in all $\alpha\beta$ expected cell frequencies which are subject to the restriction that their sum must be N. Besides, we have also estimated the $p_{i.}$ and $p_{.j}$ from the sample data. These are $\alpha - 1$ and $\beta - 1$ independent parameters respectively since

$$\sum_{i=1}^{\alpha} p_{i.} = \sum_{j=1}^{\beta} p_{.j} = 1.$$

Hence the degrees of freedom for the χ^2 testing the independence of the attributes A and B are

$$\alpha\beta - 1 - (\alpha - 1) - (\beta - 1) = (\alpha - 1)(\beta - 1).$$

The test of significance for the independence of the attributes A and B is now carried out by comparing the observed value of χ^2, as defined in (1), with the appropriate percentage point of the χ^2 distribution with $(\alpha - 1) \times (\beta - 1)$ d.f. This test of significance is valid if the expected cell frequencies are all greater than the conventional number 5.

The most convenient method for evaluating this χ^2 is to use the expression (1), but this computational procedure requires some care. The main quantity to be evaluated is

$$\sum_{i=1}^{\alpha} \sum_{j=1}^{\beta} n_{ij}^2/(n_{i.} \times n_{.j}) = \sum_{j=1}^{\beta} (1/n_{.j}) \left[\sum_{i=1}^{\alpha} n_{ij}^2/n_{i.} \right], \qquad (2)$$

and the latter summation is carried out in two steps. It is convenient to form a two-way table whose entries in its $\alpha\beta$ cells are n_{ij}^2, the squares of the observed cell frequencies. This table is then bordered by $1/n_{i.}$ and $1/n_{.j}$. These reciprocals of the row and column sums and the squares n_{ij}^2 are obtained from *Barlow's Tables*. The product sums

$$\sum_{i=1}^{\alpha} n_{ij}^2 (1/n_{i.})$$

are then obtained consecutively for each value of j and these quantities are written in the auxiliary two-way table as a row below $1/n_{.j}$. The multiplication of these two rows by a continuous machine operation gives the required value of (2). This quantity will be usually just greater than unity, and a sufficient number of significant figures should be retained in the reciprocals $1/n_{i.}$ and $1/n_{.j}$ to give reasonable accuracy for the numerical value of χ^2 obtained by using (1). These computational details are best indicated by a numerical example.

Example 17.6: Size and output of mines

We quoted in Table 5.7 of Chapter 5 the data from 894 mines pertaining to the number of wage-earners in the mines and the number of mines with different outputs per manshift. Here both the attributes are quantitative, though the classes of the two-way table are distinct. If attribute A refers to the number of wage-earners in a mine, then $\alpha = 3$. Similarly, if attribute B denotes the output per manshift, then $\beta = 4$. The null hypothesis in this problem is that there is no association between the number of wage-earners in a mine and the output per manshift. The auxiliary table for the calculation of χ^2 is given below.

Table 17.8 Showing the calculations of the contingency χ^2 of the data of Table 5.7

A-class \ B-class	B_1	B_2	B_3	B_4	$1/n_{i.}$
A_1	10609	19600	5776	1764	0·002770083
A_2	3364	17161	5776	1521	0·003289474
A_3	625	5329	6889	2304	0·004366812
$1/n_{.j}$	0·005376344	0·002906977	0·004255319	0·007751938	
$\sum_{i=1}^{4} n_{ij}^2/n_{i.}$	43·182859	134·015031	65·082969	19·950851	

Hence, from the last two rows of the above table, we have

$$\sum_{j=1}^{4} (1/n_{.j}) \times \left[\sum_{i=1}^{3} n_{ij}^2/n_{i.} \right] = 1·053351,$$

whence $\qquad \chi^2 = 894 \times 0·053351 = 47·70 \quad$ with 6 d.f.

This observed value of χ^2 is highly significant since the 0·1 per cent point of the χ^2 distribution with 6 d.f. is only 22·46. The null hypothesis can be safely rejected, and we conclude that the output per manshift is almost certainly associated with the number of wage-earners in a mine.

17.10 2 × 2 contingency tables

General contingency tables are of infrequent practical use as compared with the very important class of 2 × 2 contingency tables. The χ^2 for testing the independence of two dichotomous attributes in a 2 × 2 contingency table is formally only a particular case of the χ^2 used in the general $\alpha \times \beta$ contingency table. However, the 2 × 2 contingency table has several useful and interesting special features, and we therefore consider its analysis separately. Suppose, then, that the distinct classes of the attributes A and B are denoted by A_1, A_2 and B_1, B_2 respectively. Also, let the observed distribution of a random sample of N observations in the four cells of a 2 × 2 contingency table be as given in Table 17.9 below.

Table 17.9 Showing a schematic representation of a 2 × 2 contingency table

B-class \ A-class	A_1	A_2	Total
B_1	n_1	n_3	$n_1 + n_3$
B_2	n_2	n_4	$n_2 + n_4$
Total	$n_1 + n_2$	$n_3 + n_4$	N

It is now easy to verify that for such a table ($\alpha = \beta = 2$), the general χ^2 for testing the independence of the attributes A and B reduces to

$$\chi^2 = \frac{N(n_1 n_4 - n_2 n_3)^2}{(n_1 + n_2)(n_3 + n_4)(n_1 + n_3)(n_2 + n_4)} \quad \text{with 1 d.f.}$$

As in the case of the general contingency table, this χ^2 gives an approximate test for the independence of the attributes A and B, provided the individual cell frequencies are greater than 5. For small cell frequencies, the above χ^2 tends to over-estimate significance in the sense that the χ^2 distribution gives a smaller probability of exceeding the observed value by chance than would be given by an exact determination of the total probability of obtaining as extreme distributions of the cell frequencies as the one actually observed. We do not consider here how this exact probability can be determined because of two reasons. Firstly, the exact calculation is, in general, rather laborious. Secondly, in the majority of practical cases, it is hardly necessary to use the exact method because of a very simple and accurate modification of the χ^2 test given by Yates.

Yates's continuity correction for a 2×2 contingency table

The modification suggested by Yates is generally known as *Yates's continuity correction* because for small cell frequencies it compensates for the difference between the discrete distribution of the cell frequencies and the approximating continuous χ^2 distribution. To make the continuity correction, one cell frequency, say n_1, is replaced by $n_1 \pm \frac{1}{2}$ according as $n_1 n_4 < $ or $ > n_2 n_3$, and the other three cell frequencies are then adjusted so as to leave the marginal totals of the contingency table unchanged. If $n_1 n_4 < n_2 n_3$, then the modified contingency table becomes as shown in Table 17.10 below.

Table 17.10   Showing a schematic representation of a modified 2×2 contingency table

B-class \ A-class	A_1	A_2	Total
B_1	$n_1 + \frac{1}{2}$	$n_3 - \frac{1}{2}$	$n_1 + n_3$
B_2	$n_2 - \frac{1}{2}$	$n_4 + \frac{1}{2}$	$n_2 + n_4$
Total	$n_1 + n_2$	$n_3 + n_4$	N

The corrected χ^2 is now obtained by using the first formula but with the modified cell frequencies. Thus the corrected χ^2 is

$$\chi_c^2 = \frac{N[(n_1 + \frac{1}{2})(n_4 + \frac{1}{2}) - (n_2 - \frac{1}{2})(n_3 - \frac{1}{2})]^2}{(n_1 + n_2)(n_3 + n_4)(n_1 + n_3)(n_2 + n_4)} \quad \text{with 1 d.f.}$$

On the other hand, if $n_1 n_4 > n_2 n_3$, then

$$\chi_c^2 = \frac{N[(n_1 - \frac{1}{2})(n_4 - \frac{1}{2}) - (n_2 + \frac{1}{2})(n_3 + \frac{1}{2})]^2}{(n_1 + n_2)(n_3 + n_4)(n_1 + n_3)(n_2 + n_4)} \quad \text{with 1 d.f.}$$

It is thus seen that in either case the effect of the correction is to replace $n_1 n_4 - n_2 n_3$ by $|n_1 n_4 - n_2 n_3| - \frac{1}{2}N$. We may therefore write

$$\chi_c^2 = \frac{N[\,|\,n_1 n_4 - n_2 n_3\,| - \frac{1}{2}N\,]^2}{(n_1 + n_2)(n_3 + n_4)(n_1 + n_3)(n_2 + n_4)} \quad \text{with 1 d.f.}$$

This continuity correction invariably improves the test of significance for the independence of the attributes in a 2×2 contingency table, and it should therefore be applied unless the n_i are all quite large. There is no corresponding correction for the general $\alpha \times \beta$ contingency table.

The χ^2 approximation for the exact distribution of a 2×2 contingency table is similar to the normal approximation for the binomial distribution. We recall that for a binomial variable w, the approximate probability between two values of w, say a and b inclusive, is given by the area under the corresponding normal curve between the points $a - \frac{1}{2}$ and $b + \frac{1}{2}$, and not between a and b. Yates's continuity correction works in an analogous manner.

Example 17.7: Inoculation and the incidence of cholera

We illustrate the analysis of a 2×2 contingency table by a classic example on the effect of inoculation in the prevention of cholera. The following table gives the frequency distribution of a random sample of 818 persons according to whether or not they were attacked by cholera (attribute A), and whether or not they were inoculated against cholera (attribute B).

Table 17.11 Showing the frequency distribution of the incidence of cholera and inoculation in a sample of 818 persons

B-class \ A-class	Not attacked (A_1)	Attacked (A_2)	Total
Inoculated (B_1)	276	3	279
Not inoculated (B_2)	473	66	539
Total	749	69	818

Source: M. Greenwood and G. Udny Yule (1915), *Proceedings of the Royal Society of Medicine*, Vol.8, p.113.

The null hypothesis is that the two attributes A and B are independent, that is, inoculation does not help in preventing the incidence of cholera. Here one of the cell frequencies is evidently small and we should use Yates's continuity correction for evaluating the χ^2 for testing the independence of A and B. However, suppose we ignore this. Then

$$\chi^2 = \frac{818(276 \times 66 - 473 \times 3)^2}{279 \times 539 \times 749 \times 69} = \frac{10^6 \times 230790}{10^6 \times 7772} = 29 \cdot 70 \quad \text{with 1 d.f.}$$

Since the 0·1 per cent point of the χ^2 distribution with 1 d.f. is 10·83, the observed value of χ^2 is highly significant, and we can confidently reject the null hypothesis under test. We therefore infer that the sample data strongly suggest an association between inoculation and immunity from cholera.

On the other hand, if we use, as we should, the continuity correction, then

$$\chi_c^2 = \frac{818(16797 - 409)^2}{279 \times 539 \times 749 \times 69} = \frac{10^6 \times 219687}{10^6 \times 7772} = 28{\cdot}27 \quad \text{with 1 d.f.}$$

We thus observe that the corrected value χ_c^2 is somewhat smaller than the uncorrected χ^2, though the high significance of this value is also beyond doubt. This decrease in the value of χ_c^2 is clearly typical of the effect of the continuity correction which tends to lessen the significance of the observed value of χ^2. Of course, in this example the use of the continuity correction makes no difference in the significance of the observed result and the consequent rejection of the null hypothesis. However, in less extreme cases, the use of the continuity correction can alter materially the conclusion about the null hypothesis under test.

17.11   Goodness of fit in a 2×2 table

For a slightly different and instructive approach to the problem of the independence of the attributes A and B in a 2×2 table, we go back to the general specification of the observed distribution with frequencies n_1, n_2, n_3, n_4 $\left[\sum_{i=1}^{4} n_i = N \right]$ in the four cells $A_1 B_1$, $A_1 B_2$, $A_2 B_1$, $A_2 B_2$ respectively. Also, the marginal totals give the observed frequencies in the A_1, A_2, B_1, B_2 classes as $n_1 + n_2$, $n_3 + n_4$, $n_1 + n_3$, $n_2 + n_4$ respectively.

Next, suppose that in the population from which the sample is obtained, the probabilities of a random observation being in the A_1, A_2, B_1, B_2 classes are p_a, $1 - p_a$, p_b, $1 - p_b$ respectively. If the attributes A and B are independent, then the probabilities of the observation being in the cells $A_1 B_1$, $A_1 B_2$, $A_2 B_1$, $A_2 B_2$ are $p_a p_b$, $p_a(1 - p_b)$, $(1 - p_a)p_b$, $(1 - p_a)(1 - p_b)$ respectively. Therefore we must have

$$E(n_1) = N p_a p_b, \quad E(n_2) = N p_a(1 - p_b), \quad E(n_3) = N(1 - p_a)p_b,$$

$$E(n_4) = N(1 - p_a)(1 - p_b).$$

But these expected frequencies are unknown because p_a and p_b are population parameters. However, the maximum-likelihood estimates of p_a and p_b are

$$\hat{p}_a = (n_1 + n_2)/N, \quad \hat{p}_b = (n_1 + n_3)/N,$$

that is, the sample relative frequencies in the A_1 and B_1 classes. Hence the estimates of the expected cell frequencies are

$$N\hat{p}_a \hat{p}_b, \quad N\hat{p}_a(1 - \hat{p}_b), \quad N(1 - \hat{p}_a)\hat{p}_b, \quad N(1 - \hat{p}_a)(1 - \hat{p}_b)$$

corresponding to the observed frequencies n_1, n_2, n_3, n_4 respectively. Therefore the χ^2 for testing the independence of the attributes A and B is

$$= \frac{[n_1 - N\hat{p}_a \hat{p}_b]^2}{N\hat{p}_a \hat{p}_b} + \frac{[n_2 - N\hat{p}_a(1 - \hat{p}_b)]^2}{N\hat{p}_a(1 - \hat{p}_b)} + \frac{[n_3 - N(1 - \hat{p}_a)\hat{p}_b]^2}{N(1 - \hat{p}_a)\hat{p}_b} +$$

$$+ \frac{[n_4 - N(1 - \hat{p}_a)(1 - \hat{p}_b)]^2}{N(1 - \hat{p}_a)(1 - \hat{p}_b)}$$

$$= \frac{[Nn_1 - (n_1 + n_2)(n_1 + n_3)]^2}{N(n_1 + n_2)(n_1 + n_3)} + \frac{[Nn_2 - (n_1 + n_2)(n_2 + n_4)]^2}{N(n_1 + n_2)(n_2 + n_4)} +$$

$$+ \frac{[Nn_3 - (n_3 + n_4)(n_1 + n_3)]^2}{N(n_3 + n_4)(n_1 + n_3)} + \frac{[Nn_4 - (n_3 + n_4)(n_2 + n_4)]^2}{N(n_3 + n_4)(n_2 + n_4)}$$

$$= \frac{(n_1 n_4 - n_2 n_3)^2}{N} \left[\frac{1}{(n_1 + n_2)(n_1 + n_3)} + \frac{1}{(n_1 + n_2)(n_2 + n_4)} + \right.$$

$$\left. + \frac{1}{(n_3 + n_4)(n_1 + n_3)} + \frac{1}{(n_3 + n_4)(n_2 + n_4)} \right]$$

$$= \frac{N(n_1 n_4 - n_2 n_3)^2}{(n_1 + n_2)(n_1 + n_3)(n_2 + n_4)(n_3 + n_4)}.$$

In fact, this is the contingency χ^2 with 1 d.f., and it tests the independence of the attributes A and B by using the maximum-likelihood estimates $\hat{p}_a$ and $\hat{p}_b$ to determine the expected frequencies in the 2×2 table. The important point is that the independence of the attributes A and B can be tested only if either p_a and p_b are estimated from the sample data, or their values are specified by some other hypothesis. The first condition leads to the above contingency χ^2, and we next consider the test for the independence of A and B when there is a hypothetical specification of the values of p_a and p_b.

Suppose, then, that two specific hypotheses postulate that in the sampled population the individuals in the A_1 and A_2 classes are in the ratio of $l : 1$, and those in the B_1 and B_2 classes in the ratio of $k : 1$, where l and k are positive numbers. It is clear that these two hypotheses are equivalent to the hypotheses that $p_a = l/(l + 1)$ and $p_b = k/(k + 1)$. Therefore if the two attributes A and B are assumed to be independent, then the probabilities of a random observation from the population being in the four classes $A_1 B_1$, $A_1 B_2$, $A_2 B_1$, $A_2 B_2$ respectively are

$$p_1 = \frac{lk}{(l + 1)(k + 1)}, \qquad\qquad p_2 = \frac{l}{(l + 1)(k + 1)},$$

$$p_3 = \frac{k}{(l + 1)(k + 1)}, \qquad\qquad p_4 = \frac{1}{(l + 1)(k + 1)}.$$

Hence the expected frequencies in the four cells are Np_1, Np_2, Np_3, Np_4. Thus, using these expected frequencies, the goodness of fit χ^2 is

$$\chi^2 = \sum_{i=1}^{4} (n_i - Np_i)^2 / Np_i \quad \text{with 3 d.f.,}$$

as no parameters have been estimated from the data. This χ^2 tests jointly the three null hypotheses, namely,

(i) $H[p_a = l/(l + 1)];$ (ii) $H[p_b = k/(k + 1)];$ and

(iii) assuming (i) and (ii), the two attributes A and B are independent. Our purpose now is to partition this composite χ^2 with 3 d.f. into three independent χ^2's each with 1 d.f. for testing separately the null hypotheses (i) to (iii).

17.12 Partitioning of χ^2 in a 2×2 table

The null hypothesis (i) states that $p_a = l/(l + 1)$, whereas the maximum-likelihood estimate of p_a is $\hat{p}_a = (n_1 + n_2)/N$. Therefore, if this null hypothesis is true, then the quantity

$$\hat{p}_a - \frac{l}{l + 1} = \frac{n_1 + n_2 - l(n_3 + n_4)}{N(l + 1)}$$

must measure the difference between observation and the null hypothesis which should be ascribable wholly to errors of sampling. Also, since $N(l + 1)$ is a constant, the difference is proportional to the linear function of the cell frequencies

$$X = n_1 + n_2 - l(n_3 + n_4).$$

Similarly, if the null hypothesis (ii) is true, then

$$\hat{p}_b - \frac{k}{k + 1} = \frac{n_1 + n_3 - k(n_2 + n_4)}{N(k + 1)},$$

so that the difference between observation and expectation is proportional to the linear function

$$Y = n_1 + n_3 - k(n_2 + n_4).$$

Now a third linear function of the n_i which is orthogonal to both X and Y is

$$Z = n_1 - kn_2 - ln_3 + lkn_4.$$

This function can be written down simply by noting that the coefficient of any n_i in Z is the product of the corresponding coefficients of n_i in X and Y. We shall indicate shortly the meaning of this linear function Z. But, before we do this, it is important to recall that if the null hypotheses (i) to (iii) are true, then the n_i have a multinomial distribution such that

$$E(n_i) = Np_i, \quad \text{var}(n_i) = Np_i(1 - p_i), \quad \text{for } i = 1, 2, 3, 4;$$

and $\qquad\qquad \text{cov}(n_i, n_j) = -Np_i p_j, \quad \text{for all } j \neq i.$

By using these results, it is simple to verify that

$$E(X) = E(Y) = E(Z) = 0;$$
$$\text{var}(X) = lN, \quad \text{var}(Y) = kN, \quad \text{var}(Z) = lkN;$$

and $\qquad \text{cov}(X, Y) = \text{cov}(Y, Z) = \text{cov}(Z, X) = 0.$

Thus X, Y, and Z are a set of mutually orthogonal linear functions. Besides, it is also known that for large samples

$$\chi_1^2 = X^2/\text{var}(X) = X^2/lN,$$

$$\chi^2_2 = Y^2/\text{var}(Y) = Y^2/kN,$$

and
$$\chi^2_3 = Z^2/\text{var}(Z) = Z^2/lkN$$

are distributed as independent χ^2's each with a single degree of freedom. Finally, it can be verified that

$$\chi^2_1 + \chi^2_2 + \chi^2_3 \equiv \sum_{i=1}^{4} (n_i - Np_i)^2/Np_i.$$

We have thus partitioned the goodness of fit χ^2 into three independent component χ^2's. This partitioning ensures that χ^2_1 tests the null hypothesis $H[p_a = l/(l+1)]$, χ^2_2 the null hypothesis $H[p_b = k/(k+1)]$, and χ^2_3 the independence of the attributes A and B, assuming that $p_a = l/(l+1)$ and $p_b = k/(k+1)$.

To conclude, it is instructive to consider the difference between χ^2_3 and the usual large-sample contingency χ^2 for testing the independence of the attributes A and B. The difference is that χ^2_3 is evaluated when p_a and p_b have values assigned by hypothesis, whereas the contingency χ^2 is used when p_a and p_b are estimated from the marginal totals of the 2×2 table. Indeed, the maximum-likelihood estimates $\hat{p}_a$ and $\hat{p}_b$ are obtained by using the estimating equations $X = 0$ and $Y = 0$. Since $p_a = l/(l+1)$ we have $l = p_a/(1 - p_a)$, and so $X = 0$ gives the estimating equation

$$(1 - \hat{p}_a)(n_1 + n_2) - \hat{p}_a(n_3 + n_4) = 0, \quad \text{whence} \quad \hat{p}_a = (n_1 + n_2)/N, \text{ as before.}$$

Similarly, $k = p_b/(1 - p_b)$, and we have from $Y = 0$ the estimate $\hat{p}_b = (n_1 + n_3)/N$. This means that in estimating p_a and p_b from the data, we reduce both χ^2_1 and χ^2_2 to zero. Then χ^2_3 is simply the contingency χ^2 as obtained earlier. Thus if we write for l and k the estimated values

$$l^* = \hat{p}_a/(1 - \hat{p}_a) = (n_1 + n_2)/(n_3 + n_4),$$

$$k^* = \hat{p}_b/(1 - \hat{p}_b) = (n_1 + n_3)/(n_2 + n_4),$$

then

$$\chi^2_3 = \frac{(n_1 - k^*n_2 - l^*n_3 + l^*k^*n_4)^2}{l^*k^*N} \equiv \frac{N(n_1 n_4 - n_2 n_3)^2}{(n_1 + n_2)(n_3 + n_4)(n_1 + n_3)(n_2 + n_4)}.$$

The expression on the left is another way for evaluating the contingency χ^2 for testing the independence of the attributes A and B.

Example 17.8: Mendelian inheritance in plant breeding

We illustrate this technique of partitioning a composite χ^2 by the analysis of a simple but frequently occurring situation in the genetical study of plant breeding experiments. The attribute A refers to a gene which has two distinct and mutually exclusive forms A_1 and A_2 known as *alleles*. Similarly, the attribute B refers to another gene which also has two alleles B_1 and B_2. Generally, in a population any A_i may occur with any B_j to give plants of the constitution A_iB_j. There are thus four distinct types of plants with constitutions A_1B_1, A_1B_2, A_2B_1, A_2B_2. The alleles of the two genes occur in different proportions in a population depending upon the constitution of its parental plants.

A breeding experiment with poppy plants *Papaver Rhoeas* gave the following frequency distribution of plants in the four classes $A_1 B_1$, $A_1 B_2$, $A_2 B_1$, $A_2 B_2$.

Table 17.12 Showing the observed and expected frequency distributions of the four types of progeny in a breeding experiment with poppy plants

Frequency \\ Type	$A_1 B_1$	$A_1 B_2$	$A_2 B_1$	$A_2 B_2$	Total
Observed	72	29	40	12	153
Expected	86·0625	28·6875	28·6875	9·5625	153

Source: J. Philp (1933), *Journal of Genetics*, Vol.28, p.175.

In this case, the theory of Mendel suggests that in the population from which the sample of 153 plants was obtained, the relative frequencies of the alleles A_1, A_2, and B_1, B_2 should both be in the ratio 3 : 1. Furthermore, it is also expected that the genes A and B are inherited independently in the plants bred. Hence our theoretical analysis is applicable with $l = k = 3$. We therefore have

$$p_1 = \frac{9}{16}, \qquad p_2 = p_3 = \frac{3}{16}, \qquad p_4 = \frac{1}{16}.$$

Since $N = 153$, the expected frequencies are easily evaluated, and these are given in the last row of Table 17.12. Hence the goodness of fit χ^2 is

$$= \frac{72^2}{86\cdot0625} + \frac{29^2}{28\cdot6875} + \frac{40^2}{28\cdot6875} + \frac{12^2}{9\cdot5625} - 153$$

$$= 60\cdot2353 + 29\cdot3159 + 55\cdot7734 + 15\cdot0588 - 153 = 7\cdot38 \quad \text{with 3 d.f.}$$

The five per cent point of the χ^2 distribution with 3 d.f. is 7·82, and the observed χ^2 value is rather close to the tabular value. There is therefore some suspicion about possible disagreement between observation and expectation. We now partition this χ^2 to test separately each of the three constituent elements of the Mendelian hypothesis which gave the expected cell frequencies.

Thus, to test the null hypothesis $H(p_a = 3/4)$, we have

$$\chi_1^2 = \frac{[72 + 29 - 3(40 + 12)]^2}{3 \times 153} = 6\cdot59 \quad \text{with 1 d.f.}$$

Similarly, to test the null hypothesis $H(p_b = 3/4)$, we have

$$\chi_2^2 = \frac{[72 + 40 - 3(29 + 12)]^2}{3 \times 153} = 0\cdot26 \quad \text{with 1 d.f.}$$

Finally, to test the independence of the inheritance of the genes A and B, we have

$$\chi_3^2 = \frac{[72 - 3 \times 29 - 3 \times 40 + 9 \times 12]^2}{9 \times 153} = 0\cdot53 \quad \text{with 1 d.f.}$$

As a check, we have $\chi_1^2 + \chi_2^2 + \chi_3^2 = 7\cdot38$ in agreement with the composite χ^2 calculated above. Since the five per cent point of the χ^2 distribution with 1 d.f. is $3\cdot84$, it is clear that both χ_2^2 and χ_3^2 are not significant, whereas the significance of χ_1^2 is beyond reasonable doubt. We therefore infer that the experimental data indicate that

(i)   the alleles A_1 and A_2 are not in the expected ratio of 3 : 1;

(ii) the alleles B_1 and B_2 are in agreement with the expected ratio of 3 : 1; and

(iii) there is no association between the inheritance of the genes A and B.

The composite χ^2 was masking the possible reason for the suspected deviation of the observations from the expectation based on Mendel's theory, but the partitioned χ^2's have clearly indicated the nature of the departures from expectation.

Example 17.9: Sex and mortality in pulmonary tuberculosis

We conclude by an analysis of G. Berg's data on the incidence of open pulmonary tuberculosis in Table 5.4 of Chapter 5. The figures pertain to the number of deaths among men and women of different age-groups during the first year after the detection of infection. There are ten age-groups and, as the samples observed in each are clearly independent, they can, in the first instance, be analysed separately to determine whether attribute A (sex) is independent of attribute B (mortality within a year after the detection of infection). The cell frequencies in the ten 2×2 tables are reasonably large, and we ignore the correction for continuity in evaluating the contingency χ^2's. Also, it is adequate to retain four significant figures in the denominators of the χ^2's and to evaluate the numerators to the same accuracy. This is a useful computational point, especially when the capacity of the calculating machine is limited.

Table 17.13 gives the ten 2×2 tables and the calculation of the χ^2's for testing the independence of sex and mortality in the age-groups. Since the five per cent point of the χ^2 distribution with 1 d.f. is $3\cdot84$, it is clear that the χ^2's for the fourth, sixth, and seventh age-groups are significant, the remaining seven χ^2's being quite non-significant. This means that in these three age-groups there is evidence for sex differences in the incidence of mortality. However, the ten χ^2's are based on samples and their values are clearly dependent upon chance. The question of interest is whether these ten observed χ^2's could, as a whole, be regarded as showing a consistent pattern. In other words, we wish to evaluate a heterogeneity χ^2 to compare the differences between the male and female mortality over the age-groups. To do this, we first add the ten observed χ^2's to obtain a total χ^2 as

$$\chi_t^2 = 22\cdot23 \quad \text{with 10 d.f.}$$

Table 17.13   Showing the calculations of the χ^2's based on the data of Table 5.4

1st age-group

	Deaths	Survivals	Total
Men	156	250	406
Women	174	326	500
Total	330	576	906

$$\chi^2 = \frac{10^7 \times 4902}{10^7 \times 3859} = 1 \cdot 27 \quad \text{with 1 d.f.}$$

2nd age-group

	Deaths	Survivals	Total
Men	204	491	695
Women	246	570	816
Total	450	1061	1511

$$\chi^2 = \frac{10^7 \times 3068}{10^7 \times 27077} = 0 \cdot 11 \quad \text{with 1 d.f.}$$

3rd age-group

	Deaths	Survivals	Total
Men	169	416	585
Women	184	435	619
Total	353	851	1204

$$\chi^2 = \frac{10^7 \times 1105}{10^7 \times 10878} = 0 \cdot 10 \quad \text{with 1 d.f.}$$

4th age-group

	Deaths	Survivals	Total
Men	128	326	454
Women	150	283	433
Total	278	609	887

$$\chi^2 = \frac{10^7 \times 14252}{10^7 \times 3328} = 4 \cdot 28 \quad \text{with 1 d.f.}$$

5th age-group

	Deaths	Survivals	Total
Men	82	192	274
Women	92	165	257
Total	174	357	531

$$\chi^2 = \frac{10^6 \times 9075}{10^6 \times 4374} = 2 \cdot 07 \quad \text{with 1 d.f.}$$

6th age-group

	Deaths	Survivals	Total
Men	68	153	221
Women	83	111	194
Total	151	264	415

$$\chi^2 = \frac{10^6 \times 11011}{10^6 \times 1709} = 6 \cdot 44 \quad \text{with 1 d.f.}$$

Table 17.13 (continued)

7th age-group

	Deaths	Survivals	Total
Men	41	112	153
Women	39	55	94
Total	80	167	247

$$\chi^2 = \frac{10^5 \times 11028}{10^5 \times 1921} = 5 \cdot 74 \quad \text{with 1 d.f.}$$

8th age-group

	Deaths	Survivals	Total
Men	34	76	110
Women	20	38	58
Total	54	114	168

$$\chi^2 = \frac{10^3 \times 8733}{10^3 \times 39275} = 0 \cdot 22 \quad \text{with 1 d.f.}$$

9th age-group

	Deaths	Survivals	Total
Men	36	33	69
Women	13	16	29
Total	49	49	98

$$\chi^2 = \frac{10^3 \times 2118}{10^3 \times 4804} = 0 \cdot 44 \quad \text{with 1 d.f.}$$

10th age-group

	Deaths	Survivals	Total
Men	43	46	89
Women	28	19	47
Total	71	65	136

$$\chi^2 = \frac{10^4 \times 3017}{10^4 \times 1930} = 1 \cdot 56 \quad \text{with 1 d.f.}$$

This χ_t^2 is significant at the five per cent level since the corresponding percentage point of the χ^2 distribution with 10 d.f. is $18 \cdot 31$. Next, to derive the heterogeneity χ^2, we have to subtract from χ_t^2 the contribution of the χ^2 which tests the independence of sex and mortality irrespective of age. Thus, if we sum the ten 2×2 tables over the age-groups, we obtain the following 2×2 table.

Table 17.14   Showing the incidence of mortality for men and women
irrespective of age

	Deaths	Survivals	Total
Men	961	2095	3056
Women	1029	2018	3047
Total	1990	4113	6103

For this contingency table, the χ^2 for testing the independence of sex and mortality is

$$\chi_m^2 = \frac{10^{10} \times 28595}{10^{10} \times 7621} = 3 \cdot 75 \quad \text{with 1 d.f.}$$

This χ^2 is not significant at the five per cent level, though its value is close to the significance point $3 \cdot 84$. However, we can conclude that there is some evidence that over the whole age-range the average incidence of mortality is the same for the two sexes. The heterogeneity χ^2 is now obtained simply as the difference

$$\chi_t^2 - \chi_m^2 = 22 \cdot 23 - 3 \cdot 75 = 18 \cdot 48 \quad \text{with 9 d.f.}$$

This χ^2 is significant at the five per cent level since the corresponding percentage point of the χ^2 distribution with 9 d.f. is $16 \cdot 92$. This means that, as a whole, the ten age-groups are not showing a consistent pattern for the association between sex and mortality. Furthermore, since the heterogeneity χ^2 is significant, a more appropriate test of significance for the overall independence of sex and mortality is obtained by testing χ_m^2 with 1 d.f. against the heterogeneity χ^2 with 9 d.f. This is done by using the fact that the ratio

$$F = \frac{9\chi_m^2}{\chi_t^2 - \chi_m^2} = \frac{9 \times 3 \cdot 75}{18 \cdot 48} = 1 \cdot 83$$

has the F distribution with 1, 9 d.f. This observed F value is clearly not significant since the five per cent point of the F distribution with 1, 9 d.f. is $5 \cdot 12$. We thus observe that the significance of the heterogeneity χ^2 has reduced the significance of χ_m^2.

An alternative model

In the preceding analysis, we postulated a single population of tubercular persons, irrespective of sex, from which a random sample of 6,103 persons was observed. The sampled individuals were then classified according to sex and mortality within a year of detection of infection. The resulting frequency distribution in the four cells of Table 17.14 is multinomial and it leads to the overall contingency χ^2 test for the independence of sex and mortality. It is instructive now to look at the data of Table 17.14 from a different point of view. Suppose we postulate that there are two distinct populations of detected tubercular men and women respectively, and that we obtained samples of 3,056 men and 3,047 women from these populations. The interest is in the comparison of the true incidence of mortality in the two populations.

If p_m and p_f are the true proportions of mortality in the populations of detected tubercular men and women respectively, then the maximum-likelihood estimates of these parameters are

$$\hat{p}_m = \frac{961}{3056} = 0 \cdot 3145 \quad \text{and} \quad \hat{p}_f = \frac{1029}{3047} = 0 \cdot 3377.$$

Therefore

$$\text{estimate var}(\hat{p}_m) = \frac{0 \cdot 3145 \times 0 \cdot 6855}{3055} = 0 \cdot 00007057,$$

and $$\text{estimate var}(\hat{p}_f) = \frac{0 \cdot 3377 \times 0 \cdot 6623}{3046} = 0 \cdot 00007343.$$

Hence   estimate var$(\hat{p}_f - \hat{p}_m) = 0 \cdot 00014400,$

so that   s.e.$(\hat{p}_f - \hat{p}_m) = 0 \cdot 0120.$

Also, we have $\hat{p}_f - \hat{p}_m = 0 \cdot 0232.$

Therefore, assuming that the random variable $\hat{p}_f - \hat{p}_m$ is approximately normally distributed, we obtain the 95 per cent confidence interval for the parametric difference $p_f - p_m$ as

$$0 \cdot 0232 - 1 \cdot 96 \times 0 \cdot 0120 \leqslant p_f - p_m \leqslant 0 \cdot 0232 + 1 \cdot 96 \times 0 \cdot 0120,$$

or $$- 0 \cdot 0003 \leqslant p_f - p_m \leqslant 0 \cdot 0467.$$

Since this confidence interval contains the point zero, it follows that the null hypothesis $H(p_f - p_m = 0)$ is acceptable at the five per cent level. This test means that, on the average, the data do not give evidence for a difference between the true incidence of mortality of the two sexes. We therefore conclude that this comparison of the two binomial proportions is another way for testing the lack of association between sex and the incidence of mortality. But, and this is a fundamental point, the model underlying this analysis is that there are two distinct binomial populations instead of a single multinomial one. The choice between the two models depends upon whether the total number of men (3,056) and women (3,047) are regarded as predetermined by the experimenter or simply their sum (6,103). In the first case, the model of two binomial populations is the correct one to use, whilst in the second the multinomial model is applicable.

Finally, continuing with the model of two binomial populations, if we assume that $p_f = p_m = p$, say, then p denotes the true proportion of deaths in the combined population of detected tubercular persons irrespective of age and sex. The maximum-likelihood estimate of p is

$$\hat{p} = \frac{1990}{6103} = 0 \cdot 3261.$$

Also, $$\text{estimate var}(\hat{p}) = \frac{0 \cdot 3261 \times 0 \cdot 6739}{6102} = 0 \cdot 00003601,$$

so that   s.e.$(\hat{p}) = 0 \cdot 006001.$

Hence, assuming approximate normality for the distribution of $\hat{p}$, we have the 95 per cent confidence interval for p as

$$0 \cdot 3261 - 1 \cdot 96 \times 0 \cdot 006001 \leqslant p \leqslant 0 \cdot 3261 + 1 \cdot 96 \times 0 \cdot 006001,$$

or $$0 \cdot 3143 \leqslant p \leqslant 0 \cdot 3379.$$

These limits are rather close, and we infer that in the combined population the incidence of mortality is approximately one-third.

It is to be noted that this confidence interval for p is also applicable if the multinomial model is used, provided χ_m^2 is not significant.

APPENDIX

Table A.1
Ordinates and areas of the unit normal curve

$$\phi(z) = \frac{1}{\sqrt{2\pi}} e^{-\frac{1}{2}z^2}; \quad \Phi(z) = \int_{-\infty}^{z} \phi(u)\,du.$$

z	$\phi(z)$	$\Phi(z)$	z	$\phi(z)$	$\Phi(z)$	z	$\phi(z)$	$\Phi(z)$
0·00	0·39894	0·50000	0·47	0·35723	0·68082	0·94	0·25647	0·82639
·01	·39892	·50399	·48	·35553	·68439	·95	·25406	·82894
·02	·39886	·50798	·49	·35381	·68793	·96	·25164	·83147
·03	·39876	·51197	·50	·35207	·69146	·97	·24923	·83398
·04	·39862	·51595	·51	·35029	·69497	·98	·24681	·83646
·05	·39844	·51994	·52	·34849	·69847	·99	·24439	·83891
·06	·39822	·52392	·53	·34667	·70194	1·00	·24197	·84134
·07	·39797	·52790	·54	·34482	·70540	1·01	·23955	·84375
·08	·39767	·53188	·55	·34294	·70884	1·02	·23713	·84614
·09	·39733	·53586	·56	·34105	·71226	1·03	·23471	·84850
·10	·39695	·53983	·57	·33912	·71566	1·04	·23230	·85083
·11	·39654	·54380	·58	·33718	·71904	1·05	·22988	·85314
·12	·39608	·54776	·59	·33521	·72240	1·06	·22747	·85543
·13	·39559	·55172	·60	·33322	·72575	1·07	·22506	·85769
·14	·39505	·55567	·61	·33121	·72907	1·08	·22265	·85993
·15	·39448	·55962	·62	·32918	·73237	1·09	·22025	·86214
·16	·39387	·56356	·63	·32713	·73565	1·10	·21785	·86433
·17	·39322	·56749	·64	·32506	·73891	1·11	·21546	·86650
·18	·39253	·57142	·65	·32297	·74215	1·12	·21307	·86864
·19	·39181	·57535	·66	·32086	·74537	1·13	·21069	·87076
·20	·39104	·57926	·67	·31874	·74857	1·14	·20831	·87286
·21	·39024	·58317	·68	·31659	·75175	1·15	·20594	·87493
·22	·38940	·58706	·69	·31443	·75490	1·16	·20357	·87698
·23	·38853	·59095	·70	·31225	·75804	1·17	·20121	·87900
·24	·38762	·59483	·71	·31006	·76115	1·18	·19886	·88100
·25	·38667	·59871	·72	·30785	·76424	1·19	·19652	·88298
·26	·38568	·60257	·73	·30563	·76730	1·20	·19419	·88493
·27	·38466	·60642	·74	·30339	·77035	1·21	·19186	·88686
·28	·38361	·61026	·75	·30114	·77337	1·22	·18954	·88877
·29	·38251	·61409	·76	·29887	·77637	1·23	·18724	·89065
·30	·38139	·61791	·77	·29659	·77935	1·24	·18494	·89251
·31	·38023	·62172	·78	·29431	·78230	1·25	·18265	·89435
·32	·37903	·62552	·79	·29200	·78524	1·26	·18037	·89617
·33	·37780	·62930	·80	·28969	·78814	1·27	·17810	·89796
·34	·37654	·63307	·81	·28737	·79103	1·28	·17585	·89973
·35	·37524	·63683	·82	·28504	·79389	1·29	·17360	·90147
·36	·37391	·64058	·83	·28269	·79673	1·30	·17137	·90320
·37	·37255	·64431	·84	·28034	·79955	1·31	·16915	·90490
·38	·37115	·64803	·85	·27798	·80234	1·32	·16694	·90658
·39	·36973	·65173	·86	·27562	·80511	1·33	·16474	·90824
·40	·36827	·65542	·87	·27324	·80785	1·34	·16256	·90988
·41	·36678	·65910	·88	·27086	·81057	1·35	·16038	·91149
·42	·36526	·66276	·89	·26848	·81327	1·36	·15822	·91309
·43	·36371	·66640	·90	·26609	·81594	1·37	·15608	·91466
·44	·36213	·67003	·91	·26369	·81859	1·38	·15395	·91621
·45	·36053	·67364	·92	·26129	·82121	1·39	·15183	·91774
·46	·35889	·67724	·93	·25888	·82381	1·40	·14973	·91924

Table A.1 (continued)
Ordinates and areas of the unit normal curve

$$\phi(z) = \frac{1}{\sqrt{2\pi}}e^{-\frac{1}{2}z^2}; \quad \Phi(z) = \int_{-\infty}^{z}\phi(u)\,du.$$

z	$\phi(z)$	$\Phi(z)$	z	$\phi(z)$	$\Phi(z)$	z	$\phi(z)$	$\Phi(z)$
1·41	0·14764	0·92073	1·89	0·06687	0·97062	2·37	0·02406	0·99111
1·42	·14556	·92220	1·90	·06562	·97128	2·38	·02349	·99134
1·43	·14350	·92364	1·91	·06439	·97193	2·39	·02294	·99158
1·44	·14146	·92507	1·92	·06316	·97257	2·40	·02239	·99180
1·45	·13943	·92647	1·93	·06195	·97320	2·41	·02186	·99202
1·46	·13742	·92786	1·94	·06077	·97381	2·42	·02134	·99224
1·47	·13542	·92922	1·95	·05959	·97441	2·43	·02083	·99245
1·48	·13344	·93056	1·96	·05844	·97500	2·44	·02033	·99266
1·49	·13147	·93189	1·97	·05730	·97558	2·45	·01984	·99286
1·50	·12952	·93319	1·98	·05618	·97615	2·46	·01936	·99305
1·51	·12758	·93448	1·99	·05508	·97670	2·47	·01889	·99324
1·52	·12566	·93574	2·00	·05399	·97725	2·48	·01842	·99343
1·53	·12376	·93699	2·01	·05292	·97778	2·49	·01797	·99361
1·54	·12188	·93822	2·02	·05186	·97831	2·50	·01753	·99379
1·55	·12001	·93943	2·03	·05082	·97882	2·51	·01709	·99396
1·56	·11816	·94062	2·04	·04980	·97932	2·52	·01667	·99413
1·57	·11632	·94179	2·05	·04879	·97982	2·53	·01625	·99430
1·58	·11450	·94295	2·06	·04780	·98030	2·54	·01585	·99446
1·59	·11270	·94408	2·07	·04682	·98077	2·55	·01545	·99461
1·60	·11092	·94520	2·08	·04586	·98124	2·56	·01506	·99477
1·61	·10915	·94630	2·09	·04491	·98169	2·57	·01468	·99492
1·62	·10741	·94738	2·10	·04398	·98214	2·58	·01431	·99506
1·63	·10567	·94845	2·11	·04307	·98257	2·59	·01394	·99520
1·64	·10396	·94950	2·12	·04217	·98300	2·60	·01358	·99534
1·65	·10226	·95053	2·13	·04128	·98341	2·61	·01323	·99547
1·66	·10059	·95154	2·14	·04041	·98382	2·62	·01289	·99560
1·67	·09893	·95254	2·15	·03955	·98422	2·63	·01256	·99573
1·68	·09728	·95352	2·16	·03871	·98461	2·64	·01223	·99585
1·69	·09566	·95449	2·17	·03788	·98500	2·65	·01191	·99598
1·70	·09405	·95543	2·18	·03706	·98537	2·66	·01160	·99609
1·71	·09246	·95637	2·19	·03626	·98574	2·67	·01130	·99621
1·72	·09089	·95728	2·20	·03547	·98610	2·68	·01100	·99632
1·73	·08933	·95818	2·21	·03470	·98645	2·69	·01071	·99643
1·74	·08780	·95907	2·22	·03394	·98679	2·70	·01042	·99653
1·75	·08628	·95994	2·23	·03319	·98713	2·71	·01014	·99664
1·76	·08478	·96080	2·24	·03246	·98745	2·72	·00987	·99674
1·77	·08329	·96164	2·25	·03174	·98778	2·73	·00961	·99683
1·78	·08183	·96246	2·26	·03103	·98809	2·74	·00935	·99693
1·79	·08038	·96327	2·27	·03034	·98840	2·75	·00909	·99702
1·80	·07895	·96407	2·28	·02965	·98870	2·76	·00885	·99711
1·81	·07754	·96485	2·29	·02898	·98899	2·77	·00861	·99720
1·82	·07614	·96562	2·30	·02833	·98928	2·78	·00837	·99728
1·83	·07477	·96638	2·31	·02768	·98956	2·79	·00814	·99736
1·84	·07341	·96712	2·32	·02705	·98983	2·80	·00792	·99744
1·85	·07206	·96784	2·33	·02643	·99010	2·81	·00770	·99752
1·86	·07074	·96856	2·34	·02582	·99036	2·82	·00748	·99760
1·87	·06943	·96926	2·35	·02522	·99061	2·83	·00727	·99767
1·88	·06814	·96995	2·36	·02463	·99086	2·84	·00707	·99774

Table A.1 (*concluded*)

Ordinates and areas of the unit normal curve

$$\phi(z) = \frac{1}{\sqrt{2\pi}}e^{-\frac{1}{2}z^2}; \quad \Phi(z) = \int_{-\infty}^{z}\phi(u)\,du.$$

z	$\phi(z)$	$\Phi(z)$	z	$\phi(z)$	$\Phi(z)$	z	$\phi(z)$	$\Phi(z)$
2·85	0·00687	0·99781	3·24	0·00210	0·99940	3·63	0·00055	0·99986
2·86	·00668	·99788	3·25	·00203	·99942	3·64	·00053	·99986
2·87	·00649	·99795	3·26	·00196	·99944	3·65	·00051	·99987
2·88	·00631	·99801	3·27	·00190	·99946	3·66	·00049	·99987
2·89	·00613	·99807	3·28	·00184	·99948	3·67	·00047	·99988
2·90	·00595	·99813	3·29	·00178	·99950	3·68	·00046	·99988
2·91	·00578	·99819	3·30	·00172	·99952	3·69	·00044	·99989
2·92	·00562	·99825	3·31	·00167	·99953	3·70	·00042	·99989
2·93	·00545	·99831	3·32	·00161	·99955	3·71	·00041	·99990
2·94	·00530	·99836	3·33	·00156	·99957	3·72	·00039	·99990
2·95	·00514	·99841	3·34	·00151	·99958	3·73	·00038	·99990
2·96	·00499	·99846	3·35	·00146	·99960	3·74	·00037	·99991
2·97	·00485	·99851	3·36	·00141	·99961	3·75	·00035	·99991
2·98	·00471	·99856	3·37	·00136	·99962	3·76	·00034	·99992
2·99	·00457	·99861	3·38	·00132	·99964	3·77	·00033	·99992
3·00	·00443	·99865	3·39	·00127	·99965	3·78	·00031	·99992
3·01	·00430	·99869	3·40	·00123	·99966	3·79	·00030	·99992
3·02	·00417	·99874	3·41	·00119	·99968	3·80	·00029	·99993
3·03	·00405	·99878	3·42	·00115	·99969	3·81	·00028	·99993
3·04	·00393	·99882	3·43	·00111	·99970	3·82	·00027	·99993
3·05	·00381	·99886	3·44	·00107	·99971	3·83	·00026	·99994
3·06	·00370	·99889	3·45	·00104	·99972	3·84	·00025	·99994
3·07	·00358	·99893	3·46	·00100	·99973	3·85	·00024	·99994
3·08	·00348	·99897	3·47	·00097	·99974	3·86	·00023	·99994
3·09	·00337	·99900	3·48	·00094	·99975	3·87	·00022	·99995
3·10	·00327	·99903	3·49	·00090	·99976	3·88	·00021	·99995
3·11	·00317	·99906	3·50	·00087	·99977	3·89	·00021	·99995
3·12	·00307	·99910	3·51	·00084	·99978	3·90	·00020	·99995
3·13	·00298	·99913	3·52	·00081	·99978	3·91	·00019	·99995
3·14	·00288	·99916	3·53	·00079	·99979	3·92	·00018	·99996
3·15	·00279	·99918	3·54	·00076	·99980	3·93	·00018	·99996
3·16	·00271	·99921	3·55	·00073	·99981	3·94	·00017	·99996
3·17	·00262	·99924	3·56	·00071	·99981	3·95	·00016	·99996
3·18	·00254	·99926	3·57	·00068	·99982	3·96	·00016	·99996
3·19	·00246	·99929	3·58	·00066	·99983	3·97	·00015	·99996
3·20	·00238	·99931	3·59	·00063	·99983	3·98	·00014	·99997
3·21	·00231	·99934	3·60	·00061	·99984	3·99	·00014	·99997
3·22	·00224	·99936	3·61	·00059	·99985	4·00	·00013	·99997
3·23	·00216	·99938	3·62	·00057	·99985			

Table A.2

VALUES OF χ^2 CORRESPONDING TO GIVEN PROBABILITIES*

Degrees of freedom n	Probability of a deviation greater than χ^2						
	.01	.02	.05	.10	.20	.30	.50
1	6.635	5.412	3.841	2.706	1.642	1.074	.455
2	9.210	7.824	5.991	4.605	3.219	2.408	1.386
3	11.341	9.837	7.815	6.251	4.642	3.665	2.366
4	13.277	11.668	9.488	7.779	5.989	4.878	3.357
5	15.086	13.388	11.070	9.236	7.289	6.064	4.351
6	16.812	15.033	12.592	10.645	8.558	7.231	5.348
7	18.475	16.622	14.067	12.017	9.803	8.383	6.346
8	20.090	18.168	15.507	13.362	11.030	9.524	7.344
9	21.666	19.679	16.919	14.684	12.242	10.656	8.343
10	23.209	21.161	18.307	15.987	13.442	11.781	9.342
11	24.725	22.618	19.675	17.275	14.631	12.899	10.341
12	26.217	24.054	21.026	18.549	15.812	14.011	11.340
13	27.688	25.472	22.362	19.812	16.985	15.119	12.340
14	29.141	26.873	23.685	21.064	18.151	16.222	13.339
15	30.578	28.259	24.996	22.307	19.311	17.322	14.339
16	32.000	29.633	26.296	23.542	20.465	18.418	15.338
17	33.409	30.995	27.587	24.769	21.615	19.511	16.338
18	34.805	32.346	28.869	25.989	22.760	20.601	17.338
19	36.191	33.687	30.144	27.204	23.900	21.689	18.338
20	37.566	35.020	31.410	28.412	25.038	22.775	19.337
21	38.932	36.343	32.671	29.615	26.171	23.858	20.337
22	40.289	37.659	33.924	30.813	27.301	24.939	21.337
23	41.638	38.968	35.172	32.007	28.429	26.018	22.337
24	42.980	40.270	36.415	33.196	29.553	27.096	23.337
25	44.314	41.566	37.652	34.382	30.675	28.172	24.337
26	45.642	42.856	38.885	35.563	31.795	29.246	25.336
27	46.963	44.140	40.113	36.741	32.912	30.319	26.336
28	48.278	45.419	41.337	37.916	34.027	31.391	27.336
29	49.588	46.693	42.557	39.087	35.139	32.461	28.336
30	50.892	47.962	43.773	40.256	36.250	33.530	29.336

* For larger values of n, the quantity $(2\chi^2)^{\frac{1}{2}} - (2n-1)^{\frac{1}{2}}$ may be used as a normal variable with zero mean and unit variance.

This table is reproduced from *Statistical Methods for Research Workers*, with the kind permission of the Literary Executor of the late Sir Ronald A. Fisher, F.R.S., Cambridge, and Oliver and Boyd, Ltd, Edinburgh.

Table A.2 (*concluded*)

VALUES OF χ^2 CORRESPONDING TO GIVEN PROBABILITIES*

Degrees of freedom n	Probability of a deviation greater than χ^2					
	.70	.80	.90	.95	.98	.99
1	.148	.0642	.0158	.00393	.000628	.000157
2	.713	.446	.211	.103	.0404	.0201
3	1.424	1.005	.584	.352	.185	.115
4	2.195	1.649	1.064	.711	.429	.297
5	3.000	2.343	1.610	1.145	.752	.554
6	3.828	3.070	2.204	1.635	1.134	.872
7	4.671	3.822	2.833	2.167	1.564	1.239
8	5.527	4.594	3.490	2.733	2.032	1.646
9	6.393	5.380	4.168	3.325	2.532	2.088
10	7.267	6.179	4.865	3.940	3.059	2.558
11	8.148	6.989	5.578	4.575	3.609	3.053
12	9.034	7.807	6.304	5.226	4.178	3.571
13	9.926	8.634	7.042	5.892	4.765	4.107
14	10.821	9.467	7.790	6.571	5.368	4.660
15	11.721	10.307	8.547	7.261	5.985	5.229
16	12.624	11.152	9.312	7.962	6.614	5.812
17	13.531	12.002	10.085	8.672	7.255	6.408
18	14.440	12.857	10.865	9.390	7.906	7.015
19	15.352	13.716	11.651	10.117	8.567	7.633
20	16.266	14.578	12.443	10.851	9.237	8.260
21	17.182	15.445	13.240	11.591	9.915	8.897
22	18.101	16.314	14.041	12.338	10.600	9.542
23	19.021	17.187	14.848	13.091	11.293	10.196
24	19.943	18.062	15.659	13.848	11.992	10.856
25	20.867	18.940	16.473	14.611	12.697	11.524
26	21.792	19.820	17.292	15.379	13.409	12.198
27	22.719	20.703	18.114	16.151	14.125	12.879
28	23.647	21.588	18.939	16.928	14.847	13.565
29	24.577	22.475	19.768	17.708	15.574	14.256
30	25.508	23.364	20.599	18.493	16.306	14.953

* For larger values of n, the quantity $(2\chi^2)^{\frac{1}{2}} - (2n-1)^{\frac{1}{2}}$ may be used as a normal variable with zero mean and unit variance.

Table A.3

VALUES OF t CORRESPONDING TO GIVEN PROBABILITIES*

Degrees of freedom n	Probability of a deviation greater than t					
	.005	.01	.025	.05	.1	.15
1	63.657	31.821	12.706	6.314	3.078	1.963
2	9.925	6.965	4.303	2.920	1.886	1.386
3	5.841	4.541	3.182	2.353	1.638	1.250
4	4.604	3.747	2.776	2.132	1.533	1.190
5	4.032	3.365	2.571	2.015	1.476	1.156
6	3.707	3.143	2.447	1.943	1.440	1.134
7	3.499	2.998	2.365	1.895	1.415	1.119
8	3.355	2.896	2.306	1.860	1.397	1.108
9	3.250	2.821	2.262	1.833	1.383	1.100
10	3.169	2.764	2.228	1.812	1.372	1.093
11	3.106	2.718	2.201	1.796	1.363	1.088
12	3.055	2.681	2.179	1.782	1.356	1.083
13	3.012	2.650	2.160	1.771	1.350	1.079
14	2.977	2.624	2.145	1.761	1.345	1.076
15	2.947	2.602	2.131	1.753	1.341	1.074
16	2.921	2.583	2.120	1.746	1.337	1.071
17	2.898	2.567	2.110	1.740	1.333	1.069
18	2.878	2.552	2.101	1.734	1.330	1.067
19	2.861	2.539	2.093	1.729	1.328	1.066
20	2.845	2.528	2.086	1.725	1.325	1.064
21	2.831	2.518	2.080	1.721	1.323	1.063
22	2.819	2.508	2.074	1.717	1.321	1.061
23	2.807	2.500	2.069	1.714	1.319	1.060
24	2.797	2.492	2.064	1.711	1.318	1.059
25	2.787	2.485	2.060	1.708	1.316	1.058
26	2.779	2.479	2.056	1.706	1.315	1.058
27	2.771	2.473	2.052	1.703	1.314	1.057
28	2.763	2.467	2.048	1.701	1.313	1.056
29	2.756	2.462	2.045	1.699	1.311	1.055
30	2.750	2.457	2.012	1.697	1.310	1.055
∞	2.576	2.326	1.960	1.645	1.282	1.036

* The probability of a deviation numerically greater than t is twice the probability given at the head of the table.

This table is reproduced from *Statistical Methods for Research Workers*, with the kind permission of the Literary Executor of the late Sir Ronald A. Fisher, F.R.S., Cambridge, and Oliver and Boyd, Ltd, Eninburgh.

Table A.3 (concluded)

VALUES OF t CORRESPONDING TO GIVEN PROBABILITIES*

Degrees of freedom n	Probability of a deviation greater than t					
	.2	.25	.3	.35	.4	.45
1	1.376	1.000	.727	.510	.325	.158
2	1.061	.816	.617	.445	.289	.142
3	.978	.765	.584	.424	.277	.137
4	.941	.741	.569	.414	.271	.134
5	.920	.727	.559	.408	.267	.132
6	.906	.718	.553	.404	.265	.131
7	.896	.711	.549	.402	.263	.130
8	.889	.706	.546	.399	.262	.130
9	.883	.703	.543	.398	.261	.129
10	.879	.700	.542	.397	.260	.129
11	.876	.697	.540	.396	.260	.129
12	.873	.695	.539	.395	.259	.128
13	.870	.694	.538	.394	.259	.128
14	.868	.692	.537	.393	.258	.128
15	.866	.691	.536	.393	.258	.128
16	.865	.690	.535	.392	.258	.128
17	.863	.689	.534	.392	.257	.128
18	.862	.688	.534	.392	..257	.127
19	.861	.688	.533	.391	.257	.127
20	.860	.687	.533	.391	.257	.127
21	.859	.686	.532	.391	.257	.127
22	.858	.686	.532	.390	.256	.127
23	.858	.685	.532	.390	.256	.127
24	.857	.685	.531	.390	.256	.127
25	.856	.684	.531	.390	.256	.127
26	.856	.684	.531	.390	.256	.127
27	.855	.684	.531	.389	.256	.127
28	.855	.683	.530	.389	.256	.127
29	.854	.683	.530	.389	.256	.127
30	.854	.683	.530	.389	.256	.127
∞	.842	.674	.524	.385	.253	.126

* The probability of a deviation numerically greater than t is twice the probability given at the head of the table.

Table A.4

5% (Roman Type) and 1% (Boldface Type) Points in the Distribution of F*

n_1 degrees of freedom (for greater mean square)

n_2	1	2	3	4	5	6	7	8	9	10	11	12	14	16	20	24	30	40	50	75	100	200	500	∞
1	161 / **4,052**	200 / **4,999**	216 / **5,403**	225 / **5,625**	230 / **5,764**	234 / **5,859**	237 / **5,928**	239 / **5,981**	241 / **6,022**	242 / **6,056**	243 / **6,082**	244 / **6,106**	245 / **6,142**	246 / **6,169**	248 / **6,208**	249 / **6,234**	250 / **6,258**	251 / **6,286**	252 / **6,302**	253 / **6,323**	253 / **6,334**	254 / **6,352**	254 / **6,361**	254 / **6,366**
2	18.51 / **98.49**	19.00 / **99.01**	19.16 / **99.17**	19.25 / **99.25**	19.30 / **99.30**	19.33 / **99.33**	19.36 / **99.34**	19.37 / **99.36**	19.38 / **99.38**	19.39 / **99.40**	19.40 / **99.41**	19.41 / **99.42**	19.42 / **99.43**	19.43 / **99.44**	19.44 / **99.45**	19.45 / **99.46**	19.46 / **99.47**	19.47 / **99.48**	19.47 / **99.48**	19.48 / **99.49**	19.49 / **99.49**	19.49 / **99.49**	19.50 / **99.50**	19.50 / **99.50**
3	10.13 / **34.12**	9.55 / **30.81**	9.28 / **29.46**	9.12 / **28.71**	9.01 / **28.24**	8.94 / **27.91**	8.88 / **27.67**	8.84 / **27.49**	8.81 / **27.34**	8.78 / **27.23**	8.76 / **27.13**	8.74 / **27.05**	8.71 / **26.92**	8.69 / **26.83**	8.66 / **26.69**	8.64 / **26.60**	8.62 / **26.50**	8.60 / **26.41**	8.58 / **26.35**	8.57 / **26.27**	8.56 / **26.23**	8.54 / **26.18**	8.54 / **26.14**	8.53 / **26.12**
4	7.71 / **21.20**	6.94 / **18.00**	6.59 / **16.69**	6.39 / **15.98**	6.26 / **15.52**	6.16 / **15.21**	6.09 / **14.98**	6.04 / **14.80**	6.00 / **14.66**	5.96 / **14.54**	5.93 / **14.45**	5.91 / **14.37**	5.87 / **14.24**	5.84 / **14.15**	5.80 / **14.02**	5.77 / **13.93**	5.74 / **13.83**	5.71 / **13.74**	5.70 / **13.69**	5.68 / **13.61**	5.66 / **13.57**	5.65 / **13.52**	5.64 / **13.48**	5.63 / **13.46**
5	6.61 / **16.26**	5.79 / **13.27**	5.41 / **12.06**	5.19 / **11.39**	5.05 / **10.97**	4.95 / **10.67**	4.88 / **10.45**	4.82 / **10.27**	4.78 / **10.15**	4.74 / **10.05**	4.70 / **9.96**	4.68 / **9.89**	4.64 / **9.77**	4.60 / **9.68**	4.56 / **9.55**	4.53 / **9.47**	4.50 / **9.38**	4.46 / **9.29**	4.44 / **9.24**	4.42 / **9.17**	4.40 / **9.13**	4.38 / **9.07**	4.37 / **9.04**	4.36 / **9.02**
6	5.99 / **13.74**	5.14 / **10.92**	4.76 / **9.78**	4.53 / **9.15**	4.39 / **8.75**	4.28 / **8.47**	4.21 / **8.26**	4.15 / **8.10**	4.10 / **7.98**	4.06 / **7.87**	4.03 / **7.79**	4.00 / **7.72**	3.96 / **7.60**	3.92 / **7.52**	3.87 / **7.39**	3.84 / **7.31**	3.81 / **7.23**	3.77 / **7.14**	3.75 / **7.09**	3.72 / **7.02**	3.71 / **6.99**	3.69 / **6.94**	3.68 / **6.90**	3.67 / **6.88**
7	5.59 / **12.25**	4.74 / **9.55**	4.35 / **8.45**	4.12 / **7.85**	3.97 / **7.46**	3.87 / **7.19**	3.79 / **7.00**	3.73 / **6.84**	3.68 / **6.71**	3.63 / **6.62**	3.60 / **6.54**	3.57 / **6.47**	3.52 / **6.35**	3.49 / **6.27**	3.44 / **6.15**	3.41 / **6.07**	3.38 / **5.98**	3.34 / **5.90**	3.32 / **5.85**	3.29 / **5.78**	3.28 / **5.75**	3.25 / **5.70**	3.24 / **5.67**	3.23 / **5.65**
8	5.32 / **11.26**	4.46 / **8.65**	4.07 / **7.59**	3.84 / **7.01**	3.69 / **6.63**	3.58 / **6.37**	3.50 / **6.19**	3.44 / **6.03**	3.39 / **5.91**	3.34 / **5.82**	3.31 / **5.74**	3.28 / **5.67**	3.23 / **5.56**	3.20 / **5.48**	3.15 / **5.36**	3.12 / **5.28**	3.08 / **5.20**	3.05 / **5.11**	3.03 / **5.06**	3.00 / **5.00**	2.98 / **4.96**	2.96 / **4.91**	2.94 / **4.88**	2.93 / **4.86**
9	5.12 / **10.56**	4.26 / **8.02**	3.86 / **6.99**	3.63 / **6.42**	3.48 / **6.06**	3.37 / **5.80**	3.29 / **5.62**	3.23 / **5.47**	3.18 / **5.35**	3.13 / **5.26**	3.10 / **5.18**	3.07 / **5.11**	3.02 / **5.00**	2.98 / **4.92**	2.93 / **4.80**	2.90 / **4.73**	2.86 / **4.64**	2.82 / **4.56**	2.80 / **4.51**	2.77 / **4.45**	2.76 / **4.41**	2.73 / **4.36**	2.72 / **4.33**	2.71 / **4.31**
10	4.96 / **10.04**	4.10 / **7.56**	3.71 / **6.55**	3.48 / **5.99**	3.33 / **5.64**	3.22 / **5.39**	3.14 / **5.21**	3.07 / **5.06**	3.02 / **4.95**	2.97 / **4.85**	2.94 / **4.78**	2.91 / **4.71**	2.86 / **4.60**	2.82 / **4.52**	2.77 / **4.41**	2.74 / **4.33**	2.70 / **4.25**	2.67 / **4.17**	2.64 / **4.12**	2.61 / **4.05**	2.59 / **4.01**	2.56 / **3.96**	2.55 / **3.93**	2.54 / **3.91**
11	4.84 / **9.65**	3.98 / **7.20**	3.59 / **6.22**	3.36 / **5.67**	3.20 / **5.32**	3.09 / **5.07**	3.01 / **4.88**	2.95 / **4.74**	2.90 / **4.63**	2.86 / **4.54**	2.82 / **4.46**	2.79 / **4.40**	2.74 / **4.29**	2.70 / **4.21**	2.65 / **4.10**	2.61 / **4.02**	2.57 / **3.94**	2.53 / **3.86**	2.50 / **3.80**	2.47 / **3.74**	2.45 / **3.70**	2.42 / **3.66**	2.41 / **3.62**	2.40 / **3.60**
12	4.75 / **9.33**	3.88 / **6.93**	3.49 / **5.95**	3.26 / **5.41**	3.11 / **5.06**	3.00 / **4.82**	2.92 / **4.65**	2.85 / **4.50**	2.80 / **4.39**	2.76 / **4.30**	2.72 / **4.22**	2.69 / **4.16**	2.64 / **4.05**	2.60 / **3.98**	2.54 / **3.86**	2.50 / **3.78**	2.46 / **3.70**	2.42 / **3.61**	2.40 / **3.56**	2.36 / **3.49**	2.35 / **3.46**	2.32 / **3.41**	2.31 / **3.38**	2.30 / **3.36**
13	4.67 / **9.07**	3.80 / **6.70**	3.41 / **5.74**	3.18 / **5.20**	3.02 / **4.86**	2.92 / **4.62**	2.84 / **4.44**	2.77 / **4.30**	2.72 / **4.19**	2.67 / **4.10**	2.63 / **4.02**	2.60 / **3.96**	2.55 / **3.85**	2.51 / **3.78**	2.46 / **3.67**	2.42 / **3.59**	2.38 / **3.51**	2.34 / **3.42**	2.32 / **3.37**	2.28 / **3.30**	2.26 / **3.27**	2.24 / **3.21**	2.22 / **3.18**	2.21 / **3.16**

* Reproduced from *Statistical Methods* (5th edition, 1956), by kind permission of Professor George W. Snedecor and the Iowa State University Press, Ames.

Table A.4 (*continued*)

5% (Roman Type) and 1% (Boldface Type) Points in the Distribution of F

n_1 degrees of freedom (for greater mean square). Values shown as 5% (roman) / **1% (boldface)**.

n_2	1	2	3	4	5	6	7	8	9	10	11	12	14	16	20	24	30	40	50	75	100	200	500	∞
14	4.60 **8.86**	3.74 **6.51**	3.34 **5.56**	3.11 **5.03**	2.96 **4.69**	2.85 **4.46**	2.77 **4.28**	2.70 **4.14**	2.65 **4.03**	2.60 **3.94**	2.56 **3.86**	2.53 **3.80**	2.48 **3.70**	2.44 **3.62**	2.39 **3.51**	2.35 **3.43**	2.31 **3.34**	2.27 **3.26**	2.24 **3.21**	2.21 **3.14**	2.19 **3.11**	2.16 **3.06**	2.14 **3.02**	2.13 **3.00**
15	4.54 **8.68**	3.68 **6.36**	3.29 **5.42**	3.06 **4.89**	2.90 **4.56**	2.79 **4.32**	2.70 **4.14**	2.64 **4.00**	2.59 **3.89**	2.55 **3.80**	2.51 **3.73**	2.48 **3.67**	2.43 **3.56**	2.39 **3.48**	2.33 **3.36**	2.29 **3.29**	2.25 **3.20**	2.21 **3.12**	2.18 **3.07**	2.15 **3.00**	2.12 **2.97**	2.10 **2.92**	2.08 **2.89**	2.07 **2.87**
16	4.49 **8.53**	3.63 **6.23**	3.24 **5.29**	3.01 **4.77**	2.85 **4.44**	2.74 **4.20**	2.66 **4.03**	2.59 **3.89**	2.54 **3.78**	2.49 **3.69**	2.45 **3.61**	2.42 **3.55**	2.37 **3.45**	2.33 **3.37**	2.28 **3.25**	2.24 **3.18**	2.20 **3.10**	2.16 **3.01**	2.13 **2.96**	2.09 **2.89**	2.07 **2.86**	2.04 **2.80**	2.02 **2.77**	2.01 **2.75**
17	4.45 **8.40**	3.59 **6.11**	3.20 **5.18**	2.96 **4.67**	2.81 **4.34**	2.70 **4.10**	2.62 **3.93**	2.55 **3.79**	2.50 **3.68**	2.45 **3.59**	2.41 **3.52**	2.38 **3.45**	2.33 **3.35**	2.29 **3.27**	2.23 **3.16**	2.19 **3.08**	2.15 **3.00**	2.11 **2.92**	2.08 **2.86**	2.04 **2.79**	2.02 **2.76**	1.99 **2.70**	1.97 **2.67**	1.96 **2.65**
18	4.41 **8.28**	3.55 **6.01**	3.16 **5.09**	2.93 **4.58**	2.77 **4.25**	2.66 **4.01**	2.58 **3.85**	2.51 **3.71**	2.46 **3.60**	2.41 **3.51**	2.37 **3.44**	2.34 **3.37**	2.29 **3.27**	2.25 **3.19**	2.19 **3.07**	2.15 **3.00**	2.11 **2.91**	2.07 **2.83**	2.04 **2.78**	2.00 **2.71**	1.98 **2.68**	1.95 **2.62**	1.93 **2.59**	1.92 **2.57**
19	4.38 **8.18**	3.52 **5.93**	3.13 **5.01**	2.90 **4.50**	2.74 **4.17**	2.63 **3.94**	2.55 **3.77**	2.48 **3.63**	2.43 **3.52**	2.38 **3.43**	2.34 **3.36**	2.31 **3.30**	2.26 **3.19**	2.21 **3.12**	2.15 **3.00**	2.11 **2.92**	2.07 **2.84**	2.02 **2.76**	2.00 **2.70**	1.96 **2.63**	1.94 **2.60**	1.91 **2.54**	1.90 **2.51**	1.88 **2.49**
20	4.35 **8.10**	3.49 **5.85**	3.10 **4.94**	2.87 **4.43**	2.71 **4.10**	2.60 **3.87**	2.52 **3.71**	2.45 **3.56**	2.40 **3.45**	2.35 **3.37**	2.31 **3.30**	2.28 **3.23**	2.23 **3.13**	2.18 **3.05**	2.12 **2.94**	2.08 **2.86**	2.04 **2.77**	1.99 **2.69**	1.96 **2.63**	1.92 **2.56**	1.90 **2.53**	1.87 **2.47**	1.85 **2.44**	1.84 **2.42**
21	4.32 **8.02**	3.47 **5.78**	3.07 **4.87**	2.84 **4.37**	2.68 **4.04**	2.57 **3.81**	2.49 **3.65**	2.42 **3.51**	2.37 **3.40**	2.32 **3.31**	2.28 **3.24**	2.25 **3.17**	2.20 **3.07**	2.15 **2.99**	2.09 **2.88**	2.05 **2.80**	2.00 **2.72**	1.96 **2.63**	1.93 **2.58**	1.89 **2.51**	1.87 **2.47**	1.84 **2.42**	1.82 **2.38**	1.81 **2.36**
22	4.30 **7.94**	3.44 **5.72**	3.05 **4.82**	2.82 **4.31**	2.66 **3.99**	2.55 **3.76**	2.47 **3.59**	2.40 **3.45**	2.35 **3.35**	2.30 **3.26**	2.26 **3.18**	2.23 **3.12**	2.18 **3.02**	2.13 **2.94**	2.07 **2.83**	2.03 **2.75**	1.98 **2.67**	1.93 **2.58**	1.91 **2.53**	1.87 **2.46**	1.84 **2.42**	1.81 **2.37**	1.80 **2.33**	1.78 **2.31**
23	4.28 **7.88**	3.42 **5.66**	3.03 **4.76**	2.80 **4.26**	2.64 **3.94**	2.53 **3.71**	2.45 **3.54**	2.38 **3.41**	2.32 **3.30**	2.28 **3.21**	2.24 **3.14**	2.20 **3.07**	2.14 **2.97**	2.10 **2.89**	2.04 **2.78**	2.00 **2.70**	1.96 **2.62**	1.91 **2.53**	1.88 **2.48**	1.84 **2.41**	1.82 **2.37**	1.79 **2.32**	1.77 **2.28**	1.76 **2.26**
24	4.26 **7.82**	3.40 **5.61**	3.01 **4.72**	2.78 **4.22**	2.62 **3.90**	2.51 **3.67**	2.43 **3.50**	2.36 **3.36**	2.30 **3.25**	2.26 **3.17**	2.22 **3.09**	2.18 **3.03**	2.13 **2.93**	2.09 **2.85**	2.02 **2.74**	1.98 **2.66**	1.94 **2.58**	1.89 **2.49**	1.86 **2.44**	1.82 **2.36**	1.80 **2.33**	1.76 **2.27**	1.74 **2.23**	1.73 **2.21**
25	4.24 **7.77**	3.38 **5.57**	2.99 **4.68**	2.76 **4.18**	2.60 **3.86**	2.49 **3.63**	2.41 **3.46**	2.34 **3.32**	2.28 **3.21**	2.24 **3.13**	2.20 **3.05**	2.16 **2.99**	2.11 **2.89**	2.06 **2.81**	2.00 **2.70**	1.96 **2.62**	1.92 **2.54**	1.87 **2.45**	1.84 **2.40**	1.80 **2.32**	1.77 **2.29**	1.74 **2.23**	1.72 **2.19**	1.71 **2.17**
26	4.22 **7.72**	3.37 **5.53**	2.98 **4.64**	2.74 **4.14**	2.59 **3.82**	2.47 **3.59**	2.39 **3.42**	2.32 **3.29**	2.27 **3.17**	2.22 **3.09**	2.18 **3.02**	2.15 **2.96**	2.10 **2.86**	2.05 **2.77**	1.99 **2.66**	1.95 **2.58**	1.90 **2.50**	1.85 **2.41**	1.82 **2.36**	1.78 **2.28**	1.76 **2.25**	1.72 **2.19**	1.70 **2.15**	1.69 **2.13**

Table A.4 (*continued*)

5% (ROMAN TYPE) AND 1% (BOLDFACE TYPE) POINTS IN THE DISTRIBUTION OF F

n_1 degrees of freedom (for greater mean square)

Each cell shows the 5% point (roman) over the 1% point (boldface).

n_2	1	2	3	4	5	6	7	8	9	10	11	12	14	16	20	24	30	40	50	75	100	200	500	∞
27	4.21 / 7.68	3.35 / 5.49	2.96 / 4.60	2.73 / 4.11	2.57 / 3.79	2.46 / 3.56	2.37 / 3.39	2.30 / 3.26	2.25 / 3.14	2.20 / 3.06	2.16 / 2.98	2.13 / 2.93	2.08 / 2.83	2.03 / 2.74	1.97 / 2.63	1.93 / 2.55	1.88 / 2.47	1.84 / 2.38	1.80 / 2.33	1.76 / 2.25	1.74 / 2.21	1.71 / 2.16	1.68 / 2.12	1.67 / 2.10
28	4.20 / 7.64	3.34 / 5.45	2.95 / 4.57	2.71 / 4.07	2.56 / 3.76	2.44 / 3.53	2.36 / 3.36	2.29 / 3.23	2.24 / 3.11	2.19 / 3.03	2.15 / 2.95	2.12 / 2.90	2.06 / 2.80	2.02 / 2.71	1.96 / 2.60	1.91 / 2.52	1.87 / 2.44	1.81 / 2.35	1.78 / 2.30	1.75 / 2.22	1.72 / 2.18	1.69 / 2.13	1.67 / 2.09	1.65 / 2.06
29	4.18 / 7.60	3.33 / 5.42	2.93 / 4.54	2.70 / 4.04	2.54 / 3.73	2.43 / 3.50	2.35 / 3.33	2.28 / 3.20	2.22 / 3.08	2.18 / 3.00	2.14 / 2.92	2.10 / 2.87	2.05 / 2.77	2.00 / 2.68	1.94 / 2.57	1.90 / 2.49	1.85 / 2.41	1.80 / 2.32	1.77 / 2.27	1.73 / 2.19	1.71 / 2.15	1.68 / 2.10	1.65 / 2.06	1.64 / 2.03
30	4.17 / 7.56	3.32 / 5.39	2.92 / 4.51	2.69 / 4.02	2.53 / 3.70	2.42 / 3.47	2.34 / 3.30	2.27 / 3.17	2.21 / 3.06	2.16 / 2.98	2.12 / 2.90	2.09 / 2.84	2.04 / 2.74	1.99 / 2.66	1.93 / 2.55	1.89 / 2.47	1.84 / 2.38	1.79 / 2.29	1.76 / 2.24	1.72 / 2.16	1.69 / 2.13	1.66 / 2.07	1.64 / 2.03	1.62 / 2.01
32	4.15 / 7.50	3.30 / 5.34	2.90 / 4.46	2.67 / 3.97	2.51 / 3.66	2.40 / 3.42	2.32 / 3.25	2.25 / 3.12	2.19 / 3.01	2.14 / 2.94	2.10 / 2.86	2.07 / 2.80	2.02 / 2.70	1.97 / 2.62	1.91 / 2.51	1.86 / 2.42	1.82 / 2.34	1.76 / 2.25	1.74 / 2.20	1.69 / 2.12	1.67 / 2.08	1.64 / 2.02	1.61 / 1.98	1.59 / 1.96
34	4.13 / 7.44	3.28 / 5.29	2.88 / 4.42	2.65 / 3.93	2.49 / 3.61	2.38 / 3.38	2.30 / 3.21	2.23 / 3.08	2.17 / 2.97	2.12 / 2.89	2.08 / 2.82	2.05 / 2.76	2.00 / 2.66	1.95 / 2.58	1.89 / 2.47	1.84 / 2.38	1.80 / 2.30	1.74 / 2.21	1.71 / 2.15	1.67 / 2.08	1.64 / 2.04	1.61 / 1.98	1.59 / 1.94	1.57 / 1.91
36	4.11 / 7.39	3.26 / 5.25	2.86 / 4.38	2.63 / 3.89	2.48 / 3.58	2.36 / 3.35	2.28 / 3.18	2.21 / 3.04	2.15 / 2.94	2.10 / 2.86	2.06 / 2.78	2.03 / 2.72	1.98 / 2.62	1.93 / 2.54	1.87 / 2.43	1.82 / 2.35	1.78 / 2.26	1.72 / 2.17	1.69 / 2.12	1.65 / 2.04	1.62 / 2.00	1.59 / 1.94	1.56 / 1.90	1.55 / 1.87
38	4.10 / 7.35	3.25 / 5.21	2.85 / 4.34	2.62 / 3.86	2.46 / 3.54	2.35 / 3.32	2.26 / 3.15	2.19 / 3.02	2.14 / 2.91	2.09 / 2.82	2.05 / 2.75	2.02 / 2.69	1.96 / 2.59	1.92 / 2.51	1.85 / 2.40	1.80 / 2.32	1.76 / 2.22	1.71 / 2.14	1.67 / 2.08	1.63 / 2.00	1.60 / 1.97	1.57 / 1.90	1.54 / 1.86	1.53 / 1.84
40	4.08 / 7.31	3.23 / 5.18	2.84 / 4.31	2.61 / 3.83	2.45 / 3.51	2.34 / 3.29	2.25 / 3.12	2.18 / 2.99	2.12 / 2.88	2.07 / 2.80	2.04 / 2.73	2.00 / 2.66	1.95 / 2.56	1.90 / 2.49	1.84 / 2.37	1.79 / 2.29	1.74 / 2.20	1.69 / 2.11	1.66 / 2.05	1.61 / 1.97	1.59 / 1.94	1.55 / 1.88	1.53 / 1.84	1.51 / 1.81
42	4.07 / 7.27	3.22 / 5.15	2.83 / 4.29	2.59 / 3.80	2.44 / 3.49	2.32 / 3.26	2.24 / 3.10	2.17 / 2.96	2.11 / 2.86	2.06 / 2.77	2.02 / 2.70	1.99 / 2.64	1.94 / 2.54	1.89 / 2.46	1.82 / 2.35	1.78 / 2.26	1.73 / 2.17	1.68 / 2.08	1.64 / 2.02	1.60 / 1.94	1.57 / 1.91	1.54 / 1.85	1.51 / 1.80	1.49 / 1.78
44	4.06 / 7.24	3.21 / 5.12	2.82 / 4.26	2.58 / 3.78	2.43 / 3.46	2.31 / 3.24	2.23 / 3.07	2.16 / 2.94	2.10 / 2.84	2.05 / 2.75	2.01 / 2.68	1.98 / 2.62	1.92 / 2.52	1.88 / 2.44	1.81 / 2.32	1.76 / 2.24	1.72 / 2.15	1.66 / 2.06	1.63 / 2.00	1.58 / 1.92	1.56 / 1.88	1.52 / 1.82	1.50 / 1.78	1.48 / 1.75
46	4.05 / 7.21	3.20 / 5.10	2.81 / 4.24	2.57 / 3.76	2.42 / 3.44	2.30 / 3.22	2.22 / 3.05	2.14 / 2.92	2.09 / 2.82	2.04 / 2.73	2.00 / 2.66	1.97 / 2.60	1.91 / 2.50	1.87 / 2.42	1.80 / 2.30	1.75 / 2.22	1.71 / 2.13	1.65 / 2.04	1.62 / 1.98	1.57 / 1.90	1.54 / 1.86	1.51 / 1.80	1.48 / 1.76	1.46 / 1.72
48	4.04 / 7.19	3.19 / 5.08	2.80 / 4.22	2.56 / 3.74	2.41 / 3.42	2.30 / 3.20	2.21 / 3.04	2.14 / 2.90	2.08 / 2.80	2.03 / 2.71	1.99 / 2.64	1.96 / 2.58	1.90 / 2.48	1.86 / 2.40	1.79 / 2.28	1.74 / 2.20	1.70 / 2.11	1.64 / 2.02	1.61 / 1.96	1.56 / 1.88	1.53 / 1.84	1.50 / 1.78	1.47 / 1.73	1.45 / 1.70

Table A.4 (concluded)

5% (Roman Type) and 1% (Boldface Type) Points in the Distribution of F

n_1 degrees of freedom (for greater mean square). Each cell shows 5% (roman) / 1% (boldface).

n_2	1	2	3	4	5	6	7	8	9	10	11	12	14	16	20	24	30	40	50	75	100	200	500	∞
50	4.03/7.17	3.18/5.06	2.79/4.20	2.56/3.72	2.40/3.41	2.29/3.18	2.20/3.02	2.13/2.88	2.07/2.78	2.02/2.70	1.98/2.62	1.95/2.56	1.90/2.46	1.85/2.39	1.78/2.26	1.74/2.18	1.69/2.10	1.63/2.00	1.60/1.94	1.55/1.86	1.52/1.82	1.48/1.76	1.46/1.71	1.44/1.68
55	4.02/7.12	3.17/5.01	2.78/4.16	2.54/3.68	2.38/3.37	2.27/3.15	2.18/2.98	2.11/2.85	2.05/2.75	2.00/2.66	1.97/2.59	1.93/2.53	1.88/2.43	1.83/2.35	1.76/2.23	1.72/2.15	1.67/2.06	1.61/1.96	1.58/1.90	1.52/1.82	1.50/1.78	1.46/1.71	1.43/1.66	1.41/1.64
60	4.00/7.08	3.15/4.98	2.76/4.13	2.52/3.65	2.37/3.34	2.25/3.12	2.17/2.95	2.10/2.82	2.04/2.72	1.99/2.63	1.95/2.56	1.92/2.50	1.86/2.40	1.81/2.32	1.75/2.20	1.70/2.12	1.65/2.03	1.59/1.93	1.56/1.87	1.50/1.79	1.48/1.74	1.44/1.68	1.41/1.63	1.39/1.60
65	3.99/7.04	3.14/4.95	2.75/4.10	2.51/3.62	2.36/3.31	2.24/3.09	2.15/2.93	2.08/2.79	2.02/2.70	1.98/2.61	1.94/2.54	1.90/2.47	1.85/2.37	1.80/2.30	1.73/2.18	1.68/2.09	1.63/2.00	1.57/1.90	1.54/1.84	1.49/1.76	1.46/1.71	1.42/1.64	1.39/1.60	1.37/1.56
70	3.98/7.01	3.13/4.92	2.74/4.08	2.50/3.60	2.35/3.29	2.23/3.07	2.14/2.91	2.07/2.77	2.01/2.67	1.97/2.59	1.93/2.51	1.89/2.45	1.84/2.35	1.79/2.28	1.72/2.15	1.67/2.07	1.62/1.98	1.56/1.88	1.53/1.82	1.47/1.74	1.45/1.69	1.40/1.62	1.37/1.56	1.35/1.53
80	3.96/6.96	3.11/4.88	2.72/4.04	2.48/3.56	2.33/3.25	2.21/3.04	2.12/2.87	2.05/2.74	1.99/2.64	1.95/2.55	1.91/2.48	1.88/2.41	1.82/2.32	1.77/2.24	1.70/2.11	1.65/2.03	1.60/1.94	1.54/1.84	1.51/1.78	1.45/1.70	1.42/1.65	1.38/1.57	1.35/1.52	1.32/1.49
100	3.94/6.90	3.09/4.82	2.70/3.98	2.46/3.51	2.30/3.20	2.19/2.99	2.10/2.82	2.03/2.69	1.97/2.59	1.92/2.51	1.88/2.43	1.85/2.36	1.79/2.26	1.75/2.19	1.68/2.06	1.63/1.98	1.57/1.89	1.51/1.79	1.48/1.73	1.42/1.64	1.39/1.59	1.34/1.51	1.30/1.46	1.28/1.43
125	3.92/6.84	3.07/4.78	2.68/3.94	2.44/3.47	2.29/3.17	2.17/2.95	2.08/2.79	2.01/2.65	1.95/2.56	1.90/2.47	1.86/2.40	1.83/2.33	1.77/2.23	1.72/2.15	1.65/2.03	1.60/1.94	1.55/1.85	1.49/1.75	1.45/1.68	1.39/1.59	1.36/1.54	1.31/1.46	1.27/1.40	1.25/1.37
150	3.91/6.81	3.06/4.75	2.67/3.91	2.43/3.44	2.27/3.14	2.16/2.92	2.07/2.76	2.00/2.62	1.94/2.53	1.89/2.44	1.85/2.37	1.82/2.30	1.76/2.20	1.71/2.12	1.64/2.00	1.59/1.91	1.54/1.83	1.47/1.72	1.44/1.66	1.37/1.56	1.34/1.51	1.29/1.43	1.25/1.37	1.22/1.33
200	3.89/6.76	3.04/4.71	2.65/3.88	2.41/3.41	2.26/3.11	2.14/2.90	2.05/2.73	1.98/2.60	1.92/2.50	1.87/2.41	1.83/2.34	1.80/2.28	1.74/2.17	1.69/2.09	1.62/1.97	1.57/1.88	1.52/1.79	1.45/1.69	1.42/1.62	1.35/1.53	1.32/1.48	1.26/1.39	1.22/1.33	1.19/1.28
400	3.86/6.70	3.02/4.66	2.62/3.83	2.39/3.36	2.23/3.06	2.12/2.85	2.03/2.69	1.96/2.55	1.90/2.46	1.85/2.37	1.81/2.29	1.78/2.23	1.72/2.12	1.67/2.04	1.60/1.92	1.54/1.84	1.49/1.74	1.42/1.64	1.38/1.57	1.32/1.47	1.28/1.42	1.22/1.32	1.16/1.24	1.13/1.19
1000	3.85/6.66	3.00/4.62	2.61/3.80	2.38/3.34	2.22/3.04	2.10/2.82	2.02/2.66	1.95/2.53	1.89/2.43	1.84/2.34	1.80/2.26	1.76/2.20	1.70/2.09	1.65/2.01	1.58/1.89	1.53/1.81	1.47/1.71	1.41/1.61	1.36/1.54	1.30/1.44	1.26/1.38	1.19/1.28	1.13/1.19	1.08/1.11
∞	3.84/6.64	2.99/4.60	2.60/3.78	2.37/3.32	2.21/3.02	2.09/2.80	2.01/2.64	1.94/2.51	1.88/2.41	1.83/2.32	1.79/2.24	1.75/2.18	1.69/2.07	1.64/1.99	1.57/1.87	1.52/1.79	1.46/1.69	1.40/1.59	1.35/1.52	1.28/1.41	1.24/1.35	1.17/1.25	1.11/1.15	1.00/1.00

INDEX

(References are to pages)